GROUP THEORY

This book is in the

ADDISON-WESLEY SERIES IN PHYSICS

GROUP THEORY
AND ITS APPLICATION
TO PHYSICAL PROBLEMS

---

by

MORTON HAMERMESH

Argonne National Laboratory

ADDISON-WESLEY PUBLISHING COMPANY, INC.

READING, MASSACHUSETTS · PALO ALTO · LONDON

ISBN 0-201-02780-1
EFGHIJKLMN-CO-79876543

PREFACE

This book was written in an attempt to make group-theoretical methods a useful tool instead of an esoteric study. I have tried to reformulate the results of papers and books to make the material understandable. For most of the text, no previous knowledge of group theory is needed, but it is assumed that the reader knows quantum mechanics.

The book developed from courses of lectures presented at various times at Argonne National Laboratory. Most of the material on crystal groups and crystalline fields was presented in a course in 1953. Some of the material related to nuclear physics was discussed in a course in 1955. In 1957, I lectured on the Lorentz group. The book contains only the introduction to a treatment of the Lorentz group. I felt that the subject could not be presented properly without a full discussion of quantum field theory.

Much of the final manuscript was written in Zürich, Switzerland, in 1958–59. I am very grateful to the directors of the Argonne National Laboratory for enabling me to spend this concentrated effort on completion of the book. I am also indebted to the Council of the Royal Society, London, for permission to reproduce the tables in Chapters 10 and 11, which were originally published in *Proceedings of the Royal Society*.

The book is dedicated to my wife, Madeline, who typed the manuscript and corrected many stylistic and technical errors.

M. H.

CONTENTS

INTRODUCTION xiii

CHAPTER 1. ELEMENTS OF GROUP THEORY 1

1–1 Correspondences and transformations 1
1–2 Groups. Definitions and examples 6
1–3 Subgroups. Cayley's theorem 15
1–4 Cosets. Lagrange's theorem 20
1–5 Conjugate classes 23
1–6 Invariant subgroups. Factor groups. Homomorphism . . . 28
1–7 Direct products 30

CHAPTER 2. SYMMETRY GROUPS 32

2–1 Symmetry elements. Pole figures 32
2–2 Equivalent axes and planes. Two-sided axes 38
2–3 Groups whose elements are pure rotations: uniaxial groups,
     dihedral groups 41
2–4 The law of rational indices 45
2–5 Groups whose elements are pure rotations. Regular polyhedra . 48
2–6 Symmetry groups containing rotation reflections. Adjunction
     of reflections to $\mathfrak{C}_n$ 52
2–7 Adjunction of reflections to the groups D_n 55
2–8 The complete symmetry groups of the regular polyhedra . . 58
2–9 Summary of point groups. Other systems of notation . . . 60
2–10 Magnetic symmetry groups (color groups) 63

CHAPTER 3. GROUP REPRESENTATIONS 68

3–1 Linear vector spaces 68
3–2 Linear dependence; dimensionality 70
3–3 Basis vectors (coordinate axes); coordinates 71
3–4 Mappings; linear operators; matrix representations; equivalence 74
3–5 Group representations. 77
3–6 Equivalent representations; characters 79
3–7 Construction of representations. Addition of representations . 80
3–8 Invariance of functions and operators. Classification of
     eigenfunctions 86
3–9 Unitary spaces; scalar product; unitary matrices; Hermitian
     matrices 88
3–10 Operators: adjoint, self-adjoint, unitary 91

vii

3–11 Unitary representations   92
3–12 Hilbert space   93
3–13 Analysis of representations; reducibility; irreducible
      representations   94
3–14 Schur's lemmas   98
3–15 The orthogonality relations 101
3–16 Criteria for irreducibility. Analysis of representations . . . 104
3–17 The general theorems. Group algebra 106
3–18 Expansion of functions in basis functions of irreducible
      representations 111
3–19 Representations of direct products 114

CHAPTER 4. IRREDUCIBLE REPRESENTATIONS OF THE POINT
            SYMMETRY GROUPS 115

4–1 Abelian groups 115
4–2 Nonabelian groups 119
4–3 Character tables for the crystal point groups 125

CHAPTER 5. MISCELLANEOUS OPERATIONS WITH GROUP
            REPRESENTATIONS 128

5–1 Product representations (Kronecker products) 128
5–2 Symmetrized and antisymmetrized products 132
5–3 The adjoint representation. The complex conjugate
     representation 135
5–4 Conditions for existence of invariants 136
5–5 Real representations 138
5–6 The reduction of Kronecker products. The Clebsch-Gordan
     series 147
5–7 Clebsch-Gordan coefficients 148
5–8 Simply reducible groups 151
5–9 Three-j symbols 156

CHAPTER 6. PHYSICAL APPLICATIONS 161

6–1 Classification of spectral terms 161
6–2 Perturbation theory 162
6–3 Selection rules 166
6–4 Coupled systems 178

CHAPTER 7. THE SYMMETRIC GROUP 182

7–1 The deduction of the characters of a group from those
     of a subgroup 182

7–2   Frobenius' formula for the characters of the symmetric group . 189
7–3   Graphical methods. Lattice permutations. Young patterns.
      Young tableaux 198
7–4   Graphical method for determining characters 201
7–5   Recursion formulas for characters. Branching laws 208
7–6   Calculation of characters by means of the Frobenius formula . 212
7–7   The matrices of the irreducible representations of S_n.
      Yamanouchi symbols 214
7–8   Hund's method 231
7–9   Group algebra 239
7–10 Young operators 243
7–11 The construction of product wave functions of a given symmetry.
      Fock's cyclic symmetry conditions 246
7–12 Outer products of representations of the symmetric group . . 249
7–13. Inner products. Clebsch-Gordan series for the symmetric group 254
7–14 Clebsch-Gordan (CG) coefficients for the symmetric group.
      Symmetry properties. Recursion formulas 260

CHAPTER 8.   CONTINUOUS GROUPS 279

8–1   Summary of results for finite groups 279
8–2   Infinite discrete groups 281
8–3   Continuous groups. Lie groups 283
8–4   Examples of Lie groups 287
8–5   Isomorphism. Subgroups. Mixed continuous groups . . . 291
8–6   One-parameter groups. Infinitesimal transformations . . . 293
8–7   Structure constants 299
8–8   Lie algebras 301
8–9   Structure of Lie algebras 304
8–10 Structure of compact semisimple Lie groups and their algebras . 309
8–11 Linear representations of Lie groups 311
8–12 Invariant integration 313
8–13 Irreducible representations of Lie groups and Lie algebras.
      The Casimir operator 317
8–14 Multiple-valued representations. Universal covering group . . 319

CHAPTER 9.   AXIAL AND SPHERICAL SYMMETRY 322

9–1   The rotation group in two dimensions 322
9–2   The rotation group in three dimensions 325
9–3   Continuous single-valued representations of the three-
      dimensional rotation group 333
9–4   Splitting of atomic levels in crystalline fields (single-valued
      representations) 337
9–5   Construction of crystal eigenfunctions 342

9–6   Two-valued representations of the rotation group. The unitary
       unimodular group in two dimensions 348
9–7   Splitting of atomic levels in crystalline fields. Double-valued
       representations of the crystal point groups 357
9–8   Coupled systems. Addition of angular momenta. Clebsch-
       Gordan coefficients 367

CHAPTER 10.   LINEAR GROUPS IN n-DIMENSIONAL SPACE.
              IRREDUCIBLE TENSORS 377

10–1   Tensors with respect to $GL(n)$ 377
10–2   The construction of irreducible tensors with respect to $GL(n)$. 378
10–3   The dimensionality of the irreducible representations of $GL(n)$. 384
10–4   Irreducible representations of subgroups of $GL(n)$: $SL(n)$,
       $U(n)$, $SU(n)$ 388
10–5   The orthogonal group in n dimensions. Contraction. Traceless
       tensors 391
10–6   The irreducible representations of $O(n)$ 394
10–7   Decomposition of irreducible representations of $U(n)$ with
       respect to $O^+(n)$ 399
10–8   The symplectic group $Sp(n)$. Contraction. Traceless Tensors . 403
10–9   The irreducible representations of $Sp(n)$. Decomposition of
       irreducible representations of $U(n)$ with respect to its
       symplectic subgroup 408

CHAPTER 11.   APPLICATIONS TO ATOMIC AND NUCLEAR PROBLEMS . . 413

11–1   The classification of states of systems of identical particles
       according to $SU(n)$ 413
11–2   Angular momentum analysis. Decomposition of representations
       of $SU(n)$ into representations of $O^+(3)$ 414
11–3   The Pauli principle. Atomic spectra in Russell-Saunders
       coupling 417
11–4   Seniority in atomic spectra 423
11–5   Atomic spectra in jj-coupling 430
11–6   Nuclear structure. Isotopic spin 433
11–7   Nuclear spectra in L-S coupling. Supermultiplets 435
11–8   The L-S coupling shell model. Seniority 443
11–9   The jj-coupling shell model. Seniority in jj-coupling . . . 448

CHAPTER 12.   RAY REPRESENTATIONS. LITTLE GROUPS 458

12–1   Projective representations of finite groups 458
12–2   Examples of projective representations of finite groups . . . 463
12–3   Ray representations of Lie groups 469

12–4   Ray representations of the pseudo-orthogonal groups 478
12–5   Ray representations of the Galilean group 484
12–6   Irreducible representations of translation groups 486
12–7   Little groups 489

BIBLIOGRAPHY AND NOTES 499

INDEX 505

INTRODUCTION

The purpose of this book is to present those aspects of the theory of groups which are relevant to the treatment of problems in physics. It must be stated at the outset that all our results are obtainable without the formal methods of group theory. The alternative "simple" methods are, in fact, a physicist's rediscovery of some group-theoretical techniques. For simple problems, the formal treatment is also simple; in complex problems, the use of powerful tools can save us considerable labor. We should not deprecate formalism as such—so long as the physical ideas are not lost from sight, the formalism is valuable. In the course of our study, the "intuitive" methods will be treated as well as more formal techniques.

My hope is that this study of group theory will make the reader aware of the wide range of physical problems where the concepts of symmetry and invariance are important. Also, we shall see that many things which we have learned as isolated notions, such as parity, tensor character, spinor, angular momentum, etc., are aspects of group properties.

Before starting our presentation of group theory, let us consider some simple examples.

The Schrödinger equation for a one-dimensional problem can be written as

$$u'' + [\lambda - V(x)]u = 0, \qquad (0\text{--}1)$$

where λ is the eigenvalue, u the eigenfunction, and $V(x)$ the potential. In a one-dimensional problem the solutions are necessarily nondegenerate; i.e., to each eigenvalue λ there corresponds only one solution $u(x)$. Now suppose that our potential $V(x)$ is an even function of x,

$$V(x) = V(-x). \qquad (0\text{--}2)$$

Replacing x by $-x$, we see that if $u(x)$ is a solution belonging to the eigenvalue λ, so is $u(-x)$. The nondegeneracy then requires that $u(-x) = cu(x)$, where c is some constant:

$$u(x) = cu(-x) = c^2 u(x); \qquad c = \pm 1. \qquad (0\text{--}3)$$

Thus the eigenfunctions $u(x)$ are either even or odd. We may generalize the statement of our result as follows:

In the equation $Lu = 0$, where L is a linear operator, a symmetry property of L (in our case L was unchanged when we replaced x by $-x$) leads to a *classification* of the solutions u according to the same symmetry property.

We also note that the symmetry properties of the system (i.e., of the Hamiltonian) lead to *selection rules:*

$$\int u_n x u_m \, dx = 0$$

if u_n and u_m are both even or both odd; only terms of different symmetry combine with each other through the operator x.

This simple example indicates the general question: What properties of the eigenfunctions follow from the invariance of the Hamiltonian under various symmetry operations?

As another example, consider the motion of an electron in a spherically symmetric field (potential function $V(\mathbf{r})$ depending on r only). The invariance of the Hamiltonian under rotations leads to the result that the eigenfunctions are of the form $R(r)Y_l^m(\theta, \phi)$, where the Y_l^m are spherical harmonics. Each eigenvalue of the energy is characterized by an azimuthal quantum number l, and has $2l + 1$ eigenfunctions belonging to it ($m = l$, $l - 1, \ldots, -l$).

The eigenfunctions are classified according to their behavior under rotations. For very special forms of $V(\mathbf{r})$, say $V(r) = 1/r$, it may happen that terms with different values of l coincide. It was shown by Fock that this increased degeneracy results from the fact that the Hamiltonian in this case is invariant under a larger class of symmetry operations than just the rotations in three dimensions. A similar effect with a similar explanation occurs for the isotropic harmonic oscillator. It may also happen that, for special choices of constants in a Hamiltonian, terms can be brought into coincidence. In such a case, we speak of "accidental" degeneracy. By this we mean that the degeneracy is not a consequence of the symmetry properties, but is rather the result of a special choice of the Hamiltonian; such degeneracy can be removed *without* changing the symmetry properties of the Hamiltonian.

Another example is the motion of an electron in the periodic potential in the interior of a metal. The periodicity of the potential enables one to draw those conclusions concerning the eigenfunctions which are the content of Bloch's theorem. The classification of spectral terms for an electron in a crystal will be different from that in a free atom because of the loss of spherical symmetry.

The determination of the characteristic vibrations of a molecule requires the solution of the secular equation. Except for the simplest molecules, this is a formidable problem. The symmetry properties of the molecule can be used to reduce the secular equation to a more manageable one. They will also enable us to classify the term spectrum of the molecule and to deduce the selection rules for various processes.

For systems of identical particles, the Hamiltonian is invariant under any interchange. In atomic problems, this leads to the classification of terms according to their spin quantum number. In nuclear problems, when we consider neutron and proton as different charge states of the same particle (nucleon), we obtain an additional classification of terms according to a "charge" quantum number.

Finally, the behavior of wave functions under rotations can be used to treat problems of coupling of angular momenta, and of angular correlation between particles emitted in successive processes. We have mentioned briefly some possible applications of symmetry considerations and of group theory. We shall discuss them in detail later on.

CHAPTER 1

ELEMENTS OF GROUP THEORY

1–1 Correspondences and transformations. All of us are familiar with the concept of a *correspondence*, or *mapping:* We have a *set* of *objects* which we call *points;* these may be *finite* in number, in which case we can enumerate and label them as, say "the points a, b, c," (for a set of three objects) or the points $p_1, \ldots, p_n$ (or the points $1, 2, \ldots, n$) for a set of n objects. They may be *infinite* in number, say "the points designated by the integers 1, 2, etc."; or they may constitute a continuum (all the points in the XY-plane). By a *mapping* of the set of points on itself, we mean that we are given a recipe whereby we associate with each point p of the set an *image* point p' in the set. We say that p' is the *image* of p under the *mapping M*. We indicate this symbolically by

$$\overset{M}{p \to p'}, \quad \text{or} \quad p' = Mp. \tag{1–1}$$

We can describe Eq. (1–1) by the statement: The operator M acting on the object p changes it to the object p'. For a finite set of points the description of a mapping can be done by enumeration; e.g., for a set of three points, a, b, c, we can describe a mapping by saying: The mapping M takes the point a into its image b, the point b into a, and the point c into c, or symbolically,

$$M \equiv \begin{Bmatrix} a \to b \\ b \to a \\ c \to c \end{Bmatrix}. \tag{1–2}$$

Another possible mapping M' might be

$$M' \equiv \begin{Bmatrix} a \to a \\ b \to a \\ c \to a \end{Bmatrix}. \tag{1–3}$$

For an infinite set of points, enumeration is not possible. Instead we give a *functional law* (or recipe) for the mapping M. For example, we may consider the set of points on the X-axis, and a mapping M whose law is $x \to x' = x + 2$; i.e., each point is to be shifted two units to the right to arrive at its image.

Two mappings M, M' of a set of points are *identical* if $Mp = M'p$ for all points p. Conversely, $M = M'$ means that $Mp = M'p$ for all points p.

1

One special and important mapping is the *identity* I which maps each point on itself; $Ip = p$. We can perform mappings in succession: If M takes p into $p'(p' = Mp)$, and M' takes p' into $p''(p'' = M'p')$, then

$$p'' = M'p' = M'(Mp), \tag{1-4}$$

which we write as

$$p'' = M'Mp. \tag{1-4a}$$

In other words, there is a single mapping (which we denote by $M'M$) which produces the same effect as the successive application of M and M'. If by a sequence of mappings M_1, M_2, etc., we set up the correspondences

$$p \rightarrow p' \rightarrow p'' \cdots ,$$

$$p' = M_1 p, \quad p'' = M_2 p', \quad p''' = M_3 p'', \ldots ,$$

then

$$p''' = M_3(M_2 p') = M_3(M_2(M_1 p)) = M_3 M_2(M_1 p) = M_3 M_2 M_1 p,$$

i.e., *mappings* satisfy the *associative law.*

In our example of three points [Eqs. (1–2) and (1–3)] the mapping $M'M$ means

$$\begin{Bmatrix} a \rightarrow b \rightarrow a \\ b \rightarrow a \rightarrow a \\ c \rightarrow c \rightarrow a \end{Bmatrix} = \begin{Bmatrix} a \rightarrow a \\ b \rightarrow a \\ c \rightarrow a \end{Bmatrix} . \tag{1-5}$$

If we had performed the mappings in reverse order, we would have obtained MM',

$$\begin{Bmatrix} a \rightarrow b \\ b \rightarrow b \\ c \rightarrow b \end{Bmatrix} , \tag{1-5a}$$

so that $MM' \neq M'M$. Thus the *composition* or *product* of *mappings* gives a result which, in general, depends on the order in which the mappings are performed; mappings are *noncommutative operations.*

We shall be interested only in one-to-one mappings, or *transformations;* i.e., mappings in which no two points of the set have the same image, and every point p' of the set is the image of one (*only* one) point p. The mapping M in Eq. (1–2) was one-to-one, while M' in Eq. (1–3) was not.

Given a one-to-one mapping, we can find the *inverse* mapping which undoes its work. Thus, if the transformation M takes p into p', $p' = Mp$, its inverse M^{-1} takes p' into p, $p = M^{-1}p'$. Then $p' = Mp = MM^{-1}p'$, whence

$$MM^{-1} = I, \tag{1-6}$$

and $p = M^{-1}p' = M^{-1}Mp$, so that $M^{-1}M = I$. For example, the inverse of

$$M = \begin{Bmatrix} a \to b \\ b \to a \\ c \to c \end{Bmatrix}$$

is

$$M^{-1} = \begin{Bmatrix} a \to b \\ b \to a \\ c \to c \end{Bmatrix} ;$$

hence, in this case, M is its own inverse.

The inverse of a composite of transformations is easily found. For the mappings M, M' which appear in Eq. (1–4),

$$p = M^{-1}p', \qquad p' = M'^{-1}p'', \qquad p = M^{-1}M'^{-1}p'',$$

so that

$$(M'M)^{-1} = M^{-1}M'^{-1}. \tag{1–7}$$

In words, the *inverse* of the *composite* is obtained by carrying out the *inverse transformations* in *reverse* order.

The following cases are examples of transformations.

EXAMPLES.

(1) *Permutation.* A set of n boxes (points) are labeled 1 to n. Each box contains an object. The objects are rearranged in the boxes so that once more there is one object in each of the n boxes. For example, if the object which was in box 1 before is now in box 3, we shall say that 3 is the image of 1 under the transformation. Consider a specific example of 4 boxes. The objects in the boxes are rearranged so that the occupant of 1 goes to 4, of 2 goes to 3, of 4 goes to 2, and of 3 goes to 1. This mapping is

$$\begin{Bmatrix} 1 \to 4 \\ 2 \to 3 \\ 4 \to 2 \\ 3 \to 1 \end{Bmatrix}.$$

We may say that our transformation is a *transition* from one *arrangement* of the numbers $1, \ldots, 4$ to another arrangement. Such transformations are called *permutations*. One common notation for permutations is to write below each object its image under the transformation. In this notation, our example would be written as

$$\begin{pmatrix} 1234 \\ 4312 \end{pmatrix}.$$

We have written the numbers in the top line in natural order, but this is not necessary; the same permutation could have been written as

$$\begin{pmatrix} 1324 \\ 4132 \end{pmatrix} \quad \text{or} \quad \begin{pmatrix} 4213 \\ 2341 \end{pmatrix}.$$

So long as the object-image associations are the same, these various ways of recording represent the same permutation.

Our example is a permutation of four symbols, so we say that it is a permutation of *degree* 4. We could have carried out other rearrangements of the four symbols, e.g., the point 1 could have 1, 2, 3, or 4 as its image. This would leave only three choices for the image of 2, which would then leave only two choices for the image of 3, and finally one choice for the image of 4. Thus there are $4 \cdot 3 \cdot 2 \cdot 1 = 4! = 24$ permutations of degree 4. Similarly, the number of permutations on n symbols, i.e., the number of permutations of degree n, is $n!$.

(2) *Translations.* The points of a line are labeled by the coordinate x. The transformation is the shifting of each point two units to the right:

$$x \rightarrow x' = x + 2.$$

(3) *Projective transformations of a line.* The projective transformations of the points on the X-axis are defined by

$$x \rightarrow x'; \quad x' = \frac{ax + b}{cx + d}, \quad \text{where } ad - bc \neq 0.$$

Problem. The cross ratio of four points on a line is defined as

$$\frac{(x_1 - x_2)/(x_3 - x_2)}{(x_1 - x_4)/(x_3 - x_4)},$$

where x_1, x_2, x_3, x_4 are the coordinates of the four points. Show that the cross ratio is *invariant* under projective transformation, i.e., the cross ratio calculated for the image points has the same form as that for the object points.

EXAMPLES (*cont.*).

(4) The permutations of degree n can also be viewed as special *linear transformations* in *n-dimensional space.* The point whose νth coordinate is x_ν is mapped into the point whose p_νth coordinate (relative to the same

axes) is x_ν, where p_ν is the image of ν under the permutation. The permutation is

$$\begin{pmatrix} 1 \ 2 \ \ldots n \\ p_1 p_2 \ldots p_n \end{pmatrix},$$

while the corresponding coordinate transformation is

$$x'_{p_1} = x_1,$$

$$x'_{p_2} = x_2,$$

$$\vdots$$

$$x'_{p_n} = x_n.$$

For the special case given in Example 1, the corresponding coordinate transformation would be

$$x_1 = x'_4,$$

$$x_2 = x'_3,$$

$$x_3 = x'_1,$$

$$x_4 = x'_2.$$

(5) *Linear transformations in n-dimensional space.* The transformations of Example 4 are special cases of linear transformations in n-dimensional space. With respect to a fixed coordinate system, the linear transformation maps the point with coordinates $(x_1, \ldots, x_n)$ into the point with coordinates $(x'_1, \ldots, x'_n)$, where the x''s are given by

$$x'_i = \sum_{j=1}^{n} a_{ij}x_j \qquad (i = 1, \ldots, n). \tag{1-8}$$

The n^2 coefficients a_{ij} in these linear equations form a square array, the *matrix* **a** of the transformation. The transformation (1–8) will have an inverse if the *determinant* of the matrix **a** differs from zero, i.e., if **a** is a *nonsingular* matrix. If we follow the transformation (1–8) by a second transformation,

$$x''_i = \sum_{j=1}^{n} b_{ij}x'_j \qquad (i = 1, \ldots, n),$$

the resultant transformation (product) will be

$$x''_i = \sum_{j=1}^{n} b_{ij}x'_j = \sum_{j=1}^{n} b_{ij} \sum_{k=1}^{n} a_{jk}x_k$$

$$= \sum_{k=1}^{n} \left(\sum_{j=1}^{n} b_{ij}a_{jk} \right) x_k. \tag{1-9}$$

From (1–9) we see that the single transformation which would take us directly from the x's to the x''''s is

$$x_i'' = \sum_{k=1}^{n} c_{ik} x_k,$$

where

$$c_{ik} = \sum_{j=1}^{n} b_{ij} a_{jk} \qquad (i,\, k = 1, \ldots, n). \tag{1–10}$$

Equation (1–10) gives the rule for combination of elements of the matrices **b** and **a** to form the *product matrix* **c**:

$$\mathbf{c} = \mathbf{ba}. \tag{1–11}$$

The last equation is a symbolic statement of the n^2 equations (1–10). Equation (1–8) can also be expressed as a symbolic equation. We may look upon the coordinates $x_1, \ldots, x_n$ as elements of a matrix having n rows and *one* column:

$$\mathbf{x} = \begin{bmatrix} x_1 \\ \vdots \\ x_n \end{bmatrix}, \tag{1–12}$$

and similarly, for $\mathbf{x}'$. Then Eq. (1–8) can be expressed symbolically as

$$\mathbf{x}' = \mathbf{ax}, \tag{1–13}$$

and the inverse transformation as

$$\mathbf{x} = \mathbf{a}^{-1}\mathbf{x}', \tag{1–13a}$$

where $\mathbf{a}^{-1}$ is the matrix of the coefficients of the inverse transformation.

1–2 Groups. Definitions and examples. By a group of transformations, G, we mean an aggregate of transformations of a given point set which
(1) contains the identity transformation;
(2) for every transformation M, also contains its inverse M^{-1};
(3) if it includes M and M', also includes their composite MM'.
The euclidean (rigid-body) motions of ordinary space form a group; the set of all $n!$ permutations of n points form a group; the motions of an equilateral triangle which bring it into coincidence with itself form a group, etc.

From the notion of transformation groups, we arrive at the concept of an abstract group by dissociating points and transformations from any special pictorial significance.

An *abstract group* G is a set of *elements a, b, c,* etc., for which a law of composition or "multiplication" is given so that the *"product" ab* of any two elements is well defined, and which satisfies the following conditions:*

(1) If a and b are elements of the set, then so is ab.

(2) Multiplication is *associative;* that is, $a(bc) = (ab)c$.

(3) The set contains an element e called the *identity* such that $ae = ea = a$ for every element a of the set.

(4) If a is in the set, then so is an element b such that $ab = ba = e$. The element b is called the *inverse* of a, and is denoted by $b = a^{-1}$.

Though we frequently refer to the law of combination as "multiplication," this does not imply ordinary multiplication. The set of rational fractions (omitting zero) form a group under ordinary multiplication. The set of integers, positive, negative, and zero, form a group if the law of combination is ordinary addition. Even in these examples we are not referring to the abstract group, but rather to a concrete example (a "realization") of the abstract group. The *structure* of an abstract group is determined by stating the result of the "multiplication" of every ordered pair of elements, either by enumeration or by stating a functional law, without any particularization of the nature of the elements.

Two elements a, b of a group are said to *commute* with each other if

$$ab = ba, \tag{1–14}$$

i.e., if their product is independent of order. From the group postulates we see that the *identity* element e *commutes* with *all* the elements of the group. If all the elements of a group commute with one another, the group is said to be *commutative* or *abelian*. The use of the symbol $+$ for combination is customary for abelian groups. In such groups, the "product" of the elements a and b is written as $a + b$ (or $b + a$), the symbol 0 is used for the identity, and the inverse of a is written $(-a)$.

The number of elements in a group is called the *order* of the group.

---

* These postulates are actually redundant; (3) and (4) can be replaced by the weaker requirements:

(3′) The set contains an element e called a *left identity*, such that $ea = a$ for every element a of the set.

(4′) If a is in the set, then so is an element b, such that $ba = e$, where e is the left identity described in (3′). The element b is called a *left inverse* of a with respect to the left identity e.

The interested reader should prove that the left identity e in (3′) is also a right identity, that is, $ae = a$, and is unique, and that the element b in (4′) is also a right inverse ($ab = e$) and is uniquely determined by a. Thus (3) and (4) are consequences of the weaker requirements (3′) and (4′).

If we select an element a of the group G, we can form products of a with itself, e.g.,

$$aa,$$

which we denote by a^2. Similarly, by a^n we mean the element which results from a product of n elements each equal to a. Our symbol a^{-1} for the inverse of a was introduced with this notation in mind. Thus

$$a^{-1} \cdot a = e = a^0.$$

We can also define negative powers of a:

$$a^{-m} = (a^{-1})^m = (a^m)^{-1}, \qquad (1\text{--}15)$$

the last step being a consequence of (1–7). If all the powers of the element a are distinct, then a is said to be of *infinite order*. If this is not the case, then, as we take successive powers of a, we shall find two integers r and s $(r > s)$ such that

$$a^r = a^s.$$

Multiplying by a^{-s}, we have

$$a^{r-s} = e, \qquad r - s > 0.$$

Suppose that n is the smallest positive integer such that a^n is the identity, i.e.,

$$a^n = e, \qquad n > 0; \qquad (1\text{--}16)$$

and if $a^k = e, k > 0$, then $k \geq n$. We say that the element a is of *order n*. If the element a is of order n, then all the elements

$$a^0 = e, a, a^2, \ldots, a^{n-1} \qquad (1\text{--}17)$$

are distinct (for if $a^r = a^s$, $a^{r-s} = e$ with $r - s < n$). Thus, every other power of a will be equal to one of the elements listed in (1–17). Any integer k can be written as

$$k = sn + t, \qquad 0 \leq t < n;$$

hence

$$a^k = a^{sn+t} = a^{sn}a^t = (a^n)^s a^t = a^t.$$

From this argument, we also see that if $a^k = e$, k is a multiple of n.

The group of order 1 contains one element, the identity e: $ee = e$.

The group of order 2 contains two distinct elements. One of these must be the identity e. Call the second element a. Then $aa(= a^2)$ cannot be a

since $a^2 = a$ implies $a = e$, so we must have $a^2 = e$; that is, a is its own inverse, $a = a^{-1}$. The following examples are groups of order 2:

EXAMPLES.

(1) The elements are the integers 0, 1. The law of combination is addition modulo 2, i.e., we add the elements and find the remainder after division by 2:

$$0 + 0 = 0, \qquad 0 + 1 = 1 + 0 = 1, \qquad 1 + 1 = 0.$$

(2) The objects are the points of three-dimensional space. The element e is the identity transformation, while a is the transformation which replaces each point by its mirror image in a given plane (say the YZ-plane), so that a means changing the sign of the x-coordinate of every point.

(3) The same as (2), but now a is the inversion in the origin of coordinates (i.e., a replaces x, y, z by $-x$, $-y$, $-z$).

(4) The same, but with a the rotation through 180° about, say the Z-axis.

(5) The same, but now a is the reciprocal transformation in the unit sphere: The point with spherical coordinates r, θ, ϕ is imaged in the point $1/r$, θ, ϕ.

(6) The elements are the integers 1, -1, and the law of combination is ordinary multiplication.

(7) The elements are the permutations on two symbols,

$$e = \begin{Bmatrix} 1 \to 1 \\ 2 \to 2 \end{Bmatrix}, \qquad a = \begin{Bmatrix} 1 \to 2 \\ 2 \to 1 \end{Bmatrix}.$$

Here we have seven different realizations of the abstract group of order 2. All these groups have precisely the same structure, and are said to be *isomorphic*.

In general, we shall say that two groups G, G' are *isomorphic* $(G \approx G')$ if their elements can be put into one-to-one correspondence which is preserved under combination. For example, consider the groups (1), (6), and (7) cited above. We make the correspondence:

Group (1): $\quad 0 \qquad\qquad 1 \qquad\qquad\qquad 1 + 1 = 0$,

Group (6): $\quad 1 \qquad\qquad -1 \qquad\qquad\quad (-1)(-1) = 1$,

Group (7): $\begin{pmatrix} 1 \to 1 \\ 2 \to 2 \end{pmatrix} \quad \begin{pmatrix} 1 \to 2 \\ 2 \to 1 \end{pmatrix} \quad \begin{pmatrix} 1 \to 2 \\ 2 \to 1 \end{pmatrix}\begin{pmatrix} 1 \to 2 \\ 2 \to 1 \end{pmatrix} = \begin{pmatrix} 1 \to 1 \\ 2 \to 2 \end{pmatrix}.$

If we indicate the correspondent of a (in G) by a' (in G'), so that the prime on an element signifies its correspondent, then G and G' are isomorphic if

$$a'b' = (ab)'.$$

Note that $a'b'$ means the product of a', b' according to the law of combination of G', while ab means the product of a and b according to the law of combination of the group G. For example, the laws of combination in the three groups above were addition, multiplication, and successive transformation, respectively.

If two groups are isomorphic, they have the same structure; the symbols or words may differ, but the abstract groups are the same.

Next consider the abstract group of order 3. Call the distinct elements a, b, e. The product ab cannot equal a (or b) since this would imply $b = e$ (or $a = e$). So ab must equal e. Similarly, we see that $a^2 = b$. Thus the group of order 3 consists of a, a^2, $a^3 = e$. This is an example of a *cyclic* group, one which consists of the powers of a single element.

Some realizations of this group are:

(1) The rotations in the plane of the equilateral triangle ABC which bring it into coincidence with itself.

(2) The rotations of three-dimensional space through angles of $0°$, $120°$, and $240°$ about the Z-axis.

(3) The cube roots of unity with ordinary multiplication as the law of combination.

For groups of higher order, this process of enumeration would become quite painful. The law of combination can be presented conveniently in the form of a *group table:* We label the rows and columns of a square array according to the elements of the group. In the box in the nth row and mth column, we record the product of the element labeling the nth row by the element labeling the mth column:

For example, the table for the group of order 2 is

	e	a
e	$e^2 = e$	$ea = a$
a	$ae = a$	$a^2 = e$

or briefly,

	e	a
e	e	a
a	a	e

For the group of order 3, we have

	e	a	$b(= a^2)$
e	e	a	b
a	a	b	e
b	b	e	a

We shall usually omit the first row and first column since all they do is to repeat the labeling. Note that the group tables for these examples are symmetric about the main diagonal. This will be the case if, and only if, the group is abelian. Also, we see that in each column (and in each row) all the elements of the group are displayed once and only once. The reason for this is that if we multiply all elements of the group from the left (or right) by a particular element a, then all the products obtained must be distinct: If $ab = ac$, then $b = c$.

There are two distinct structures for abstract groups of order 4:

(A)

e	a	b	c
a	b	c	e
b	c	e	a
c	e	a	b

$a^2 = b, \qquad ab = c = a^3,$
$a^4 = b^2 = e,$

i.e., the cyclic group a, a^2, a^3, $a^4 = e$.

(B)

e	a	b	c
a	e	c	b
b	c	e	a
c	b	a	e

$a^2 = b^2 = c^2 = e,$
$ab = c, \qquad ac = b, \qquad bc = a.$

This group is called the *four-group*, and is often denoted by the symbol V. From the symmetry of the tables we see that both groups are abelian.

Problems. (1) Show that (A) and (B) are the only possible structures for the group of order 4.

(2) Show directly that the group of order 4 must be abelian.

(3) Give a realization of each of the groups of order 4.

Let us consider some further examples of groups.

EXAMPLES.

(1) The elements are the integers

$$\ldots -3, -2, -1, 0, 1, 2, \ldots ,$$

and the law of combination is ordinary addition. The sum of any two integers is again an integer, 0 is the identity, and the inverse of n is $-n$. This group is of *infinite order*. The identity, 0, has order one; the order of all other elements is infinite.

(2) The elements are all rational numbers with addition as the law of combination.

(3) The elements are all complex numbers with addition as the law of combination.

Note that the non-negative integers do *not* form a group under addition. The "products" are in the set, the set contains the identity (0), but the inverses are not in the set.

(4) The group of even integers under addition: This group is of infinite order. Note that group (1) is isomorphic to group (4). To each element n of group (1) we may associate the element $2n$ in group (4). This example shows that a group can be isomorphic with another group which is only a part of the first. (Of course, this *cannot* occur for groups of *finite* order.)

(5) The powers of 2 under ordinary multiplication, that is,

$$\ldots 2^{-2}, 2^{-1}, 2^0, 2^1, 2^2, \ldots .$$

If we make n in group (1) correspond to 2^n in group (5), we see that the two groups are isomorphic.

(6) The rational numbers, excluding zero, under multiplication.

(7), (8) Similarly, the real numbers (complex numbers), excluding zero, under multiplication.

(9) The nth roots of unity under multiplication. These are the numbers $e^{2\pi mi/n}$, $m = 0, 1, \ldots, n - 1$. This is a cyclic group, all the elements being powers of $e^{2\pi i/n}$. It is clear that there is only one structure of the cyclic group of order n; i.e., all cyclic groups of order n are isomorphic to one another, for if the elements of the two cyclic groups are powers of a and b, respectively, we make the correspondence $a^r \leftrightarrow b^r$. This example also shows that there are groups of *any finite order* n. Moreover, if we have any cyclic group of infinite order, i.e., a group with elements

$$\ldots a^{-3}, a^{-2}, a^{-1}, a^0 = e, a^1, a^2, \ldots ,$$

it is isomorphic to group (1), and hence there is only one structure for the cyclic group of *denumerable* order.

(10) The permutations of degree n,

$$\begin{pmatrix} 1 & 2 & \ldots n \\ p_1 p_2 & \ldots p_n \end{pmatrix},$$

form a group called the *symmetric group* of degree n, which we denote by the symbol S_n. The number of elements in S_n is easily found: 1 can be replaced by any of the symbols 1 to n; after this is done, 2 can be replaced by any of the remaining $n - 1$ symbols, etc. Thus the order of the symmetric group S_n is $n!$

Consider the permutation of degree 8,

$$\begin{pmatrix} 12345678 \\ 23154768 \end{pmatrix}.$$

Starting with the symbol 1, we see that the permutation takes 1 into 2. We then look for 2 in the top line and see that the permutation takes 2 into 3, and 3 into 1, closing a *cycle*, which we write as (123). We now start with some other symbol in the top line, say 8. The permutation takes 8 into 8, giving a cycle (8). Continuing, we find the cycles (45) and (67). We may write our permutation alternatively as

$$(123)(45)(67)(8).$$

Note that the cycles have no letters in common. The cycle (123) may be looked upon as an abbreviation for the following permutation of degree 8:

$$\begin{pmatrix} 12345678 \\ 23145678 \end{pmatrix}.$$

We may shorten the notation and omit the unpermuted symbol 8, writing our original permutation as

$$(123)(45)(67), \tag{1–18}$$

but we must keep in mind the degree of the permutation. Since the cycles have no elements in common, they commute with one another, e.g.,

$$(123)(45) = (45)(123),$$

so the order in which we write the cycles is irrelevant. Moreover, in writing an individual cycle, we can start at any point in the chain:

$$(123) = (231) = (312).$$

A cycle containing *two* symbols (2-cycle) is called a *transposition*. Any cycle can be written as a product of transpositions (having elements in common). Thus

$$(123) = (13)(12),$$

and, in general,

$$(12 \ldots n) = (1n) \cdots (13)(12).$$

The cycle on 3 letters is written as a product of 2 transpositions, the cycle on n letters as a product of $n - 1$ transpositions.

Proceeding in this fashion, one can resolve any permutation into a product of transpositions. In our example,

$$(123)(45)(67) = (13)(12)(45)(67). \tag{1-19}$$

In (1–18), the number of permuted symbols is 7, and the number of *independent* cycles is 3. The difference of these numbers, $7 - 3 = 4$, is called the *decrement* of the permutation. The reader should supply the proof that if the decrement of a permutation is even (odd), its resolution into a product of transpositions will have an even (odd) number of factors. Permutations with even (odd) decrement are said to be even (odd) permutations.

We now show that if we multiply a given permutation by a transposition, we change the decrement by ± 1, i.e., we go from an even permutation to an odd one, or vice versa. Consider the transposition (ab) by which we multiply the given permutation. We resolve the permutation into independent cycles. Suppose a and b occur in the same cycle:

$$(a \ldots xb \ldots y).$$

Then

$$(ab)(a \ldots xb \ldots y) = (a \ldots x)(b \ldots y),$$

so that the decrement is decreased by 1. Conversely, if a and b are in different independent cycles, we see (by reversing the steps) that multiplying by (ab) increases the decrement by 1. The product of two even (or odd) permutations is an even permutation. The product of an odd and an even permutation is odd. The inverse of an even (odd) permutation is even (odd). The odd permutations of degree n do not form a group (since their products are even permutations). The even permutations of degree n do form a group, the *alternating group* α_n, which has $n!/2$ elements.

(11) Start with the set of all positive integers. Consider permutations in which any *finite* number of the symbols are permuted. The set of permutations constructed in this way is called the *denumerable symmetric group.*

(12) The group of nonsingular n-by-n matrices, with matrix multiplication as the law of combination.

(13) The unimodular group: the same as (12), except that the elements have determinant $= \pm 1$.

(14) The special unimodular group: the same as (13), except that the determinants are $+1$.

1–3 Subgroups. Cayley's theorem. If we select from the elements of the group G a subset $\mathcal{K}$, we use the notation $\mathcal{K} \subset G$ to symbolize that $\mathcal{K}$ is contained in G. If the subset $\mathcal{K}$ forms a group (under the *same* law of combination that was used in G), $\mathcal{K}$ is said to be a *subgroup* of the group G. Every group has two trivial subgroups: the group consisting of the identity element alone, and the whole group G itself. These are said to be *improper* subgroups. The problem of finding all other (*proper*) subgroups of a given group G is one of the main problems of group theory.

We must emphasize the requirement that $\mathcal{K}$ form a group under the same law of combination as G. The rational numbers form a group G under addition. The positive rational numbers form a group G' under multiplication; but G' is not a subgroup of G, even though the elements of G' are a subset of the elements of G.

To test a subset $\mathcal{K}$ of elements of the group G to see whether $\mathcal{K}$ is a subgroup, we require that

(1) the product of any pair of elements of $\mathcal{K}$ be in $\mathcal{K}$;

(2) $\mathcal{K}$ contain the inverse of each of its elements.

The other requirements for a group are taken care of because $\mathcal{K}$ is contained in the group G. Thus, the associative law held for G and thus holds for $\mathcal{K}$. The group G contained an identity element, and from (2) and (1) it follows that $\mathcal{K}$ contains it.

For a finite group (or any group, all of whose elements are of finite order), only requirement (1) is needed. In this case, for each element a in $\mathcal{K}$, there exists a power of a (also in $\mathcal{K}$), say a^{n-1}, such that $aa^{n-1} = a^n = e$, so that requirement (2) is a consequence of (1).

In general, both requirements are needed for infinite groups. For example, the positive integers under addition satisfy the first and not the second requirement, and hence do not form a subgroup of all integers under addition.

The alternating group $\mathcal{A}_n$ is a subgroup of the symmetric group S_n on the same symbols ($\mathcal{A}_n \subset S_n$). For $n = 3$, S_3 consists of the elements e; (123), (132); (12), (13), (23), and $\mathcal{A}_3$ contains e; (123), (132).

If $\mathcal{K}$ is a subgroup of G, and K is a subgroup of $\mathcal{K}$, then K is a subgroup of G. This *transitive* relation leads to the idea of sequences of subgroups, each containing all those preceding it in the sequence. For example, we may take for G all integers under addition, for $\mathcal{K}$ all even integers under addition, for K all multiples of $2^2 = 4$ under addition, next all multiples of 2^3, etc.:

$$G: \quad \ldots -2, -1, 0, 1, 2, \ldots,$$

$$\mathcal{K}: \quad \ldots -4, -2, 0, 2, 4, \ldots,$$

$$K: \quad \ldots -8, -4, 0, 4, 8, \ldots,$$

and so on.

Each group contains all the successive subgroups listed below it, that is,

$$G \supset \mathcal{H} \supset K \supset \ldots;$$

and in this case, moreover, all these groups are isomorphic:

$$G \approx \mathcal{H} \approx K \approx \cdots.$$

Thus a group can be isomorphic with one of its proper subgroups, but this is not possible for groups of finite order.

In the case of S_3 given above,

$$S_3 \supset \mathcal{Q}_3 \supset e,$$

where e is the group consisting of the identity only. For all groups of finite order, sequences such as those presented in this special example start with G and end with the group containing only e. Other sequences in S_3 would be:

$$S_3 \supset \mathcal{H}_1 \supset e,$$

$$S_3 \supset \mathcal{H}_2 \supset e,$$

$$S_3 \supset \mathcal{H}_3 \supset e,$$

where $\mathcal{H}_1$ contains two elements, e and (23), $\mathcal{H}_2$ contains e and (13), and $\mathcal{H}_3$ contains e and (12).

A group G' may be isomorphic with one or more subgroups of another group G; for example, S_2, the permutation group on two symbols, is isomorphic with the subgroups $\mathcal{H}_1$, $\mathcal{H}_2$, and $\mathcal{H}_3$ of S_3. Clearly this implies that

$$\mathcal{H}_1 \approx \mathcal{H}_2 \approx \mathcal{H}_3.$$

In general, S_{n-1} is isomorphic with the n subgroups of S_n which are obtained by leaving first the symbol 1, then 2, $\ldots$, then n unchanged. We may also note that the elements of G which are common to two subgroups F_1, F_2 of G form a set D (the *intersection* of F_1 and F_2) which is again a subgroup of G, for if a and b are in D, then a, b are in F_1 and in F_2; therefore, ab and a^{-1} are in F_1 and F_2, and hence are in D, so that D satisfies our requirements for a subgroup. Since every subgroup of G contains the identity e, the intersection of any number of subgroups (i.e., the elements common to all) forms a subgroup which contains at least the identity.

The symmetric groups S_n are especially important because they actually exhaust the possible structures of finite groups. This is shown by *Cayley's theorem:*

THEOREM. Every group G of order n is isomorphic with a subgroup of the symmetric group S_n.

Before proving the theorem, we wish to note its consequences. Cayley's theorem implies that the number of possible nonisomorphic groups of order n is *finite*, for these groups are isomorphic with subgroups of S_n. Since S_n is a finite group, it has only a finite number of subgroups, so that our statement is proved. This is an important result because it at least limits the task of finding independent structures of the group of order n. Now for Cayley's theorem: Let the elements of the group G be $a_1, a_2, \ldots, a_n$. Take any element b; its products with each of the elements in group G are $ba_1, ba_2, \ldots, ba_n$, all of which are different since $ba_i = ba_j$ means $a_i = a_j$. Thus the products ba_i give the elements of G in some new order. We associate with the element b the permutation π_b,

$$b \rightarrow \pi_b = \begin{pmatrix} a_1 \ldots a_n \\ ba_1 \ldots ba_n \end{pmatrix},$$

of the n objects $a_1, \ldots, a_n$. Another element c of G is associated with the permutation π_c:

$$c \rightarrow \pi_c = \begin{pmatrix} a_1 \ldots a_n \\ ca_1 \ldots ca_n \end{pmatrix}.$$

By this same rule, the element cb is associated with the permutation

$$cb \rightarrow \pi_{cb} = \begin{pmatrix} a_1 \ldots a_n \\ cba_1 \ldots cba_n \end{pmatrix}.$$

To prove the isomorphism we must show that π_{cb} is the product of the permutations π_c and π_b.

As we have seen before, only the object-image relation is essential in describing a permutation, while the order in which we write the symbols to be permuted is irrelevant. Thus, π_c can also be written as

$$\pi_c = \begin{pmatrix} ba_1 \ldots ba_n \\ c(ba_1) \ldots c(ba_n) \end{pmatrix}.$$

Taking the product of the images of c and b,

$$\pi_c \pi_b = \begin{pmatrix} ba_1 \ldots ba_n \\ cba_1 \ldots cba_n \end{pmatrix} \begin{pmatrix} a_1 \ldots a_n \\ ba_1 \ldots ba_n \end{pmatrix},$$

we obtain

$$\begin{pmatrix} a_1 \ldots a_n \\ cba_1 \ldots cba_n \end{pmatrix} = \pi_{cb}.$$

The permutation π_e associated with the identity e of G is the identity permutation

$$\pi_e \pi_b = \pi_{eb = b}.$$

Also,

$$\pi_{a^{-1}}\pi_a = \pi_{a^{-1}a} = \pi_e,$$

so that the permutations associated with the elements of G are a subgroup of S_n, and the theorem is proved.

The permutations which we have associated with each group element are just those we would obtain by looking at the group table. For example, in the four-group V, the structure table is

	e	a	b	c
e	e	a	b	c
a	a	e	c	b
b	b	c	e	a
c	c	b	a	e

To find the permutation corresponding to, say the element a, we write in the top line the symbols

$$eabc,$$

and enter below them the symbols as they appear in the row where we multiply by a on the left, that is,

$$aecb;$$

so

$$a \rightarrow \pi_a = \begin{pmatrix} eabc \\ aecb \end{pmatrix}.$$

If we label the elements e, a, b, c as 1, 2, 3, 4, the permutations corresponding to e, a, b, c are:

$$\pi_e = \begin{pmatrix} 1234 \\ 1234 \end{pmatrix},$$

$$\pi_a = \begin{pmatrix} 1234 \\ 2143 \end{pmatrix} = (12)(34),$$

$$\pi_b = \begin{pmatrix} 1234 \\ 3412 \end{pmatrix} = (13)(24),$$

$$\pi_c = \begin{pmatrix} 1234 \\ 4321 \end{pmatrix} = (14)(23).$$

The permutation groups formed in this way have some special features: (1) They are subgroups of order n of the symmetric group S_n. (2) We see that, except for the identity (which permutes *none* of the symbols), the permutations leave *no* symbol unchanged, for π_b takes a_i into ba_i which can equal a_i only if b is the identity. Permutations in S_n having these two properties are called *regular permutations*. Subgroups which contain only regular permutations have other properties which are consequences of the first two. If two of the permutations, π_1, π_2, took the same symbol, say 3, into the same symbol, say 4, then the permutation $\pi_1\pi_2^{-1}$, which is also in the group, would have to leave 3 and 4 unchanged. On the other hand, so long as $\pi_1 \neq \pi_2$, $\pi_1\pi_2^{-1}$ cannot be the identity, and hence it does permute some of the symbols. But this contradicts our previous result that all the permutations (except the identity) leave no symbol unchanged. Since there are n permutations on the n letters, we conclude that the symbol 1 is taken into a different symbol $(1, \ldots, n)$ by each of the permutations in the group. (The same applies to each of the other symbols.) For example, in V, π_e takes 1 into 1, π_a takes 1 into 2, π_b takes 1 into 3, and π_c takes 1 into 4.

Another interesting property of the regular permutation subgroups is the following: Consider any one of the permutations, π_b, and resolve it into independent cycles. We claim that *all* the cycles must have the *same length*. If the resolution of π_b contained two cycles of differing lengths l_1, l_2 $(l_1 < l_2)$, then $\pi_b^{l_1}$, which is also in the group, would leave all the symbols in the first cycle unchanged, while it would *not* leave all the symbols in the second cycle unchanged; this contradicts our definition of the regular permutation subgroup. For example, in a regular subgroup, a permutation like

$$(12)(345)$$

cannot occur, since its square is

$$(1)(2)(354).$$

The four-group illustrates our statement. Each of its permutations is a product of two independent 2-cycles.

The cyclic group of order 4 gives us the regular subgroup of S_4:

$$e; \quad (1234); \quad (13)(24); \quad (1432).$$

The first element has four 1-cycles, the second and fourth have one 4-cycle, and the third has two 2-cycles.

Cayley's theorem can be used to determine possible group structures. For example, suppose that n is a prime number. Then the group of order n is isomorphic to a regular subgroup of S_n. Since the permutations are regular, their resolution into independent cycles must have all cycles equal in length; therefore, the cycle length must be a divisor of n. Since n is prime, the cycle lengths can only be n or 1 (for the identity). So the regular subgroup is just the cyclic group containing the permutation $(12 \ldots n)$ and its powers. Thus we show that if the order n of a group is *prime*, then the group is *cyclic*. We shall give an alternative proof later in this chapter. Our result solves the problem of finding the possible structures of groups of *prime order*.

Now let us use Cayley's theorem to find the possible group structures of order 4. According to the theorem, our problem is equivalent to finding the regular subgroups of S_4. The permutations of the subgroup must have either one 4-cycle or two 2-cycles (the identity, of course, has four 1-cycles). First suppose that one of the permutations is a 4-cycle, say (1234). Taking successive powers, we obtain (1234), (13)(24), (1432), e, that is, a total of four elements. This is the cyclic group of order 4. Next, suppose that there are no 4-cycles, but only permutations with two 2-cycles. The square of such a permutation is e, and there are just three such permutations,

$$(12)(34); \qquad (13)(24); \qquad (14)(23),$$

which, together with e, give us the four-group.

Problems. (1) Give the elements of the regular subgroup of S_6 which is isomorphic with the cyclic group of order 6.

(2) Use Cayley's theorem to find the possible structures of groups of order 6.

1–4 Cosets. Lagrange's theorem. The abstract groups of order 1, 2, 3 have no proper subgroups. The cyclic group of order 4 has a subgroup of order 2, consisting of a^2 and e, since $(a^2)(a^2) = a^4 = e$. The four-group has three proper subgroups of order 2: a, e; b, e; c, e, all of which have the same structure. We note that the order of the subgroups, 2, is a divisor of the order of the group, 4. We shall now show that this result (due to Lagrange) is generally valid:

The *order* of a *subgroup* of a *finite group* is a *divisor* of the *order* of the group.

Let the group G of order g contain the subgroup $\mathcal{K}$ of order h. If $\mathcal{K}$ exhausts the group G, then $\mathcal{K} = G$, and the result is trivial. (Note that the equals sign here means that the two sets contain the same elements.) If not, then let a be an element of G which is not contained in $\mathcal{K}$. Denoting the elements of $\mathcal{K}$ by $e, H_2, H_3, \ldots, H_h$, we form the products $ae = a$, $aH_2, aH_3, \ldots, aH_h$. We shall denote this aggregate of products symbolically as $a\mathcal{K}$. The products aH_ν are all different, for if $aH_\nu = aH_\mu$, then $H_\nu = H_\mu$. Also, none of them are contained in $\mathcal{K}$, for if $aH_\nu = H_\mu$, then $a = H_\mu H_\nu^{-1}$, and a would be contained in $\mathcal{K}$ contrary to our assumption.

Now we have two sets of h distinct elements each, $\mathcal{H}$ and $a\mathcal{H}$, which are contained in G. If the group G has not been exhausted, we proceed as before, choosing some element b of G which is not contained in either $\mathcal{H}$ or $a\mathcal{H}$. The set $b\mathcal{H}$ will again yield h new elements of the group G, for if $bH_\nu = aH_\mu$, then $b = aH_\mu H_\nu^{-1}$, and b would be contained in $a\mathcal{H}$ contrary to our assumption. Continuing this process, we can exhibit the group G as the sum of a finite number of distinct sets of h elements each:

$$G = \mathcal{H} + a\mathcal{H} + b\mathcal{H} + \cdots + k\mathcal{H}.* \qquad (1\text{–}20)$$

Thus the order g of the group G is a multiple of the order h of its subgroup $\mathcal{H}$, that is,

$$g = mh, \qquad (1\text{–}21)$$

where m is called the *index* of the subgroup $\mathcal{H}$ under the group G. The sets (complexes) of elements of the form $a\mathcal{H}$ in (1–20) are called *left cosets* (*residue classes, nebenklassen, nebengruppen*) of $\mathcal{H}$ in G. They are certainly not subgroups since they do not contain the identity element. We could, of course, also have carried out our proof by multiplying $\mathcal{H}$ on the right. This would lead to the resolution of G into *right cosets* with respect to the subgroup $\mathcal{H}$:

$$G = \mathcal{H} + \mathcal{H}a' + \mathcal{H}b' + \cdots + \mathcal{H}k'. \qquad (1\text{–}20\text{a})$$

From Eq. (1–21) we see that a group whose order is a prime number can have no proper subgroups.

Starting with any element a (of order h) of the group G, we can generate a cyclic group by taking successive powers of a. The cyclic group $\{a\}$: $a^0 = e, a, a^2, \ldots, a^{h-1}$, which is generated by a is called the *period* of a. It is the smallest subgroup of G which contains the element a. From Eq. (1–21), h is a divisor of g, so the *orders* of all *elements* of a *finite group* must be *divisors* of the *order* of the group. We conclude that a group of prime order is necessarily cyclic, and can be generated from any of its elements other than the identity element. This gives us the answer to the problem of finding the structure of the group of order 5: It is generated from an element a such that $a^5 = e$; and similarly for the groups of order 7, 11, 13, etc.

---

* The plus sign in Eq. (1–20) denotes the summation of sets: The set G contains all the elements in $\mathcal{H}$ and all the elements in $a\mathcal{H}$, etc. In this case, the sets $\mathcal{H}$, $a\mathcal{H}$, ... have no elements in common. In general, the *sum* of two *sets* A and B is the set of all objects which are contained in at least one of the sets A, B.

The following are examples of the resolution of a group into cosets:

EXAMPLES.

(1) Let G be the group of integers under addition, and $\mathcal{K}$ the group of multiples of 4 under addition. Then

$$G = \mathcal{K} + (1 + \mathcal{K}) + (2 + \mathcal{K}) + (3 + \mathcal{K}),$$

so that $\mathcal{K}$ has the index 4 in G. Here the coset $(1 + \mathcal{K})$ is the set of integers of the form $4n + 1$ (n integral). Two elements a and b of G are in the same coset if $a - b$ is divisible by 4.

(2) Let $G = S_3$ and $\mathcal{K}$ be the subgroup containing e and (12). Then

$$S_3 = \mathcal{K} + (13)\mathcal{K} + (23)\mathcal{K} = \mathcal{K} + \mathcal{K}(13) + \mathcal{K}(23).$$

Problem. The cyclic permutations on four symbols form a subgroup $\mathcal{K}$ of S_4. Resolve S_4 into left cosets with respect to $\mathcal{K}$. Compare this resolution with one into right cosets.

Lagrange's theorem can be used for finding the possible structures of groups of a given order. As an example, we find all structures of order 6. Since the order of the group is 6, the order of each of its elements is a divisor of 6, i.e., 1, 2, 3, or 6. If the group contains an element a of order 6, then the group is the cyclic group $a, a^2, \ldots, a^6 = e$. To find other possible structures, we suppose that the group contains no element of order 6, but has an element a of order 3. Thus the group contains the subgroup a, $a^2, a^3 = e$. If the group also contains another element, b, then it contains the six distinct elements e, a, a^2, b, ba, ba^2. The element b has order 2 or 3. If the order of b is 3, that is, $b^3 = e$, the element b^2 must be one of the six elements listed. We cannot have $b^2 = e$ (since we assumed that b is of order 3), and $b^2 = b$, ba, or ba^2 implies $b = e$, a, or a^2, respectively, which contradicts our assumption that b is distinct from these elements. Furthermore, $b^2 = a$ implies $ba = e$, and $b^2 = a^2$ implies $ba^2 = e$, both of which contradict our assumptions. Thus, the order of b cannot be 3, and we must have $b^2 = e$. The product ab cannot be e, a, a^2, or b. If $ab = ba$, then

$$(ab)^2 = (ab)(ab) = (ab)ba = ab^2a = a^2,$$

$$(ab)^3 = a^2(ab) = b; \qquad (ab)^4 = a, \qquad (ab)^5 = ba^2, \qquad (ab)^6 = e,$$

and therefore the group would contain an element, ab, of order 6, contrary to assumption. (This can be done more briefly: a is of order 2, b of order 3, and $ab = ba$, that is, a and b commute. Taking powers, we see that the order of ab is the least common multiple of the orders of a and b.) We are now left with

$$b^2 = e, \qquad ab = ba^2;$$

$ab = ba^2$ implies $(ab)^2 = (ab)ba^2 = e$.

This last assumption leads to no contradictions. The group of order 6 thus found can be characterized briefly. It is constructed by taking all possible products of a and b, where

$$a^3 = b^2 = (ab)^2 = e.$$

The group table is easily constructed:

e	a	a^2	b	ba	ba^2
a	a^2	e	ba^2	b	ba
a^2	e	a	ba	ba^2	b
b	ba	ba^2	e	a	a^2
ba	ba^2	b	a^2	e	a
ba^2	b	ba	a	a^2	e

This group is isomorphic with S_3.

Problem. Find the possible structures of groups of order 8.

1–5 Conjugate classes. An element b of the group G is said to be *conjugate* to the element a if we can find an element u in G such that

$$uau^{-1} = b. \tag{1–22}$$

Sometimes, we say that b is the *transform* of a by u. By choosing $u = e$, we see that a is conjugate to itself. Also, if b is conjugate to a and c is conjugate to b, then c is conjugate to a, for if $c = vbv^{-1}$, $b = uau^{-1}$, then $c = vuau^{-1}v^{-1} = (vu)a(vu)^{-1}$. Furthermore, from (1–22), $a = u^{-1}b(u^{-1})^{-1}$; so if b is conjugate to a, a is conjugate to b. Thus we have a relation between elements which fulfills all the requirements for an *equivalence relation* (symbol $\equiv$):

(1) $a \equiv a$.

(2) If $a \equiv b$, then $b \equiv a$.

(3) If $a \equiv b$, and $b \equiv c$, then $a \equiv c$.

An equivalence relation can be used to separate a set into *classes*—in our case, to separate the group into classes of elements which are conjugate to one another. In an abelian group each element forms a class by itself, since for any a, b, $bab^{-1} = a$. In every group, the identity element e forms a class by itself. We note that all elements in the same class have the same order; if $a^h = e$ and $b = uau^{-1}$, then $b^h = (uau^{-1})^h = uau^{-1}uau^{-1} \ldots = ua^h u^{-1} = ueu^{-1} = e$.

Conjugate transformations. For groups of transformations, the notion of conjugate elements has a simple physical meaning. Suppose that a is the reflection in the plane P, and c is a rotation about some axis. Since a leaves every point x of P unchanged, $ax = x$. Then

$$(cac^{-1})(cx) = cax = cx.$$

So the transformation cac^{-1} leaves the set of points cx unchanged. In other words, cac^{-1} is the reflection in the plane obtained from P by the rotation c. Similarly, if a represents a rotation about one axis, then cac^{-1} represents a rotation through the same angle about the axis obtained from the first by transforming with c.

For example, suppose a group contains an element a which is the rotation through angle ϕ about the line l. Let c be some other transformation of the group, say a translation. Then c takes l into another line l'. The transformation cac^{-1} has the following properties: c^{-1} takes the line l' into the line l; then a is a rotation through angle ϕ about l and leaves the points of l fixed; and finally, c takes l back into l'. The net result leaves the points of l' fixed; hence cac^{-1} is a rotation (through angle ϕ) about the line l'.

Conjugate permutations. Now let us apply the notion of conjugate elements to permutations. Suppose that a is the permutation

$$\begin{pmatrix} 1 & \ldots & n \\ a_1 & \ldots & a_n \end{pmatrix}$$

and b the permutation

$$\begin{pmatrix} 1 & \ldots & n \\ b_1 & \ldots & b_n \end{pmatrix} = \begin{pmatrix} a_1 & \ldots & a_n \\ a_{b_1} & \ldots & a_{b_n} \end{pmatrix}.$$

Then

$$bab^{-1} = \begin{pmatrix} a_1 & \ldots & a_n \\ a_{b_1} & \ldots & a_{b_n} \end{pmatrix} \begin{pmatrix} 1 & \ldots & n \\ a_1 & \ldots & a_n \end{pmatrix} \begin{pmatrix} b_1 & \ldots & b_n \\ 1 & \ldots & n \end{pmatrix}$$

$$= \begin{pmatrix} b_1 & \ldots & b_n \\ a_{b_1} & \ldots & a_{b_n} \end{pmatrix}.$$

Comparing with a, we see that to obtain bab^{-1} we apply the permutation b to the top row of a and to the bottom row of a individually. For example, let

$$a = \begin{pmatrix} 1234 \\ 4321 \end{pmatrix}, \qquad b = \begin{pmatrix} 1234 \\ 2134 \end{pmatrix};$$

we apply the permutation b to the top row of a, so that 1234 becomes 2134, and to the bottom row of a, so that 4321 becomes 4312; thus bab^{-1} is

$$\begin{pmatrix} 2134 \\ 4312 \end{pmatrix} = \begin{pmatrix} 1234 \\ 3412 \end{pmatrix}.$$

The procedure works equally well if the permutations are given in cycle form; e.g., $a = (12)(345)$ and $b = (24135)$. Then $bab^{-1} = (34)(512)$. We see that conjugate permutations have the same cycle structure so that the permutations in a given class are either all even or all odd. For example, in S_3, e forms a class by itself, the elements (12), (13), (23) form a class, and the elements (123), (213) form a class; so S_3 contains three conjugate classes. In S_4, the distinct classes are:

1. e;
2.   (12), (13), (14), (23), (24), (34);
3.   (12)(34), (13)(24), (14)(23);
4.   (123), (132), (124), (142), (134), (143), (234), (243);
5.   (1234), (1243), (1324), (1342), (1423), (1432).

Problem. Separate the elements of S_5 into conjugate classes.

From the procedure used in the example of S_4, the general method should be clear. The permutations of S_n operate on a total of n symbols. Suppose that we resolve the permutations into independent cycles and let the number of 1-cycles be ν_1, of 2-cycles be ν_2, ..., of n-cycles be ν_n. Since the total number of symbols is n, we must have

$$\nu_1 + 2\nu_2 + \cdots + n\nu_n = n. \tag{1–23}$$

A permutation which when resolved into independent cycles has ν_1 1-cycles, ν_2 2-cycles, ..., ν_n n-cycles is said to have the cycle structure $(1^{\nu_1}, 2^{\nu_2}, \ldots, n^{\nu_n})$, or briefly, (ν). As we saw above, all the permutations of S_n which have the same cycle structure (ν) form a conjugate class in S_n. Each solution of (1–23) for positive integers $\nu_1, \ldots, \nu_n$ gives a class in S_n, and hence the number of classes is just the number of such solutions. If we let

$$\begin{aligned} \nu_1 + \nu_2 + \cdots + \nu_n &= \lambda_1, \\ \nu_2 + \nu_3 + \cdots + \nu_n &= \lambda_2, \\ &\;\;\vdots \\ \nu_n &= \lambda_n, \end{aligned} \tag{1–24}$$

then

$$\lambda_1 + \lambda_2 + \cdots + \lambda_n = n \quad \text{and} \quad \lambda_1 \geq \lambda_2 \geq \cdots \geq \lambda_n \geq 0. \tag{1–25}$$

The splitting-up of n into a sum of n integers as in (1–25) is called a *partition* of n. From (1–23), (1–24), and (1–25) each solution of (1–23) (in nonnegative integers) corresponds to a partition of n into $(\lambda_1, \ldots, \lambda_n)$. Conversely, given a partition as in (1–25), there is a corresponding cycle structure, namely:

$$\nu_1 = \lambda_1 - \lambda_2,$$
$$\nu_2 = \lambda_2 - \lambda_3,$$
$$\vdots$$
$$\nu_n = \lambda_n. \tag{1-26}$$

Usually we do not bother to record those λ's in (1–25) which are 0; e.g., the partition of 5,

$$(22100),$$

i.e., $5 = 2 + 2 + 1 + 0 + 0 = \lambda_1 + \lambda_2 + \cdots + \lambda_5$ is written as (221) or, even more briefly, as $(2^2 1)$. From (1–26), the corresponding cycle structure is

$$\nu_1 = \lambda_1 - \lambda_2 = 2 - 2 = 0,$$
$$\nu_2 = \lambda_2 - \lambda_3 = 2 - 1 = 1,$$
$$\nu_3 = \lambda_3 = 1,$$

i.e., one 2-cycle and one 3-cycle.

Similarly in S_6, the partition (31^3) has

$$\nu_1 = 2, \qquad \nu_2 = 0, \qquad \nu_3 = 0, \qquad \nu_4 = 1;$$

so the cycle structure corresponding to (31^3) is two 1-cycles and one 4-cycle.

A transposition in S_n contains one 2-cycle and $(n-2)$ 1-cycles; hence the partition corresponding to this cycle structure is $(n-1, 1)$.

We see that the problem of finding the number of conjugate classes in S_n is reduced to the problem of partitioning n. We list the partitions for the first few symmetric groups and the total number, r, of classes:

$$S_1: \qquad (1); \quad r = 1,$$
$$S_2: \qquad (2), (1^2); \quad r = 2,$$
$$S_3: \qquad (3), (21), (1^3); \quad r = 3,$$
$$S_4: \qquad (4), (31), (2^2), (21^2), (1^4); \quad r = 5.$$

Note the order in which we record the partitions. The partition having maximum λ_1 is recorded first, and if two partitions have the same $\lambda_1, \ldots, \lambda_i$, then the one with larger λ_{i+1} is listed first.

Problem. Continue the table to $n = 5, 6, 7$. For $n = 5$, give the cycle structure corresponding to each partition.

An important datum is the number of permutations of S_n in a given conjugate class. This number $n_{(\nu)}$ is easily found as follows: The class with cycle structure (ν) contains the permutations with ν_1 1-cycles, ν_2 2-cycles, $\ldots$, ν_n n-cycles. We imagine the structure to be written out with no symbols inserted:

$$\underbrace{(.)\cdots(.)}_{\nu_1 \text{ 1-cycles,}} \qquad \underbrace{(..)(..)\cdots(..)}_{\nu_2 \text{ 2-cycles,}} \qquad \underbrace{(...)\cdots(...)}_{\nu_3 \text{ 3-cycles,}}\cdots \quad \text{etc.}$$

There is a total of n places in the various cells into which the n symbols 1 to n are to be placed. This can be done in $n!$ ways. However, duplications will result; e.g., if 1 and 2 appear in 1-cycles, (1)(2) is the same as (2)(1). All the ν_1 1-cycles can be permuted (in $\nu_1!$ ways), all the ν_2 2-cycles can be permuted among themselves in $\nu_2!$ ways, etc., so that a given permutation will be repeated $\nu_1! \nu_2! \cdots \nu_n!$ times. In addition, a 2-cycle like (12) can also appear as (21); a 3-cycle like (123) can also appear as (231) or (312), etc.; thus each permutation is repeated $1^{\nu_1} 2^{\nu_2} \ldots n^{\nu_n}$ times. Hence the number of *distinct* permutations of S_n having the cycle structure (ν) is

$$n_{(\nu)} = \frac{n!}{\nu_1! \cdot 2^{\nu_2} \cdot \nu_2! \cdot 3^{\nu_3} \cdot \nu_3! \cdots n^{\nu_n} \cdot \nu_n!} \tag{1-27}$$

Problem. Find the number of permutations in each class of S_5.

It is important to remember that what we have done above applies *only* to the full symmetric group S_n. For example, in S_4 the permutations

$$(12)(34), \quad (13)(24), \quad (14)(23)$$

are all in the same class because S_4 contains a permutation $b = (23)$ such that

$$b(12)(34)b^{-1} = (13)(24).$$

If we consider the four-group V, the permutations listed above are not in the same class in V, since V does not contain the transpositions. In fact, since V is abelian, each of its elements forms a class by itself.

Problem. By taking successive products of the permutations

$$(1234)(5678) \quad \text{and} \quad (1537)(2846),$$

show that one generates a group of order 8. Separate the elements of the group into conjugate classes. Show that this group (the *quaternion group*) is isomorphic to the group with elements

$$1, -1, i, -i, j, -j, k, -k;$$

$$i^2 = j^2 = k^2 = -1, \quad ij = k, \quad jk = i, \quad ki = j.$$

1–6 Invariant subgroups. Factor group. Homomorphism. Starting from a subgroup $\mathfrak{K}$ of the group G, we can form the set of elements $a\mathfrak{K}a^{-1}$, where a is any element of G. (By $a\mathfrak{K}a^{-1}$ we mean the set of elements aha^{-1}, where h runs through all the elements of $\mathfrak{K}$.) This set of elements (or *complex*) is again a subgroup of G, (reader, please verify), and is said to be a *conjugate subgroup* of $\mathfrak{K}$ in G. By choosing various elements a from G, we may obtain different conjugate subgroups. It may happen that for *all* a,

$$a\mathfrak{K}a^{-1} = \mathfrak{K}, \tag{1–28}$$

i.e., all the conjugate subgroups of $\mathfrak{K}$ in G are identical with $\mathfrak{K}$. In this case we say that $\mathfrak{K}$ is an *invariant subgroup* (*self-conjugate subgroup*, *normal divisor*) in G. Written out, (1–28) states that, given an element h_1 in $\mathfrak{K}$, we can, for any a, find an element h_2 in $\mathfrak{K}$ such that $ah_1a^{-1} = h_2$ or $ah_1 = h_2a$. The last equation can be written as

$$a\mathfrak{K} = \mathfrak{K}a, \tag{1–29}$$

and leads to a second definition of an invariant subgroup: The subgroup $\mathfrak{K}$ is invariant in G if the left and right cosets formed with any element a of G are the same. We may say from (1–29) that the subgroup $\mathfrak{K}$, taken as an entity, commutes with all elements of G. From the statements following (1–28) we also see that (1–28) implies the result: If h_1 is in $\mathfrak{K}$, then all elements ah_1a^{-1} are in $\mathfrak{K}$. In other words, a subgroup $\mathfrak{K}$ of G is invariant if, and only if, it contains elements of G in *complete classes;* i.e., for any class of G, $\mathfrak{K}$ contains either *all* or *none* of the elements in the class. The identity element and the whole group G are trivial invariant subgroups of G. A group which has no invariant (proper) subgroups is said to be *simple.* A group is said to be *semisimple* if none of its invariant subgroups are abelian. All the subgroups of an abelian group are clearly invariant.

Invariant subgroups have some very special properties. If $\mathfrak{K}$ is invariant in G, then

$$(a\mathfrak{K})(b\mathfrak{K}) = a(\mathfrak{K}b)\mathfrak{K} = a(b\mathfrak{K})\mathfrak{K} = ab(\mathfrak{K}\mathfrak{K}) = (ab)\mathfrak{K}, \tag{1–30}$$

since HH simply gives the set of elements in H. Thus the product of two cosets of an invariant subgroup is again a coset. Also, we note that

$$H(aH) = (Ha)H = (aH)H = a(HH) = aH, \qquad (1\text{--}31)$$

so that multiplying any coset of H by H yields the coset; the invariant subgroup H behaves like an identity element in this multiplication of cosets. Again, if we have a coset aH, we can find the coset which is its inverse (under our new "coset multiplication"), namely, the coset $a^{-1}H$ which contains a^{-1}:

$$(a^{-1}H)(aH) = a^{-1}HaH = a^{-1}aHH = H. \qquad (1\text{--}32)$$

When we consider the cosets of H as elements, and define product as the result of coset multiplication, the cosets of the invariant subgroup form a group which is called the *factor group* (or *quotient group*), symbolized by G/H, whose order is the index of H in G.

Let us find the conjugate subgroups of the subgroup H [e, $(12)(34)$] in the group S_4. The subgroup H is not invariant since it does not contain all elements of the third class listed for S_4 in Section 1–5. When we form aHa^{-1} with elements a from S_4, we must get cycle structures which are the same as those of the elements of H. Thus the conjugate subgroups are:

$$
\begin{aligned}
H: \quad & e,\ (12)(34), \\
H': \quad & e,\ (13)(24), \\
H'': \quad & e,\ (14)(23).
\end{aligned}
$$

Each is obtained 8 times:

H for $a = e, (12), (34), (12)(34), (13)(24), (14)(23), (1324), (1423)$;
H' for $a = (14), (23), (132), (124), (143), (234), (1243), (1342)$;
H'' for $a = (13), (24), (123), (142), (134), (243), (1234), (1432)$.

On the other hand, the subgroup

$$V: e,\ (12)(34),\ (13)(24),\ (14)(23)$$

is invariant in S_4 since it contains elements of S_4 in complete classes.

Problems. (1) Without enumerating, find the conjugate subgroups of S_3 in S_4; of S_2 in S_4; of the cyclic group [e, (123), (132)] in S_4.

(2) Prove that a subgroup of index 2 is necessarily invariant.

(3) Show that all subgroups of the quaternion group are invariant.

In the cyclic group G of order 4 $(a, a^2, a^3, a^4 = e)$, the subgroup $\mathcal{H}$ (e, a^2) is invariant (G is abelian!). The factor group $G/\mathcal{H}$ contains two elements,

$$E = (e, a^2), \qquad A = (a, a^3),$$

with $A^2 = E$.

Earlier we introduced the concept of isomorphism for groups which had the same structure. A one-to-one correspondence was set up between elements a of a group G and elements a' in a group G', such that $(ab)' = a'b'$. By a *homomorphic* mapping of G on G' we mean a similar correspondence which preserves products, but now *several* elements of G may have the *same image* in G'.

For example, let G be the cyclic group of order 4, and let G' consist of an identity element alone. We map all the elements of G into the identity of G'. Similarly, if G' is the group of order 2, with elements b and $b^2 = e$, we obtain a homomorphism of G on G' if we map (a^2, e) into $b^2 = e$, and (a, a^3) into b.

In a homomorphic mapping of G on G', the image of the identity element e of G is the identity e' of G', for if $ab = a$ (so that b is the identity of G), then $a'b' = a'$, so that b' is the identity in G'. If the set of elements a_1, $a_2, \ldots, a_h$ of G are mapped on e', then choosing some other element b of G, we find h distinct elements $ba_1, ba_2, \ldots, ba_h$ of G which are mapped on the same element b' of G', for $(ba)' = b'a' = b'e' = b'$. Thus equal numbers of elements of G are mapped on each element of G'. If a is mapped on e', so is a^{-1}, as one can verify by taking images in the equation $aa^{-1} = e$. Hence the elements of G which are mapped on e' form a group, in fact an *invariant subgroup* of G, for if a is mapped on e', then so are all the elements in its class, since $(uau^{-1})' = u'a'(u^{-1})' = u'e'(u')^{-1} = e'$. Calling this invariant subgroup $\mathcal{H}$, we find that any coset $a\mathcal{H}$ is mapped on a single element of G', since $(a\mathcal{H})' = a'\mathcal{H}' = a'e' = a'$. Thus we see that G' is isomorphic to the factor group $G/\mathcal{H}$.

1–7 Direct products. A group G is said to be the *direct product* of its subgroups $\mathcal{H}_1, \mathcal{H}_2, \ldots, \mathcal{H}_n$ if:

(1) The elements of different subgroups commute.

(2) Every element of g is expressible in one and only one way as

$$g = h_1 \ldots h_n,$$

where h_1 is in $\mathcal{H}_1, \ldots,$ and h_n is in $\mathcal{H}_n$.

It is assumed that none of the subgroups consists of the identity element only. Symbolically, we write

$$G = \mathcal{H}_1 \times \mathcal{H}_2 \times \cdots \times \mathcal{H}_n. \tag{1–33}$$

The subgroups $\mathcal{H}_1, \mathcal{H}_2, \ldots, \mathcal{H}_n$ are said to be *direct factors* of G.

From requirements (1) and (2) it follows that the subgroups $\mathcal{K}_i$ have only the identity in common. It also follows that all the $\mathcal{K}_i$ are invariant subgroups in G. According to (2), any element g of G can be written as

$$g = h_1 h_2 \ldots h_n,$$

where, according to (1), the h_i all commute with one another. Suppose h_i' belongs to $\mathcal{K}_i$. Then any conjugate of h_i' is of the form

$$gh_i'g = (h_1 h_2 \ldots h_n) \, h_i' \, (h_1 h_2 \ldots h_n)^{-1} = h_i h_i' h_i^{-1}, \qquad (1\text{–}34)$$

and also belongs to $\mathcal{K}_i$. [The factors in (1–34) which belong to other subgroups than $\mathcal{K}_i$ commute through and cancel.] Thus $\mathcal{K}_i$ contains elements of G in complete classes, and is therefore invariant in G.

As an example of the resolution of a group into direct factors, consider the cyclic group G of order 6 ($a^6 = e$). It can be written as the direct product of the subgroups

$$A: \quad e, a^2, a^4,$$

$$B: \quad e, a^3.$$

We have

$$G = A \times B.$$

Every element of G is expressible in only one way as a product of an element of A by an element of B. (Reader, check!)

In our work we shall also deal with a direct product $G \times G'$ of two groups G, G'. To obtain it, form all pairs

$$(a, a'),$$

where a is in G and a' in G'. The product of pairs is defined by

$$(a, a')(b, b') = (ab, a'b'). \qquad (1\text{–}35)$$

If e, e' are the identity elements of G, G', then the pairs (a, e') form a subgroup Γ which is isomorphic to G, the pairs (e, a') form a subgroup Γ' which is isomorphic to G', and the group of element pairs defined above is the direct product of Γ and Γ' according to our previous definition. Usually we shall say simply that it is the direct product of G and G'. It is clear that the order of the direct product of two groups is the product of their orders.

CHAPTER 2

SYMMETRY GROUPS

2–1 Symmetry elements. Pole figures. A large class of groups which are important in physics and chemistry are the so-called symmetry groups. In this chapter, we shall discuss them in some detail as a preliminary to later use and to provide concrete examples of the concepts introduced in Chapter 1.

The classification of the spectral terms of a polyatomic molecule is related to its symmetry. So, too, is the problem of finding the vibration spectrum of the molecule. The underlying microscopic structure of crystals is related to the symmetry of their external macroscopic form. The classification of electron energy levels in a crystal will be related to the symmetry of the field in the crystal. For all these problems it is essential first to enumerate in a systematic fashion the possible types of symmetry which a molecule or crystal can possess.

The symmetry of a body is described by giving the set of all those transformations which preserve the distances between all pairs of points of the body and bring the body into coincidence with itself. Any such transformation is called a *symmetry transformation*. It is clear that this set forms a group, the *symmetry group* of the body. The distance-preserving transformations can all be built up from three fundamental types:

(1) Rotation through a definite angle about some axis.

(2) Mirror reflection in a plane.

(3) Parallel displacement (translation).

The last symmetry element, translation, can occur *only* if the body is infinite in extent (e.g., an infinite crystal lattice). It is worth pointing out that when one uses translational symmetry in a physical argument, one implies that the solution to the problem is not affected by far-distant points of the body, since the necessarily finite body is replaced by its infinite extrapolation. For example, in the theory of solids one must later consider separately the problem of surface states.

For a body of finite extension, a molecule or the macroscopic form of a mineral, only the first two symmetry types are possible. In fact, all transformations of the symmetry group of a finite body must leave *at least one* point of the body fixed. In other words, all axes of rotation and all planes of reflection must intersect in (at least) one point. Clearly, successive rotations about nonintersecting axes or reflections in nonintersecting planes will result in the introduction of translation and continual shift of the body. The symmetry groups of finite bodies (which must leave at least one point of the body fixed) are called *point groups*. We shall restrict ourselves in this chapter to point groups of finite order.

First, let us suppose that the body is brought into coincidence with itself if we rotate it through an angle $\psi = 2\pi/n$ (n integral) about a certain axis. Such an axis is said to be an *n-fold rotation axis*. If $n = 1$, we have coincidence after rotation through 2π, which is the identity transformation. We shall denote the operation of rotation through $2\pi/n$ by the symbol C_n. Successive applications of this transformation, C_n^2, C_n^3, etc., (i.e., rotations through $4\pi/n$, $6\pi/n$, etc.), must also bring the body into coincidence with itself. If n is divisible by the integer l, then

$$(C_n)^l = C_{n/l}. \tag{2–1}$$

Thus a 6-fold axis is at the same time a 2-fold and a 3-fold axis. The *largest* n (or the smallest angle ψ) characterizes the axis. It is also clear that n successive rotations through $2\pi/n$ about the same axis bring us back to the initial position and produce the identity transformation. Thus

$$C_n^n = E, \tag{2–2}$$

where we have introduced the symbol E (*einheit*) for the identity transformation.

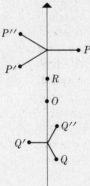

FIGURE 2–1

Since it is essential to visualize what is happening under symmetry transformations, we shall draw numerous diagrams to try to help the reader. One type of diagram is shown in Fig. 2–1. The vertical line is a 3-fold axis passing through the center (fixed point) O of the molecule. The fact that the axis is 3-fold is indicated by the small triangle ▲ affixed to the end of the axis. Similarly, a 2-fold axis would have ◆, and a 4-fold axis ■ at its end, etc. If there is an atom of the molecule at the point P, the presence of the 3-fold axis requires that identical atoms be at P', P'', where $OP = OP' = OP''$. Similarly, if an atom is at Q, we must have identical atoms at Q', Q''. For special positions, like R, on the axis of rotation, the image points are coincident with R.

This method of presentation is satisfactory for molecules. When we consider the external symmetry of crystals, the problem is a little different. The areas of the crystal faces of a given substance may vary considerably, depending on the conditions under which the crystal was grown. The main content of all the investigations on morphology of crystals is the *law of constancy of crystal angles:* The angles which the crystal faces make with one another are fixed characteristics of the crystal. This means that the crystal is characterized by the directions of the normals to the crystal faces and the angles which the normals make with one another. Note that the distances of the crystal faces from the "center" are not the same for all samples. Thus, the simplest method for describing the crystal is to indicate on the surface of a unit sphere the *poles* of a polyhedron, i.e., the points in which the normals to the crystal faces intersect the surface of the sphere. To obtain the shape of the crystal from the pole diagram, we merely draw the tangent planes to the unit sphere at all poles and obtain a closed polyhedron.

To present the *pole figure* in two dimensions, we make use of *stereographic projection.* In Fig. 2–2 we join the pole P on the unit sphere to the south pole S. The intersection P' of the line PS with the equatorial plane is the stereographic projection of P. To show the poles in stereographic projection, we draw the equatorial circle AB and mark the projections of the poles on it by open circles (see Fig. 2–3). Note that all points in the northern hemisphere project into points in the interior of the equatorial circle, and points on the equator project into points on the periphery of the circle. If we treated the points in the southern hemisphere in the same way, their projections would be outside the unit circle, and the projection of the south pole S would be at infinity. To avoid treating the

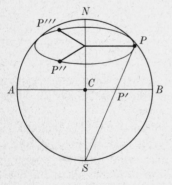

FIGURE 2–2

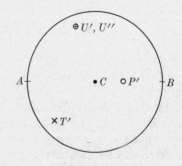

FIGURE 2–3

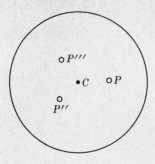

FIGURE 2–4

two halves of the sphere unsymmetrically, we use S for projecting the points of the upper hemisphere, while the points in the lower hemisphere are projected from the north pole N. To distinguish between these, one can either use two separate circles, one for each hemisphere, or, as we shall do, mark projections of poles in the lower hemisphere by crosses. Thus, in Fig. 2–3, T' is the image of a point in the lower hemisphere, and U', U'' (with both cross and circle) means that there are two poles, one in each hemisphere, which are at mirror-image positions relative to the equatorial plane.

From our 3-fold axis of Fig. 2–1, the picture of P, P', P'' in Fig. 2–2 would be three points on a latitude circle, as shown. The stereographic projection of the same points is shown in Fig. 2–4.

If we draw the tangent planes to the sphere of Fig. 2–2 at P, P'', P''', we obtain a triangular pyramid with vertex along the vertical 3-fold axis. If we add a triplet of poles Q, Q'', Q''' along the equator, we obtain a triangular prism. To close the polyhedron, we need another triplet of points R, R'', R''' in the lower hemisphere. We note that the points P and R alone suffice to give a closed polyhedron, but we can add triplets like the points Q to get more complex figures. To avoid giving the figure higher symmetry than the 3-fold axis, we must be sure that there is no simple relation between the positions of P and R. This is more easily accomplished in the molecular problem since, as we see in Fig. 2–1, we can vary the distance of the atom from the axis.

We should note that this first symmetry operation, rotation, is one which can move a body rigidly from its initial to its final position. The second fundamental operation, *reflection in a plane*, is different. It does not correspond to a physically possible rigid displacement. Figures which can be superposed only after mirror reflection are said to be *enantiomorphic*.

We shall use the symbol σ for reflection in a plane. Since two reflections in the same plane return us to the initial position, we have

$$\sigma^2 = E, \tag{2-3}$$

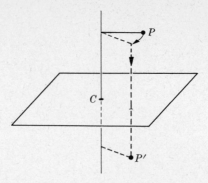

FIGURE 2–5

so that reflection in a plane is an element of order 2. If it is necessary to specify the plane of reflection, this will be indicated by a subscript. Generally, there will be one axis of rotational symmetry which will be the main symmetry of the body. If this is the case, then we shall denote reflection in a plane perpendicular to this principal axis by σ_h (h for horizontal), while we use σ_v for reflection in a plane passing through the axis (v for vertical).

The combined application of our two fundamental operations (rotation about an axis and reflection in a perpendicular plane) leads to a symmetry transformation which we shall call a *rotation-reflection symmetry* (*drehspiegelung*). A body is said to have an n-fold rotation-reflection axis if it is brought into coincidence with itself under the combined operation of rotation through $2\pi/n$ about the axis and reflection in the perpendicular plane (see Fig. 2–5). We denote this operation by S_n. Clearly

$$S_n = C_n \sigma_h = \sigma_h C_n, \qquad (2\text{--}4)$$

where we note that rotation about an axis and reflection in the perpendicular plane commute with each other. From Eqs. (2–2), (2–3), and (2–4) we have

$$(S_n)^n = (C_n \sigma_h)^n = (C_n)^n (\sigma_h)^n; \qquad (2\text{--}5)$$

so, for even n, $(S_n)^n = E$. But for odd n, $(S_n)^n = \sigma_h$. Thus for odd n, if the body has the symmetry S_n, it also has the symmetries σ_h and C_n as independent symmetry elements.

A very important case is that of a rotation-reflection axis of order 2: S_2 is the rotation through π about an axis followed by reflection in the plane perpendicular to the axis. The result of this combined operation, which we call *inversion*, is to move each point P of the body to the point P' which is its inverse relative to the fixed point O: P and P' are on opposite sides of O on the straight line POP', and are equidistant from O. Denoting the inversion by I, we have

$$I = S_2 = C_2\sigma_h. \tag{2-6}$$

From Eq. (2–6) it follows that

$$C_2 = I\sigma_h, \qquad \sigma_h = IC_2. \tag{2-6a}$$

All three elements I, σ_h, and C_2 commute. If any two of these elements belong to the symmetry group, then so does the third.

We now consider a few simple properties of successive reflections in different planes, or of rotations about different axes. First, we note that the product of rotations about two axes passing through the point O is again a rotation about some axis passing through O. Next consider the product of reflections in two planes intersecting in a line. In Fig. 2–6, let v and v' be the traces of the two planes, and O the trace of their line of intersection. Let ϕ be the angle between the planes. Reflecting first in v' and then in v, we see that P goes to P' and then to P'', where $OP = OP' = OP''$. From the construction, the angle between OP and OP'' equals 2ϕ. Thus,

$$\sigma_v\sigma_{v'} = C(2\phi), \tag{2-7}$$

where $C(2\phi)$ is a rotation through 2ϕ about the line of intersection of the planes, in the direction from v' to v. Similarly, $\sigma_{v'}\sigma_v$ is a rotation through the same angle about the same axis, but in opposite direction. So σ_v and $\sigma_{v'}$ do not commute except for the special cases where $\phi = \pi/2$ and the product is C_2, and the trivial case of $\phi = \pi$, when $\sigma_v = \sigma_{v'}$. Multiplying Eq. (2–7) on the left by σ_v, we find

$$\sigma_{v'} = \sigma_v C(2\phi). \tag{2-7a}$$

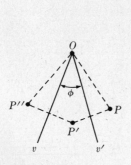

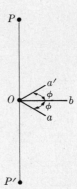

FIGURE 2–6   FIGURE 2–7

Thus the product of a rotation through a given angle about an axis and reflection in a plane passing through the axis is a reflection in a second plane passing through the axis, making an angle with the first plane equal to *half* the angle of the rotation. Again we have three mutually dependent elements. Any two of the symmetry elements σ_v, $\sigma_{v'}$, and $C(2\phi)$ imply the presence of the third.

Another important property is shown in Fig. 2–7: OP is vertical, while Oa and Ob lie in a horizontal plane. Consider the effect of successive rotations through π, first about Oa and then about Ob. The rotation about Oa leaves Oa fixed and moves P to P' ($OP = OP'$). If next we rotate through π about Ob, P' moves back to P, so the product must be a rotation about POP'. Also, the point a which was left fixed by the first rotation now moves to the position a' in Fig. 2–7. Thus successive rotations through angle π about two axes making an angle ϕ with each other result in a rotation through 2ϕ about an axis perpendicular to the first two. In particular, if the X- and Y-axes are 2-fold rotation axes, then so is the Z-axis.

2–2 Equivalent axes and planes. Two-sided axes.

In our enumeration of the possible point-symmetry groups, we shall find the distribution of the group elements in classes. One important means for obtaining conjugate transformations was already mentioned in Chapter 1, Section 1–5, and will now be discussed in detail. If, as in Fig. 2–2, we have a 3-fold axis in the vertical direction, and if the line OP is an n-fold axis, then we must also have n-fold axes along OP'' and OP'''; rotations through the same angle around these three axes all belong to the same class. All the axes (or planes) which can be brought into coincidence by some operation of the group are said to be *equivalent* axes (or planes). Rotations through the same angle about equivalent axes are in the same class. The same rule applies to rotation-reflection operations about equivalent axes.

This result does not help us in the special case of rotations or rotation reflections about the same axis. To see how we treat this case, let us look again at what we did in obtaining equivalent axes. When we describe the rotation about an axis, we first assign a polarity to the axis and a positive sense of rotation as given, say, by the right-hand rule. In Fig. 2–8, if the axis $A'A$ can be shifted to the position $B'B$ by one of the rotations of the symmetry group, the directions of rotation around the equivalent axes will be related as shown in the figure. If, on the other hand, by a rotation of the group, $A'A$ is brought not into $B'B$ but into BB', the diagram would be that shown in Fig. 2–9. In the two cases, the rotations around $B'B$ which are conjugate to a given rotation around $A'A$ are equal rotations in opposite directions. Now let us consider Fig. 2–10. Suppose that there is

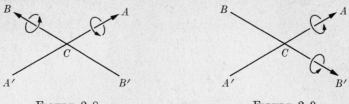

FIGURE 2–8   FIGURE 2–9

a rotation ρ in the group which takes $A'A$ into AA', e.g., a rotation through π about the line CR (so that $\rho^2 = E, \rho = \rho^{-1}$). If the rotation $C(\phi)$ through angle ϕ around $A'A$ takes the point P into P', then the transform of $C(\phi)$ by ρ, namely $\rho C(\phi)\rho$, takes P into the point Q'' and represents a rotation around $A'A$ through angle ϕ in the reverse direction. To see this, carry out $\rho C(\phi)\rho$ step by step: Thus ρ takes P to Q, then $C(\phi)$ takes Q to Q', and finally ρ takes Q' to Q''. Whenever there is a rotation in the group which carries $A'A$ into AA', we say that $A'A$ is a *two-sided axis*, and as we see from Fig. 2–10, equal rotations about $A'A$ in opposite directions are conjugate to one another. If $A'A$ is an n-fold symmetry axis, then C_n^k and $C_n^{n-k} = C_n^{-k}$ are conjugates. So we have the result that for a two-sided axis, any rotation and its inverse belong to the same class.

Next, suppose there is a plane of symmetry perpendicular to AA', namely, the equatorial plane in Fig. 2–11. If $C(\phi)$ takes P to P', then the

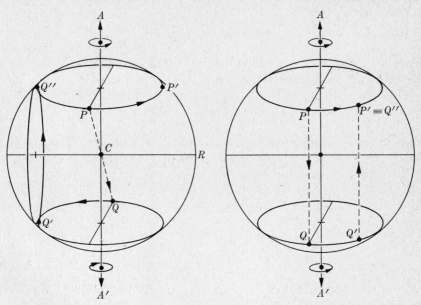

FIGURE 2–10   FIGURE 2–11

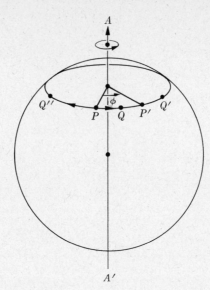

FIGURE 2–12

transform by σ_h, $\sigma_h C(\phi) \sigma_h^{-1}$, gives the following sequence of operations: $\sigma_h^{-1} = \sigma_h$ takes P to Q, then $C(\phi)$ takes Q to Q', and finally σ_h takes Q' to P' so that

$$\sigma_h C(\phi) \sigma_h^{-1} = C(\phi), \tag{2-8}$$

which is evident from our earlier statement that σ_h commutes with $C(\phi)$. The operation σ_h reverses the polarity of the axis and at the same time reverses the sense of rotation as given by the right-hand rule, so that we get the same rotation as before, and the axis $A'A$ is not two-sided (see Fig. 2–11).

On the other hand, suppose that there is a plane of symmetry passing through the axis AA' (perpendicular to the paper) as shown in Fig. 2–12. The operation σ_v does not change the polarity of the axis (it takes $A'A$ into $A'A$), but it does reverse the sense of rotation. In Fig. 2–12, $C(\phi)$ takes P into P', while $\sigma_v C(\phi) \sigma_v^{-1}$ takes P first to Q, then to Q', and finally to Q'' so that

$$\sigma_v C(\phi) \sigma_v^{-1} = \sigma_v C(\phi) \sigma_v = C(-\phi). \tag{2-9}$$

In this case also the axis $A'A$ is two-sided.

After these preliminaries, we can proceed to our problem. We can divide our symmetry operations into two types: pure rotations, and rotation reflections (including reflection and inversion as special cases). We first build up the point groups which contain only rotations.

2–3 Groups whose elements are pure rotations: uniaxial groups, dihedral groups. In this category we first consider the case where we have only one axis of rotational symmetry. We choose the Z-axis as this principal axis. The number of elements in the group is equal to the order of the axis. If the axis is n-fold, we designate the symmetry group as $\mathfrak{C}_n$.

I. *Groups having a single n-fold rotation axis: $\mathfrak{C}_n$.*

The group $\mathfrak{C}_1$ contains only the identity E, and corresponds to complete absence of symmetry. All groups of this type are cyclic, and each element forms a class by itself. To illustrate the symmetry, we draw a set of equivalent poles in stereographic projection. These are shown for $n = 1$, 2, 3, 4, 6 in Figs. 2–13 through 2–17.

We must emphasize that Figs. 2–13 through 2–17 merely show one set of equivalent poles. They are clearly not sufficient to bound the volume of a crystal. (At least four faces are needed to bound a volume.) To show a complete set of poles having the desired symmetry (and no higher symmetry) would make the diagrams extremely complicated. Figure 2–18 shows a complete set of poles for a crystal with symmetry $\mathfrak{C}_1$. Molecular configurations having the desired symmetries are somewhat easier to

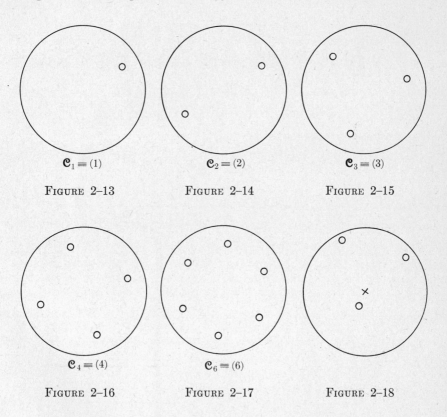

$\mathfrak{C}_1 \equiv (1)$ $\mathfrak{C}_2 \equiv (2)$ $\mathfrak{C}_3 \equiv (3)$

FIGURE 2–13   FIGURE 2–14   FIGURE 2–15

$\mathfrak{C}_4 \equiv (4)$ $\mathfrak{C}_6 \equiv (6)$

FIGURE 2–16   FIGURE 2–17   FIGURE 2–18

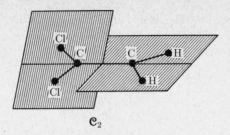

$\mathcal{C}_2$

FIGURE 2–19

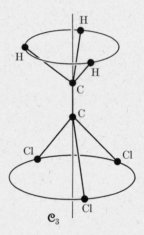

$\mathcal{C}_3$

FIGURE 2–20

illustrate, since we can vary distances from the center and use nuclei of different species. Figures 2–19 and 2–20 show molecules having the symmetry groups $\mathcal{C}_2$ and $\mathcal{C}_3$. A molecule containing four different nuclei not all in the same plane would "belong" to $\mathcal{C}_1$ (i.e., have no symmetry). Figure 2–19 shows the molecule $H_2C{=}CCl_2$ in a twisted configuration having the symmetry group $\mathcal{C}_2$. (If the molecule were plane, it would have higher symmetry, as we shall see later.) Figure 2–20 shows twisted $H_3C{-}CCl_3$ which "belongs" to $\mathcal{C}_3$ (i.e., has the symmetry group $\mathcal{C}_3$).

Now we proceed to add other axes of rotational symmetry. First we impose a restriction by requiring that there shall be no more than one rotation axis of order greater than 2. Thus, if we start from the group $\mathcal{C}_n$, we can adjoin only a 2-fold axis. Moreover, this axis must be at right angles to the n-fold axis; otherwise we could obtain a second n-fold axis by rotation through π about the 2-fold axis. Thus the only possible adjunction is a 2-fold axis perpendicular to the n-fold axis. If we now take products of these symmetry operations, we generate a total of n 2-fold axes in the

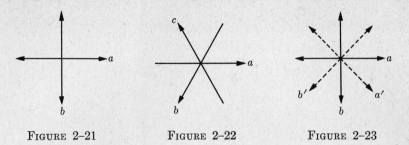

FIGURE 2-21   FIGURE 2-22   FIGURE 2-23

horizontal plane and arrive at:

II. *Groups having an n-fold axis and a system of 2-fold axes at right angles to it: the dihedral groups D_n.*

These groups contain $2n$ elements. The principal (n-fold) axis is two-sided, so C_n^k and C_n^{-k} are in the same class. Figure 2-21 shows the horizontal plane with a single 2-fold axis a; the n-fold axis is perpendicular to the paper. Suppose first that $n = 2$. Applying C_2 to the axis a merely reverses it. But if we let C_a be the rotation through π about a, then the product $C_2 C_a = C_b$, where C_b is a rotation through π about the axis b in Fig. 2-21. Thus, the group D_2 has three mutually perpendicular axes of second order. All the axes are two-sided. The group is an abelian group of order 4, and is isomorphic to the four-group. Sometimes the symbol V is used for this group.

For $n > 2$, the dihedral group D_n is not abelian.

Figure 2-22 shows the case of $n = 3$. When C_3 and C_3^2 are applied to the axis a, they generate the equivalent axes b and c. The group D_3 contains six elements, E, C_3, C_3^2, and the three 2-fold rotations. There are three classes:

$$E; \quad C_3, C_3^2; \quad C_a, C_b, C_c.$$

The case of $n = 4$ is shown in Fig. 2-23. The 4-fold rotations applied to the axis a generate only one new axis b equivalent to a. The product $C_4 C_a$ is a 2-fold rotation about an axis a' midway between a and b. Applying C_4 to a', we now generate the axis b'. Thus for $n = 4$, the two-fold axes split into two sets of equivalent axes. The group D_4 has eight elements in the following five classes:

$$E; \quad C_4, C_4^3; \quad C_4^2; \quad C_a, C_b; \quad C_{a'}, C_{b'}.$$

The results for the general case should now be clear. The group D_n contains $2n$ elements. If n is even ($n = 2p$), then E and $C_n^p = C_2$ each form a class; this leaves $(n - 2) = (2p - 2)$ n-fold rotations which fall into classes by pairs, giving $(p - 1)$ classes. The 2-fold rotations fall into two

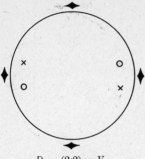

$D_2 \equiv (2:2) \equiv V$

FIGURE 2–24

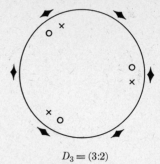

$D_3 \equiv (3:2)$

FIGURE 2–25

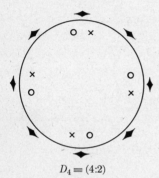

$D_4 \equiv (4:2)$

FIGURE 2–26

$D_6 \equiv (6:2)$

FIGURE 2–27

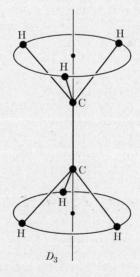

FIGURE 2–28

classes of p elements each. The total number of classes is $p + 3 = (n + 6)/2$.

If n is odd, $(n = 2p + 1)$, all 2-fold axes are equivalent so that the $(2p + 1)$ 2-fold rotations are all in the same class. The element E forms a class by itself, while the rotations about the n-fold axis give $(n - 1)/2 = p$ classes. The total number of classes is $p + 2 = (n + 3)/2$.

Problem. Show that the dihedral group D_n is generated by two elements a, b, such that $a^n = b^2 = (ab)^2 = e$.

The pole figures for $D_2 \equiv V$, D_3, D_4, D_6 are shown in Figs. 2–24 through 2–27. A possible example of D_2 would be obtained if we replaced the Cl-atoms in Fig. 2–19 by H-atoms (giving C_2H_4) and made all the CH-distances equal. An example of D_3 would be C_2H_6 in the configuration shown in Fig. 2–28. (The angle of twist of the CH_3-groups relative to one another must not be 60° since the resulting symmetry would be higher than D_3.)

2–4 The law of rational indices. We have now exhausted the possible point groups containing only rotations and having at most one axis of order greater than two. The reader may have noted that we only went up to $n = 6$ and omitted $n = 5$. The reason for this is that, for molecules, the cases which we have omitted are uncommon. In the case of crystals the nonexistence of axes of order 5, 7, 8, etc., follows from the empirical "law of rational indices." Consider the various faces and edges of a crystal polyhedron. Choose any three noncoplanar edges and use lines parallel to them through the origin as a system of coordinate axes. Any face of the polyhedron will have intercepts on these axes equal to u, v, w. The law of rational indices states that for any two crystal faces,

$$\frac{u'}{u} : \frac{v'}{v} : \frac{w'}{w} = n_1 : n_2 : n_3, \qquad (2\text{–}10)$$

where n_1, n_2, n_3 are integers. Note that parallel displacement of a plane or use of reciprocals of the intercepts makes no change in Eq. (2–10). This law is the expression of the results of measurements on crystals. One thing worth checking is whether the law is mathematically self-consistent. The planes considered in Eq. (2–10) form new edges which can be used as axes. For these new axes, we may use planes formed by the old axes to give an equation similar to (2–10). For self-consistency we must require that the ratios be again integers. A complete proof in three dimensions is rather lengthy. Instead we shall sketch a typical part of the proof

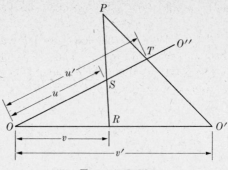

FIGURE 2–29

in the simpler two-dimensional case. In Fig. 2–29, let OO' and OO'' be lines parallel to crystal edges, while SR and TO' are two crystal faces of the lattice. The intercepts of these planes on OO' and OO'' must satisfy

$$\frac{u/u'}{v/v'} = r,$$

where r is rational. From the figure,

$$\frac{\sin OTO'}{\sin OO'T} = \frac{v'}{u'}, \qquad \frac{\sin OSR}{\sin ORS} = \frac{v}{u}.$$

If we now use the axes PR and PO' and consider the faces OO' and OO'', the ratio of indices is

$$\frac{PS/PR}{PT/PO'} = \frac{PS/PT}{PR/PO'} = \frac{\sin PTO}{\sin PST} \cdot \frac{\sin PRO'}{\sin PO'R} = \frac{\sin OTO'}{\sin OSR} \cdot \frac{\sin ORS}{\sin OO'T}$$

$$= \frac{v'/u'}{v/u} = r.$$

This proof of consistency is essentially the statement of cross-ratio theorems of projective geometry.

Now using the law of rational indices, we prove that for $n \geq 5$ the only possible order of a rotation (or rotation-reflection) axis is $n = 6$. Suppose that the vertical axis in Fig. 2–30 is an n-fold axis with $n \geq 5$. Then starting with the pole 1, we can generate a total of at least five poles. All these poles lie on a latitude circle. The radii from the center of the sphere to the poles are normals to crystal faces, and all make the same angle θ with the vertical. Let $\mathbf{n}_1, \mathbf{n}_2, \mathbf{n}_3, \mathbf{n}_4, \mathbf{n}_5$ be unit vectors along the normals. All of them have the same vertical projection, $\cos \theta$. Their horizontal projections have magnitude $\sin \theta$ and make successive angles $\psi = 2\pi/n$, as shown in the figure. (For a rotation-reflection axis, normals 1 and 3,

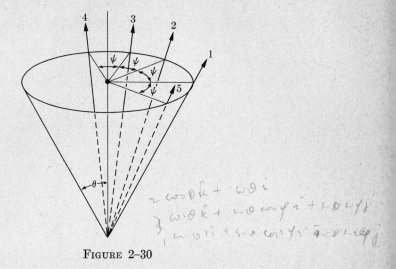

FIGURE 2–30

say, would be reversed in direction, but this would not affect our proof.)
Any two crystal faces determine a crystal edge which lies in both planes,
and therefore is perpendicular to the normal to each plane. Thus the edge
produced by the faces at poles 1 and 2 has a direction given by the cross
product $\mathbf{n}_1 \times \mathbf{n}_2$. We choose as our axes the crystal edges $\mathbf{n}_1 \times \mathbf{n}_2$,
$\mathbf{n}_2 \times \mathbf{n}_3$, $\mathbf{n}_1 \times \mathbf{n}_3$. We now apply the law of rational indices to the crystal
faces at the poles 4 and 5 (or to planes parallel to them). In Fig. 2–31, if

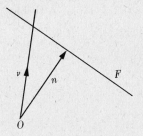

FIGURE 2–31

$\mathbf{n}$ is a unit vector normal to the crystal face F, the intercept of F on the
axis $\boldsymbol{\nu}$ is $1/\boldsymbol{\nu} \cdot \mathbf{n}$. We shall work with the reciprocal intercepts, so we want
$\mathbf{n}_4 \cdot \mathbf{n}_1 \times \mathbf{n}_2$, $\mathbf{n}_4 \cdot \mathbf{n}_2 \times \mathbf{n}_3$, $\mathbf{n}_4 \cdot \mathbf{n}_1 \times \mathbf{n}_3$, and similarly for $\mathbf{n}_5$. The values
of all these quantities can be obtained from Fig. 2–30. One finds that the
intercepts have a common factor $\sin^2 \theta \cos \theta$ which drops out when we
take ratios. The law of rational indices then states:

$$\frac{\sin 3\psi - \sin 2\psi - \sin \psi}{\sin 2\psi - \sin \psi - \sin \psi} : \frac{\sin 2\psi - \sin \psi - \sin \psi}{\sin 3\psi - \sin 2\psi - \sin \psi} :1 = a:b:c,$$

where a, b, c are integers, or

$$\frac{\sin 3\psi - \sin 2\psi - \sin \psi}{\sin 2\psi - \sin \psi - \sin \psi} = r, \qquad \text{where } r \text{ is a rational number,}$$

$$= \frac{2 \cos (5\psi/2) \sin (\psi/2) - 2 \sin (\psi/2) \cos (\psi/2)}{2 \cos (3\psi/2) \sin (\psi/2) - 2 \sin (\psi/2) \cos (\psi/2)}$$

$$= \frac{\cos (5\psi/2) - \cos (\psi/2)}{\cos (3\psi/2) - \cos (\psi/2)}$$

$$= \frac{\sin 3(\psi/2)}{\sin (\psi/2)} = 1 + 2 \cos \psi = r.$$

Hence $\cos \psi = \cos (2\pi/n)$ is a rational number. For $n \geq 5$, the only possible solution is $n = 6$ ($\cos \pi/3 = \frac{1}{2}$), and thus our theorem is proved.

2–5 Groups whose elements are pure rotations. Regular polyhedra.
Now we return to groups containing only rotations. Our next step is to consider groups having more than one n-fold axis with $n > 2$. Among all these axes of order n, choose the two which make the smallest angle with each other. If we draw the points (poles) where the axes intersect the unit sphere as in Fig. 2–32, let us suppose that P_0, P_1 are the poles of these closest axes. In other words, the great circle P_0P_1 is assumed to be the shortest great circle joining any two poles of nth-order axes. If we now perform an n-fold rotation about the axis P_0, we generate from P_1 $(n - 1)$ new n-fold axes, among which is the one through A shown in the figure. The spherical angle between P_0P_1 and P_0A is $\psi = 2\pi/n$, and $P_0P_1 = P_0A$. Rotating through ψ around A, we now generate from P_0 another n-fold axis through B. The points P_1, P_0, A, B, etc., obtained by this procedure all lie in the same plane, and therefore the figure must close again at P_1. (Otherwise we would obtain an n-fold axis nearer to P_1

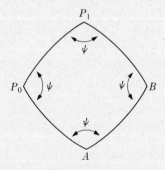

FIGURE 2–32

than P_0.) Now starting again from P_0, generate another axis from P_1, and continue to form a second regular spherical polygon. By this process we cover the whole surface of the unit sphere with similar regular spherical polygons. Thus we obtain a regular spherical polyhedron with F faces, V vertices, and E edges. Each face of the polyhedron is a regular spherical polygon of s sides. The number of edges which meet at a vertex is n. Since each edge has a vertex at its two ends, $E = nV/2$. Since each edge is shared by two faces, we have $E = Fs/2$. Combining these, we have $Fs = nV$. The area of the polygon of s sides on the unit sphere is equal to the sum of the angles of the polygon minus $(s - 2)\pi$. For our regular polygon, the area is $s(2\pi/n) - (s - 2)\pi$. Multiplying by the number of faces F, we obtain the total surface of the unit sphere, 4π. Thus,

$$F\left[s\left(\frac{2\pi}{n}\right) - (s - 2)\pi\right] = 4\pi,$$

or

$$\frac{Fs}{n} - \frac{Fs}{2} + F = 2. \tag{2–11}$$

Using the relations obtained above, we can rewrite (2–11) as

$$V - E + F = 2. \tag{2–12}$$

This is Euler's theorem, which in the form (2–12) is valid for any network on the surface of the sphere. Rewrite (2–11) as

$$\frac{2s}{n} - (s - 2) = \frac{4}{F}. \tag{2–13}$$

We see that $2s/n$ must be greater than $s - 2$, and therefore, since $n > 2$, the possible values of s are restricted.

For $n = 3$, $s < 6$: $s = 3$ gives $F = 4$; $s = 4$ gives $F = 6$; $s = 5$ requires $F = 12$.
For $n = 4$, $s < 4$: $s = 3$ gives $F = 8$.
For $n = 5$, only $s = 3$ is possible, with $F = 20$.
There is no solution for $n \geq 6$.

We have the following possibilities:

n	s	F	
3	3	4	tetrahedron
3	4	6	cube
4	3	8	octahedron
3	5	12	dodecahedron
5	3	20	icosahedron

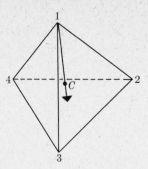

FIGURE 2–33

The first two have the same symmetry elements—four 3-fold axes which join one of the vertices 1, 2, 3, 4 of the tetrahedron to the opposite face as shown in Fig. 2–33. Starting with the pole of one face of a crystal, we find that rotations about these axes generate a total of 12 poles, so the number of elements in the group is 12. Rotations about any one axis bring the other three into coincidence, so all four axes are equivalent. The axes are one-sided, so we have two classes of four elements each: $C_3(4)$; $C_3^2(4)$. If we rotate about axis 1 so that $2 \to 3$, $3 \to 4$, $4 \to 2$, and follow with a rotation about axis 2 which takes $3 \to 4$, $4 \to 1$, $1 \to 3$, the product takes $2 \to 4$, $4 \to 2$; $3 \to 1$, $1 \to 3$, and is therefore a rotation through π around the line joining the midpoints of the opposite edges 24 and 13. In this way we find three 2-fold axes which are equivalent to one another. Hence the 2-fold rotations form a class of three elements. Finally, the tetrahedral group T has 12 elements in 4 classes,

$$T: E; \quad C_2(3); \quad C_3(4); \quad C_3^2(4),$$

where the number in parentheses is the number of conjugate elements in the class. Another way of representing these axes is shown in Fig. 2–34.

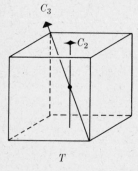

FIGURE 2–34

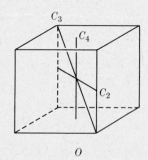

FIGURE 2–35

We also note that the group T can be obtained from the group V by adjoining to it a 3-fold axis placed symmetrically with respect to the three 2-fold axes of V. For this reason the symbol $C_x \sim C_y \sim C_z$ is sometimes used for T.

Next we consider the octahedron. Here we start with three mutually perpendicular 4-fold axes. By taking products of these rotations one obtains the total rotational symmetry of a cube, as shown in Fig. 2–35. There are four 3-fold axes (space diagonals), three 4-fold axes (joining centers of opposite faces), and six 2-fold axes (joining midpoints of opposite edges).

By applying 3-fold and 4-fold rotations to any one of the 2-fold axes, we can generate them all, and hence all six 2-fold rotations are in the same class. One can easily see that all axes are two-sided, that all 3-fold axes are equivalent to one another, and that all 4-fold axes are equivalent to one another. So the group O has 24 elements in 5 classes:

$$O: \quad E; \quad C_2(6); \quad C_3, C_3^2(8); \quad C_4, C_4^3(6); \quad C_4^2(3).$$

The icosahedral group Y, which is obtained for both of the last entries in our table of possibilities, has no physical interest, since for crystals 5-fold axes cannot occur, and no examples of molecules with this symmetry are known. Y is the group of 60 rotations about the symmetry axes of an icosahedron, which has six 5-fold axes, ten 3-fold axes, and fifteen 2-fold axes.

To summarize: By permitting chains of higher-order axes, we obtain three new groups of rotations:

III. T, O, Y. The pole figures for T and O are shown in Figs. 2–36 and 2–37. With regard to all these pole figures, we must point out that one set of equivalent poles may not be enough to form a closed polyhedron or to fix the symmetry unambiguously. In such cases one must add a second equivalent set.

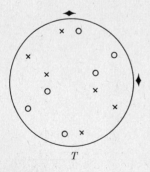

T

FIGURE 2–36

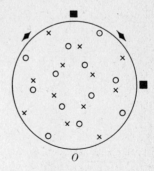

O

FIGURE 2–37

Problems. (1) Show that all the elements of the group O are generated from rotations around the 4-fold axes. (2) Enumerate all subgroups of O. Which of these are invariant?

2–6 Symmetry groups containing rotation reflections. Adjunction of reflections to $\mathfrak{C}_n$.

Now that we have all the possible symmetry groups containing only rotations, we must add elements of the second type, i.e., rotation reflections. Let S be a rotation reflection which is an element of the group. The product of S with any rotation is again a rotation reflection. The product of two rotation reflections is a rotation. We note first that the pure rotations form a subgroup $\mathfrak{K}$ of the group G. Moreover, this subgroup is necessarily of index 2, and is therefore an invariant subgroup. In other words, we claim that every rotation reflection S_i of the group is contained in the coset $S\mathfrak{K}$ formed with any one of them. In fact,

$$S_i = (SS^{-1})S_i = S(S^{-1}S_i),$$

and $S^{-1}S_i$, being the product of two rotation reflections, is a rotation, and is therefore in $\mathfrak{K}$. The factor group $G/\mathfrak{K}$ is always of order 2.

We now proceed as follows: We choose for $\mathfrak{K}$ any one of the pure rotation groups. In adjoining a rotation reflection S, we do not permit the appearance of new rotations (since by the above argument they should already be in $\mathfrak{K}$). What choices do we have for S? One possibility is that $S^2 = E$, for which there are two solutions: $S = \sigma$, $S = I$. Thus we can add a reflection or an inversion. If $S^2 \neq E$, then S^2 must be one of the elements of $\mathfrak{K}$ other than the identity. The axis of the rotation reflection must be one of the axes of rotation of the group $\mathfrak{K}$. If C_n is an element of $\mathfrak{K}$, the only other choice is $S^2 = C_n$, for if we take $S^2 = C_n^{2p}$, then $S^2 C_n^{-2p} = E$, or $(SC_n^{-p})^2 = E$; by adjoining $S' = SC_n^{-p}$ instead of S, we would have $(S')^2 = E$. Or if we take $S^2 = C_n^{2p+1}$, then we could instead adjoin $S' = SC_n^{-p}$ and obtain $(S')^2 = C_n$. So we have three possible adjunctions: σ, I, and S, such that $S^2 = C_n$.

Adjunction of reflections to the group $\mathfrak{C}_n$.

Case 1. Adjoin $S^2 = C_n$.

For $n = 1$, $S^2 = E$, and we have cases 2 and 3.

For $n = 2$, $S^2 = C_2$, $S = S_4$. We obtain the abelian group $\mathcal{S}_4$ containing 4 elements in 4 classes.

For $n = 3$, $S^2 = C_3$, $S = S_6$. Again we obtain an abelian group $\mathcal{S}_6$ with 6 elements in 6 classes.

For $n \geq 4$, we would obtain a rotation-reflection axis of order greater than 6, which was excluded earlier.

Case 2. Adjoin σ. If we are not to introduce new rotations, the plane of reflection must either be perpendicular to, or pass through the principal axis.

(a) If we adjoin σ_h to $\mathcal{C}_n$, the group obtained is called $\mathcal{C}_{nh}$. All these groups are abelian, and contain $2n$ elements in $2n$ classes. If n is even, the group contains $(C_n^{n/2} \cdot \sigma_h) = C_2\sigma_h = I$, so that the body has a center of symmetry. For $n = 1$, the group $\mathcal{C}_{1h}$ contains the two elements E and σ; this group is usually denoted by $\mathcal{C}_s$. In this way we obtain the new groups

$$\mathcal{C}_s, \ \mathcal{C}_{2h}, \ \mathcal{C}_{3h}, \ \mathcal{C}_{4h}, \ \mathcal{C}_{6h}.$$

(b) If we adjoin σ_v to $\mathcal{C}_n$, then by combining σ_v with rotations around the vertical axis we generate a set of n vertical planes. Excluding $n = 1$, which gives $\mathcal{C}_s$, we obtain new groups which we label $\mathcal{C}_{2v}, \ \mathcal{C}_{3v}, \ \mathcal{C}_{4v}, \ \mathcal{C}_{6v}$. The group $\mathcal{C}_{nv}$ contains $2n$ elements. As shown earlier, the presence of σ_v makes the rotation axis two-sided. If n is odd $(n = 2p + 1)$, all planes are equivalent, and all reflections fall into one class. At the same time, the rotations about the two-sided principal axis yield $(p + 1)$ classes: E; C_{2p+1}^k, C_{2p+1}^{-k} (for $k = 1, 2, \ldots, p$). We have a total of $p + 2 = (n + 3)/2$ classes. If n is even $(n = 2p)$, the reflections form two classes with p elements in each. The rotations give the class E, the class C_2, and $(p - 1)$ classes C_{2p}^k, C_{2p}^{-k} for $k = 1, 2, \ldots, (p - 1)$. We have a total of $p + 3 = (n + 6)/2$ classes.

Case 3. Finally, adjoining the inversion I to $\mathcal{C}_n$, we obtain only one new group, namely from $\mathcal{C}_1$. This is the group $\mathcal{C}_i$ (or $\mathcal{S}_2$) having two elements E and I.

Summary: Adjunctions to $\mathcal{C}_n$ yield the groups

$$\mathcal{S}_4, \mathcal{S}_6; \quad \mathcal{C}_i; \quad \mathcal{C}_s; \quad \mathcal{C}_{2h}, \mathcal{C}_{3h}, \mathcal{C}_{4h}, \mathcal{C}_{6h}; \quad \mathcal{C}_{2v}, \mathcal{C}_{3v}, \mathcal{C}_{4v}, \mathcal{C}_{6v}.$$

The pole figures are shown in Figs. 2–38 to 2–49. An example of   would be the trans-ClBrHC—CHBrCl shown in Fig. 2–50. Any planar nonlinear molecule, all of whose atoms are different (for example, NOCl) has the symmetry group $\mathcal{C}_{2h}$. Figure 2–51 shows the plane molecule trans-$C_2H_2Cl_2$ which has the symmetry group $\mathcal{C}_{2h}$. Examples of the symmetry group $\mathcal{C}_{2v}$ are the molecules H_2O, SO_2, H_2S, as shown in Fig. 2–52; examples of $\mathcal{C}_{3v}$ are NH_3, CH_3Cl, PCl_3, as shown in Fig. 2–53.

Problem. What group is obtained if we adjoin I to $\mathcal{C}_3$?

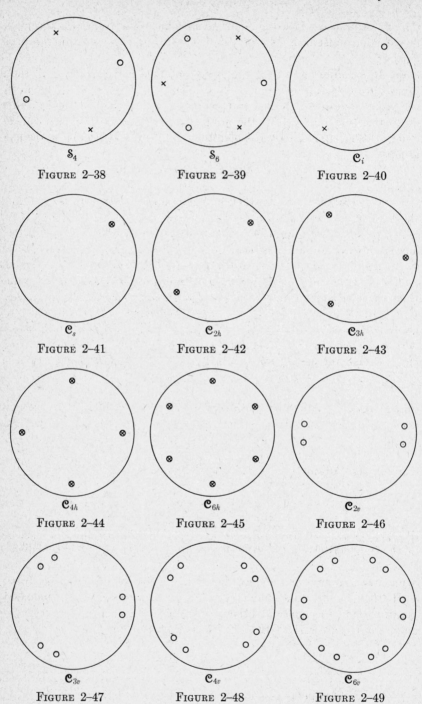

$\mathcal{S}_4$

FIGURE 2–38

$\mathcal{S}_6$

FIGURE 2–39

$\mathcal{C}_i$

FIGURE 2–40

$\mathcal{C}_s$

FIGURE 2–41

$\mathcal{C}_{2h}$

FIGURE 2–42

$\mathcal{C}_{3h}$

FIGURE 2–43

$\mathcal{C}_{4h}$

FIGURE 2–44

$\mathcal{C}_{6h}$

FIGURE 2–45

$\mathcal{C}_{2v}$

FIGURE 2–46

$\mathcal{C}_{3v}$

FIGURE 2–47

$\mathcal{C}_{4v}$

FIGURE 2–48

$\mathcal{C}_{6v}$

FIGURE 2–49

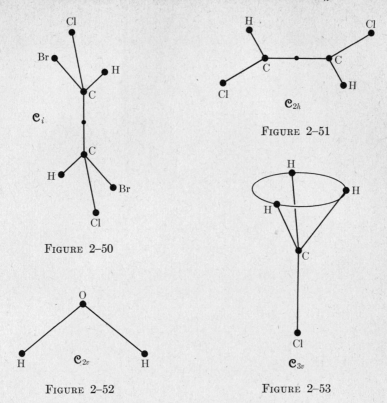

FIGURE 2–51

FIGURE 2–50

FIGURE 2–52

FIGURE 2–53

2-7 Adjunction of reflections to the groups D_n. We proceed similarly with the groups D_n. Consider the adjunction of a reflection plane. If we adjoin σ_h, the product of σ_h with the rotation around any 2-fold axis gives the reflection in the vertical plane through the axis. Thus the adding of a horizontal plane of symmetry gives n vertical reflection planes with n corresponding operations σ_v. The new group D_{nh} contains $4n$ elements: $2n$ pure rotations from D_n, n reflections σ_v in the n vertical planes, and n rotation reflections $C_n^k \sigma_h$. Note that σ_h commutes with all the elements of the group. We may therefore (see Section 1–7) write D_{nh} as the direct product of D_n and $\mathcal{C}_s$,

$$D_{nh} = D_n \times \mathcal{C}_s,$$

(If n is even, $n = 2p$, the group contains the inversion, and we can also write $D_{2p,h} = D_{2p} \times \mathcal{C}_i$.) The number of classes in D_{nh} is just twice the number in D_n. First we have all the classes of D_n, and then the classes obtained by multiplying each element by σ_h. We find, just as in our discussion of D_n, that if n is odd, all the reflections are in the same class, while they form two classes if n is even. The rotation reflections $C_n^k \sigma_h$ and

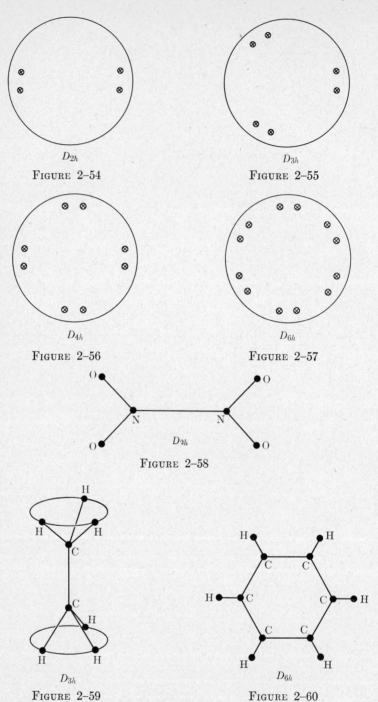

D_{2h}

FIGURE 2–54

D_{3h}

FIGURE 2–55

D_{4h}

FIGURE 2–56

D_{6h}

FIGURE 2–57

D_{2h}

FIGURE 2–58

D_{3h}

FIGURE 2–59

D_{6h}

FIGURE 2–60

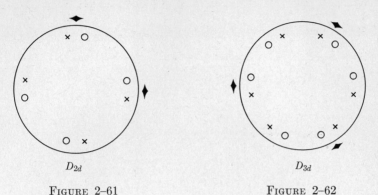

FIGURE 2–61   FIGURE 2–62

$C_n^{-k}\sigma_h$ fall into classes in pairs. The new groups are D_{2h}, D_{3h}, D_{4h}, D_{6h}. The pole figures are shown in Figs. 2–54 through 2–57.

The planar molecule N_2O_4 shown in Fig. 2–58 has the symmetry group D_{2h}; the "eclipsed" form of C_2H_6 shown in Fig. 2–59 belongs to D_{3h}; the benzene molecule C_6H_6 has the symmetry of D_{6h}, and is shown in Fig. 2–60.

A plane of reflection can be adjoined to the groups D_n in still another way: We can add a vertical reflection plane which bisects the angle between two neighboring 2-fold axes. As soon as we add one vertical plane, the 2-fold rotations generate a total of n vertical reflection planes. The group obtained in this way is D_{nd} (d for diagonal), and contains $4n$ elements. Of these $4n$ elements, $2n$ are the pure rotations of D_n. In addition, we have n mirror reflections, σ_d, in the n vertical planes. The remaining n elements are rotation reflections around the principal axis, of the form S_{2n}^{2k+1}, where $k = 0, 1, 2, \ldots, (n-1)$. The simplest way of showing this is by looking at the pole figures (see Figs. 2–61 and 2–62). Thus the principal axis is not just an n-fold rotation axis, but is rather a $2n$-fold rotation-reflection axis. As a result, we can only form these groups for $n = 2$ or 3. (For $n > 3$, the order of the rotation-reflection axis would be greater than 6, and this was ruled out earlier.) The two new groups obtained are D_{2d} and D_{3d}. (See Figs. 2–61 and 2–62.)

Problem. By taking products of σ_d and rotations show that the principal axis of D_{nd} is a $2n$-fold rotation-reflection axis.

The 2-fold axes are all equivalent since each axis can be made to coincide with its neighbor by reflecting in the plane midway between them. Similarly, all the reflection planes are equivalent. (Apply 2-fold rotations.)

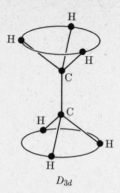

D_{3d}

FIGURE 2–63

Finally, S_{2n}^{2k+1} and $S_{2n}^{-(2k+1)}$ are conjugates, since

$$\sigma_d S_{2n}^{2k+1} \sigma_d^{-1} = \sigma_d \sigma_h C_{2n}^{2k+1} \sigma_d = \sigma_h \sigma_d C_{2n}^{2k+1} \sigma_d = \sigma_h C_{2n}^{-(2k+1)} = S_{2n}^{-(2k+1)}.$$

Thus, in general, for even $n = 2p$, $D_{2p,d}$ has $n + 3 = 2p + 3$ classes: E; rotation $C_{2p}^{p} = C_2$ around the principal axis; $(p - 1)$ classes of pairs C_{2p}^{k}, C_{2p}^{-k} for $k = 1, 2, \ldots, p - 1$; the class of $2p$ 2-fold rotations around horizontal axes; the class of $2p$ reflections σ_d; and p classes of pairs of rotation reflections S_{2n}^{2k+1}, $S_{2n}^{-(2k+1)}$ for $k = 0, 1, 2, \ldots, p - 1$. For odd n, $n = 2p + 1$, D_{nd} contains the inversion I, and we can therefore write it as the direct product $D_{2p+1,d} = D_{2p+1} \times \mathcal{C}_i$. Thus it has $(2p + 4)$ classes, just double the number in D_{2p+1}. The new classes in $D_{2p+1,d}$ are obtained from those of D_{2p+1} by multiplying with I.

The "staggered" form of C_2H_6 shown in Fig. 2–63 has the symmetry group D_{3d}.

We could now try to adjoin the inversion or a rotation reflection to D_n. If we adjoin I, we obtain the direct product $D_n \times \mathcal{C}_i$, which we have shown to be D_{nh} if n is even, and D_{nd} if n is odd. Similarly, the adjunction of a rotation reflection gives no new groups.

2–8 The complete symmetry groups of the regular polyhedra.

The last step in the enumeration is to add rotation reflections to the groups T, O, Y.

Suppose we wish to adjoin a plane of reflection to T. Since reflections in this plane must not produce any new rotation axes, the plane must either pass through two opposite edges of the cube in Fig. 2–34 or be parallel to two faces and midway between them.

The first alternative gives the group T_d with the typical axes and planes of symmetry shown in Figs. 2–64 and 2–65. The group T_d exhibits all the

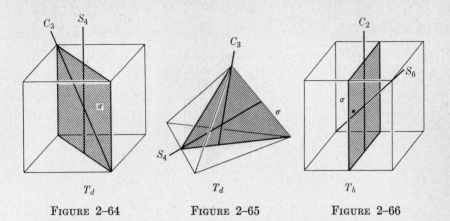

FIGURE 2–64   FIGURE 2–65   FIGURE 2–66

symmetries of a tetrahedron. The reason for the designation T_d is that the plane which we have added bisects the angle between the two horizontal 2-fold axes. Just as in the case of the group D_{2d}, this means that the 2-fold rotation axes become 4-fold rotation-reflection axes, as indicated in the figures. The symmetry planes pass through the 3-fold axes, so that these axes are two-sided. All reflection planes are equivalent, and all S_4-axes are equivalent. Therefore, the 24 elements of T_d are distributed in 5 classes:

$$T_d: \quad E; \quad C_3, C_3^2(8); \quad S_4, S_4^3(6); \quad S_4^2 = C_2(3); \quad \sigma_d(6).$$

The second alternative gives the group T_h with typical axes as shown in Fig. 2–66. The subscript h means that the plane is horizontal relative to 2-fold axes, but the planes bisect the angles between the 3-fold axes, and thus convert them to 6-fold rotation-reflection axes. Since the group contains S_6, it contains I, and hence we may write $T_h = T \times \mathcal{C}_i$. The group T_h therefore has 24 elements in 8 classes which are obtainable from the classes of T. The other possible adjunctions are already included in T_h and T_d.

In the case of the group O, the position of an added reflection plane is restricted in the same way as for T. But now the addition of one type of plane immediately generates the other. Just as in the case of T_h, the C_3-axes become S_6-axes, and the group contains the inversion I. Thus the group O_h can be expressed as $O_h = O \times \mathcal{C}_i$, and O_h is the group of all symmetry transformations of a cube (see Fig. 2–67). It contains 48 elements distributed in 10 classes (twice the number of classes in O):

$$O_h: \quad E; \quad C_2(6); \quad C_3, C_3^2(8); \quad C_4, C_4^3(6); \quad C_4^2(3);$$
$$I; \quad \sigma_v(6); \quad S_6, S_6^5(8); \quad C_4\sigma_h, C_4^3\sigma_h(6); \quad \sigma_h(3).$$

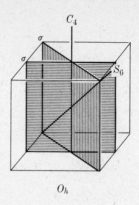

O_h

FIGURE 2–67

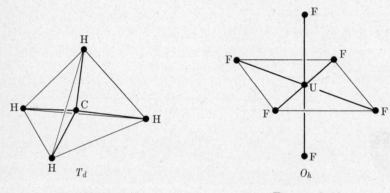

FIGURE 2–68   FIGURE 2–69

Finally, we can adjoin the inversion to the group Y to yield $Y_h = Y \times \mathcal{C}_i$. This is the complete symmetry group of the icosahedron. Since it has no physical interest, we shall not discuss it further.

Because the group T_d has the complete symmetry of the tetrahedron, all tetrahedral molecules like CH_4 and CCl_4 belong to this group; the methane molecule is shown in Fig. 2–68. Uranium hexafluoride, UF_6, has the symmetry group O_h and is shown in Fig. 2–69.

2–9 Summary of point groups. Other systems of notation. We have now found all possible point groups. Of these there are only 32 which are consistent with the law of rational indices. We tabulate them under different systems, stating in each case the number of elements in the group.

    I. *Triclinic system.*

        1. $\mathcal{C}_1(1)$; 2. $\mathcal{C}_i(2)$.

II. *Monoclinic system.*

    3. $C_{2h}(4)$; 4. $C_2(2)$; 5. $C_s(2)$.

III. *Rhombic system.*

    6. $D_{2h}(8)$; 7. $D_2 \equiv V(4)$; 8. $C_{2v}(4)$.

IV. *Trigonal system.*

    9. $D_{3h}(12)$; 10. $D_3(6)$; 11. $C_{3v}(6)$; 12. $S_6(6)$; 13. $C_3(3)$.

V. *Tetragonal system.*

    14. $D_{4h}(16)$; 15. $D_4(8)$; 16. $C_{4v}(8)$; 17. $C_{4h}(8)$; 18. $C_4(4)$.

VI. *Hexagonal system.*

    19. $D_{6h}(24)$; 20. $D_6(12)$; 21. $C_{6v}(12)$; 22. $C_{6h}(12)$;
    23. $C_6(6)$.

VII. *Regular system.*

    24. $O_h(48)$; 25. $O(24)$; 26. $T_d(24)$; 27. $T_h(24)$; 28. $T(12)$.

Of the remaining groups, 29 [$C_{3h}(6)$] and 30 [$D_{3d}(12)$] are usually included in VI, while 31 [$S_4(4)$] and 32 [$D_{2d}(8)$] are included in V.

Note that, in each system, the group listed first has the highest symmetry. In systems IV, V, VI, and VII, the symmetry of the first group in the system is said to be "holohedric" (having its full quota of crystal faces). This group is followed by three groups which have half the number of elements (hemihedry), while the fifth group in the set has only one-quarter the number of elements of the first group (tetartohedry).

The notation which we have used for the point groups is that of Schoenflies. There have been many different notations, all with some advantages and some disadvantages.

A second system of notation is one which we shall designate as the Shubnikov system. In this notation, the presence of an n-fold rotation axis is indicated by the symbol n. Thus the group C_1 is denoted by 1, C_2 by 2, etc. Rotation-reflection axes are indicated by placing a bar over the corresponding symbol, so that $C_i \equiv S_2$ is now $\bar{2}$, S_4 is $\bar{4}$, etc. The presence of a plane of symmetry is indicated by the letter m, so that the group $C_s \equiv C_{1h}$ is denoted by m, since its only symmetry element is the reflection.

When the point group contains *more* than one symmetry element, this is indicated by giving the symbols for the individual symmetries, together with an indication of relative orientation. A single dot is used to indicate that two symmetry elements are parallel (e.g., a rotation axis and a symmetry plane passing through the axis). In this notation, the group C_{2v} would be $2 \cdot m$.

TABLE 2–1

Schoenflies	C_1	C_i	C_{2h}	C_2	C_s	D_{2h}	D_2	C_{2v}
Shubnikov	1	$\bar{2}$	$2:m$	2	m	$m\cdot2:m$	$2:2$	$2\cdot m$
International	1	$\bar{1}$	$2/m$	2	m	mmm	222	$mm2 \equiv 2mm$

Schoenflies	D_{3h}	D_3	C_{3v}	S_6	C_3	D_{3d}	D_{4h}	D_4	C_{4v}	C_{4h}	D_{2d}	S_4	C_4
Shubnikov	$m\cdot3:m$	$3:2$	$3\cdot m$	$\bar{6}$	3	$\bar{6}\cdot m$	$m\cdot4:m$	$4:2$	$4\cdot m$	$4:m$	$\bar{4}\cdot m$	$\bar{4}$	4
International	$\bar{6}m2 \equiv \bar{6}2m$	32	$3m$	$\bar{3}$	3	$\bar{3}m$	$4/mmm$	$422 \equiv 42$	$4mm$	$4/m$	$\bar{4}2m$	$\bar{4}$	4

Schoenflies	D_{6h}	D_6	C_{6v}	C_{6h}	C_{3h}	C_6	O_h	O	T_d	T_h	T
Shubnikov	$m\cdot6:m$	$6:2$	$6\cdot m$	$6:m$	$3:m$	6	$\bar{6}/4$	$3/4$	$3/\bar{4}$	$\bar{6}/2$	$3/2$
International	$6/mmm$	$622 \equiv 62$	$6mm$	$6/m$	$\bar{6}$	6	$m3m$	$432 \equiv 43$	$\bar{4}3m$	$m3$	23

A double dot or colon is used to indicate that two symmetry elements are perpendicular. For example, 3:2 means that the group contains a 3-fold axis and a 2-fold axis which are at right angles. This is precisely the group D_3 in the Schoenflies notation. Similarly the symbol 3:m indicates that there is a plane of symmetry perpendicular to the 3-fold axis, so that this is the group which we called C_{3h}.

A diagonal line is used to indicate that the group contains two axes which are not at right angles to one another. Thus 3/2 denotes the group which has a 3-fold axis and a 2-fold axis at some angle other than 90°. From Fig. 2–34 and our discussion we see that this is the group T.

The system of notation in most common use by crystallographers is the *international system*. Here the bar over a symbol means that one takes the product of the operation with the *space inversion*. Thus $\bar{1}$ is the group C_i. The symbol m is used to denote a symmetry plane, so C_s is m in this notation. Note that 2 would be the same as m. In this notation, the group S_6 is $\bar{3}$, while C_{3h} is $\bar{6}$.

A diagonal line between the symbols indicates that the group contains a plane of symmetry perpendicular to a rotation axis. Thus C_{2h} is 2/m, C_{4h} is 4/m. Symmetry planes which are not perpendicular to the rotation axis are recorded without any additional marking. The group C_{3v} is 3m, but C_{4v} and C_{6v} are 4mm and 6mm, because the vertical reflections are all in one class in the first group, while they are in *two* classes in the other groups.

The groups D_{4h} and D_{6h} are 4/mmm and 6/mmm, because they have one symmetry plane perpendicular to the rotation axis; the other reflections are in *two* classes.

The group $D_2 \equiv V$ is denoted by 222 since it contains a (principal) 2-fold axis and two 180° rotations which are not in the same class. Similarly, D_4 and D_6 are 422 and 622. On the other hand, D_3 is 32, since all 180° rotations are in the *same* class.

For reference we present the notation for the 32 point groups according to the three systems in Table 2–1.

Problem. Explain the significance of the symbol $m{\cdot}6$:m. Why is this the group which we called D_{6h}? Explain the significance of the symbols $\bar{3}$, 23, $m3m$ in the international system.

2–10 Magnetic symmetry groups (color groups).

We have considered the point-symmetry groups mainly as describing the external symmetry of crystals. For a microscopic description, we should examine the symmetry of the distribution of atoms (or ions) in the crystal. When we do this, the symmetry we are describing is that of the time-averaged distribution of the matter in the crystal. We may regard the point groups as describing the possible point symmetry of the time-averaged charge density ρ in the equilibrium state of the crystal. In this equilibrium state there may also be present a time-averaged distribution of current density $\mathbf{j}$. Since there must be no sources or sinks of charge in the equilibrium state, $\mathbf{j}$ must satisfy div $\mathbf{j} = 0$. For most substances, $\mathbf{j} = 0$, but in ferromagnets or antiferromagnets we must have $\mathbf{j} \neq 0$.

If we reverse $\mathbf{j}$ at every point, the equilibrium state remains an equilibrium state. We now consider a new symmetry operation R which reverses the sign of $\mathbf{j}$ at each point in space, but *does not* act on the space coordinates. (R is the "time-reversal" operator.) The element R has order 2 ($R^2 = E$), and commutes with all space rotations and reflections.

We can now consider the possible point symmetry groups of crystals in which $\mathbf{j} \neq 0$. Such groups can contain the ordinary rotation and rotation-reflection operations A, but, in addition, they may contain elements of the form RA, i.e., a combination of the geometric transformation A and the reversal R of the current $\mathbf{j}$. The 32 point groups which we found previously are all admissible symmetry groups when $\mathbf{j} \neq 0$. We now wish to find further symmetry groups containing at least one element of the form RA. We see at once that such groups are not to be obtained by simply adjoining R to one of the ordinary point groups. In fact, if the group contains the element R, this means that $\mathbf{j} = -\mathbf{j}$, so that $\mathbf{j} = 0$ everywhere. Thus our new groups should contain one or more elements RA, but they must not contain R.

Before proceeding to determine all these new groups, we consider other interpretations of them. Suppose that the faces of a crystal can be colored white (W) or black (B), and that R is the operation which changes the color (from W to B or from B to W). In addition to the geometric symmetry operations A, which shift the crystal faces but do not change their color, we now consider elements RA. For example, if A is a rotation which takes face F into face F', then RA will take F into F' and color F' opposite to F. We also require that no face receive both colors, so the element R itself must not appear in the group. We call these groups "color groups."

We note first that if the group G contains an element $M = RA$, A must not be of odd order, since G will contain all powers of M, among which is R itself. Hence, elements like RC_3 or RS_3 cannot occur.

Both A and $M = RA$ cannot be in G since G would then contain $MA^{-1} = R$. In accordance with this result, we label the group elements as A_k ($k = 1, 2, \ldots, m$), $M_i = RA_i$ ($i = m + 1, m + 2, \ldots, n$), where all the geometric operations A are different. It is clear that if we replaced R by the identity, the n elements A_k ($k = 1, \ldots, m$) and A_i ($i = m + 1, \ldots, m$) would form one of the 32 point groups. Also the elements A_k form a subgroup $\mathcal{K}$ in G, which is one of the 32 point groups.

We could now proceed as follows: Take any one of the point groups G. Find a subgroup $\mathcal{3C}$ with elements A_k. Multiply all elements A_i in the set $G - \mathcal{3C}$ (i.e., all elements of G which are not in the subgroup $\mathcal{3C}$) by R, obtaining $M_i = RA_i$. If the elements M_i and A_k form a group, it is one of the type we are seeking. This method would be quite tedious. But the whole problem resolves itself since we now prove the following:

A necessary and sufficient condition for the elements M_i and A_k to form a group is that $\mathcal{3C}$ have index 2 in G.

If $\mathcal{3C}$ has index 2 in G, then

$$G = \mathcal{3C} + A_i\mathcal{3C},$$

where A_i is one of the elements of $G - \mathcal{3C}$. Our new set G' is

$$G' = \mathcal{3C} + RA_i\mathcal{3C}.$$

Since $\mathcal{3C} \cdot \mathcal{3C} = \mathcal{3C}$, $RA_i\mathcal{3C} \cdot \mathcal{3C} = RA_i\mathcal{3C}$, $RA_i\mathcal{3C} \cdot RA_i\mathcal{3C} = \mathcal{3C}$, G' is a group.

Conversely, if the elements A_k and $M_i = RA_i$ form a group G', then multiplying the m elements in $\mathcal{3C}$ by any one of the elements M_i yields m different elements of the type M_i; if we multiply the M_i's by any one of their number, we obtain n different elements of the type A_k. Thus $\mathcal{3C}$ has index 2 in G' and, consequently, also has index 2 in G.

Our procedure for finding the new groups is as follows: Select a point group G. Choose any subgroup $\mathcal{3C}$ of index 2 in G. Multiply the elements of $G - \mathcal{3C}$ by R. Then $G' = \mathcal{3C} + R(G - \mathcal{3C})$ is the new group.

For each of the 32 point groups G, all possible subgroups $\mathcal{3C}$ of index 2 are listed in Table 2–2. Each of these gives rise to a new group (magnetic class) G'. We use the international notation which is particularly useful for the magnetic classes. The bar below a symbol means that one should take the product of the corresponding element with the time-reversal operator R.

We have found 58 new groups which, together with the 32 point groups, give us a total of 90 "magnetic symmetry groups." Our discussion was based on the distribution of current density $\mathbf{j}$, but we could equally well have considered the distribution of magnetization $\boldsymbol{\mu}$. (The only precaution necessary is to remember that $\boldsymbol{\mu}$ is an axial vector, whereas $\mathbf{j}$ is a polar vector.) The magnetic symmetry groups (and the space groups based on them) have been applied to the analysis of antiferromagnetic structures.

Generalizations of these groups can be made in several ways. For example, we may assign n different independent two-valued attributes, and introduce n operators R_i such that R_i changes the value of the ith attribute, and all R_i commute with one another and with all geometric operations; or we can consider an n-valued attribute (multi-color groups)

TABLE 2–2

MAGNETIC CLASSES

G	$\mathcal{H}$	G'		G	$\mathcal{H}$	G'
1	—	—		$\bar{4}2m$	$2mm$	$\bar{4}2m$
$\bar{1}$	1	$\bar{1}$		$4/mmm$	42	$4/\underline{m}mm$
2	1	$\underline{2}$		$4/mmm$	$4mm$	$4/\underline{mm}m$
m	1	$\underline{m}$		$4/mmm$	mmm	$\underline{4}/mmm$
$2/m$	2	$2/\underline{m}$		$4/mmm$	$\bar{4}2m$	$\underline{4}/mmm$
$2/m$	m	$\underline{2}/m$		$4/mmm$	$4/m$	$4/m\underline{m}m$
$2/m$	$\bar{1}$	$2/\underline{m}$		6	3	$\underline{6}$
222	2	$2\underline{22}$		$\bar{3}$	3	$\underline{3}$
$2mm$	2	$2\underline{mm}$		$\bar{3}m$	$\bar{3}$	$\bar{3}\underline{m}$
$2mm$	m	$\underline{2}mm$		$\bar{3}m$	$3m$	$\bar{3}\underline{m}$
mmm	222	$\underline{mmm}$		$\bar{3}m$	32	$\bar{3}\underline{m}$
mmm	$2mm$	$mm\underline{m}$		62	6	$6\underline{2}$
mmm	$2/m$	$m\underline{m}m$		62	32	$6\underline{2}$
3	—	—		$6/m$	6	$6/\underline{m}$
32	3	$3\underline{2}$		$6/m$	$\bar{3}$	$\underline{6}/m$
$3m$	3	$3\underline{m}$		$6/m$	$3/m$	$\underline{6}/m$
$\bar{6}$	3	$\underline{\bar{6}}$		$6mm$	6	$6\underline{mm}$
$\bar{6}m2$	$\bar{6}$	$\bar{6}\underline{m2}$		$6mm$	$3m$	$6\underline{mm}$
$\bar{6}m2$	$3m$	$\bar{6}\underline{m2}$		$6/mmm$	$\bar{6}2m$	$\underline{6}/m\underline{m}m$
$\bar{6}m2$	32	$\bar{6}\underline{m2}$		$6/mmm$	$\bar{3}m$	$\underline{6}/mmm$
4	2	$\underline{4}$		$6/mmm$	62	$6/\underline{m}mm$
$\bar{4}$	2	$\underline{\bar{4}}$		$6/mmm$	$6mm$	$6/m\underline{m}m$
42	4	$4\underline{2}$		$6/mmm$	$6/m$	$6/m\underline{m}m$
42	222	$\underline{4}2$		23	—	—
$4/m$	4	$4/\underline{m}$		$m3$	23	$\underline{m}3$
$4/m$	$\bar{4}$	$\underline{4}/m$		$\bar{4}3m$	23	$\bar{4}3\underline{m}$
$4/m$	$2/m$	$4/m$		43	23	$\underline{4}3$
$4mm$	4	$4\underline{mm}$		$m3m$	43	$m3\underline{m}$
$4mm$	$2mm$	$\underline{4}mm$		$m3m$	$\bar{4}3m$	$\underline{m}3\underline{m}$
$\bar{4}2m$	$\bar{4}$	$\bar{4}\underline{2m}$		$m3m$	$m3$	$m3\underline{m}$
$\bar{4}2m$	222	$\bar{4}\underline{2m}$				

by introducing an operator R such that $R^n = E$ (R commutes with all geometric operations) and requiring that no elements of the form R^m shall appear. Extensive work in this field has been done by the Russian school. ———————————————

Problems. (1) Derive all possible point symmetry groups in two dimensions. (2) Derive all possible two-color point groups in two dimensions.

CHAPTER 3

GROUP REPRESENTATIONS

In the introduction, we described briefly how the symmetry of the Hamiltonian of a physical system leads to a classification of the eigenfunctions (eigenvectors) of the system. In this chapter, we take up this problem once again, and develop the mathematical apparatus for its complete solution.

3–1 Linear vector spaces. Our intuitive notion of a vector space is based on a picture of directed lines in a plane or in three-dimensional space. The vectors in such a space are described by their magnitude and direction. The vectors can be multiplied by any real number. We introduce coordinate axes in the space by drawing any two noncollinear vectors in the plane (or any three noncoplanar vectors in three-space). The vectors can then be described in terms of coordinates relative to the particular axes.

For applications to physics we must generalize from this intuitive picture. We consider a set of objects $\mathbf{x}$, $\mathbf{y}$, . . . in which the elements can be "multiplied" by a complex number α or "added" to one another to give members of the same set. Such a set is called a *linear vector space L:*

If $\mathbf{x}$ and $\mathbf{y}$ are in L, then

$$\alpha\mathbf{x} \quad \text{and} \quad \mathbf{x} + \mathbf{y} = \mathbf{y} + \mathbf{x} \tag{3–1}$$

are also in L. The "multiplication" and "addition" must satisfy the conditions:

$$(\alpha + \beta)\mathbf{x} = \alpha\mathbf{x} + \beta\mathbf{x}, \tag{3–2}$$

$$(\alpha\beta)\mathbf{x} = \alpha(\beta\mathbf{x}), \tag{3–2a}$$

$$1\mathbf{x} = \mathbf{x}, \tag{3–2b}$$

$$\alpha(\mathbf{x} + \mathbf{y}) = \alpha\mathbf{x} + \alpha\mathbf{y}. \tag{3–2c}$$

The space L will contain a zero vector (null vector), $\mathbf{0}$, such that

$$\mathbf{x} + \mathbf{0} = \mathbf{x} \quad \text{for all} \quad \mathbf{x}. \tag{3–2d}$$

Thus the linear vector space L forms an abelian group under the "addition" operation, and its elements can be multiplied by complex numbers.

If we restrict ourselves to real multipliers, we obtain a real space. Examples of such spaces are the plane and the three-space described above.

68

In the complex plane, we may speak in an intuitive way about vectors starting from the origin. Multiplication of the vector $\mathbf{x}$ by a complex number α multiplies the magnitude of $\mathbf{x}$ by the factor $|\alpha|$ and turns the vector $\mathbf{x}$ through an angle equal to the argument of α.

The set of all n-by-n matrices forms a linear vector space. The sum of two vectors $\mathbf{x}$ and $\mathbf{y}$ (with matrix elements x_{ik} and y_{ik}) is the matrix $\mathbf{x} + \mathbf{y}$ (with matrix elements $x_{ik} + y_{ik}$), and the matrix $\alpha\mathbf{x}$ has matrix elements αx_{ik}. All elements of the null matrix are equal to zero.

The infinite sequences

$$\mathbf{x} = (x_1, x_2, \ldots) \equiv (x_i), \qquad i = 1, \ldots, \infty, \tag{3-3}$$

form a linear vector space with

$$\mathbf{x} + \mathbf{y} = (x_i + y_i), \qquad \alpha\mathbf{x} = (\alpha x_i).$$

The set of all polynomials in the "variable" ζ of the form

$$\mathbf{x} = x_0 + x_1\zeta, \tag{3-4}$$

where x_0, x_1 are complex numbers, is a linear vector space in which

$$\alpha\mathbf{x} = \alpha x_0 + \alpha x_1\zeta,$$
$$\mathbf{x} + \mathbf{y} = (x_0 + y_0) + (x_1 + y_1)\zeta.$$

An immediate extension of the last example is the vector space of all polynomials of degree n in the variable ζ:

$$\mathbf{x} = x_0 + x_1\zeta + x_2\zeta^2 + \cdots + x_n\zeta^n \equiv \sum_{r=0}^{n} x_r\zeta^r. \tag{3-5}$$

In particular, we can allow n to go to infinity, thus yielding the vector space of polynomials of the form

$$\mathbf{x} = \sum_{r=0}^{\infty} x_r\zeta^r. \tag{3-6}$$

We can also consider spaces in which the vectors are functions of a real (or complex) variable z. For example, we may have a space of all functions of the form

$$\mathbf{x} = x_1 + x_2 e^z, \tag{3-7}$$

or the space of all functions of the form

$$\mathbf{x} = x_1 \cos z + x_2 \sin z, \tag{3-8}$$

or the space

$$\mathbf{x} = x_1 f_1(z) + x_2 f_2(z), \tag{3-9}$$

where f_1 and f_2 are given functions of z. As the general case, we can consider the function space

$$\mathbf{x} = \sum_{r=1}^{n} x_r f_r(z),$$ (3–10)

where $f_1, f_2, \ldots, f_n(z)$ are given functions of z. Again, we can allow n to go to infinity in Eq. (3–10).

We can generalize still further and consider the space of all functions $f(z)$ of the real (or complex) variable z, defined in some range of values of the variable z, and subject to some conditions of continuity, integrability, etc. For example, we may have a space of all continuous functions of the real variable z in the interval 0 to 1; or the space of all square-integrable functions, i.e., functions $f(z)$ for which

$$\int_{-\infty}^{+\infty} |f(z)|^2 \, dz$$

converges.

Problem. Show that the square-integrable functions form a linear vector space.

3–2 Linear dependence; dimensionality. A *linear combination* of vectors $\mathbf{x}_1, \mathbf{x}_2, \ldots, \mathbf{x}_n$ is a vector $\mathbf{x}$ of the form

$$\mathbf{x} = \alpha_1 \mathbf{x}_1 + \cdots + \alpha_n \mathbf{x}_n,$$ (3–11)

where $\alpha_1, \ldots, \alpha_n$ are complex numbers.

The vectors $\mathbf{x}_1, \mathbf{x}_2, \ldots, \mathbf{x}_n$ are said to be linearly dependent if we can construct the null vector as a linear combination of $\mathbf{x}_1, \ldots, \mathbf{x}_n$ (excluding the trivial possibility of setting $\alpha_1 = \alpha_2 = \cdots = \alpha_n = 0$),

$$\alpha_1 \mathbf{x}_1 + \alpha_2 \mathbf{x}_2 + \cdots + \alpha_n \mathbf{x}_n = 0.$$ (3–12)

If Eq. (3–12) has no nontrivial solution, we say that the vectors $\mathbf{x}_1, \mathbf{x}_2, \ldots, \mathbf{x}_n$ are *linearly independent*.

We now wish to build up sets of linearly independent vectors in our vector space. We first try this with a single vector $\mathbf{x}$. If

$$\alpha \mathbf{x} = 0, \qquad \alpha \neq 0,$$

for all the vectors in the space, then $\mathbf{x} = 0$, and we have the null space consisting solely of the null vector. If the space contains a vector $\mathbf{x}_1 \neq 0$, we try to find a second vector $\mathbf{x}_2$ such that

$$\alpha_1 \mathbf{x}_1 + \alpha_2 \mathbf{x}_2 = 0 \qquad \text{only if} \quad \alpha_1 = \alpha_2 = 0.$$

Continuing this procedure, we arrive at our definition of the dimensionality of the space L_n. In an n-*dimensional vector space* L_n we can find n vectors $\mathbf{u}_1, \mathbf{u}_2, \ldots, \mathbf{u}_n$ which are linearly independent, while $n + 1$ vectors in the space are always linearly dependent.

In the plane, two collinear vectors are linearly dependent. If the two vectors are not collinear, they are linearly independent; but any three vectors in the plane are linearly dependent, so the plane is a two-dimensional vector space.

The space of all n-by-n matrices has dimension n^2. To show this we consider a matrix $\mathbf{e}^{(jk)}$ which has all its elements equal to zero except for the (jk)-element, which we set equal to any nonzero value $\alpha^{(jk)}$. Letting j and k run from 1 to n, we obtain a set of n^2 matrices which are linearly independent, whereas any larger number of matrices will be linearly dependent.

The space of the polynomials in Eq. (3–4) is two-dimensional; e.g., the polynomials 1, ζ are linearly independent, while any three polynomials are linearly dependent. (We can always find a nontrivial solution of the equation $\alpha \mathbf{x} + \beta \mathbf{y} + \gamma \mathbf{z} = 0$ since this expression yields two equations,

$$\alpha x_0 + \beta y_0 + \gamma z_0 = 0,$$
$$\alpha x_1 + \beta y_1 + \gamma z_1 = 0,$$

in the three unknowns α, β, γ.) Similarly the space defined by Eq. (3–5) is $(n + 1)$-dimensional.

The vector space defined by Eq. (3–6) is *infinite-dimensional* since the infinite set of polynomials 1, ζ, ζ^2, $\ldots$ is linearly independent.

Problems. (1) What is the dimensionality of the vector spaces given by Eqs. (3–7), (3–8), (3–9), and (3–10)?

(2) Discuss the dimensionality of the space of square-integrable functions.

3–3 Basis vectors (coordinate axes); coordinates.

In an n-dimensional space L_n, any n linearly independent vectors $\mathbf{u}_1, \ldots, \mathbf{u}_n$ are said to form a set of *basis vectors*, or to provide a *basis* (coordinate system) in L_n.

If the vectors $\mathbf{u}_1, \ldots, \mathbf{u}_n$ form a basis in L_n, we can prove that any vector $\mathbf{x}$ in L_n is expressible as a linear combination of the vectors $\mathbf{u}_i$. The equation

$$\beta \mathbf{x} + \alpha_1 \mathbf{u}_1 + \alpha_2 \mathbf{u}_2 + \cdots + \alpha_n \mathbf{u}_n = 0$$

must have a nontrivial solution ($n + 1$ vectors in an n-dimensional space!), and in this solution, $\beta \neq 0$, for if $\beta = 0$, we would obtain a nontrivial solution,

$$\alpha_1 \mathbf{u}_1 + \cdots + \alpha_n \mathbf{u}_n = 0,$$

in contradiction to our assumption that the vectors $\mathbf{u}_i$ form a basis. Since $\beta \neq 0$, we can solve for $\mathbf{x}$:

$$\mathbf{x} = -\frac{1}{\beta}(\alpha_1 \mathbf{u}_1 + \cdots + \alpha_n \mathbf{u}_n) = x_1 \mathbf{u}_1 + \cdots + x_n \mathbf{u}_n \equiv x_i \mathbf{u}_i. \quad (3\text{--}13)$$

Thus the arbitrary vector $\mathbf{x}$ is expressed as a linear combination of the basis vectors $\mathbf{u}_i$. The coefficients x_i in the linear combination (3–13) are called the *coordinates* of the vector $\mathbf{x}$ in the basis (or coordinate system) $\mathbf{u}_1, \ldots, \mathbf{u}_n$.

In the space of n-by-n matrices, the matrices $\mathbf{e}^{(jk)}$ described above form a basis. The matrix vector $\mathbf{x}$ (matrix elements x_{jk}) is expressible in terms of these basis vectors as

$$\mathbf{x} = \sum_{jk} \frac{x_{jk}}{\alpha^{(jk)}} \mathbf{e}^{(jk)},$$

so that the coordinates of $\mathbf{x}$ are the numbers $x_{jk}/\alpha^{(jk)}$ in this basis.

The polynomials 1 and ζ are a set of basis vectors for the two-dimensional space of Eq. (3–4). The vector $\mathbf{x}$ of Eq. (3–4) has the coordinates x_0, x_1 in this basis.

The coordinates of the polynomial $\mathbf{x}$ in Eq. (3–6) are the infinite sequence $(x_0, x_1, \ldots)$ in the basis provided by the linearly independent polynomials $1, \zeta, \zeta^2, \ldots$

Problem. Find a set of basis vectors for the spaces described by Eqs. (3–7), (3–8), (3–9), and (3–10), and give the coordinates of a general vector relative to this basis.

It should be clear that the basis vectors are *not* uniquely given and in fact can be selected in infinitely many ways. In the plane, any two non-collinear vectors provide a basis. In the space of polynomials of the form $x_0 + x_1 \zeta$, we can choose, in addition to the basis $(1, \zeta)$, a basis $(1 + \zeta, 1 - \zeta)$ or $(1 + \alpha\zeta, 1 - \alpha\zeta)$ with $\alpha \neq 0$ or, in general, $(\alpha + \beta\zeta, \gamma + \delta\zeta)$ so long as $\mathbf{u}_1 = \alpha + \beta\zeta$ and $\mathbf{u}_2 = \gamma + \delta\zeta$ are linearly independent $(\alpha\delta - \beta\gamma \neq 0)$.

A given vector $\mathbf{x}$ will have *different* coordinates with respect to *different* systems of basis vectors. The vector $\mathbf{x}$ itself has an *intrinsic* significance, whereas its description in terms of coordinates changes with the coordinate axes. For example, the polynomial

$$\mathbf{x} = x_0 + x_1 \zeta = \frac{x_0 + x_1/\alpha}{2} (1 + \alpha\zeta) + \frac{x_0 - x_1/\alpha}{2} (1 - \alpha\zeta)$$

$$= \frac{x_0\delta - x_1\gamma}{\alpha\delta - \beta\gamma} (\alpha + \beta\zeta) + \frac{x_1\alpha - x_0\beta}{\alpha\delta - \beta\gamma} (\gamma + \delta\zeta)$$

has the coordinates

$$(x_0, x_1), \qquad \left(\frac{x_0 + x_1/\alpha}{2}, \frac{x_0 - x_1/\alpha}{2} \right), \qquad \left(\frac{x_0\delta - x_1\gamma}{\alpha\delta - \beta\gamma}, \frac{x_1\alpha - x_0\beta}{\alpha\delta - \beta\gamma} \right)$$

with respect to these three coordinate systems.

Once we have found a basis $\mathbf{u}_1, \ldots, \mathbf{u}_n$, we can easily determine *all* possible systems of basis vectors. Every vector is expressible as a linear combination of $\mathbf{u}_1, \ldots, \mathbf{u}_n$ as in Eq. (3–13). The n vectors $\mathbf{u}'_1, \ldots, \mathbf{u}'_n$ can be written as

$$\mathbf{u}'_i = \sum_{j=1}^{n} a_{ij}\mathbf{u}_j \equiv a_{ij}\mathbf{u}_j \qquad (i = 1, \ldots, n), \qquad (3\text{–}14)$$

where, in the last step, we have introduced the convention of summing over repeated indices. The vectors $\mathbf{u}'_i$ will be a basis if they are linearly independent, which will be the case if and only if the determinant of the coefficients a_{ij} is not zero. The coefficients form a matrix $\mathbf{a}$. The new vectors $\mathbf{u}'_i$ in Eq. (3–14) will form a basis if the matrix $\mathbf{a}$ is nonsingular (determinant $\neq 0$). All possible bases in L_n are obtained from any one basis by allowing the matrix $\mathbf{a}$ to run through the full set of nonsingular matrices of degree n.

When we change basis from the vectors $\mathbf{u}_i$ to the vectors $\mathbf{u}'_i$, the *fixed* vector $\mathbf{x}$ changes its coordinates from x_i to x'_i:

$$\mathbf{x} = x_i\mathbf{u}_i = x'_i\mathbf{u}'_i. \qquad (3\text{–}15)$$

Using Eq. (3–14), we have

$$x_j\mathbf{u}_j = x'_i a_{ij}\mathbf{u}_j,$$

and, since the vectors $\mathbf{u}_j$ are linearly independent,

$$x_j = x'_i a_{ij} = \tilde{a}_{ji}x'_i, \qquad (3\text{–}16)$$

where $\tilde{\mathbf{a}}$ is the transpose of the matrix $\mathbf{a}$, that is,

$$\tilde{a}_{ji} = a_{ij}. \qquad (3\text{–}17)$$

If we regard the vectors $\mathbf{u}_i(\mathbf{u}'_i)$ and the coordinate sets $x_i(x'_i)$ as matrices with n rows and a single column, Eqs. (3–14) and (3–16) can be written in matrix form:

$$\mathbf{u}' = \mathbf{a}\mathbf{u}, \tag{3–14a}$$

$$\mathbf{x} = \tilde{\mathbf{a}}\mathbf{x}'. \tag{3–16a}$$

The transpose of a product of matrices is given by

$$\widetilde{\mathbf{a}\mathbf{b}} = \tilde{\mathbf{b}}\tilde{\mathbf{a}}, \tag{3–18}$$

i.e., it is equal to the product of the transposed matrices in reverse order. The transpose of the column matrix $\mathbf{x}$ is a matrix $\tilde{\mathbf{x}}$ having one row and n columns:

$$\tilde{\mathbf{x}} \equiv (x_1, \ldots, x_n). \tag{3–19}$$

With this notation, Eq. (3–15) can be written as

$$\mathbf{x} = \tilde{\mathbf{x}}\mathbf{u} = \tilde{\mathbf{x}}'\mathbf{u}'. \tag{3–15a}$$

From Eq. (3–16a) we obtain

$$\mathbf{x}' = \tilde{\mathbf{a}}^{-1}\mathbf{x}. \tag{3–16b}$$

We can easily verify the correctness of Eq. (3–15):

$$\tilde{\mathbf{x}}' = \tilde{\mathbf{x}}\mathbf{a}^{-1}, \qquad \mathbf{u}' = \mathbf{a}\mathbf{u},$$

so that

$$\tilde{\mathbf{x}}'\mathbf{u}' = \tilde{\mathbf{x}}\mathbf{a}^{-1}\mathbf{a}\mathbf{u} = \tilde{\mathbf{x}}\mathbf{u}.$$

Problems. (1) Prove that the polynomials

$$f_n' = e^{-\zeta}\frac{d^n}{d\zeta^n}(\zeta^{2n}e^{\zeta}), \qquad n = 0, 1, 2, \ldots \tag{3–19}$$

form a basis in the vector space of all polynomials [Eq. (3–6)]. Find the expansion coefficients of the f_n' in terms of the basis functions $f_n = \zeta^n$.

(2) Do the same for the polynomials

$$f_n'' = \frac{d^n}{d\zeta^n}[(\zeta^2 - 1)^n]. \tag{3–20}$$

3–4 Mappings; linear operators; matrix representatives; equivalence. As in Chapter 1, we can define a mapping of the vector space L onto itself.

The mapping T associates with each vector $\mathbf{x}$ a new vector $\mathbf{y}$ in L,

$$\mathbf{y} = T\mathbf{x}. \tag{3-21}$$

If the mapping T is one-to-one there exists an *inverse* mapping T^{-1} such that

$$\mathbf{x} = T^{-1}\mathbf{y}. \tag{3-22}$$

The mapping T may be regarded as an *operator* which acts on vectors $\mathbf{x}$ in L to produce vectors $\mathbf{y}$ in L. If the mapping is one-to-one, the inverse operator T^{-1} exists. For every vector $\mathbf{x}$ in L,

$$T^{-1}T\mathbf{x} = TT^{-1}\mathbf{x} = \mathbf{x},$$

that is,

$$T^{-1}T = TT^{-1} = 1, \tag{3-23}$$

where the operator 1 is the identity operator, which leaves all vectors unchanged.

The operator T is said to be a *linear operator* if

$$T(\mathbf{x} + \mathbf{z}) = T\mathbf{x} + T\mathbf{z}, \tag{3-24}$$

$$T(\alpha\mathbf{x}) = \alpha T\mathbf{x}. \tag{3-24a}$$

We should emphasize that no coordinate system is specified in the definition of operators, so that operators have an *intrinsic* significance.

If we choose a particular basis $\mathbf{u}_i$ in the space L, we can describe the action of the operator T in Eq. (3–21) by saying that the coordinates y_i of the image are certain functions of the coordinates x_i:

$$y_i = T_i(x_1, \ldots, x_n), \qquad i = 1, \ldots, n. \tag{3-25}$$

If the mapping is one-to-one, these equations are solvable for the x_i in terms of y_i,

$$x_i = T_i^{-1}(y_1, \ldots, y_n), \qquad i = 1, \ldots, n. \tag{3-25a}$$

The necessary and sufficient condition for the mapping to be one-to-one is that the Jacobian

$$J \equiv \left(\frac{\partial T_i}{\partial x_j}\right) \tag{3-26}$$

be different from zero. The functions T_i and T_i^{-1} relate the coordinates x_i and y_i in a *given* basis $\mathbf{u}_i$. If we change to a new basis, $\mathbf{u}_i'$, the new coordinates x_i', y_i' can be related by using Eqs. (3–16a) and (3–16b):

$$y_i' = \tilde{a}_{ij}^{-1} y_j = \tilde{a}_{ij}^{-1} T_j(x_1, \ldots, x_n)$$

$$= \tilde{a}_{ij}^{-1} T_j(\tilde{a}_{1k} x_k', \ldots, \tilde{a}_{nk} x_k'). \tag{3-27}$$

In the case of a nonsingular linear operator T, the coordinates x_i, y_i are related more simply. Applying the linear operator T to Eq. (3–13), we obtain

$$\mathbf{y} = T\mathbf{x} = T(x_i \mathbf{u}_i) = x_i T\mathbf{u}_i. \tag{3-28}$$

Equation (3–28) states that the coordinates of $\mathbf{y}$ with respect to the basis vectors $\mathbf{v}_i = T\mathbf{u}_i$ are the *same* as the coordinates of $\mathbf{x}$ with respect to the basis vectors $\mathbf{u}_i$. The vectors $\mathbf{v}_i$ are linear combinations of the $\mathbf{u}_i$, that is,

$$\mathbf{v}_i = T\mathbf{u}_i = \mathbf{u}_j T_{ji}, \tag{3-29}$$

where the T_{ji} are complex numbers forming a matrix T. Substituting in (3–28), we find that in the *fixed basis* $\mathbf{u}_i$,

$$y_j \mathbf{u}_j = \mathbf{u}_j T_{ji} x_i, \tag{3-30}$$

and, since the $\mathbf{u}_j$ are linearly independent,

$$y_j = T_{ji} x_i, \tag{3-31}$$

or, in matrix notation,

$$\mathbf{y} = \mathbf{T}\mathbf{x}. \tag{3-32}$$

In Eq. (3–32), $\mathbf{T}$ is the *matrix representative* of the linear operator T in the basis $\mathbf{u}_i$. If we change to the *new* basis $\mathbf{u}_i'$ given by Eq. (3–14a), we find by means of (3–16a) and (3–16b) that

$$\mathbf{y}' = \tilde{\mathbf{a}}^{-1}\mathbf{y} = \tilde{\mathbf{a}}^{-1}\mathbf{T}\mathbf{x} = \tilde{\mathbf{a}}^{-1}\mathbf{T}\tilde{\mathbf{a}}\mathbf{x}'. \tag{3-33}$$

Thus the matrix representative of the linear operator T in the new basis $\mathbf{u}_i'$ is

$$\mathbf{T}' = \tilde{\mathbf{a}}^{-1}\mathbf{T}\tilde{\mathbf{a}}. \tag{3-34}$$

The *transform* of a matrix $\mathbf{A}$ by a (nonsingular) matrix $\mathbf{S}$ is defined to be the matrix

$$\mathbf{A}' = \mathbf{S}\mathbf{A}\mathbf{S}^{-1}. \tag{3-35}$$

Thus Eq. (3–34) states that $\mathbf{T}'$ is the transform of $\mathbf{T}$ by the matrix $\tilde{\mathbf{a}}^{-1}$.

The matrix representatives (in different bases) of the fixed linear operator T are thus transforms of one another. Matrices which bear this relation to one another are said to be *equivalent*. Thus the matrix representatives of the linear operator T in different bases are *equivalent matrices*.

We can also consider one-to-one mappings of *one* space L on *another* space L'. (Clearly, if the mapping is one-to-one, the two spaces have the same dimensionality.) In such a mapping the operator S acts on vectors $\mathbf{x}$ in L to give vectors $\mathbf{x}'$ in L':

$$\mathbf{x}' = S\mathbf{x}, \qquad \mathbf{x} = S^{-1}\mathbf{x}'.$$

If a linear operator T is defined in the space L, the mapping S *induces* a linear operator T' in L':

$$T' = STS^{-1}. \tag{3–36}$$

The operator T' is well defined: S^{-1} takes vectors $\mathbf{x}'$ in L' into vectors $\mathbf{x}$ in L, T transforms vectors $\mathbf{x}$ in L into vectors $\mathbf{y}$ in L, and finally S takes vectors $\mathbf{y}$ in L into vectors $\mathbf{y}'$ in L'. The net result is that T' takes vectors $\mathbf{x}'$ in L' into vectors $\mathbf{y}'$ in L'. The operator T' is the *transform* of T by the operator S.

Problems. (1) Show that $d/d\zeta$ is a linear operator in the polynomial space of Eq. (3–5). Choose a basis and give the matrix representative of the operator in this basis. Does the matrix have an inverse?

(2) Discuss the corresponding problem for the operator $d/d\zeta$ in the space of Eq. (3–6).

(3) The Fourier transform of the function $f(x)$ is

$$g(k) = (2\pi)^{-1/2} \int_{-\infty}^{+\infty} f(x)e^{-ikx}\, dx; \qquad f(x) = (2\pi)^{-1/2} \int_{-\infty}^{+\infty} g(k)e^{ikx}\, dk. \tag{3–37}$$

Show that the operator $(1/i)(\partial/\partial x)$ in the space of functions $f(x)$ induces the operator of multiplication by k in the space of functions $g(k)$.

3–5 Group representations. A set of operators $A, B, \ldots$ in a vector space L forms a group if the operators satisfy the postulates given in Chapter 1. Here the product of the operators A and B means the single operator C such that

$$C\mathbf{x} = A(B\mathbf{x}) \qquad \text{for all } \mathbf{x} \text{ in } L. \tag{3–38}$$

The identity of the group is the unit operator which leaves all vectors in L unchanged. All operators of the group possess inverses.

If we map the space L on another space L', using an operator S, we get an isomorphic group of operators in the space L' which are the transforms of A, B, ... by the operator S:

$$A' = SAS^{-1}, \qquad B' = SBS^{-1}, \ldots \qquad (3\text{-}39)$$

If we map an arbitrary group G homomorphically on a group of operators $D(G)$ in the vector space L, we say that the operator group $D(G)$ is a *representation* of the group G in the *representation space* L. If the dimensionality of L is n, we say that the representation has *degree* n (or is an *n-dimensional representation*). The operator corresponding to the element R of G will be denoted by $D(R)$. If R and S are elements of the group G, then

$$D(RS) = D(R)D(S), \qquad (3\text{-}40)$$

$$D(R^{-1}) = [D(R)]^{-1}, \qquad (3\text{-}40\text{a})$$

and

$$D(E) = 1. \qquad (3\text{-}40\text{b})$$

A *linear representation* is a representation in terms of *linear operators*. We shall restrict ourselves almost entirely to such linear representations. (All representations which appear should be assumed to be linear unless a specific statement is made to the contrary.)

If we choose a basis in the n-dimensional space L, the linear operators of the representation can be described by their matrix representatives. We then obtain a homomorphic mapping of the group G on a group of n-by-n matrices $D(G)$, i.e., a *matrix representation* of the group G. From Eqs. (3–40), (3–40a), and (3–40b) we see that the matrices are nonsingular, and that

$$D_{ij}(E) = \delta_{ij} = \begin{cases} 1 & \text{for } i = j \\ 0 & \text{for } i \neq j \end{cases}, \qquad i, j = 1, \ldots, n; \qquad (3\text{-}41)$$

$$D_{ij}(RS) = \sum_k D_{ik}(R)D_{kj}(S) \equiv D_{ik}(R)D_{kj}(S). \qquad (3\text{-}41\text{a})$$

If we deal with several different representations, we distinguish among them by using superscripts: $D_{ij}^{(\mu)}(R)$. Another notation is $[_{ij}^{\mu R}]$. The dimension of the μth representation will be denoted by n_μ or $[\mu]$.

If the homomorphic mapping of G on $D(G)$ reduces to an isomorphism, then the representation is said to be "faithful"; in this case the order of the group of matrices $D(G)$ is equal to the order g of the group G. In general, there will be several elements in G which are mapped on the unit matrix $D(E) = 1$. As we saw in Chapter 1, the set of elements $\mathfrak{K}$ of G which are mapped on 1 form an invariant subgroup of G, and the group of matrices $D(R)$ is a faithful representation of the factor group $G/\mathfrak{K}$. From this it follows that if we have found a representation for the factor group relative to an invariant subgroup, we automatically have a representation for the whole group G. In this representation, all elements in a coset of $\mathcal{K}$ in G are mapped into the same matrix.

3-6 Equivalent representations; characters. If we change the basis in the n-dimensional space L, the matrices $D(R)$ will be replaced by their transforms by some matrix C [cf. Eq. (3-34)]. The matrices

$$D'(R) = CD(R)C^{-1} \tag{3-42}$$

also provide a representation of the group G, which is equivalent to the representation $D(R)$. From our previous discussion it is clear that equivalent representations have the same structure, even though the matrices look different.

What we wish to find are quantities which are intrinsic properties of $D(R)$, i.e., are invariant under a change of coordinate axes. One such invariant is easily found, for if we take the sum of the diagonal elements of the matrix, we obtain

$$\sum_i (CD(R)C^{-1})_{ii} = \sum_{ikl} C_{ik}D_{kl}(R)C_{li}^{-1}$$

$$= \sum_{kl} \delta_{kl}D_{kl}(R) = \sum_k D_{kk}(R). \tag{3-43}$$

Thus the sum of the diagonal elements, or *trace*, of a matrix $D(R)$ is invariant under a transformation of the coordinate axes. When we deal with group representations, the trace $\sum_i D_{ii}(R)$ is called the *character* of R in the *representation* D and is denoted by

$$\chi(R) = \sum_i D_{ii}(R). \tag{3-44}$$

We see that equivalent representations have the same set of characters. To indicate the representation, we shall use a superscript. Thus $\chi^{(\mu)}(R)$ (or $[\mu; R]$) means the character of R in the μ-representation.

If we consider two conjugate elements S and R of G so that $S = URU^{-1}$, then $D(S) = D(U)D(R)[D(U)]^{-1}$. Since $D(U)$ is a possible axis transformation in our space, we see that $\sum_i D_{ii}(S) = \sum_i D_{ii}(R)$, or $\chi(S) = \chi(R)$. In other words, conjugate elements in G always have the same character. Hence, when we describe a group by listing the characters of its elements in a given representation, the same number (character) is assigned to all the elements in a given class. Labeling the classes of G by $K_1, K_2, \ldots$, etc., the representation will be described by the set of characters $\chi_1, \ldots, \chi_\nu$, where ν is the number of classes in G. Again, if we have several different representations, the characters for the various representations will be distinguished by a superscript, e.g., χ_1^μ is the character of the class K_1 in the μ-representation. Thus each representation gives us a set of ν numbers which we can consider as a vector in a ν-dimensional space, namely the vector χ^μ with components $\chi_1^\mu, \ldots, \chi_\nu^\mu$.

3–7 Construction of representations. Addition of representations.

Before proceeding any further with this formal development of the theory, let us look at the problem from a slightly different point of view. The theory as presented above is necessary if we start from an abstract group. In physical problems we start not from an abstract group, but from a group of transformations of the configuration space of a physical system. For example, our symmetry groups in the previous chapter were groups of transformations in three-dimensional space. The group elements *themselves* give a representation of the group in three dimensions. For example, the operation $C(\theta)$ was the transformation

$$x_1' = x_1 \cos\theta - x_2 \sin\theta;$$
$$x_2' = x_1 \sin\theta + x_2 \cos\theta; \qquad (3\text{–}45)$$
$$x_3' = x_3,$$

and the operation I was the transformation

$$x_1' = -x_1, \qquad x_2' = -x_2, \qquad x_3' = -x_3. \qquad (3\text{–}46)$$

One of our problems is to determine how to go about constructing representations of a group. Another is to see what connection representations have with physics. Suppose that we are given a group G of transformations like the symmetry groups of the previous chapter, or a representation like the ones discussed earlier in this chapter. If we have a transformation T belonging to the group of transformations G [or to the group of transformations associated with the matrix representation $D(G)$], we can construct new representations as follows: The transformation T takes x into x': $x' = Tx$. We now associate with T a linear operator O_T which acts on functions $\psi(x)$: Given any function $\psi(x)$, the effect of the operator O_T on ψ is to change it to the function $O_T\psi \equiv \psi'$ such that

$$\psi'(x') \equiv O_T\psi(x') = \psi(x) \qquad \text{if } x' = Tx. \qquad (3\text{–}47)$$

In other words, the transformed function $\psi' \equiv O_T\psi$ takes the same value at the image point x' that the original function ψ had at the object point x. Or we may say that the point P (with coordinates x) moves to its image point P' (with coordinates x') under the transformation T, carrying with it the numerical value of ψ at P. For example, if T is $x' = x + a$ (one dimension!), then $\psi'(x)$ is obtained from $\psi(x)$ by sliding the graph of $\psi(x)$ a units to the right; so $\psi'(x) = \psi(x - a)$. We can rewrite (3–47) as

$$O_T\psi(Tx) = \psi(x) \tag{3-48}$$

or

$$O_T\psi(x) = \psi(T^{-1}x). \tag{3-48a}$$

If we follow the transformation T by a transformation S such that

$$x'' = Sx',$$

then the associated operator O_S is defined as in (3–48); acting on any function ϕ, this operator yields a new function $O_S\phi$ such that

$$O_S\phi(Sx') = \phi(x').$$

If ϕ is the function ψ' defined by (3–47), then

$$\begin{aligned}
O_S\psi'(Sx') &= \psi'(x'), \\
O_SO_T\psi(Sx') &= \psi(x), \\
O_SO_T\psi(STx) &= \psi(x).
\end{aligned} \tag{3-49}$$

On the other hand,

$$x'' = Sx' = STx,$$

and we can therefore define an operator O_{ST} such that

$$O_{ST}\psi(x'') = \psi(x). \tag{3-50}$$

Comparing with (3–49), we see that

$$O_{ST} = O_SO_T, \tag{3-51}$$

so the operators satisfy the same relations as the group elements. To each element S there corresponds an operator O_S acting on functions ψ, and to the element S^{-1} there corresponds the operator

$$O_{S^{-1}} = (O_S)^{-1}. \tag{3-51a}$$

If we can find a representation for the operators, we automatically obtain a representation of G.

To see how this method works in a simple case, we consider the symmetry group $\mathcal{C}_i$ with two elements E and I; E is the identity transformation

$x' = Ex = x$, I is the inversion $x' = Ix = -x$ (all in three dimensions!). We choose a function $\psi(x)$. From (3–48),

$$\psi(x) = O_E\psi(Ex) = O_E\psi(x),$$
$$\psi(-x) = O_E\psi(-Ex) = O_E\psi(-x); \tag{3–52}$$

that is, O_E is the identity operator. Similarly,

$$\psi(x) = O_I\psi(Ix) = O_I\psi(-x), \tag{3–53}$$

or

$$O_I\psi(x) = \psi(-x); \tag{3–53a}$$

so the operator I changes the sign of x in ψ. Equations (3–52), (3–53), and (3–53a) state that $O_E\psi(\pm x)$, $O_I\psi(\pm x)$ are expressible as linear combinations of $\psi(x)$ and $\psi(-x)$:

$$\begin{cases} O_E\psi(x) = \psi(x) + 0 \cdot \psi(-x), \\ O_E\psi(-x) = 0 \cdot \psi(x) + \psi(-x); \end{cases}$$

$$\begin{cases} O_I\psi(x) = 0 \cdot \psi(x) + \psi(-x), \\ O_I\psi(-x) = \psi(x) + 0 \cdot \psi(-x). \end{cases} \tag{3–54}$$

If we let $\psi(x) = f_1$, $\psi(-x) = f_2$, Eqs. (3–54) state that

$$\begin{cases} O_E f_1 = f_1 + 0 \cdot f_2, \\ O_E f_2 = 0 \cdot f_1 + f_2, \end{cases} \qquad \begin{cases} O_I f_1 = 0 \cdot f_1 + f_2, \\ O_I f_2 = f_1 + 0 \cdot f_2. \end{cases} \tag{3–54a}$$

The operators transform the functions f_i among themselves; we may write

$$O_R f_i = \sum_{j=1}^{2} f_j D_{ji}(R), \qquad i = 1, 2, \tag{3–55}$$

where, comparing with (3–54a), we see that

$$D(E) = \begin{bmatrix} 1 & 0 \\ 0 & 1 \end{bmatrix}; \qquad D(I) = \begin{bmatrix} 0 & 1 \\ 1 & 0 \end{bmatrix}. \tag{3–56}$$

It is easily verified that the matrices (3–56) give a two-dimensional representation of the group:

$$I^2 = E, \qquad \text{and} \qquad [D(I)]^2 = \begin{bmatrix} 0 & 1 \\ 1 & 0 \end{bmatrix}\begin{bmatrix} 0 & 1 \\ 1 & 0 \end{bmatrix} = \begin{bmatrix} 1 & 0 \\ 0 & 1 \end{bmatrix} = D(E).$$

Suppose that we had chosen $\psi(x)$ to be an *even* function, $\psi(x) \equiv \psi(-x)$.

Then we would obtain only the two relations

$$O_E \psi = \psi, \qquad O_I \psi = \psi,$$

that is, the even function ψ is transformed into a multiple of itself by all the operators; we have a single basis function $f = \psi$, and our representation is one-dimensional:

$$D^{(1)}(E) = (1), \qquad D^{(1)}(I) = (1), \tag{3–57}$$

where (1) is a 1×1 matrix whose single element is equal to unity.

If we had chosen ψ to be an *odd* function, $\psi(x) \equiv -\psi(-x)$, we would also obtain only two relations:

$$O_E \psi = \psi, \qquad O_I \psi = -\psi.$$

Again ψ is transformed into a multiple of itself by all the operators; we have a single basis function $f = \psi$, and our representation is one-dimensional:

$$D^{(2)}(E) = (1), \qquad D^{(2)}(I) = (-1). \tag{3–58}$$

Suppose that we now start with two functions, an even function f_1 and an odd function f_2. Then

$$\begin{aligned} O_E f_1 &= f_1, & O_E f_2 &= f_2, \\ O_I f_1 &= f_1, & O_I f_2 &= -f_2. \end{aligned} \tag{3–59}$$

We may adopt the apparently perverse viewpoint that the functions f_1 and f_2 are transformed into linear combinations of f_1 and f_2, and say that we have obtained a two-dimensional representation:

$$D^{(3)}(E) = \begin{bmatrix} 1 & 0 \\ 0 & 1 \end{bmatrix}, \qquad D^{(3)}(I) = \begin{bmatrix} 1 & 0 \\ 0 & -1 \end{bmatrix}. \tag{3–60}$$

Comparing with the previous two cases, we see that, in fact, f_1 and f_2 separately went into multiples of f_1 and f_2, respectively, and all we did was to consider together two functions, f_1 and f_2, each of which transformed into a multiple of itself. The matrices $D^{(3)}$ can be written as

$$D^{(3)}(E) = \begin{bmatrix} D^{(1)}(E) & 0 \\ 0 & D^{(2)}(E) \end{bmatrix}; \quad D^{(3)}(I) = \begin{bmatrix} D^{(1)}(I) & 0 \\ 0 & D^{(2)}(I) \end{bmatrix}. \tag{3–61}$$

To go to extremes, suppose we had chosen three even functions ψ_1, ψ_2, ψ_3 which are linearly independent (for example, x^2, y^2, z^2). Then we would have obtained a three-dimensional representation D such that

$$D(E) = \begin{bmatrix} 1 & 0 & 0 \\ 0 & 1 & 0 \\ 0 & 0 & 1 \end{bmatrix} = D(I) = \begin{bmatrix} D^{(1)} & 0 & 0 \\ 0 & D^{(1)} & 0 \\ 0 & 0 & D^{(1)} \end{bmatrix}. \quad (3\text{-}62)$$

Similarly, if we had chosen ψ_2 as an even function and ψ_1, ψ_3 as odd functions, we would have obtained the representation

$$D(E) = \begin{bmatrix} 1 & 0 & 0 \\ 0 & 1 & 0 \\ 0 & 0 & 1 \end{bmatrix} = \begin{bmatrix} D^{(2)}(E) & 0 & 0 \\ 0 & D^{(1)}(E) & 0 \\ 0 & 0 & D^{(2)}(E) \end{bmatrix},$$

$$\quad (3\text{-}63)$$

$$D(I) = \begin{bmatrix} -1 & 0 & 0 \\ 0 & 1 & 0 \\ 0 & 0 & -1 \end{bmatrix} = \begin{bmatrix} D^{(2)}(I) & 0 & 0 \\ 0 & D^{(1)}(I) & 0 \\ 0 & 0 & D^{(2)}(I) \end{bmatrix}.$$

As a last example we return to Eq. (3–56). If we had started with two linearly independent functions ψ and ϕ (neither of which is purely odd or even), we could have found a set of four functions:

$$f_1 = \psi(x), \quad f_2 = \psi(-x), \quad f_3 = \phi(x), \quad f_4 = \phi(-x),$$

and thus obtained the four-dimensional representation D':

$$D'(E) = \begin{bmatrix} 1 & 0 & 0 & 0 \\ 0 & 1 & 0 & 0 \\ \cdots & \cdots & \cdots & \cdots \\ 0 & 0 & 1 & 0 \\ 0 & 0 & 0 & 1 \end{bmatrix}; \quad D'(I) = \begin{bmatrix} 0 & 1 & 0 & 0 \\ 1 & 0 & 0 & 0 \\ \cdots & \cdots & \cdots & \cdots \\ 0 & 0 & 0 & 1 \\ 0 & 0 & 1 & 0 \end{bmatrix}. \quad (3\text{-}64)$$

The process we have used in these three examples is called addition of representations. In each case the matrices consist of submatrices along the main diagonal bordered by zeros, the subdivision being the same for all matrices. In general, if we have a set of functions $f_1, \ldots, f_n$ such that

$$O_R f_i = \sum_{j=1}^{n} f_j D_{ji}^{(1)}(R); \quad i = 1, \ldots, n, \quad (3\text{-}65)$$

giving a representation $D^{(1)}(G)$, and a second set of functions, $f_{n+1}, \ldots, f_{n+m}$, linearly independent of $f_1, \ldots, f_n$ such that

$$O_R f_i = \sum_{j=n+1}^{n+m} f_j D_{ji}^{(2)}(R), \quad i = n+1, \ldots, n+m, \quad (3\text{-}65a)$$

then we may look upon Eqs. (3–65) and (3–65a) as linear transformations among the composite set of $(n + m)$ functions. Written in this form, Eq. (3–65) for $i = 1$ would read

$$O_R f_1 = D_{11}^{(1)}(R)f_1 + D_{21}^{(1)}(R)f_2 + \cdots + D_{n1}^{(1)}(R)f_n + 0 \cdot f_{n+1} + \cdots$$
$$+ 0 \cdot f_{n+m}.$$

Using the composite set as a basis, we would obtain a representation $D(R)$ of degree $(n + m)$ such that

$$D(R) = \left[\begin{array}{c|c} D^{(1)}(R) & 0 \\ \hline 0 & D^{(2)}(R) \end{array} \right] \begin{array}{l} {}^{}\}n \\ {}^{}\}m \end{array} \quad \text{rows}$$

$$\underbrace{\phantom{D^{(1)}(R)}}_{n} \quad \underbrace{\phantom{D^{(2)}(R)}}_{m}$$
$$\text{columns}$$

whence

$$D = D^{(1)} + D^{(2)}.$$

According to this terminology, (3–61) means

$$D^{(3)} = D^{(1)} + D^{(2)},$$

while (3–62) states that

$$D = D^{(1)} + D^{(1)} + D^{(1)} = 3D^{(1)},$$

and (3–63) that

$$D = D^{(2)} + D^{(1)} + D^{(2)} = D^{(1)} + 2D^{(2)}.$$

The rearrangement of terms in the sum of representations is clearly permissible; it amounts to a relabeling of the basis functions in a different order.

The general procedure for constructing representations should now be clear. We start from any set of linearly independent functions, and apply to each of the functions all the operators O_R corresponding to elements R of the transformation group G. We then get a set of functions which, though they may not be linearly independent, can all be expressed linearly in terms of n of them, $\psi_1, \psi_2, \ldots, \psi_n$. If we now apply to these functions any operator O_R, the resulting function can be expressed as a linear combination of these same n functions:

$$O_R \psi_\nu = \sum_{\mu=1}^{n} \psi_\mu D_{\mu\nu}(R), \qquad \nu = 1, \ldots, n. \tag{3–66}$$

The correspondent of the element R in the representation is then the matrix $D(R)$. We must show that we get a proper homomorphism of G on $D(G)$. From (3–51) and (3–66) we have

$$O_{SR}\psi_\nu = O_S O_R \psi_\nu = O_S \sum_{\mu=1}^{n} \psi_\mu D_{\mu\nu}(R)$$

$$= \sum_{\mu,\sigma=1}^{n} \psi_\sigma D_{\sigma\mu}(S) D_{\mu\nu}(R) \tag{3–67}$$

$$= \sum_{\sigma=1}^{n} \psi_\sigma \left[\sum_{\mu=1}^{n} D_{\sigma\mu}(S) D_{\mu\nu}(R) \right].$$

But

$$O_{SR}\psi_\nu = \sum_{\sigma=1}^{n} \psi_\sigma D_{\sigma\nu}(SR),$$

so

$$D_{\sigma\nu}(SR) = \sum_{\mu=1}^{n} D_{\sigma\mu}(S) D_{\mu\nu}(R),$$

or

$$D(SR) = D(S)D(R).$$

3–8 Invariance of functions and operators. Classification of eigenfunctions. Now we examine further the operators O_R which we introduced earlier. The operator O_R applied to the function $\psi(x)$ changed it to a function which we called $O_R\psi$ such that

$$O_R\psi(x') = \psi(x) \qquad \text{if } x' = Rx, \tag{3–47}$$

that is,

$$O_R\psi(Rx) = \psi(x) \qquad \text{for all } x,$$

or

$$O_R\psi(x) = \psi(R^{-1}x) \qquad \text{for all } x. \tag{3–68}$$

This last form is most useful: O_R operating on ψ replaces x by $R^{-1}x$. It may happen that $O_R\psi$ is identical with ψ, i.e.,

$$O_R\psi(x) \equiv \psi(x), \tag{3–69}$$

so that

$$\psi(x) \equiv \psi(R^{-1}x) \qquad \text{or} \qquad \psi(Rx) \equiv \psi(x), \tag{3–69a}$$

and the function ψ takes on the same value at the image point Rx as at the point x. In this case we say that the function ψ is *invariant* under the operator O_R or, more briefly, under the transformation R. For example, the function $\psi(\mathbf{x}) = x^4 + y^2$ is invariant under inversion; the function $\psi = x^2 + y^2$ is invariant under rotations. To test for invariance of a function we replace the arguments x by their images Rx and see whether we get the same expression again.

We have already noted that the operators O_R are linear:

$$O_R[\psi(x) + \phi(x)] = O_R\psi(x) + O_R\phi(x). \tag{3–70}$$

From (3–68) it is clear that

$$O_R[\psi(x) \cdot \phi(x)] = O_R\psi(x) \cdot O_R\phi(x). \tag{3–71}$$

If we have an operator $H(x)$ which acts on the function $\psi(x)$, giving the function $\phi(x) = H(x)\psi(x)$, then

$$O_R[H(x)\psi(x)] = O_R\phi(x) = \phi(R^{-1}x) = H(R^{-1}x)\,\psi(R^{-1}x),$$
$$O_R H(x) O_R^{-1} O_R \psi(x) = H(R^{-1}x) O_R \psi(x) = H'(x) O_R \psi(x), \tag{3–72}$$

where $H'(x) = H(R^{-1}x) = O_R H(x) O_R^{-1}$, or $H'(Rx) = H(x)$. The transformed operator H' at the point Rx is the same as the operator H at the point x. The operators H and H' are in general *not* the same at a *given* point x. If $H'(x) = H(x)$, so that $H(Rx) = H(x)$ and

$$O_R H(x) O_R^{-1} = H(x), \tag{3–73}$$

the operator $H(x)$ is said to be *invariant* under the transformation R [$H(x)$ commutes with the operator O_R].

Problems. (1) Show that the operator $\partial^2/\partial x_1^2 + \partial^2/\partial x_2^2$ is invariant under the transformations of Eqs. (3–45) and (3–46).

(2) The Fourier transform operator F is defined by

$$F\psi(\mathbf{x}) \equiv \bar{\psi}(\mathbf{y}) = (2\pi)^{-3/2} \int_{-\infty}^{+\infty} d\mathbf{x}\, e^{i\mathbf{y}\cdot\mathbf{x}} \psi(\mathbf{x}).$$

Show that F is a linear operator. Show that the operator $H(x) = \nabla^2 + \mathbf{x}^2$ commutes with F.

We are now in a position to see the connection of our work with physics. Let us look at the eigenfunctions of a problem of the form

$$H\psi_\nu = \epsilon\psi_\nu, \tag{3–74}$$

and consider the n linearly independent eigenfunctions ψ_ν belonging to a given eigenvalue ϵ. If the Hamiltonian H is invariant under a symmetry transformation R, then applying O_R to (3–74) yields

$$O_R[H\psi_\nu] = O_R H O_R^{-1} O_R \psi_\nu = H[O_R \psi_\nu] = \epsilon O_R \psi_\nu, \qquad (3\text{–}75)$$

and $O_R\psi_\nu$ is an eigenfunction belonging to the same eigenvalue ϵ. Thus $O_R\psi_\nu$ can be expressed as a linear combination,

$$O_R\psi_\nu = \sum_{\mu=1}^{n} \psi_\mu D_{\mu\nu}(R). \qquad (3\text{–}76)$$

Carrying out this procedure for all operators in the symmetry group of the Hamiltonian, we obtain a representation $D_{\mu\nu}(R)$ in n dimensions. If S is another transformation of the symmetry group, then

$$O_S\psi_\nu = \sum_{\mu=1}^{n} \psi_\mu D_{\mu\nu}(S),$$

and

$$O_S(O_R\psi_\nu) = O_S \sum_\mu \psi_\mu D_{\mu\nu}(R) = \sum_{\lambda,\mu} \psi_\lambda D_{\lambda\mu}(S) D_{\mu\nu}(R)$$

$$= \sum_\lambda \psi_\lambda [D(S)D(R)]_{\lambda\nu} = \sum_\lambda \psi_\lambda D_{\lambda\nu}(SR); \qquad (3\text{–}77)$$

so to the symmetry transformation SR there corresponds the matrix $D(SR) = D(S)D(R)$.

The eigenfunctions of each degenerate level provide the basis for a representation of the symmetry group. If we can find some way of characterizing the possible representations of the symmetry group, we shall be able to classify the eigenfunctions.

3–9 Unitary spaces; scalar product; unitary matrices; Hermitian matrices. In quantum theories, we associate numerical values with pairs of vectors ("state vectors"). In order to bring the theory of representations into closer contact with physics, we define a *metric* in the n-dimensional space L. For this purpose we associate with each pair of vectors $\mathbf{x}$, $\mathbf{y}$ in L a complex number $(\mathbf{x}, \mathbf{y})$. The complex number $(\mathbf{x}, \mathbf{y})$ is called the *scalar product* of $\mathbf{x}$ and $\mathbf{y}$ and is required to satisfy the conditions:

$$(\mathbf{x}, \mathbf{y}) = (\mathbf{y}, \mathbf{x})^*, \qquad \text{where * denotes the complex conjugate,} \qquad (3\text{–}78)$$

$$(\mathbf{x}, \alpha\mathbf{y}) = \alpha(\mathbf{x}, \mathbf{y}), \qquad (3\text{–}78a)$$

$$(\mathbf{x}_1 + \mathbf{x}_2, \mathbf{y}) = (\mathbf{x}_1, \mathbf{y}) + (\mathbf{x}_2, \mathbf{y}), \qquad (3\text{–}78b)$$

$$(\mathbf{x}, \mathbf{x}) \geq 0, \qquad (3\text{–}78c)$$

and $(\mathbf{x}, \mathbf{x}) = 0$ only if $\mathbf{x} = 0$. [Setting $\mathbf{y} = \mathbf{x}$ in (3–78) shows that $(\mathbf{x}, \mathbf{x})$

is real, so that (3–78c) is meaningful.] The quantity $(\mathbf{x}, \mathbf{x})$ is the square of the *length* of the vector $\mathbf{x}$. A space L in which a scalar product is defined is called a *unitary space*.

The scalar product $(\mathbf{x}, \mathbf{y})$ is a function which is defined for any pair of vectors $\mathbf{x}$, $\mathbf{y}$ in L, and whose values are complex numbers. In defining the scalar product, we make no mention of a basis in L, so the scalar product $(\mathbf{x}, \mathbf{y})$ is an intrinsic property of $\mathbf{x}$, $\mathbf{y}$, independent of basis.

Problems. (1) Prove the Schwartz inequality

$$|(\mathbf{x}, \mathbf{y})|^2 \leq (\mathbf{x}, \mathbf{x})(\mathbf{y}, \mathbf{y}), \tag{3–79}$$

where the equality sign applies only if $\mathbf{x}$ and $\mathbf{y}$ are linearly dependent. What does Eq. (3–79) mean for the case of ordinary geometrical vectors in three-space?

(2) In the space of square-integrable functions, we define the scalar product of functions f, g to be

$$(f, g) = \int dz f^*(z) g(z). \tag{3–80}$$

Show that (3–80) is finite for all f and g in the space.

Any function satisfying conditions (3–78) through (3–78c) can be used to define a scalar product in the space L. *Different* definitions of the scalar product in the same space L yield *different* unitary spaces.

The vectors in a unitary space can be *normalized* (adjusted to have length equal to unity) by multiplying them by a complex number; for any $\mathbf{x}$, if

$$\mathbf{x}' = \frac{1}{\sqrt{(\mathbf{x}, \mathbf{x})}} \mathbf{x}, \quad \text{then} \quad (\mathbf{x}', \mathbf{x}') = 1. \tag{3–81}$$

One way of assigning the scalar product $(\mathbf{x}, \mathbf{y})$ is to write it as a function of the coordinates x_i, y_i in a particular basis. If the basis vectors are $\mathbf{u}_i$, the scalar product is determined by assigning the numbers

$$m_{ij} = (\mathbf{u}_i, \mathbf{u}_j). \tag{3–82}$$

From Eq. (3–78) we see that

$$m_{ij} = m_{ji}^*. \tag{3–83}$$

The numbers m_{ij} defined by Eq. (3–82) form a matrix $\mathbf{m}$, the *metric matrix*. According to Eq. (3–83),

$$\mathbf{m} = \mathbf{m}^\dagger, \tag{3–83a}$$

$$(\mathbf{m}^\dagger)_{ij} = m_{ji}^*, \tag{3–84}$$

$\mathbf{m}^\dagger$ is the *adjoint* (or "Hermitian conjugate," or "conjugate transposed") of the matrix $\mathbf{m}$. We see from (3–84) that

$$(\mathbf{AB})^\dagger = \mathbf{B}^\dagger \mathbf{A}^\dagger. \tag{3–85}$$

A matrix $\mathbf{A}$ which is identical with its adjoint is said to be *self-adjoint* or *Hermitian*. According to Eq. (3–83a) the metric matrix $\mathbf{m}$ must be Hermitian.

We expand the vectors $\mathbf{x}$, $\mathbf{y}$ in the basis $\mathbf{u}_i$ by means of Eq. (3–13), and use Eqs. (3–78) through (3–78b) and (3–82):

$$(\mathbf{x}, \mathbf{y}) = (x_i\mathbf{u}_i, y_j\mathbf{u}_j) = x_i^* y_j m_{ij} = \mathbf{x}^\dagger \mathbf{m} \mathbf{y}, \tag{3–86}$$

where $\mathbf{x}$ and $\mathbf{y}$ are one-column matrices and $\mathbf{x}^\dagger$ is the adjoint of $\mathbf{x}$. The quantity $\mathbf{x}^\dagger \mathbf{m} \mathbf{y}$, with $\mathbf{m}$ Hermitian, is called a *Hermitian form*. Equation (3–78c) requires that (for $\mathbf{y} = \mathbf{x}$)

$$\mathbf{x}^\dagger \mathbf{m} \mathbf{x} \geq 0. \tag{3–87}$$

A form satisfying such a condition is said to be *positive definite*. Thus the scalar product must be a positive definite Hermitian form in the coordinates x_i, y_i. The matrix $\mathbf{m}$ of such a form is also said to be a *positive definite Hermitian matrix*.

Two vectors $\mathbf{x}$, $\mathbf{y}$ in a unitary space are *orthogonal* (or perpendicular) if

$$(\mathbf{x}, \mathbf{y}) = 0. \tag{3–88}$$

If we have a basis $\mathbf{v}_i$ in the unitary space L, we can always (by taking linear combinations of the $\mathbf{v}_i$) construct a new set of basis vectors $\mathbf{u}_i$, all of which have length unity and are mutually perpendicular, i.e.,

$$(\mathbf{u}_i, \mathbf{u}_j) = \delta_{ij}. \tag{3–89}$$

The basis vectors $\mathbf{u}_i$ constitute an *orthonormal basis*.

Problem. In the two-dimensional space of the basis vectors $\mathbf{v}_1$, $\mathbf{v}_2$, construct linear combinations $\mathbf{u}_1$, $\mathbf{u}_2$ which form an orthonormal basis.

In an orthonormal basis, the metric matrix (3–82) reduces to the unit matrix, and the scalar product $(\mathbf{x}, \mathbf{y})$ in (3–86) assumes the simple form

$$(\mathbf{x}, \mathbf{y}) = \sum_i x_i^* y_i = \mathbf{x}^\dagger \mathbf{y}, \tag{3–90}$$

while
$$(\mathbf{x}, \mathbf{x}) = \sum_i |x_i|^2. \tag{3-91}$$

If we shift from the orthonormal basis $\mathbf{u}_i$ to another basis $\mathbf{u}_i'$, using Eq. (3-14), the new basis $\mathbf{u}_i'$ will also be orthonormal if

$$\delta_{ij} = (\mathbf{u}_i', \mathbf{u}_j') = (a_{ik}\mathbf{u}_k, a_{jl}\mathbf{u}_l)$$
$$= a_{ik}^* a_{jl}(\mathbf{u}_k, \mathbf{u}_l) = a_{ik}^* a_{jl}\, \delta_{kl} = a_{ik}^* a_{jk}; \tag{3-92}$$

that is,

$$\mathbf{a}\mathbf{a}^\dagger = \mathbf{1} = \mathbf{a}^\dagger \mathbf{a}, \qquad \mathbf{a}^\dagger = \mathbf{a}^{-1}, \tag{3-93}$$

since the transformation $\mathbf{a}$ has an inverse $\mathbf{a}^{-1}$.

A matrix $\mathbf{A}$ is said to be unitary if

$$\mathbf{A}^\dagger = \mathbf{A}^{-1}. \tag{3-94}$$

Thus the transformation from one orthonormal basis to another is accomplished by a *unitary matrix*.

Problems. Prove that the rows (columns) of a unitary matrix are orthonormal.

3-10 Operators: adjoint, self-adjoint, unitary. The *adjoint* $T^\dagger$ of a linear operator T is defined by

$$(T\mathbf{x}, \mathbf{y}) = (\mathbf{x}, T^\dagger\mathbf{y}) \qquad \text{for all } \mathbf{x}, \mathbf{y}. \tag{3-95}$$

In an orthonormal basis we find, using (3-31) and (3-90),

$$(T\mathbf{x}, \mathbf{y}) = T_{ij}^* x_j^* y_i = x_j^* \widetilde{T}_{ji}^* y_i,$$
$$(\mathbf{x}, T^\dagger\mathbf{y}) = x_j^* T_{ji}^\dagger y_i,$$

and since x_j, y_i are arbitrary, it follows from (3-95) that

$$(T^\dagger)_{ji} = \widetilde{T}_{ji}^* \qquad \text{or} \qquad \mathbf{T}^\dagger = \widetilde{\mathbf{T}}^*; \tag{3-96}$$

i.e., in an orthonormal basis the matrix representative of the adjoint operator $T^\dagger$ is the conjugate transpose of the matrix representing T.

Problem. In a basis with metric matrix m, prove that the matrix representation of the adjoint operator $T^\dagger$ is $\mathbf{m}^{-1}\mathbf{T}\mathbf{m}$.

An operator is *self-adjoint* or *Hermitian* if it is identical with its adjoint:

$$(T\mathbf{x}, \mathbf{y}) = (\mathbf{x}, T\mathbf{y}) \qquad \text{for all } \mathbf{x}, \mathbf{y}, \quad \text{or} \quad T^\dagger = T. \qquad (3\text{-}97)$$

In an orthonormal coordinate system a Hermitian operator is represented by a Hermitian matrix.

The operator U is called a *unitary operator* if

$$(U\mathbf{x}, U\mathbf{y}) = (\mathbf{x}, \mathbf{y}) \qquad \text{for all } \mathbf{x}, \mathbf{y}, \qquad (3\text{-}98)$$

i.e., if the scalar product of the image vectors $\mathbf{x}' = U\mathbf{x}$, $\mathbf{y}' = U\mathbf{y}$ is the same as the scalar product of $\mathbf{x}$, $\mathbf{y}$ for all vectors $\mathbf{x}$, $\mathbf{y}$. In an orthonormal system the matrix representative of U is a unitary matrix:

$$\mathbf{U}^\dagger\mathbf{U} = \mathbf{U}\mathbf{U}^\dagger = \mathbf{1}. \qquad (3\text{-}99)$$

3–11 Unitary representations. If the operators of a representation of the group G are unitary operators (or if the matrices of the representation are unitary matrices), the representation is called a *unitary representation*.

In Section 3–6 we saw that there are infinitely many representations of the group G equivalent to any given representation $D(G)$. Because unitary matrices have especially useful properties (e.g., see the problem at the end of Section 3–9), it is important to determine whether a given representation $D(G)$ is equivalent to a unitary representation. This is *not true* in general. For *finite groups* G we can prove that *every representation* is *equivalent* to a *unitary representation:*

For an arbitrary pair of vectors $\mathbf{x}$, $\mathbf{y}$ we construct the expression

$$\{\mathbf{x}, \mathbf{y}\} = \frac{1}{g} \sum_{R \subset G} (D(R)\mathbf{x}, D(R)\mathbf{y}). \qquad (3\text{-}100)$$

The sum in Eq. (3–100) is extended over all elements R of the group G. The quantity $\{\mathbf{x}, \mathbf{y}\}$ defined by (3–100) satisfies all the requirements for a scalar product. Furthermore, for any element S of G,

$$\{D(S)\mathbf{x}, D(S)\mathbf{y}\} = \frac{1}{g} \sum_{R \subset G} (D(R)D(S)\mathbf{x}, D(R)D(S)\mathbf{y})$$

$$= \frac{1}{g} \sum_{R \subset G} (D(RS)\mathbf{x}, D(RS)\mathbf{y}). \qquad (3\text{-}101)$$

But for fixed S, as R runs through the elements of G, so does RS; hence the expressions on the right sides of (3–100) and (3–101) are identical, and

$$\{D(S)\mathbf{x}, D(S)\mathbf{y}\} = \{\mathbf{x}, \mathbf{y}\}. \qquad (3\text{-}102)$$

In other words, the operators of our representation are unitary with respect to the scalar product $\{\mathbf{x}, \mathbf{y}\}$ [but not with respect to $(\mathbf{x}, \mathbf{y})$].

Next we consider a set of vectors $\mathbf{u}_i$ which are orthonormal with respect to the original scalar product, and a second set $\mathbf{v}_i$ which are orthonormal with respect to the new scalar product:

$$(\mathbf{u}_i, \mathbf{u}_j) = \delta_{ij} = \{\mathbf{v}_i, \mathbf{v}_j\}. \tag{3-103}$$

We define the operator T which takes the $\mathbf{u}$'s into the $\mathbf{v}$'s:

$$\mathbf{v}_i = T\mathbf{u}_i. \tag{3-104}$$

Since $T\mathbf{x} = Tx_i\mathbf{u}_i = x_iT\mathbf{u}_i = x_i\mathbf{v}_i$,

$$\{T\mathbf{x}, T\mathbf{y}\} = x_i^* y_i = (\mathbf{x}, \mathbf{y}). \tag{3-105}$$

We now consider the equivalent representation defined by

$$D'(S) = T^{-1}D(S)T, \tag{3-106}$$

and find that

$$(T^{-1}D(S)T\mathbf{x},\, T^{-1}D(S)T\mathbf{y}) = \{D(S)T\mathbf{x},\, D(S)T\mathbf{y}\}$$
$$\text{[from (3–105)],}$$
$$= \{T\mathbf{x}, T\mathbf{y}\} \qquad \text{[from (3–102)],} \quad (3\text{–}107)$$
$$= (\mathbf{x}, \mathbf{y}) \qquad \text{[again from (3–105)].}$$

The last equation shows that the equivalent representation $D'(G)$ defined by (3–106) is unitary. So for finite groups we can *always* choose our representation to be unitary. For infinite groups we shall have to investigate later the meaning to be assigned to the summation over group elements which occurs in Eq. (3–101).

3–12 Hilbert space. We have defined unitary spaces in Section 3–9 by introducing a scalar product in the linear vector space. We can then define the *distance* $|\mathbf{x} - \mathbf{y}|$ between two vectors (points, functions) from

$$|\mathbf{x} - \mathbf{y}|^2 = (\mathbf{x} - \mathbf{y}, \mathbf{x} - \mathbf{y}). \tag{3-108}$$

A sequence of vectors $\mathbf{x}_n$ $(n = 1, \ldots, \infty)$ in L is said to *converge* to the vector $\mathbf{x}$ in L if

$$\lim_{n \to \infty} |\mathbf{x}_n - \mathbf{x}| = 0, \tag{3-109}$$

i.e., if for any $\epsilon > 0$ there is an integer $n(\epsilon)$ such that $|\mathbf{x}_m - \mathbf{x}| < \epsilon$ for $m > n(\epsilon)$.

The sequence of vectors $\mathbf{x}_n$ is said to *converge in itself* or to be a *fundamental sequence* if

$$\lim_{m,n \to \infty} |\mathbf{x}_m - \mathbf{x}_n| = 0. \tag{3-110}$$

The space L is said to be *complete* if *every* fundamental sequence converges to a vector in L, i.e., if Eq. (3–110) implies the existence of a vector $\mathbf{x}$ in L for which (3–109) holds.

A *complete unitary* space is called a *Hilbert space*. The unitary spaces of finite dimension are necessarily complete.

When we consider infinite-dimensional representations, we shall restrict ourselves to representations by *linear operators* in a *Hilbert space*. Here we impose the requirement that the linear operators be *continuous*, i.e., if

$$|\mathbf{x}_n - \mathbf{x}| \to 0,$$

then

$$|A\mathbf{x}_n - A\mathbf{x}| \to 0. \tag{3-111}$$

3–13 Analysis of representations; reducibility; irreducible representations. In Section 3–7, we discussed the addition of representations. Now we wish to consider the reverse process. Given a representation D, is it possible to describe it in terms of "simpler" representations? A crude criterion of simplicity would be that our representations have as low a dimensionality as possible. For example, if all matrices of the three-dimensional representation D are of the form

$$\begin{bmatrix} a_i & b_i & \vdots & e_i \\ c_i & d_i & \vdots & f_i \\ \cdots & \cdots & \vdots & \cdots \\ 0 & 0 & \vdots & g_i \end{bmatrix}, \tag{3-112}$$

the products will have the same form, that is,

$$\begin{bmatrix} a_1a_2 + b_1c_2 & a_1b_2 + b_1d_2 & \vdots & a_1e_2 + b_1f_2 + e_1g_2 \\ c_1a_2 + d_1c_2 & c_1b_2 + d_1d_2 & \vdots & c_1e_2 + d_1f_2 + f_1g_2 \\ \cdots & \cdots & \vdots & \cdots \\ 0 & 0 & \vdots & g_1g_2 \end{bmatrix}. \tag{3-112a}$$

We see that the matrices in the upper left corner provide us with a two-dimensional representation:

$$\begin{bmatrix} a_i & b_i \\ c_i & d_i \end{bmatrix}, \tag{3-112b}$$

while the matrices in the lower right corner provide us with a one-dimensional representation:

$$[g_i]. \tag{3-112c}$$

When this occurs, we say that the original representation is *reducible*.

The matrices of the representation may not have the simple form (3–112), but if we can find a basis transformation which brings *all* matrices of the representation to the (equivalent) form (3–112), we say that the representation is reducible. In general, if we can find a basis in which *all* matrices $D(R)$ of an n-dimensional representation can be brought simultaneously to the form

$$D(R) = \begin{bmatrix} D^{(1)}(R) & \vdots & A(R) \\ \cdots\cdots & \vdots & \cdots\cdots \\ 0 & \vdots & D^{(2)}(R) \end{bmatrix}, \tag{3-113}$$

where the $D^{(1)}(R)$ are m-by-m matrices, the $D^{(2)}(R)$ are $(n - m)$-by-$(n - m)$, $A(R)$ is a rectangular matrix with m rows and $(n - m)$ columns, and 0 denotes a matrix with $(n - m)$ rows and m columns all of whose elements are 0, then we say that the representation $D(R)$ is *reducible*. Clearly the matrix product

$$D(RS) = D(R)D(S) = \begin{bmatrix} D^{(1)}(R)D^{(1)}(S) & \vdots & D^{(1)}(R)A(S) + A(R)D^{(2)}(S) \\ \cdots\cdots\cdots\cdots & \vdots & \cdots\cdots\cdots\cdots\cdots\cdots \\ 0 & \vdots & D^{(2)}(R)D^{(2)}(S) \end{bmatrix}$$

has the same form as (3–113), and hence the matrices

$$D^{(1)}(RS) = D^{(1)}(R)D^{(1)}(S)$$

provide us with an m-dimensional representation, and the matrices

$$D^{(2)}(RS) = D^{(2)}(R)D^{(2)}(S)$$

give an $(n - m)$-dimensional representation.

We can now continue as follows: We transform the basis in the m-dimensional space of $D^{(1)}$ and try to bring *all* matrices $D^{(1)}(R)$ to a form like that in (3–113), i.e.,

$$D^{(1)}(R) = \begin{bmatrix} D^{(3)}(R) & \vdots & A'(R) \\ \cdots\cdots & \vdots & \cdots\cdots \\ 0 & \vdots & D^{(4)}(R) \end{bmatrix},$$

where $D^{(3)}$ is p-dimensional and $D^{(4)}$ is $(m - p)$-dimensional, and apply the same procedure to the matrices $D^{(2)}(R)$. This process clearly comes to an end, and we then have all the matrices of the representation D expressed in the form

$$D(R) = \begin{bmatrix} D^{(1)}(R) & \vdots & & & & A^{(1)}(R) \\ \cdots & \vdots & D^{(2)}(R) & \vdots & & A^{(2)}(R) \\ & \vdots & \cdots & \vdots & D^{(3)}(R) & \vdots & A^{(3)}(R) \\ & \vdots & & \vdots & \cdots & & \\ & \vdots & & \vdots & & \vdots & D^{(k-1)}(R) & \vdots & A^{(k-1)}(R) \\ 0 & \vdots & 0 & \vdots & 0 & \vdots & 0 & \cdots \\ & & & & & & 0 & \vdots & D^{(k)}(R) \end{bmatrix},$$

(3–114)

where the k sets of matrices $D^{(1)}(R) \ldots D^{(i)}(R) \ldots D^{(k)}(R)$ are irreducible representations of dimension m_i ($n = \sum_{i=1}^{k} m_i$).

We can give an *intrinsic* criterion of reducibility as follows: In the case of (3–112) we consider those vectors which are in the two-dimensional subspace of the first two components. The matrix (3–112) applied to vectors $\mathbf{x}$ with $x_3 = 0$ gives

$$\begin{bmatrix} a & b & e \\ c & d & f \\ 0 & 0 & g \end{bmatrix} \begin{bmatrix} x_1 \\ x_2 \\ 0 \end{bmatrix} = \begin{bmatrix} ax_1 + bx_2 \\ cx_1 + dx_2 \\ 0 \end{bmatrix},$$

which is again in the subspace $x_3 = 0$. In other words, the two-dimensional subspace is invariant under all transformations of the group. On the other hand, the vectors with $x_1 = x_2 = 0$ are transformed into

$$\begin{bmatrix} a & b & e \\ c & d & f \\ 0 & 0 & g \end{bmatrix} \begin{bmatrix} 0 \\ 0 \\ x_3 \end{bmatrix} = \begin{bmatrix} ex_3 \\ fx_3 \\ gx_3 \end{bmatrix},$$

so the (complementary) one-dimensional space of the third component is *not* invariant.

In the case of (3–113), the subspace of the first m components is invariant:

$$\begin{bmatrix} D^{(1)}(R) & \vdots & A(R) \\ \cdots & \cdots & \cdots \\ 0 & \vdots & D^{(2)}(R) \end{bmatrix} \begin{bmatrix} \mathbf{x} \\ \cdots \\ 0 \end{bmatrix} = \begin{bmatrix} D^{(1)}(R)\mathbf{x} \\ \cdots \\ 0 \end{bmatrix},$$

while the complementary $(n - m)$-dimensional subspace is not invariant:

$$\begin{bmatrix} D^{(1)}(R) & \vdots & A(R) \\ \cdots\cdots & \vdots & \cdots\cdots \\ 0 & \vdots & D^{(2)}(R) \end{bmatrix} \begin{bmatrix} 0 \\ \cdots \\ \mathbf{x} \end{bmatrix} = \begin{bmatrix} A(R)\mathbf{x} \\ \cdots\cdots \\ D^{(2)}(R)\mathbf{x} \end{bmatrix}.$$

Similarly, in the case of (3–114), the m_1-dimensional subspace of the first m_1 components is invariant.

In general, if there exists some subspace of dimension $m < n$ which is invariant under all transformations of the group, the representation is reducible. We can choose m new basis functions $\phi_1, \ldots, \phi_m$ in this subspace, and complete the set with $(n - m)$ other basis functions $\phi_{m+1}, \ldots, \phi_n$ to give n basis functions for the whole n-dimensional space. In this basis, the matrices of the representation will assume the reduced form (3–112).

If there is no proper subspace which is invariant, the representation is *irreducible*.

If it is possible to find a basis in which all the matrices of the representation assume the form (3–113), but with $A(R) = 0$, i.e.,

$$D(R) = \begin{bmatrix} D^{(1)}(R) & 0 \\ 0 & D^{(2)}(R) \end{bmatrix}, \qquad (3\text{–}115)$$

we say that the representation is *fully reducible*. Both the m-dimensional subspace of $D^{(1)}$ and the $(n - m)$-dimensional subspace of $D^{(2)}$ are invariant in this case. In other words, the basis functions $\phi_1, \ldots, \phi_m$ transform among themselves, and the basis functions $\phi_{m+1}, \ldots, \phi_n$ transform among themselves, so the transformations of the group do not couple the two sets of functions. The space L is *decomposed* into the *direct sum* of $L^{(1)}$ and $L^{(2)}$,

$$L = L^{(1)} + L^{(2)},$$

and the representation D is the sum of $D^{(1)}$ and $D^{(2)}$,

$$D = D^{(1)} + D^{(2)}. \qquad (3\text{–}116)$$

This is precisely the converse of the process of adding representations which we used in Section 3–7. Assuming for the moment that we are dealing with fully reducible representations, we can now examine the representations $D^{(1)}$ and $D^{(2)}$ to see whether they in turn are reducible. Notice that we can treat $D^{(1)}$ and $D^{(2)}$ separately, since we can treat the subspaces independently of one another. Thus, a transformation of coordinates of the form

$$C = \begin{bmatrix} C_1 & \vdots & 0 \\ \cdots & \vdots & \cdots \\ 0 & \vdots & 1 \end{bmatrix} \begin{matrix} \} & m \\ \\ \} & n - m \end{matrix}$$

transforms only the subspace of the representation $D^{(1)}$ and leaves the matrices of $D^{(2)}$ unchanged. By continuing this procedure, we can finally exhibit D in the form

$$
D(R) = \begin{bmatrix} D^{(1)}(R) & 0 & \cdots\cdots\cdots & 0 \\ 0 & D^{(2)}(R) & \cdots\cdots\cdots & 0 \\ 0 & 0 & \cdots\cdots & \begin{matrix}0 \\ 0\end{matrix} \\ 0 & 0 & 0 & D^{(k)}(R) \end{bmatrix}, \quad (3\text{--}117)
$$

that is, $D = D^{(1)} + D^{(2)} + \cdots + D^{(k)}$, where all the $D^{(\nu)}$ are irreducible representations. We say then that D is fully reducible to the sum

$$
D^{(1)} + \cdots + D^{(k)}.
$$

Now among the irreducible representations $D^{(\nu)}$ there may be several which are equivalent to one another. (For this, they must, of course, have the same dimensions.) Equivalent irreducible representations are not counted as distinct, and we can use the same symbol for them. So the representation D may contain a particular irreducible representation $D^{(\nu)}$ several times. We express this symbolically as

$$
D = a_1 D^{(1)} + \cdots + a_r D^{(r)} = \sum_{\nu} a_{\nu} D^{(\nu)}, \quad (3\text{--}118)
$$

where the a_{ν} are *positive integers*.

For the most part we shall deal with fully reducible representations and will use the term *reducible* to mean *decomposable*.

If the matrices of a representation are unitary (unitary representation), then reducibility implies full reducibility; if the matrices $D(R)$ in (3–113) are unitary, $A(R) = 0$. In Section 3–11, we showed that for finite groups the representations can always be chosen to be unitary; hence, for finite groups, reducible representations *always* decompose into a sum of representations.

3–14 Schur's lemmas. We now proceed to the general theorems. The purpose of these theorems is (1) to find some simpler criteria for irreducibility, and (2) to give some restrictions on the number of nonequivalent representations. We start by proving two fundamental lemmas (Schur's lemmas).

Suppose that we have a set of n functions $\psi_\nu(\nu = 1, \ldots, n)$ which serve as a basis for a representation of the group G. Then

$$O_R\psi_\nu = \sum_\mu \psi_\mu D_{\mu\nu}(R) \tag{3-119}$$

for all R in G. If the representation is reducible, then, by definition, we can find m functions ϕ_l, not identically zero ($m < n$), which are linear combinations of the ψ_ν,

$$\phi_l = \sum_\rho \psi_\rho a_{\rho l}, \tag{3-120}$$

such that

$$O_R\phi_l = \sum_k \phi_k D'_{kl}(R) \tag{3-121}$$

for all R in G. Using (3–119) through (3–121), we obtain

$$O_R\phi_l = \sum_\rho O_R\psi_\rho a_{\rho l} = \sum_{\sigma,\rho} \psi_\sigma D_{\sigma\rho}(R)a_{\rho l}$$

$$= \sum_k \phi_k D'_{kl}(R) = \sum_{\sigma,k} \psi_\sigma a_{\sigma k} D'_{kl}(R). \tag{3-122}$$

So,

$$\sum_{\sigma,\rho} \psi_\sigma D_{\sigma\rho}(R)a_{\rho l} = \sum_{\sigma,k} \psi_\sigma a_{\sigma k} D'_{kl}(R). \tag{3-123}$$

Since the ψ_σ are linearly independent,

$$\sum_\rho D_{\sigma\rho}(R)a_{\rho l} = \sum_k a_{\sigma k} D'_{kl}(R) \tag{3-124}$$

or

$$D(R)A = AD'(R) \tag{3-125}$$

for all R in G. Thus if D is reducible, we can find a nonzero matrix A such that Eq. (3–125) is valid for all R in G.

Conversely, if Eq. (3–125) is valid, i.e., if there exists a matrix A such that

$$D(R)A = AD'(R), \tag{3-125}$$

where the $D'(R)$ are a set of matrices of degree $m < n$, then

$$\sum_\rho D_{\sigma\rho}(R)a_{\rho l} = \sum_k a_{\sigma k} D'_{kl}(R). \tag{3-124}$$

Multiply by ψ_σ and sum:

$$\sum_{\rho,\sigma} \psi_\sigma D_{\sigma\rho}(R)a_{\rho l} = \sum_\rho O_R(\psi_\rho a_{\rho l}) = \sum_{\sigma,k} \psi_\sigma a_{\sigma k} D'_{kl}(R). \quad (3\text{-}123)$$

Thus the m functions

$$\phi_l = \sum_\rho \psi_\rho a_{\rho l} \quad\quad\quad\quad\quad (3\text{-}120)$$

form a basis for a representation D', and D is reducible. If D is assumed to be irreducible, then the only way to avoid a contradiction is for $A = 0$, i.e., all elements of A are zero. So we can state our result as follows:

LEMMA I. If D and D' are two irreducible representations of a group G, having different dimensions, then if the matrix A satisfies

$$D(R)A = AD'(R) \quad\quad\quad\quad (3\text{-}125)$$

for all R in G, it follows that $A = 0$.

Next we consider the special case where $n = m$. If Eq. (3-125) holds, we repeat the same argument and arrive at

$$\sum_\rho O_R(\psi_\rho a_{\rho l}) = \sum_{\sigma,k} (\psi_\sigma a_{\sigma k}) D'_{kl}(R); \quad\quad k = 1, \ldots, n. \quad (3\text{-}123)$$

The n functions $\sum_\rho \psi_\rho a_{\rho l}$ must be linearly independent; otherwise D would be reducible. But this means that the determinant of A is not zero, and A has an inverse, so that from (3-125), $D(R) = AD'(R)A^{-1}$, and the two representations are equivalent. If D and D' are irreducible and not equivalent, the only way to avoid a contradiction is for $A = 0$. So we state this special case:

LEMMA Ia. If D and D' are irreducible representations of the group G, having the same dimensions, and if the matrix A satisfies

$$D(R)A = AD'(R) \quad\quad\quad\quad (3\text{-}125)$$

for all R in G, then either D and D' are equivalent or $A = 0$.

Finally, we consider a single irreducible representation of the group G.

LEMMA II. If the matrices $D(R)$ are an irreducible representation of a group G, and if

$$AD(R) = D(R)A \quad\quad\quad\quad (3\text{-}126)$$

for all R in G, then $A = $ constant $\cdot$ 1.

In other words, if a matrix commutes with all the matrices of an irreducible representation, the matrix must be a multiple of the unit matrix.

Consider the equation

$$A\mathbf{x} = \lambda\mathbf{x}, \tag{3–127}$$

where $\mathbf{x}$ is some vector in the space. The solutions give the eigenvalues and eigenvectors of A. If $\mathbf{x}$ is an eigenvector belonging to λ, then using (3–126), we find that $D(R)\mathbf{x}$ is also an eigenvector belonging to λ. So the subspace of eigenvectors of A belonging to a given λ is invariant under all transformations of the group G. But this would mean that D is reducible unless this subspace is the whole space or the zero vector. The first possibility implies that A has only one eigenvalue λ, that is, $A = \lambda \cdot 1$, the second that $A = 0$.

This lemma is very useful as a test of irreducibility. Given a representation D, we try to find A such that (3–126) will be satisfied. If we can show that A must be a multiple of the unit matrix, then D is irreducible.

Problem. Apply Lemma II to prove that all irreducible representations of an abelian group are one-dimensional.

3–15 The orthogonality relations.   Now we use the lemmas to prove our general theorems. Suppose that in Lemma II we are given an irreducible representation of degree n for the group G of order g. Construct the matrix

$$A = \sum_S D(S)XD(S^{-1}), \tag{3–128}$$

where X is an arbitrary matrix and the sum $\sum_S$ runs through all the elements of the group G. We claim that A satisfies the conditions of Lemma II, since

$$D(R)A = \sum_S D(R)D(S)XD(S^{-1})$$

$$= \sum_S D(R)D(S)XD(S^{-1})D(R^{-1}) \cdot D(R)$$

$$= \left[\sum_S D(RS)XD(\{RS\}^{-1})\right]D(R). \tag{3–129}$$

Now we note that for fixed R, as S runs through the elements of the group, so does RS (remember the group table), and therefore,

$$\sum_S D(RS)XD(\{RS\}^{-1}) = \sum_S D(S)XD(S^{-1}) \tag{3–130}$$

and $$D(R)A = AD(R). \tag{3-131}$$

But then, according to the lemma, $A = \lambda \cdot 1$. The value of the constant λ will depend on our choice of the arbitrary matrix X. We choose X to have all its elements zero, except $X_{lm} = 1$, and let the constant λ be λ_{lm}. Then from (3–128),

$$\sum_S D_{il}(S)D_{mj}(S^{-1}) = \lambda_{lm}\,\delta_{ij}, \tag{3-132}$$

or, if D is unitary,

$$\sum_S D_{il}(S)D^*_{jm}(S) = \lambda_{lm}\,\delta_{ij}. \tag{3-133}$$

To evaluate λ_{lm}, we set $i = j$, and sum over i:

$$\sum_S \sum_i D_{il}(S)D_{mi}(S^{-1}) = n\lambda_{lm} = \sum_S D_{ml}(SS^{-1})$$

$$= \sum_S D_{ml}(E) = \sum_S \delta_{ml} = g\,\delta_{ml}. \tag{3-134}$$

So,

$$\lambda_{lm} = \frac{g}{n}\,\delta_{lm}, \tag{3-135}$$

and

$$\sum_S D_{il}(S)D_{mj}(S^{-1}) = \frac{g}{n}\,\delta_{lm}\,\delta_{ij}, \tag{3-136}$$

or, if D is unitary,

$$\sum_S D_{il}(S)D^*_{jm}(S) = \frac{g}{n}\,\delta_{lm}\,\delta_{ij}. \tag{3-137}$$

Now we construct in similar fashion a matrix A satisfying the conditions of Lemma I. Given any two nonequivalent irreducible representations of $G, D^{(1)}$ and $D^{(2)}$ (of dimensions n_1 and n_2), let

$$A = \sum_S D^{(2)}(S)XD^{(1)}(S^{-1}), \tag{3-138}$$

where X is an arbitrary matrix. Then

$$D^{(2)}(R)A = \sum_S D^{(2)}(R)D^{(2)}(S)XD^{(1)}(S^{-1})$$

$$= \sum_S D^{(2)}(R)D^{(2)}(S)XD^{(1)}(S^{-1})D^{(1)}(R^{-1}) \cdot D^{(1)}(R)$$

$$= \sum_S D^{(2)}(RS)XD^{(1)}(\{RS\}^{-1}) \cdot D^{(1)}(R) = AD^{(1)}(R).$$

$$\tag{3-139}$$

Thus, according to Lemma I, $A = 0$. Choosing X as before, we find

$$\sum_S D_{il}^{(2)}(S)D_{mj}^{(1)}(S^{-1}) = 0, \qquad \text{for all } i, j, l, m; \qquad (3\text{-}140)$$

or, if $D^{(1)}$ and $D^{(2)}$ are unitary,

$$\sum_S D_{il}^{(2)}(S)D_{jm}^{(1)*}(S) = 0. \qquad (3\text{-}141)$$

Together, Eqs. (3-136) and (3-140) [or (3-137) and (3-141)] state the following: If we consider all nonequivalent irreducible representations of a group G, then the quantities $D_{ij}^{(\mu)}(R)$, for *fixed* μ, i, j, form a vector in a g-dimensional space, such that

$$\sum_R D_{il}^{(\mu)}(R)D_{mj}^{(\nu)}(R^{-1}) = \frac{g}{n_\mu}\,\delta_{\mu\nu}\,\delta_{ij}\,\delta_{lm} \qquad (3\text{-}142)$$

or, if the representation is unitary,

$$\sum_R D_{il}^{(\mu)}(R)D_{jm}^{(\nu)*}(R) = \frac{g}{n_\mu}\,\delta_{\mu\nu}\,\delta_{ij}\,\delta_{lm}. \qquad (3\text{-}143)$$

Thus each irreducible representation $D^{(\mu)}$ provides us with n_μ^2 vectors $D_{ij}^{(\mu)}(R)$ $(i, j = 1, \ldots, n_\mu)$ which are orthogonal to one another and to the vectors $D_{lm}^{(\nu)}(R)$ formed for all nonequivalent representations. Since the number of orthogonal vectors in a g-dimensional space cannot exceed g, we have the result

$$\sum_\mu n_\mu^2 \leq g, \qquad (3\text{-}144)$$

or, the *number* of *nonequivalent* irreducible representations of a *finite* group is *finite*. So far we have no guarantee that the orthogonal vectors thus obtained will fill out the entire g-dimensional space, but we shall later prove that they do, i.e., that the equality sign holds in Eq. (3-144).

Starting from Eq. (3-142) or (3-143), we can derive similar orthogonality relations for the characters. If we set $i = l$ and $j = m$ in (3-142), we obtain

$$\sum_R D_{ii}^{(\mu)}(R)D_{jj}^{(\nu)}(R^{-1}) = \frac{g}{n_\mu}\,\delta_{\mu\nu}\,\delta_{ij}^2 = \frac{g}{n_\mu}\,\delta_{\mu\nu}\,\delta_{ij}. \qquad (3\text{-}145)$$

Now sum over all i and j:

$$\sum_R \chi^{(\mu)}(R)\chi^{(\nu)}(R^{-1}) = g\,\delta_{\mu\nu}, \qquad (3\text{-}146)$$

or

$$\sum_R \chi^{(\mu)}(R)\chi^{(\nu)*}(R) = g\,\delta_{\mu\nu} \qquad (3\text{-}147)$$

if the representations are unitary. We restrict ourselves to unitary representations from now on.

Here we must be careful. Remember that all the elements in a given class in G have the same character. Let $K_1, \ldots, K_k$ be the k classes of G, and let the number of elements in K_i be g_i. Then for all the elements of K_i, the character in the μ-representation is the same: $\chi^{(\mu)}(R) = \chi_i^{(\mu)}$. Thus Eq. (3-147) becomes

$$\sum_i \chi_i^{(\mu)} \chi_i^{(\nu)*} \cdot g_i = g \, \delta_{\mu\nu}. \tag{3-148}$$

For given μ, the numbers $\chi_i^{(\mu)} g_i^{1/2}$ form a vector in a k-dimensional space (where k is the number of classes in G). The vectors thus obtained from nonequivalent irreducible representations are *orthogonal* (and *none* of the vectors are *zero*). Therefore the *number* of *nonequivalent* irreducible representations of a group must be *less than* or *equal* to the number of classes in the group. Again, we shall prove later that the equality holds.

Problem. Show that the reciprocals of all the elements in class K_i form a class $K_{i'}$. For general irreducible representations derive the equation

$$\sum_i g_i \chi_i^{(\mu)} \chi_{i'}^{(\nu)} = g \, \delta_{\mu\nu}. \tag{3-148a}$$

3-16 Criteria for irreducibility. Analysis of representations. The characters of irreducible representations are usually called *primitive characters* or *simple characters*. We now consider an arbitrary representation D. According to Eq. (3-118), D can be expressed in terms of irreducible representations as

$$D(R) = \sum_\nu a_\nu D^{(\nu)}(R), \tag{3-118}$$

where the a_ν are integers ≥ 0. Now take the trace of this equation for an element R in the class K_i of the group G:

$$\chi_i = \sum_\nu a_\nu \chi_i^{(\nu)}. \tag{3-149}$$

The character χ_i of a *reducible* representation D is called a *compound character*. From (3-149), we see that a compound character is a linear combination of simple characters with *positive integral* coefficients. Multiply (3–149) by $\chi_i^{(\mu)*} g_i$ and sum over i:

$$\sum_i \chi_i^{(\mu)*} \chi_i g_i = \sum_\nu a_\nu \sum_i g_i \chi_i^{(\mu)*} \chi_i^{(\nu)}$$

$$= \sum_\nu a_\nu g \, \delta_{\mu\nu} = g a_\mu,$$

or

$$a_\mu = \frac{1}{g} \sum_i g_i \chi_i^{(\mu)*} \chi_i. \qquad (3\text{–}150)$$

Thus, to find the number of times a given irreducible representation is contained in D, we use Eq. (3–150). Notice that if two representations have the *same* set of *characters*, they are *equivalent* since, according to (3–150), the coefficients a_μ are the same for both.

If we multiply (3–149) by g_i times the complex conjugate equation and sum over i, we obtain

$$\sum_i \chi_i \chi_i^* g_i = \sum_{\mu,\nu} a_\mu a_\nu \sum_i g_i \chi_i^{(\nu)} \chi_i^{(\mu)*} = \sum_{\mu,\nu} a_\mu a_\nu g \, \delta_{\mu\nu};$$

$$\sum_i \chi_i \chi_i^* g_i = g \sum_\mu a_\mu^2. \qquad (3\text{–}151)$$

In particular, if the representation is irreducible, all the a_μ are zero except for one which is unity; so if the original representation is irreducible, its characters must satisfy

$$\sum_i g_i |\chi_i|^2 = g. \qquad (3\text{–}152)$$

This gives us a very simple criterion for irreducibility.

Equation (3–151) and its special case, (3–152), are extremely useful tools. If by some means we find a representation of the group G, we calculate the compound character χ_i and evaluate

$$\frac{1}{g} \sum_i g_i |\chi_i|^2 = \sum_\mu a_\mu^2.$$

If this quantity is unity, then the representation was irreducible. Suppose $\sum_\mu a_\mu^2 = 2$; since the a_μ must be integers, the only possibility is $a_\rho = a_\sigma = 1$ for two *different* irreducible representations $D^{(\rho)}$ and $D^{(\sigma)}$. Similarly, if $\sum_\mu a_\mu^2 = 3$, the representation D is a sum of three *different* irreducible representations, each occurring once. In the terminology of characters, we would say that the compound character χ_i is the sum of three different simple characters:

$$\chi_i = \chi_i^{(\rho)} + \chi_i^{(\sigma)} + \chi_i^{(\tau)}, \qquad \rho \neq \sigma \neq \tau, \qquad (3\text{–}153)$$

with all coefficients equal to unity. If $\sum_\mu a_\mu^2 = 4$, then either

$$\chi_i = \chi_i^{(\rho)} + \chi_i^{(\sigma)} + \chi_i^{(\tau)} + \chi_i^{(\mu)}, \qquad \rho \neq \sigma \neq \tau \neq \mu, \qquad (3\text{--}154)$$

or

$$\chi_i = 2\chi_i^{(\rho)}. \qquad (3\text{--}154a)$$

The following extension of these results is also very useful. If we are given two representations D and Δ of the group G, with the compound characters χ_i and ϕ_i, respectively, we can first use Eq. (3–151) on each character separately to try to deduce how many different simple characters each contains. We would then like to find out whether χ_i and ϕ_i have any simple characters in common. We start from equations like (3–149):

$$\chi_i = \sum_\nu a_\nu \chi_i^{(\nu)}; \qquad \phi_i = \sum_\mu b_\mu \chi_i^{(\mu)},$$

where the a_ν, b_μ are positive integers. We multiply χ_i by $\phi_i^* g_i/g$ and sum over i:

$$\sum_i \frac{g_i}{g} \chi_i \phi_i^* = \sum_{\mu,\nu} a_\nu b_\mu \sum_i \frac{g_i}{g} \chi_i^{(\nu)} \chi_i^{(\mu)*}$$

$$= \sum_{\mu,\nu} a_\nu b_\mu \, \delta_{\mu\nu} = \sum_\nu a_\nu b_\nu, \qquad (3\text{--}155)$$

where we have used Eq. (3–148). If $\sum a_\nu b_\nu$ is zero, then the two representations have *no* irreducible constituents in common. If $\sum a_\nu b_\nu = 1$, they have precisely *one* irreducible constituent in common, and each contains that constituent just *once*. For $\sum a_\nu b_\nu > 1$, the possibilities increase rapidly, but one may still be able to derive useful information.

Problem. For the case of general representations derive the equations:

$$a_\mu = \frac{1}{g} \sum_i g_i \chi_i \chi_{i'}^{(\mu)}, \qquad (3\text{--}150a)$$

$$\sum_i g_i \chi_i \chi_{i'} = g \sum_\mu a_\mu^2, \qquad (3\text{--}151a)$$

$$\sum_i \frac{g_i}{g} \chi_i \phi_{i'} = \sum_\nu a_\nu b_\nu. \qquad (3\text{--}155a)$$

3–17 The general theorems; group algebra. We may now return to complete the proofs of our general theorems. First we show that the equality sign holds in Eq. (3–144). To do this, we consider a special representation of the group G, the *regular* representation, which we used in Chapter 1 in discussing Cayley's theorem. If we label the elements of the group as $S_1, \ldots, S_g$, then multiplying from the left by any element S_ν permutes the $S_1, \ldots, S_g$ among themselves Considering $S_1, \ldots, S_g$ as coordinates in a g-dimensional space, we can represent the element S_ν by a permutation of the g coordinates. Thus if $S_\nu S_i = S_{j_i} (i = 1, \ldots, g)$, then $D_{il}(S_\nu) = \delta_{lj_i}$. In this representation, the regular representation, the diagonal elements of all matrices are zero except for the element S_ν, such that $S_\nu S_i = S_i$, that is, for the unit element E. So in the regular representation,

$$\chi(R) = \begin{cases} 0 & \text{for} \quad R \neq E, \\ g & \text{for} \quad R = E. \end{cases} \tag{3–156}$$

Expressing the regular representation in terms of nonequivalent irreducible representations, we have

$$\chi_i = \sum_\nu a_\nu \chi_i^{(\nu)}. \tag{3–149}$$

Consider the class of the identity. For this class, $\chi_i = \chi_1 = g$, while in the νth irreducible representation, $\chi_1^{(\nu)} = n_\nu$. Thus

$$g = \sum_\nu a_\nu n_\nu. \tag{3–157}$$

The a_ν are given by Eq. (3–150), where in the sum only the term $i = 1$ gives a contribution:

$$a_\nu = \frac{1}{g} \sum_i g_i \chi_i \chi_i^{(\nu)*} = \frac{1}{g} \cdot g n_\nu = n_\nu = \chi_1^{(\nu)}. \tag{3–158}$$

Equation (3–158) states that the *number* of *times* each irreducible representation is *contained* in the *regular representation* is *equal* to the *degree* of the representation. Substituting in (3–157), we obtain

$$g = \sum_\nu n_\nu^2, \tag{3–159}$$

which proves our theorem. Substituting (3–158) in (3–149), we also have

$$\chi_i = \sum_\nu n_\nu \chi_i^{(\nu)} = \sum_\nu \chi_1^{(\nu)} \chi_i^{(\nu)}. \tag{3–160}$$

If $i = 1$ (the class of E), then $\chi_i = g$, while for all other classes $\chi_i = 0$; hence

$$\sum_\nu \chi_1^{(\nu)} \chi_i^{(\nu)} = \begin{cases} 0 & \text{for } i \neq 1, \\ g & \text{for } i = 1. \end{cases} \tag{3–161}$$

We note that our procedure here is the complement of what we did earlier in the orthogonality theorem of Eq. (3–148). There, for fixed ν, the $g_i^{1/2}\chi_i^{(\nu)}$ formed a vector in the k-dimensional space (k equals the number of classes). For different ν, ($\nu = 1, \ldots, r$) the vectors were orthogonal, so that $r \leq k$. Now Eq. (3–161) suggests that for fixed i we can form vectors $\chi_i^{(\nu)}$ in the r-dimensional space (i.e., for a given class, we write the characters in the r different irreducible representations and form a vector). Equation (3–161) states the orthogonality of these vectors only for the case where one of the classes is the identity. If we could extend Eq. (3–161) to any pair of classes, we could conclude that $k \leq r$, and this conclusion, combined with our previous result that $r \leq k$, would prove that $r = k$; i.e., the number of nonequivalent irreducible representations is equal to the number of classes in the group. We now describe the proof.

Starting from a group G, we can construct a system which is called an *algebra*. The quantities in the algebra are

$$\sum_R a_R R, \tag{3–162}$$

where the coefficients a_R are any complex numbers. By the sum of two quantities we understand

$$\sum_R a_R R + \sum_R b_R R = \sum_R (a_R + b_R) R, \tag{3–163}$$

and the product implies

$$\sum_R a_R R \cdot \sum_S b_S S = \sum_{R,S} a_R b_S RS = \sum_R \left(\sum_S a_R s b_{S^{-1}} \right) R. \tag{3–164}$$

For example, in the group $\mathcal{C}_i$ with elements E, I, we set up an algebra of quantities

$$a_1 E + a_2 I.$$

Thus

$$(2E + 4I) + (E - 6I) = (3E - 2I),$$

$$(2E + 4I)(E - 6I) = 2E^2 - 12EI + 4IE - 24I^2$$

$$= 2E - 12I + 4I - 24E$$

$$= -22E - 8I.$$

Let the elements in the class K_i of G be $A_1^{(i)}, \ldots, A_{g_i}^{(i)}$. We construct the algebraic quantity

$$\mathcal{K}_i = \sum_{l=1}^{g_i} A_l^{(i)}. \tag{3–165}$$

If we multiply the quantities $\mathcal{K}_i$ for two classes,

$$\mathcal{K}_i\mathcal{K}_j = \sum_{l=1}^{g_i} \sum_{m=1}^{g_j} A_l^{(i)} A_m^{(j)}, \tag{3–166}$$

the sum on the right again consists of complete classes, for if we transform $\mathcal{K}_i\mathcal{K}_j$ by any element of the group, we merely permute the terms. Thus

$$\mathcal{K}_i\mathcal{K}_j = \sum_l c_{ijl}\mathcal{K}_l, \tag{3–167}$$

where the c_{ijl} are positive integers. Consider any irreducible representation of G. If we add all the matrices corresponding to elements of the ith class, we obtain a matrix which we denote by D_i. This matrix commutes with all matrices of the representation since

$$D_i = \sum_{R \subset K_i} D(R), \tag{3–168}$$

and transforming D_i merely permutes the terms in the sum. But this means that D_i commutes with all the matrices of the representation and, according to Lemma II, $D_i = \lambda_i \cdot 1$. Taking the characters in this equation, we have

$$g_i\chi_i = \lambda_i n = \lambda_i\chi_1, \tag{3–169}$$

where n is the degree of the irreducible representation. So

$$\lambda_i = \frac{g_i\chi_i}{\chi_1}. \tag{3–170}$$

Now go over to the matrices in Eq. (3–167):

$$D_iD_j = \sum_l c_{ijl}D_l, \tag{3–171}$$

or

$$\lambda_i\lambda_j = \sum_l c_{ijl}\lambda_l, \tag{3–172}$$

and using (3–170), we obtain

$$\frac{g_i\chi_i}{\chi_1} \cdot \frac{g_j\chi_j}{\chi_1} = \sum_l c_{ijl}\frac{g_l\chi_l}{\chi_1},$$

or

$$g_ig_j\chi_i\chi_j = \chi_1 \sum_l c_{ijl}g_l\chi_l. \tag{3–173}$$

All the characters refer to a given irreducible representation, while the g's and c_{ijl} are properties of the group and are *independent* of the representation. Equation (3–173) is useful in checking computations of characters.

If two elements of the group G are conjugate to each other, then their inverses are also conjugate to each other. Thus, given a class K_i, there exists another class $K_{i'}$, consisting of the inverses of the elements in K_i. Note that K_i and $K_{i'}$ have the same number of elements, $g_i = g_{i'}$. If we take the product of these two classes K_i and $K_{i'}$, ($K_{i'}$ being made up of the inverses of elements in K_i), we obtain the identity E precisely g_i times. For $j \neq i'$, the product $K_i K_j$ does not contain the identity, that is,

$$c_{ij1} = \begin{cases} 0 & \text{for } j \neq i', \\ g_i & \text{for } j = i'. \end{cases} \tag{3-174}$$

Now rewrite (3–173) and include a superscript to indicate a particular representation:

$$g_i g_j \chi_i^{(\nu)} \chi_j^{(\nu)} = \sum_l c_{ijl} \, g_l \chi_1^{(\nu)} \chi_l^{(\nu)}.$$

Sum this expression over all ν from 1 to r:

$$g_i g_j \sum_{\nu=1}^r \chi_i^{(\nu)} \chi_j^{(\nu)} = \sum_l c_{ijl} \, g_l \sum_{\nu=1}^r \chi_1^{(\nu)} \chi_l^{(\nu)},$$

$$= \sum_l c_{ijl} \, g_1 g \, \delta_{l1} \quad \text{[using Eq. (3-161)]}, \tag{3-175}$$

$$= c_{ij1} g_1 g = c_{ij1} g.$$

Then from (3–174),

$$g_i g_j \sum_{\nu=1}^r \chi_i^{(\nu)} \chi_j^{(\nu)} = \begin{cases} 0 & \text{for } j \neq i', \\ g g_i & \text{for } j = i', \end{cases} \tag{3-176}$$

or

$$\sum_{\nu=1}^r \chi_i^{(\nu)} \chi_j^{(\nu)} = \frac{g}{g_j} \, \delta_{ji'}. \tag{3-177}$$

For a unitary representation, $\chi_{i'} = \chi_i^*$, so (3–177) can be written as

$$\sum_{\nu=1}^r \chi_i^{(\nu)} \chi_j^{(\nu)*} = \frac{g}{g_j} \, \delta_{ij}. \tag{3-178}$$

Equation (3–178) shows that the k r-dimensional nonzero vectors $\chi_i^{(\nu)}$ are *mutually orthogonal*, so $k \leq r$, and we have completed the proof of our theorem:

The number of nonequivalent irreducible representations of a group is equal to the number of classes in the group.

We summarize our results as follows. Set up a table of characters like Table 3–1. Then Eqs. (3–148) and (3–178) (for unitary representations) state that the scalar product (with weight factors g_i) of any two rows or any two columns is zero. In addition, Eq. (3–173) states that the product of two numbers in the same row is expressible as a linear combination of the terms in the same row, with coefficients which are independent of the row.

TABLE 3–1

	K_1	K_2	$\cdots$	K_i	$\cdots$	K_k
$D^{(1)}$	$\chi_1^{(1)}$	$\chi_2^{(1)}$	$\cdots$	$\chi_i^{(1)}$	$\cdots$	$\chi_k^{(1)}$
$D^{(2)}$	$\chi_1^{(2)}$	$\cdots$	$\cdots$	$\cdots$	$\cdots$	$\cdots$
$\vdots$						$\vdots$
$D^{(\mu)}$	$\chi_1^{(\mu)}$	$\cdots$	$\cdots$	$\chi_i^{(\mu)}$	$\cdots$	$\chi_k^{(\mu)}$
$\vdots$						$\vdots$
$D^{(k)}$	$\chi_1^{(k)}$	$\chi_2^{(k)}$	$\cdots$	$\chi_i^{(k)}$	$\cdots$	$\chi_k^{(k)}$

3–18 Expansion of functions in basis functions of irreducible representations. As pointed out earlier in this chapter, starting with any function ψ, we can obtain a representation of the group G by applying to ψ all the transformations of the group G. Then ψ itself will be one of the base functions, or it will be expressible linearly in terms of the base functions. This statement remains true if we now split the representation into its irreducible constituents. So we conclude that any function ψ is expressible as a sum of functions which can act as base functions in the various irreducible representations:

$$\psi = \sum_\nu \sum_{i=1}^{n_\nu} \psi_i^{(\nu)}. \tag{3–179}$$

(We note that the most general function ψ will be a function for which the derived functions $O_R\psi$ are linearly independent. In this case we obtain the regular representation which we discussed earlier.)

We recall the definition of $\psi_i^{(\nu)}$ in Eq. (3–66). The base functions for the νth irreducible (unitary) representation satisfy the equations

$$O_R\psi_i^{(\nu)} = \sum_j \psi_j^{(\nu)} D_{ji}^{(\nu)}(R). \tag{3–66}$$

The function $\psi_i^{(\nu)}$ is said to belong to the ith row of the νth irreducible representation. We now try to find the condition that a given function must satisfy in order that it may belong to the ith row of a given representation. In other words, given a function $\psi_k^{(\nu)}$, we wish to associate with it

$(n_\nu - 1)$ other functions such that the set satisfies (3–66). We take (3–66), multiply by $D_{lm}^{(\mu)*}(R)$, and sum over the group:

$$\sum_R D_{lm}^{(\mu)*}(R)O_R\psi_i^{(\nu)} = \sum_j \psi_j^{(\nu)} \sum_R D_{lm}^{(\mu)*}(R)D_{ji}^{(\nu)}(R)$$

$$= \frac{g}{n_\nu} \sum_j \psi_j^{(\nu)}\, \delta_{lj}\, \delta_{mi}\, \delta_{\mu\nu} = \frac{g}{n_\nu} \psi_l^{(\nu)}\, \delta_{mi}\, \delta_{\mu\nu}. \quad (3\text{–}180)$$

In particular, for $m = l$, $\mu = \nu$,

$$\sum_R D_{ll}^{(\nu)*}(R)O_R\psi_i^{(\nu)} = \frac{g}{n_\nu} \psi_l^{(\nu)}\, \delta_{li}, \quad (3\text{–}181)$$

and setting $l = i$ yields

$$\sum_R D_{ii}^{(\nu)*}(R)O_R\psi_i^{(\nu)} = \frac{g}{n_\nu} \psi_i^{(\nu)}. \quad (3\text{–}182)$$

Equation (3–182) is a necessary condition on $\psi_i^{(\nu)}$. We now show that it is a sufficient condition, i.e., if a function $\psi_k^{(\nu)}$ satisfies

$$\sum_R D_{kk}^{(\nu)*}(R)O_R\psi_k^{(\nu)} = \frac{g}{n_\nu} \psi_k^{(\nu)}, \quad (3\text{–}183)$$

then we can find $(n_\nu - 1)$ "partners" so that the set satisfies (3–66). We use Eq. (3–180) with $\mu = \nu$, $m = k = i$ to define a set of functions:

$$\psi_l^{(\nu)} = \frac{n_\nu}{g} \sum_R D_{lk}^{(\nu)*}(R)O_R\psi_k^{(\nu)}. \quad (3\text{–}184)$$

In particular for $l = k$, we get back Eq. (3–183). Thus Eq. (3–184) is a satisfactory definition of n_ν functions $\psi_l^{(\nu)}$ in terms of $\psi_k^{(\nu)}$, provided that $\psi_k^{(\nu)}$ satisfies Eq. (3–183). We now show that the $\psi_l^{(\nu)}$ defined according to Eq. (3–184) satisfy (3–66), i.e., they form a basis for the νth irreducible representation. To prove this, we substitute Eq. (3–184) in the right-hand side of (3–66):

$$\sum_j \psi_j^{(\nu)} D_{ji}^{(\nu)}(S) = \frac{n_\nu}{g} \sum_R \sum_j D_{ji}^{(\nu)}(S)D_{jk}^{(\nu)*}(R)O_R\psi_k^{(\nu)}$$

$$= \frac{n_\nu}{g} \sum_R \left(\sum_j D_{ij}^{(\nu)*}(S^{-1})D_{jk}^{(\nu)*}(R) \right) O_R\psi_k^{(\nu)}$$

$$= \frac{n_\nu}{g} O_S \sum_R D_{ik}^{(\nu)*}(S^{-1}R)O_{S^{-1}}O_R\psi_k^{(\nu)}$$

$$= \frac{n_\nu}{g} O_S \sum_R D_{ik}^{(\nu)*}(R)O_R\psi_k^{(\nu)} = O_S\psi_i^{(\nu)}. \quad (3\text{–}185)$$

We now return to Eq. (3–179) and ask how to find the $\psi_i^{(\nu)}$ if we are given the function ψ. In other words, how do we resolve the given function into a sum of functions, each of which belongs to a particular row of some irreducible representation? In Eq. (3–180), for $m = l$,

$$\sum_R D_{ll}^{(\mu)*}(R)O_R\psi_i^{(\nu)} = \frac{g}{n_\mu}\psi_l^{(\nu)}\,\delta_{li}\,\delta_{\mu\nu}. \tag{3–186}$$

Thus the operator

$$P_i^{(\mu)} = \frac{n_\mu}{g}\sum_R D_{ii}^{(\mu)*}(R)O_R \tag{3–187}$$

is a *projection operator;* i.e.,

$$P_i^{(\mu)}\psi_j^{(\nu)} = \psi_i^{(\mu)}\,\delta_{\mu\nu}\,\delta_{ij}. \tag{3–188}$$

So, if we apply the operator $P_i^{(\mu)}$ to Eq. (3–179), we obtain the result

$$\psi_i^{(\mu)} = \frac{n_\mu}{g}\sum_R D_{ii}^{(\mu)*}(R)O_R\psi. \tag{3–189}$$

In analogy to the above, we say that a function "belongs to the νth irreducible representation" if it is a sum of functions belonging to the various rows of that representation, i.e.,

$$\psi^{(\nu)} = \sum_{i=1}^{n_\nu}\psi_{i\cdot}^{(\nu)} \tag{3–190}$$

If we sum over l in Eq. (3–186), we obtain

$$\sum_R \chi^{(\mu)*}(R)O_R\psi_i^{(\nu)} = \frac{g}{n_\mu}\psi_i^{(\nu)}\,\delta_{\mu\nu}, \tag{3–191}$$

and summing over i from 1 to n_ν [using (3–190)] yields

$$\sum_R \chi^{(\mu)*}(R)O_R\psi^{(\nu)} = \frac{g}{n_\mu}\,\delta_{\mu\nu}\psi^{(\mu)}. \tag{3–192}$$

We see that

$$P^{(\mu)} = \frac{n_\mu}{g}\sum_R \chi^{(\mu)*}(R)O_R \tag{3–193}$$

is a projection operator; i.e.,

$$P^{(\mu)}\psi^{(v)} = \psi^{(\mu)}\,\delta_{\mu\nu}. \tag{3–194}$$

From (3–179) and (3–190),

$$\psi = \sum_\nu \psi^{(\nu)}, \tag{3–195}$$

where

$$\psi^{(\nu)} = P^{(\nu)}\psi. \tag{3–196}$$

3–19 Representations of direct products. In Section 1–7 we introduced the concept of the direct product: The group G is said to be the direct product of two of its subgroups G_1 and G_2 $(G = G_1 \times G_2)$ if all elements of G_1 commute with all elements of G_2, G_1 and G_2 have only the identity in common, and every element of G is expressible as the product of an element of G_1 and an element of G_2. Some examples of direct products were given in Chapter 2. If a group G can be expressed as the direct product of two groups G_1 and G_2, the characters of the irreducible representations of G are easily determined from those of G_1 and G_2: Let $\psi_i^{(\mu)}$ $(i = 1, \ldots, n_\mu)$ and $\phi_k^{(\nu)}$ $(j = 1, \ldots, n_\nu)$ be sets of functions which form bases for irreducible representations of G_1 and G_2, respectively. Then the $n_\mu n_\nu$ functions $\psi_i^{(\mu)}\phi_j^{(\nu)}$ form the basis for an irreducible representation of G. We distinguish the elements and representations of G_1 and G_2 by subscripts 1 and 2. Then

$$O_{R_1}\psi_i^{(\mu)} = \sum_k \psi_k^{(\mu)} D_{1ki}^{(\mu)}(R_1), \tag{3–197}$$

$$O_{R_2}\phi_j^{(\nu)} = \sum_l \phi_l^{(\nu)} D_{2lj}^{(\nu)}(R_2), \tag{3–198}$$

$$O_{R_1 R_2}\psi_i^{(\mu)}\phi_j^{(\nu)} = O_{R_1}\psi_i^{(\mu)} O_{R_2}\phi_j^{(\nu)}$$

$$= \sum_{kl} \psi_k^{(\mu)}\psi_l^{(\nu)} D_{1ki}^{(\mu)}(R_1) D_{2lj}^{(\nu)}(R_2); \tag{3–199}$$

$$D_{kl;ij}^{(\mu \times \nu)}(R_1 R_2) = D_{1ki}^{(\mu)}(R_1) D_{2lj}^{(\nu)}(R_2) \tag{3–200}$$

or, symbolically,

$$D^{(\mu \times \nu)}(R_1 R_2) = D_1^{(\mu)}(R_1) \times D_2^{(\nu)}(R_2). \tag{3–200a}$$

To find the character of $R_1 R_2$, we must sum the diagonal elements in Eq. (3–200):

$$\chi^{(\mu \times \nu)}(R_1 R_2) = \chi_1^{(\mu)}(R_1)\chi_2^{(\nu)}(R_2). \tag{3–201}$$

Thus to find the characters of the irreducible representations of the direct product of G_1 and G_2, we take the products of the characters for G_1 and G_2.

CHAPTER 4

IRREDUCIBLE REPRESENTATIONS OF THE
POINT-SYMMETRY GROUPS

We shall now apply the theorems of Chapter 3 to the point-symmetry groups derived in Chapter 2. We shall discuss various methods for finding the irreducible representations of these groups. One of our aims is to become familiar with the practical application of the various theorems on representations.

4-1 Abelian groups. The problem of finding the irreducible representations is simple for abelian groups. Since the number of classes in an abelian group is equal to the order of the group, we see that all irreducible representations must be one-dimensional. For this case, then, matrix and character coincide, and we deal with simple multiplication of numbers. Furthermore, since the order of any element is finite, the characters of all elements of the group are roots of unity. Thus if the order of the element A is 2, that is, $A^2 = E$, then $D(A) = \chi(A) = \pm 1$. If $A^3 = E$, then $D(A) = \chi(A) = e^{2\pi i l/3}$ ($l = 1, 2, 3$). In general, if the order of A is h, $A^h = E$, then

$$D(A) = \chi(A) = e^{2\pi i l/h} \qquad (l = 1, \ldots, h). \qquad (4\text{-}1)$$

If the group is cyclic, some element A generates the group, $A^g = E$, so

$$\chi(A) = e^{2\pi i l/g} \qquad (l = 1, \ldots, g), \qquad (4\text{-}2)$$

and the characters of all the elements are determined by taking powers of $\chi(A)$. For example, $\chi(A^m) = e^{2\pi i l m/g}$.

First we consider cyclic groups:

$\mathcal{C}_1$: This is a trivial group consisting only of the identity E (i.e., the system has no symmetry). There is one irreducible representation:

$$D(E) = \chi(E) = 1.$$

$\mathcal{C}_2$: The group is generated from the element C_2; $C_2^2 = E$. Then

$$D(C_2) = \chi(C_2) = e^{2\pi i l/2} \qquad (l = 1, 2).$$

The characters of the two irreducible representations are given in Table 4-1.

In this case (one-dimensional representation), the characters coincide with

115

TABLE 4-1

$\mathcal{C}_2$:	E	C_2
A; z	1	1
B; x, y	1	-1

the matrices. We have introduced the following notation for the representations: One-dimensional representations are denoted by the symbols A or B, depending on whether the basis function is symmetric or antisymmetric with respect to rotation about the principal axis. We choose the principal axis along the z-direction. The basis functions for any of the groups $\mathcal{C}_n$ can be taken as

$$\psi = e^{il\phi} \qquad (l = 1, \ldots, n). \tag{4-3}$$

Each of these ψ's gives a one-dimensional representation, since rotation about the z-axis merely multiplies the function by a factor. For example, possible basis functions for the representation A of $\mathcal{C}_2$ are 1, z, $f(z)$, $f(x^2, y^2)$, etc. For B, we can take $\psi = e^{i\phi}$. We apply the operators O_R to the basis functions ψ. (Throughout this chapter we shall often write the group element R itself to denote the operator.) Then $C_2 e^{i\phi} = e^{i(\phi - \pi)} = -e^{i\phi}$. Or we can take any odd power of x or of y. In the character table (Table 4-1), we have indicated the representation to which the coordinates x, y, z belong; thus z belongs to the symmetric representation A, while x and y belong to the antisymmetric representation B. This information will be useful later when we consider selection rules. The representations to which x, y, z belong will determine the selection rules for electric dipole transitions. For other multipoles, we shall see how to derive the selection rules from the information in the character table.

We know two other groups which are isomorphic to $\mathcal{C}_2$, namely, $\mathcal{C}_s$ and $\mathcal{C}_i$. All these groups must have the same irreducible representations. To present information on isomorphic groups in compact form, we combine their character tables as shown in Table 4-2. In groups containing the inversion, a one-dimensional representation which is symmetric (or antisymmetric) with respect to inversion has a subscript g (gerade) or u (ungerade) attached to the symbol A or B. Similarly, in a group contain-

TABLE 4-2

$\mathcal{C}_i$:			E	I
	$\mathcal{C}_2$:		E	C_2
		$\mathcal{C}_s$:	E	σ_h
A_g	A; z	A'; x, y	1	1
A_u; x, y, z	B; x, y	A''; z	1	-1

ing σ_h, symmetry or antisymmetry under reflection is indicated by one or two primes, respectively. In the group $\mathcal{C}_s$, x and y are symmetric under reflection (in the XY-plane) and therefore belong to A', while z is antisymmetric under reflection and belongs to A''. In the group $\mathcal{C}_i$, x, y, and z are all antisymmetric under inversion and belong to A_u.

We note that the orthogonality theorems, Eqs. (3–147) and (3–178), are satisfied.

The theorems of the last chapter concerning the expansion of an arbitrary function in terms of functions belonging to different irreducible representations are of interest. For example, according to Eq. (3–192), a function $f(x, y, z)$ will belong to the representation A_g of the group $\mathcal{C}_i$ if

$$O_E f + O_I f = \frac{g}{n_\mu} f = \frac{2}{1} f;$$

$$f(x, y, z) + f(-x, -y, -z) = 2f(x, y, z);$$

$$f(x, y, z) = f(-x, -y, -z); \tag{4-4}$$

while $g(x, y, z)$ belongs to A_u if

$$O_E g - O_I g = 2g;$$

$$g(x, y, z) = -g(-x, -y, -z). \tag{4-5}$$

So for the group $\mathcal{C}_i$ the expansion theorem states that an arbitrary function can be expressed as a sum of functions which are even or odd under inversion.

Similar statements apply to the groups $\mathcal{C}_s$ and $\mathcal{C}_2$. For $\mathcal{C}_2$ we may also state our result as follows: The azimuth ϕ runs from 0 to 2π. Any function $f(\phi)$ can be continued periodically outside this range. In the Fourier expansion

$$f(\phi) = \sum_{-\infty}^{+\infty} a_n e^{in\phi}, \tag{4-6}$$

we can split the sum into terms with even and odd n, respectively, that is,

$$f(\phi) = \sum a_{2m} e^{2im\phi} + \sum a_{2m+1} e^{i(2m+1)\phi}. \tag{4-7}$$

The first term in (4–7) is even under the operation C_2, while the second is odd.

Next we consider the group $\mathcal{C}_3$. There are three one-dimensional representations, with basis functions $e^{il\phi}$ ($l = 1, 2, 3$), respectively, or i, $e^{i\phi}$, $e^{-i\phi}$. Then

$$C_3 e^{i\phi} = e^{i(\phi - 2\pi/3)} = e^{-2\pi i/3} e^{i\phi}.$$

TABLE 4–3

$\mathcal{C}_3$:	E	C_3	C_3^2	
$A; z$	1	1	1	
$E; x \pm iy$	$\begin{cases} 1 \\ 1 \end{cases}$	$\begin{matrix} \epsilon \\ \epsilon^2 \end{matrix}$	$\begin{matrix} \epsilon^2 \\ \epsilon \end{matrix}$	$\epsilon = e^{-2\pi i/3}.$

The character table is shown in Table 4–3. The orthogonality relations all reduce to $1 + \epsilon + \epsilon^2 = 0$, a special case of the theorem that the sum of the nth roots of unity is zero.

We shall use the symbol E throughout to denote two-dimensional representations. The brace in front of the last two lines of the table and our treatment of these entries, as if they were a two-dimensional representation, require an explanation. These two one-dimensional representations are complex conjugate. Now in quantum mechanics (in the absence of magnetic fields) the Hamiltonian is invariant under time reversal, and the complex conjugate of an eigenfunction is an eigenfunction for the same energy. Thus, even though the group $\mathcal{C}_3$ has only one-dimensional representations, symmetry under time reversal means that we have an additional operator which leaves the Hamiltonian invariant. This operator interchanges the base functions of the two representations, so that the two complex-conjugate representations correspond physically to a doubly degenerate level.

The cyclic groups $\mathcal{C}_4$, $\mathcal{S}_4$, $\mathcal{C}_6$, and $\mathcal{S}_6$ can be treated in similar fashion.

Next we consider the abelian group $\mathcal{C}_{2h}$. It contains 4 elements and has therefore 4 irreducible representations. The square of any element is the identity E, so all the characters are ± 1. The product of any two non-identity elements yields the third, and hence either all characters are $+1$, or two are 1 and two -1. Table 4–4 is the character table. The isomorphic groups $\mathcal{C}_{2v}$ and $D_2 \equiv V$ have the same character table.

TABLE 4–4

$\mathcal{C}_{2h}$:	E	C_2	σ_h	I
A_g	1	1	1	1
$A_u; z$	1	1	-1	-1
B_g	1	-1	-1	1
$B_u; x, y$	1	-1	1	-1

Problems. (1) Give examples of functions which belong to the representations A_g and B_g of the group $\mathcal{C}_{2h}$. Classify the spherical harmonics $P_l^m(\theta)e^{im\phi}$ according to the representations of $\mathcal{C}_{2h}$.

(2) For the groups $\mathcal{C}_3$ and $\mathcal{S}_4$, find the irreducible representations to which quadratic expressions in x, y, z belong.

(3) Classify the components of an axial vector according to the representations of $\mathcal{C}_{2h}$.

4–2 Nonabelian groups.

We now go on to the nonabelian groups. First we consider $\mathcal{C}_{3v}$ (and the isomorphic group D_3). The group $\mathcal{C}_{3v}$ contains 6 elements in 3 classes. There are 3 irreducible representations of dimensions n_1, n_2, n_3, and

$$n_1^2 + n_2^2 + n_3^2 = 6.$$

The only solution is $n_1 = n_2 = 1$, $n_3 = 2$. So we have two one-dimensional representations, and one two-dimensional representation. The simplest procedure for finding the representations is to start from those previously determined for $\mathcal{C}_3$, which is a subgroup of $\mathcal{C}_{3v}$. The eigenvalues of σ_v are ± 1, since $\sigma_v^2 = E$. Hence, taking the basis function ψ of the representation A of $\mathcal{C}_3$, we can have either $\sigma_v \psi = \pm \psi$. Thus we get the two one-dimensional representations of $\mathcal{C}_{3v}$ whose characters are given in Table 4–5.

Equation (3–148) is satisfied:

$$\sum g_i |\chi_i|^2 = 1 \cdot (1)^2 + 2(1)^2 + 3(\pm 1)^2 = 6 = g,$$

$$\sum g_i \chi_i^{(\mu)} \chi_i^{(\nu)*} = 1(1 \cdot 1) + 2(1 \cdot 1) + 3(1)(-1) = 0. \tag{4–8}$$

The characters of the two-dimensional representation can now be found by using Eq. (3–178); $\chi(E) = 2$ for a two-dimensional representation. Applying Eq. (3–178) to the first two classes gives

$$(1)(1) + (1)(1) + 2\chi(C_3) = 0; \qquad \chi(C_3) = -1, \tag{4–9}$$

and similarly for the first and third classes,

$$(1)(1) + (1)(-1) + 2\chi(\sigma_v) = 0; \qquad \chi(\sigma_v) = 0. \tag{4–10}$$

Equation (3–178) with $i = j$ is also satisfied:

$$(1)^2 + (1)^2 + (2)^2 = 6 = \frac{g}{g_1}; \qquad (1)^2 + (1)^2 + (-1)^2 = 3 = \frac{g}{g_2};$$

$$\tag{4–11}$$

$$(1)^2 + (-1)^2 + (0)^2 = 2 = \frac{g}{g_3}.$$

Thus Table 4–5 represents the complete character table for the group $\mathcal{C}_{3v}$. The numbers in parentheses in the top row give the quantities g_i, the number of elements in class K_i.

TABLE 4-5

$\mathcal{C}_{3v}$:	E	$C_3, C_3^2\,(2)$	$\sigma_v\,(3)$
$A_1; z$	1	1	1
A_2	1	1	-1
$E; x, y$	2	-1	0

An alternative procedure, which also gives the matrices of the E-representation, starts from the base functions $e^{\pm i\phi}$ of the complex conjugate representations of $\mathcal{C}_3$. Choose the ZX-plane as reflection plane for one of the σ_v's (ϕ is measured from the x-axis). Then $\sigma_v e^{\pm i\phi} = e^{\mp i\phi}$, while C_3 and C_3^2 are diagonal matrices. The matrices for the other reflections can be obtained from the products $\sigma_v C_3$ and $\sigma_v C_3^2$. Hence the matrices of the E-representation are

$$E:\begin{bmatrix} 1 & 0 \\ 0 & 1 \end{bmatrix}; \quad C_3:\begin{bmatrix} \epsilon & 0 \\ 0 & \epsilon^2 \end{bmatrix}; \quad C_3^2:\begin{bmatrix} \epsilon^2 & 0 \\ 0 & \epsilon \end{bmatrix};$$

$$\sigma_v:\begin{bmatrix} 0 & 1 \\ 1 & 0 \end{bmatrix}; \quad \sigma_{v'}:\begin{bmatrix} 0 & \epsilon^2 \\ \epsilon & 0 \end{bmatrix}; \quad \sigma_{v''}:\begin{bmatrix} 0 & \epsilon \\ \epsilon^2 & 0 \end{bmatrix}.$$

We can verify Eq. (3-143) for the matrices of the three representations. Take the representation A_1 and the ij-element of the representation E. We find

$$\sum_R D_{ij}^{(\mu)}(R)\,D_{lm}^{(\nu)*}(R) = \sum_R D_{ij}^{(\mu)}(R) = 1 + \epsilon + \epsilon^2 = 0 \qquad \text{for all}\quad i, j.$$
$$(4\text{-}12)$$

When we take A_2 and E, we get the same result. Taking the E-representation and setting $i = j = 1$ and $l = m = 2$, we have

$$\sum_R D_{11}(R)\,D_{22}^*(R) = 1 \cdot 1 + \epsilon \cdot \epsilon^{2*} + \epsilon^2 \cdot \epsilon^* = 1 + \epsilon + \epsilon^2 = 0.$$
$$(4\text{-}13)$$

To find functions which belong to the A_2-representation, we can use the following method, which is long-winded but might be instructive. If we expand $f(\phi)$ in Fourier series, the symmetry under C_3 and C_3^2 requires that only those harmonics which are multiples of 3 appear, i.e.,

$$f(\phi) = a_0 + a_3 e^{3i\phi} + a_6 e^{6i\phi} + \cdots$$
$$+ a_{-3} e^{-3i\phi} + a_{-6} e^{-6i\phi} + \cdots \qquad (4\text{-}14)$$

The antisymmetry under σ_v now requires $a_0 = 0$, $a_3 = -a_3$ etc. Hence functions like $\sin 3\phi$ [$\sum a_n \sin (3n\phi)$ in general] belong to A_2.

The reader will recall that we pointed out in Chapter 2 that σ_v reverses the sense of rotation about the z-axis. So the z-component of an *axial* vector will belong to the representation A_2. In terms of the coordinates of two points x, y, z and x', y', z', a typical function would be $xy' - yx'$ or $\sin(\phi - \phi')$. Note that a function which depends on the difference of two azimuths is invariant under rotations about the z-axis.

The method we have used for $\mathcal{C}_{3v}$ can also be applied to $\mathcal{C}_{4v}$ and $\mathcal{C}_{6v}$.

Problem. Find the matrices of the irreducible representations of $\mathcal{C}_{3v}$. Find functions belonging to each of the representations.

There is an alternative procedure which can be used to find the characters, which we shall work out in detail for $\mathcal{C}_{3v}$. We first determine as we did above that the three irreducible representations have $n_1 = n_2 = 1$, $n_3 = 2$. Then we use Eq. (3–173) to find relations between the characters for a given representation. The coefficients c_{ijl} in (3–173) are determined from Eq. (3–171). We label the classes of $\mathcal{C}_{3v}$ as $K_1: E$; K_2: C_3, C_3^2; $K_3: \sigma_v(3)$. Apply Eq. (3–171) to $\mathcal{K}_2^2$:

$$\mathcal{K}_2 = C_3 + C_3^2; \qquad \mathcal{K}_2^2 = (C_3 + C_3^2)^2 = 2E + C_3 + C_3^2 = 2\mathcal{K}_1 + \mathcal{K}_2,$$

$$(4\text{--}15)$$

whence $c_{221} = 2$, $c_{222} = 1$. Similarly,

$$\mathcal{K}_3^2 = (\sigma_v + \sigma_{v'} + \sigma_{v''})^2 = 3E + 3(C_3 + C_3^2) = 3\mathcal{K}_1 + 3\mathcal{K}_2, \quad (4\text{--}16)$$

so $c_{331} = 3$, $c_{332} = 3$. Finally,

$$\mathcal{K}_2\mathcal{K}_3 = \mathcal{K}_3\mathcal{K}_2 = (C_3 + C_3^2)(\sigma_v + \sigma_{v'} + \sigma_{v''})$$

$$= 2(\sigma_v + \sigma_{v'} + \sigma_{v''}) = 2\mathcal{K}_3, \quad (4\text{--}17)$$

so $c_{233} = c_{323} = 2$.

All other c_{ijl} are zero. Now we use these coefficients in (3–173):

$i = j = 2$:

$$g_2^2 \chi_2^2 = \chi_1 \sum_l c_{22l} g_l \chi_l,$$

$$4\chi_2^2 = n(2n + 2\chi_2); \qquad \chi_2 = n \quad \text{or} \quad \chi_2 = -\frac{n}{2}. \qquad (4\text{--}18)$$

$i = j = 3$:

$$g_3^2 \chi_3^2 = \chi_1 \sum_l c_{33l} g_l \chi_l, \qquad g\chi_3^2 = n(3n + 6\chi_2). \qquad (4\text{--}19)$$

$i = 2, j = 3$:

$$g_2 g_3 \chi_2 \chi_3 = \chi_1 \sum_l c_{23l} g_l \chi_l,$$

$$6\chi_2 \chi_3 = 6n\chi_3; \qquad \chi_3 = 0 \quad \text{or} \quad \chi_2 = n. \qquad (4\text{-}20)$$

For a one-dimensional representation, the characters cannot be zero (the matrices must be nonsingular). So for $n = 1$,

$$\chi_2 = 1, \qquad \chi_3^2 = (\tfrac{1}{3})(1 + 2) = 1, \qquad \chi_3 = \pm 1,$$

and we obtain the two solutions A_1 and A_2. For $n = 2$, if we take the solution $\chi_2 = 2$, we find $\chi_3 = 2$. But for an irreducible representation, $\sum |\chi(R)|^2 = g = 6$; this solution would give $\sum |\chi(R)|^2 = 24$. The other solution is $\chi_3 = 0, \chi_2 = -n/2 = -1$, which yields

$$\sum |\chi(R)|^2 = 1(2)^2 + 2(-1)^2 + 3(0)^2 = 6 = g.$$

This is a rather laborious method for finding characters. It was used extensively by Bethe (*Ann. Physik*, 1929).

The tetrahedral group T contains 12 elements in 4 classes, so there are 4 irreducible representations, with

$$n_1^2 + n_2^2 + n_3^2 + n_4^2 = 12.$$

The only solution is $n_1 = n_2 = n_3 = 1, n_4 = 3$; i.e., three representations are one-dimensional and one is three-dimensional. (We shall use the symbol F for three-dimensional representations. The symbol T is also sometimes used for this purpose.) Remember that the group T is obtained from the group $V \equiv D_2$ by adjoining 3-fold rotations about the space diagonals (see Fig. 2–34), and we can therefore obtain the representations of T by starting from those of V. When operated upon by a rotation about a 3-fold axis, the totally symmetric function which is the basis of the A_1-representation of V can be multiplied by 1, ϵ, or ϵ^2. (Remember we wish to obtain one-dimensional representations, so the basis function must be an eigenfunction of C_3.) Thus we get three different one-dimensional representations, as shown in Table 4–6. On the other hand, if we look at the basis functions which we found for the other three representations of V, namely x, y, and z, we see that they are transformed into one another by rotations about the space diagonals. Thus these three one-dimensional representations of V are coalesced into a three-dimensional representation of T. The characters of F can be found from the orthogonality relations. Since $\chi(E) = 3$, the orthogonality of the column vectors in the table requires $\chi(C_2) = -1, \chi(C_3) = 0 = \chi(C_3^2)$.

TABLE 4–6

T:	E	$C_2(3)$	$C_3(4)$	$C_3^2(4)$
A	1	1	1	1
E $\begin{cases} \\ \\ \end{cases}$	1	1	ϵ	ϵ^2
	1	1	ϵ^2	ϵ
$F; x, y, z$	3	−1	0	0

The matrices of the F-representation are based on x, y, z, and are merely the usual three-dimensional matrices for the various rotations.

Problems. (1) Construct the matrices for the F-representation of T.
(2) Find functions which belong to the E-representation of T.

Finally we consider the group T_d (and the isomorphic group O). The group contains 24 elements in 5 classes, so there are 5 irreducible representations, and

$$\sum_{i=1}^{5} n_i^2 = 24.$$

The only solution is $n_1 = n_2 = 1$, $n_3 = 2$, $n_4 = n_5 = 3$. The group T_d is obtained from T by adjoining reflections in planes through opposite edges of the cube, as shown in Fig. 2–64. Each of these planes contains two of the 3-fold axes. The base function of the representation A of T can be either symmetric or antisymmetric with respect to σ_d, and we therefore obtain two one-dimensional representations A_1, A_2 for T_d. The pair of complex-conjugate representations E of T have basis functions which go into one another upon reflection in a plane passing through the 3-fold axis, and hence we get a two-dimensional representation E of T_d. Finally consider the three-dimensional representation F of T. If we take the basis functions x, y, z and apply a reflection in the plane $x = y$, then $O_\sigma(z) = z$, $O_\sigma(x) = y$, $O_\sigma(y) = x$, giving for the matrix of σ

$$\begin{bmatrix} 0 & 1 & 0 \\ 1 & 0 & 0 \\ 0 & 0 & 1 \end{bmatrix},$$

whence $\chi(\sigma) = 1$. [All the reflections are in the same class, so $\chi(\sigma_d) = 1$ for all σ_d.] If instead we take the components of an axial vector as basis functions for the representation F of T, then, under the pure rotations of the group T, they serve as well as a polar vector (for pure rotations their

TABLE 4–7

T_d:	E	$C_3(8)$	$C_2(3)$	$\sigma_d(6)$	$S_4(6)$
A_1	1	1	1	1	1
A_2	1	1	1	−1	−1
E	2	−1	2	0	0
$F_2; x, y, z$	3	0	−1	1	−1
F_1	3	0	−1	−1	1

transformation properties are the same). But now reflection in the plane $x = y$ changes the sign of the z-component of the axial vector, while the other two components are transformed into one another, so $\chi(\sigma_d) = -1$. Thus we obtain two different three-dimensional representations F_1 and F_2 of the group T_d. The characters of the elements S_4 can be determined from the orthogonality of the row vectors in the character table (Table 4–7).

Problems. (1) Construct the matrices for the representations E, F_1, and F_2 of the group T_d.

(2) Find the characters of representations of T by using Eqs. (3–154) and (3–156).

We have now essentially obtained the primitive characters for all crystal point groups. Tables 4–9 through 4–19 assemble all these data in compact form for reference.

Groups (such as $\mathcal{C}_{5v}$ and $\mathcal{S}_8$) which sometimes occur in molecular problems, but are not included in the crystal groups, can be treated easily by our methods.

We have not included tables for those crystal groups which are expressible as direct products, i.e., for the groups

$$\mathcal{C}_{3h} = \mathcal{C}_3 \times \mathcal{C}_s, \qquad \mathcal{C}_{4h} = \mathcal{C}_4 \times \mathcal{C}_i, \qquad \mathcal{C}_{6h} = \mathcal{C}_6 \times \mathcal{C}_i,$$

$$D_{2h} = D_2 \times \mathcal{C}_i, \qquad D_{4h} = D_4 \times \mathcal{C}_i, \qquad \mathcal{S}_6 = \mathcal{C}_3 \times \mathcal{C}_i,$$

$$D_{3d} = D_3 \times \mathcal{C}_i, \qquad D_{6h} = D_6 \times \mathcal{C}_i, \qquad T_h = T \times \mathcal{C}_i,$$

$$O_h = O \times \mathcal{C}_i.$$

The reason for this omission is that, as pointed out in the previous chapter, the characters of the irreducible representations of these direct products can be obtained from the representations of the factors. Thus all groups which are obtained by taking a direct product with the group $\mathcal{C}_i$ have double the number of classes. Each of the representations of the original

TABLE 4–8

$\mathfrak{C}_{3h}$:	E	C_3	C_3^2	σ_h	$\sigma_h C_3$	$\sigma_h C_3^2$
A'	1	1	1	1	1	1
A''	1	1	1	-1	-1	-1
E' $\begin{cases} \\ \\ \end{cases}$	1	ϵ	ϵ^2	1	ϵ	ϵ^2
	1	ϵ^2	ϵ	1	ϵ^2	ϵ
E'' $\begin{cases} \\ \\ \end{cases}$	1	ϵ	ϵ^2	-1	$-\epsilon$	$-\epsilon^2$
	1	ϵ^2	ϵ	-1	$-\epsilon^2$	$-\epsilon$

group gives rise to two representations of the direct product, one which is symmetric and one which is antisymmetric with respect to the inversion I. The same remarks apply to a direct product with $\mathfrak{C}_s$. As an example we give the character table for the group $\mathfrak{C}_{3h}$ (Table 4–8).

Problem. Construct the character table for $\mathfrak{S}_6$.

4–3 Character tables for the crystal point groups.

TABLE 4–10

TABLE 4–9

$\mathfrak{C}_1$:	E
A	1

$\mathfrak{C}_i$:	$\mathfrak{C}_2$:	$\mathfrak{C}_s$:	E	I
			E	C_2
			E	σ
A_g	$A; z$	$A'; x, y$	1	1
$A_u; x, y, z$	$B; x, y$	$A''; z$	1	-1

TABLE 4–11

$\mathfrak{C}_{2h}$:	$\mathfrak{C}_{2v}$:	$V \equiv D_2$	E	C_2	σ_h	I
			E	C_2	σ_v	$\sigma_{v'}$
			E	C_z	C_y	C_x
A_g	$A_1; z$	A_1	1	1	1	1
B_g	$B_2; y$	$B_3; x$	1	-1	-1	1
$A_u; z$	A_2	$B_1; z$	1	1	-1	-1
$B_u; x, y$	$B_1; x$	$B_2; y$	1	-1	1	-1

TABLE 4–12

$\mathcal{C}_4$:		E	C_4	C_4^2	C_4^3
	$\mathcal{S}_4$:	E	S_4	S_4^2	S_4^3
$A;z$	A	1	1	1	1
B	$B;z$	1	-1	1	-1
$E;x\pm iy$	$E;x\pm iy$ $\left\{\begin{array}{c} \\ \\ \end{array}\right.$	1	i	-1	$-i$
		1	$-i$	-1	i

TABLE 4–13

$\mathcal{C}_3$:	E	C_3	C_3^2
$A;z$	1	1	1
$E;x\pm iy$ $\left\{\begin{array}{c} \\ \\ \end{array}\right.$	1	ϵ	ϵ^2
	1	ϵ^2	ϵ

$$\epsilon = e^{-2\pi i/3}$$

TABLE 4–14

$\mathcal{C}_{3v}$:		E	$C_3(2)$	$\sigma_v(3)$
	D_3:	E	$C_3(2)$	$C_x(3)$
$A_1;z$	A_1	1	1	1
A_2	$A_2;z$	1	1	-1
$E;x,y$	$E;x,y$	2	-1	0

TABLE 4–15

$\mathcal{C}_6$	E	C_6	C_6^2	C_6^3	C_6^4	C_6^5
$A;z$	1	1	1	1	1	1
B	1	-1	1	-1	1	-1
E_1 $\left\{\begin{array}{c} \\ \\ \end{array}\right.$	1	ω^2	$-\omega$	1	ω^2	$-\omega$
	1	$-\omega$	ω^2	1	$-\omega$	ω^2
$E_2;x\pm iy$ $\left\{\begin{array}{c} \\ \\ \end{array}\right.$	1	ω	ω^2	-1	$-\omega$	$-\omega^2$
	1	$-\omega^2$	$-\omega$	-1	ω^2	ω

$$\omega = e^{2\pi i/6}$$

TABLE 4–16

$\mathcal{C}_{4v}$:			E	C_4^2	$C_4(2)$	$\sigma_v(2)$	$\sigma_{v'}(2)$
	D_4:		E	C_4^2	$C_4(2)$	$C_2(2)$	$C_{2'}(2)$
		D_{2d}:	E	C_2	$S_4(2)$	$C_2(2)$	$\sigma_d(2)$
$A_1;z$	A_1	A_1	1	1	1	1	1
A_2	$A_2;z$	A_2	1	1	1	-1	-1
B_1	B_1	B_1	1	1	-1	1	-1
B_2	B_2	$B_2;z$	1	1	-1	-1	1
$E;x,y$	$E;x,y$	$E;x,y$	2	-2	0	0	0

TABLE 4–17

D_6:			E	C_6^3	$C_6^2(2)$	$C_6(2)$	$C_2(3)$	$C_{2'}(3)$
	C_{6v}:		E	C_6^3	$C_6^2(2)$	$C_6(2)$	$\sigma_v(3)$	$\sigma_{v'}(3)$
		D_{3h}:	E	σ_h	$S_6^2(2)$	$S_6(2)$	$C_2(3)$	$\sigma_v(3)$
A_1	$A_1; z$	A_1'	1	1	1	1	1	1
$A_2; z$	A_2	A_2'	1	1	1	1	-1	-1
B_1	B_2	A_1''	1	-1	1	-1	1	-1
B_2	B_1	$A_2''; z$	1	-1	1	-1	-1	1
E_2	E_1	$E'; x, y$	2	2	-1	-1	0	0
$E_1; x, y$	$E_2; x, y$	E''	2	-2	-1	1	0	0

TABLE 4–18

T:	E	$C_2(3)$	$C_3(4)$	$C_3^2(4)$
A	1	1	1	1
E $\Big\{$	1	1	ϵ	ϵ^2
	1	1	ϵ^2	ϵ
$F; x, y, z$	3	-1	0	0

TABLE 4–19

O:		E	$C_3(8)$	$C_4^2(3)$	$C_2(6)$	$C_4(6)$
	T_d:	E	$C_3(8)$	$S_4^2(3)$	$\sigma_d(6)$	$S_4(6)$
A_1	A_1	1	1	1	1	1
A_2	A_2	1	1	1	-1	-1
E	E	2	-1	2	0	0
F_2	$F_2; x, y, z$	3	0	-1	1	-1
$F_1; x, y, z$	F_1	3	0	-1	-1	1

CHAPTER 5

MISCELLANEOUS OPERATIONS WITH GROUP REPRESENTATIONS

5–1 Product representations (Kronecker products). In Chapter 3 we discussed the addition of representations of a group G to form new representations. Another method of deriving new representations, which is of basic importance in physical applications, is the construction of the *product representation (Kronecker product)*. This procedure is similar to the one used in Section 3–19 to obtain the irreducible representations of the direct product of two groups.

Suppose that we have found all irreducible representations of the group G. Let $D^{(\mu)}$ be an irreducible representation in the n_μ-dimensional space of vectors $\mathbf{x}$ (having components $x_1, \ldots, x_{n_\mu}$ in a particular basis), and let $D^{(\nu)}$ be another irreducible representation in the n_ν-dimensional space of vectors $\mathbf{y}$ (components $y_1, \ldots, y_{n_\nu}$):

$$x'_i = \sum_{j=1}^{n_\mu} D_{ij}^{(\mu)}(R) x_j \qquad (i = 1, \ldots, n_\mu); \tag{5–1}$$

$$y'_k = \sum_{l=1}^{n_\nu} D_{kl}^{(\nu)}(R) y_l \qquad (k = 1, \ldots, n_\nu). \tag{5–1a}$$

Multiplying the two equations, we obtain

$$x'_i y'_k = \sum_{j=1}^{n_\mu} \sum_{l=1}^{n_\nu} D_{ij}^{(\mu)}(R)\, D_{kl}^{(\nu)}(R) x_j y_l \qquad \begin{pmatrix} i = 1, \ldots, n_\mu \\ k = 1, \ldots, n_\nu \end{pmatrix}. \tag{5–2}$$

We can regard the quantities $x_i y_k$ as the components of a vector in the $(n_\mu \cdot n_\nu)$-dimensional product space. Equation (5–2) associates a transformation in this space with each of the elements R of the group G:

$$x'_i y'_k = \sum_{j,l} D_{ik,jl}^{(\mu \times \nu)}(R) x_j y_l \qquad \begin{pmatrix} i, j = 1, \ldots, n_\mu \\ k, l = 1, \ldots, n_\nu \end{pmatrix}, \tag{5–3}$$

where

$$D_{ik,jl}^{(\mu \times \nu)}(R) = \left(D^{(\mu)}(R) \times D^{(\nu)}(R) \right)_{ik,jl} = D_{ij}^{(\mu)}(R) \cdot D_{kl}^{(\nu)}(R). \tag{5–4}$$

The matrices $D_{ik,jl}^{(\mu \times \nu)}(R)$ form a representation of the group G:

$$
\begin{aligned}
D_{ik,jl}^{(\mu \times \nu)}(RS) &= D_{ij}^{(\mu)}(RS) D_{kl}^{(\nu)}(RS) \\
&= \sum_{\alpha} D_{i\alpha}^{(\mu)}(R) D_{\alpha j}^{(\mu)}(S) \sum_{\beta} D_{k\beta}^{(\nu)}(R) D_{\beta l}^{(\nu)}(S) \\
&= \sum_{\alpha, \beta} [D_{i\alpha}^{(\mu)}(R) D_{k\beta}^{(\nu)}(R)][D_{\alpha j}^{(\mu)}(S) D_{\beta l}^{(\nu)}(S)] \\
&= \sum_{\alpha \beta} D_{ik,\alpha \beta}^{(\mu \times \nu)}(R) D_{\alpha \beta, jl}^{(\mu \times \nu)}(S),
\end{aligned}
\tag{5–5}
$$

that is,

$$
D^{(\mu \times \nu)}(RS) = D^{(\mu \times \nu)}(R) \cdot D^{(\mu \times \nu)}(S). \tag{5–6}
$$

The representation (5–4) is called the *Kronecker product* of the representations $D^{(\mu)}$ and $D^{(\nu)}$.

Equation (5–6) also gives us the general rule for multiplication of Kronecker products of matrices. If A_1, A_2, ... are matrices of degree n, and B_1, B_2, ... are matrices of degree m, the Kronecker products $(A_1 \times B_1)$, $(A_2 \times B_2)$, ... will be matrices of degree nm, in which

$$
(A_\nu \times B_\nu)_{ik,jl} = (A_\nu)_{ij}(B_\nu)_{kl}. \tag{5–4a}
$$

As in Eq. (5–6),

$$
(A_1 \times B_1) \cdot (A_2 \times B_2) = (A_1 A_2) \times (B_1 B_2) \tag{5–6a}
$$

or, in general,

$$
(A_1 \times B_1) \cdot (A_2 \times B_2) \cdots (A_s \times B_s) = (A_1 A_2 \cdots A_s) \times (B_1 B_2 \cdots B_s).
\tag{5–6b}
$$

The above definition is given in terms of abstract vector spaces. We come closer to the physical problem if we restate our definition in terms of wave functions of a physical system. In Section 3–7 we associated an operator O_R with each transformation R, where O_R is a transformation of the arguments of the wave function:

$$
O_R \psi(x) = \psi(R^{-1} x). \tag{3–68}
$$

The set of transformations R which leave the Hamiltonian H invariant form a group G such that

$$
O_R H O_R^{-1} = H \tag{3–73}
$$

for all elements R in G. If ψ is an eigenfunction of H belonging to a given eigenvalue λ, then $O_R\psi$ is also an eigenfunction with eigenvalue λ. In the absence of "accidental" degeneracy, the set of n_μ eigenfunctions $\psi_i^{(\mu)}(i = 1, \ldots, n_\mu)$ belonging to a given eigenvalue λ_μ will provide the basis for an irreducible representation of the symmetry group G of the Hamiltonian:

$$O_R\psi_j^{(\mu)} = \sum_i \psi_i^{(\mu)} D_{ij}^{(\mu)}(R). \tag{5-7}$$

Similarly the set of n_ν degenerate eigenfunctions $\phi_i^{(\nu)}$ belonging to the eigenvalue λ_ν form the basis of the irreducible representation $D^{(\nu)}$:

$$O_R\phi_l^{(\nu)} = \sum_k \phi_k^{(\nu)} D_{kl}^{(\nu)}(R). \tag{5-7a}$$

From Eqs. (5-7) and (5-7a),

$$O_R(\psi_j^{(\mu)}\phi_l^{(\nu)}) = \sum_{ik} \psi_i^{(\mu)}\phi_k^{(\nu)} D_{ij}^{(\mu)}(R)D_{kl}^{(\nu)}(R)$$

$$= \sum_{ik} \psi_i^{(\mu)}\phi_k^{(\nu)} D_{ik,jl}^{(\mu \times \nu)}(R), \tag{5-8}$$

so the products $\psi_i^{(\mu)}\phi_j^{(\nu)}$ form the basis for the product representation $D^{(\mu)} \times D^{(\nu)}$.

The notation with double subscripts is quite complicated. To gain some familiarity with it, we give the explicit results for the simple case where $n_\mu = n_\nu = 2$. Then

$$O_R\psi_1^{(\mu)} = \psi_1^{(\mu)}D_{11}^{(\mu)}(R) + \psi_2^{(\mu)}D_{21}^{(\mu)}(R),$$

$$O_R\psi_2^{(\mu)} = \psi_1^{(\mu)}D_{12}^{(\mu)}(R) + \psi_2^{(\mu)}D_{22}^{(\mu)}(R),$$

$$O_R\phi_1^{(\nu)} = \phi_1^{(\nu)}D_{11}^{(\nu)}(R) + \phi_2^{(\nu)}D_{21}^{(\nu)}(R),$$

$$O_R\phi_2^{(\nu)} = \phi_1^{(\nu)}D_{12}^{(\nu)}(R) + \phi_2^{(\nu)}D_{22}^{(\nu)}(R). \tag{5-9}$$

From (5-9) we have

$$O_R(\psi_1^{(\mu)}\phi_1^{(\nu)}) = [\psi_1^{(\mu)}D_{11}^{(\mu)}(R) + \psi_2^{(\mu)}D_{21}^{(\mu)}(R)][\phi_1^{(\nu)}D_{11}^{(\nu)}(R) + \phi_2^{(\nu)}D_{21}^{(\nu)}(R)]$$

$$= \psi_1^{(\mu)}\phi_1^{(\nu)}D_{11}^{(\mu)}(R)D_{11}^{(\nu)}(R) + \psi_1^{(\mu)}\phi_2^{(\nu)}D_{11}^{(\mu)}(R)D_{21}^{(\nu)}(R)$$

$$+ \psi_2^{(\mu)}\phi_1^{(\nu)}D_{21}^{(\mu)}(R)D_{11}^{(\nu)}(R) + \psi_2^{(\mu)}\phi_2^{(\nu)}D_{21}^{(\mu)}(R)D_{21}^{(\nu)}(R). \tag{5-10}$$

In the expansion of $O_R(\psi_1^{(\mu)}\phi_1^{(\nu)})$, the coefficient of $\psi_1^{(\mu)}\phi_2^{(\nu)}$ is $D_{11}^{(\mu)}(R)D_{21}^{(\nu)}(R)$, so

$$(D^{(\mu)} \times D^{(\nu)}(R))_{12,11} = D_{11}^{(\mu)}(R)D_{21}^{(\nu)}(R),$$

and

$$(D^{(\mu)} \times D^{(\nu)}(R))_{22,11} = D_{21}^{(\mu)}(R)D_{21}^{(\nu)}(R).$$

Similarly, we find

$$O_R(\psi_1^{(\mu)}\phi_2^{(\nu)}) = \psi_1^{(\mu)}\phi_1^{(\nu)}D_{11}^{(\mu)}(R)D_{12}^{(\nu)}(R) + \psi_1^{(\mu)}\phi_2^{(\nu)}D_{11}^{(\mu)}(R)D_{22}^{(\nu)}(R)$$
$$+ \psi_2^{(\mu)}\phi_1^{(\nu)}D_{21}^{(\mu)}(R)D_{12}^{(\nu)}(R) + \psi_2^{(\mu)}\phi_2^{(\nu)}D_{21}^{(\mu)}(R)D_{22}^{(\nu)}(R);$$

$$(5–11)$$

$$O_R(\psi_2^{(\mu)}\phi_1^{(\nu)}) = \psi_1^{(\mu)}\phi_1^{(\nu)}D_{12}^{(\mu)}(R)D_{11}^{(\nu)}(R) + \psi_1^{(\mu)}\phi_2^{(\nu)}D_{12}^{(\mu)}(R)D_{21}^{(\nu)}(R)$$
$$+ \psi_2^{(\mu)}\phi_1^{(\nu)}D_{22}^{(\mu)}(R)D_{11}^{(\nu)}(R) + \psi_2^{(\mu)}\phi_2^{(\nu)}D_{22}^{(\mu)}(R)D_{21}^{(\nu)}(R);$$

$$(5–12)$$

$$O_R(\psi_2^{(\mu)}\phi_2^{(\nu)}) = \psi_1^{(\mu)}\phi_1^{(\nu)}D_{12}^{(\mu)}(R)D_{12}^{(\nu)}(R) + \psi_1^{(\mu)}\phi_2^{(\nu)}D_{12}^{(\mu)}(R)D_{22}^{(\nu)}(R)$$
$$+ \psi_2^{(\mu)}\phi_1^{(\nu)}D_{22}^{(\mu)}(R)D_{12}^{(\nu)}(R) + \psi_2^{(\mu)}\phi_2^{(\nu)}D_{22}^{(\mu)}(R)D_{22}^{(\nu)}(R).$$

$$(5–13)$$

From (5–10) through (5–13), the matrix of R in the product representation is

$$D^{(\mu)} \times D^{(\nu)}(R) =$$

$$\begin{bmatrix} D_{11}^{(\mu)}(R)\,D_{11}^{(\nu)}(R) & D_{11}^{(\mu)}(R)\,D_{12}^{(\nu)}(R) & D_{12}^{(\mu)}(R)\,D_{11}^{(\nu)}(R) & D_{12}^{(\mu)}(R)\,D_{12}^{(\nu)}(R) \\ D_{11}^{(\mu)}(R)\,D_{21}^{(\nu)}(R) & D_{11}^{(\mu)}(R)\,D_{22}^{(\nu)}(R) & D_{12}^{(\mu)}(R)\,D_{21}^{(\nu)}(R) & D_{12}^{(\mu)}(R)\,D_{22}^{(\nu)}(R) \\ D_{21}^{(\mu)}(R)\,D_{11}^{(\nu)}(R) & D_{21}^{(\mu)}(R)\,D_{12}^{(\nu)}(R) & D_{22}^{(\mu)}(R)\,D_{11}^{(\nu)}(R) & D_{22}^{(\mu)}(R)\,D_{12}^{(\nu)}(R) \\ D_{21}^{(\mu)}(R)\,D_{21}^{(\nu)}(R) & D_{21}^{(\mu)}(R)\,D_{22}^{(\nu)}(R) & D_{22}^{(\mu)}(R)\,D_{21}^{(\nu)}(R) & D_{22}^{(\mu)}(R)\,D_{22}^{(\nu)}(R) \end{bmatrix},$$

$$(5–14)$$

where the rows and columns are numbered in dictionary order: 11, 12, 21, 22. We have been assuming that all the product functions are linearly independent, for otherwise we should have combined the terms and would obtain a representation of lower degree. So we are assuming that the μ- and ν-representations have different degree or, if $\mu = \nu$, that the ψ's and ϕ's are not related to one another.

We denote the characters of the product representation by

$$\chi^{(\mu)} \times \chi^{(\nu)}(R) = \chi^{(\mu \times \nu)}(R).$$

As we see from the special case (5–14) or the general equation (5–4),

$$\chi^{(\mu \times \nu)}(R) = \sum_{i,j} D_{ii}^{(\mu)}(R) D_{jj}^{(\nu)}(R) = \chi^{(\mu)}(R) \cdot \chi^{(\nu)}(R). \qquad (5\text{--}15)$$

Thus the character of an element in the product representation is the product of its characters in the "factor" representations.

Problem. Show that the Kronecker product of unitary representations is a unitary representation.

5–2 Symmetrized and antisymmetrized products. In the previous section it was assumed that $\mu \neq \nu$. Let us now assume that $\mu = \nu$ and that the $\psi_i^{(\mu)}$ and $\phi_i^{(\nu)}$ are independent sets of functions. Again, for simplicity, we first treat the case where $n_\mu = n_\nu = 2$. We rewrite Eqs. (5–10) through (5–13), omitting the superscripts $\mu = \nu$:

$$O_R(\psi_1\phi_1) = \psi_1\phi_1[D_{11}(R)]^2 + \psi_1\phi_2 D_{11}(R)D_{21}(R)$$
$$+ \psi_2\phi_1 D_{21}(R)D_{11}(R) + \psi_2\phi_2[D_{21}(R)]^2; \qquad (5\text{--}16)$$

$$O_R(\psi_1\phi_2) = \psi_1\phi_1 D_{11}(R)D_{12}(R) + \psi_1\phi_2 D_{11}(R)D_{22}(R)$$
$$+ \psi_2\phi_1 D_{21}(R)D_{12}(R)$$
$$+ \psi_2\phi_2 D_{21}(R)D_{22}(R); \qquad (5\text{--}17)$$

$$O_R(\psi_2\phi_1) = \psi_1\phi_1 D_{12}(R)D_{11}(R) + \psi_1\phi_2 D_{12}(R)D_{21}(R)$$
$$+ \psi_2\phi_1 D_{22}(R)D_{11}(R)$$
$$+ \psi_2\phi_2 D_{22}(R)D_{21}(R); \qquad (5\text{--}18)$$

$$O_R(\psi_2\phi_2) = \psi_1\phi_1[D_{12}(R)]^2 + \psi_1\phi_2 D_{12}(R)D_{22}(R)$$
$$+ \psi_2\phi_1 D_{22}(R)D_{12}(R) + \psi_2\phi_2[D_{22}(R)]^2. \qquad (5\text{--}19)$$

If we take the sum and difference of (5–17) and (5–18), we can rewrite the four equations as

$$O_R(\psi_1\phi_1) = \psi_1\phi_1[D_{11}(R)]^2 + (\psi_1\phi_2 + \psi_2\phi_1)D_{11}(R)D_{21}(R)$$
$$+ \psi_2\phi_2[D_{21}(R)]^2; \qquad (5\text{--}20)$$

$$O_R(\psi_1\phi_2 + \psi_2\phi_1) = 2\psi_1\phi_1 D_{11}(R)D_{12}(R)$$
$$+ (\psi_1\phi_2 + \psi_2\phi_1)[D_{11}(R)D_{22}(R) + D_{12}(R)D_{21}(R)]$$
$$+ 2\psi_2\phi_2 D_{21}(R)D_{22}(R); \tag{5–21}$$

$$O_R(\psi_1\phi_2 - \psi_2\phi_1) = (\psi_1\phi_2 - \psi_2\phi_1)[D_{11}(R)D_{22}(R) - D_{12}(R)D_{21}(R)]; \tag{5–22}$$

$$O_R(\psi_2\phi_2) = \psi_1\phi_1[D_{12}(R)]^2 + (\psi_1\phi_2 + \psi_2\phi_1)D_{12}(R)D_{22}(R)$$
$$+ \psi_2\phi_2[D_{22}(R)]^2. \tag{5–23}$$

Note that (5–22) contains only the antisymmetric combination $\psi_1\phi_2 - \psi_2\phi_1$, while (5–20), (5–21), and (5–23) couple the three symmetric combinations $\psi_1\phi_1$, $\psi_1\phi_2 + \psi_2\phi_1$, and $\psi_2\phi_2$. We conclude that the product of a representation (of degree > 1) with itself is always reducible into the sum of a symmetric and an antisymmetric product representation. We show this now for the general case. In Eq. (5–8) take $\mu = \nu$:

$$O_R(\psi_j^{(\mu)}\phi_l^{(\mu)}) = \sum_{ik} \psi_i^{(\mu)}\phi_k^{(\mu)} D_{ij}^{(\mu)}(R)D_{kl}^{(\mu)}(R). \tag{5–24}$$

Interchange j and l:

$$O_R(\psi_l^{(\mu)}\phi_j^{(\mu)}) = \sum_{ik} \psi_i^{(\mu)}\phi_k^{(\mu)} D_{il}^{(\mu)}(R)D_{kj}^{(\mu)}(R). \tag{5–25}$$

Taking the sum and difference of (5–24) and (5–25) yields

$$O_R(\psi_j^{(\mu)}\phi_l^{(\mu)} + \psi_l^{(\mu)}\phi_j^{(\mu)}) = \sum_{ik} \psi_i^{(\mu)}\phi_k^{(\mu)}[D_{ij}^{(\mu)}(R)D_{kl}^{(\mu)}(R) + D_{il}^{(\mu)}(R)D_{kj}^{(\mu)}(R)]$$
$$= \tfrac{1}{2}\sum_{ik} (\psi_i^{(\mu)}\phi_k^{(\mu)} + \psi_k^{(\mu)}\phi_i^{(\mu)})[D_{ij}^{(\mu)}(R)D_{kl}^{(\mu)}(R) + D_{il}^{(\mu)}(R)D_{kj}^{(\mu)}(R)]; \tag{5–26}$$

$$O_R(\psi_j^{(\mu)}\phi_l^{(\mu)} - \psi_l^{(\mu)}\phi_j^{(\mu)}) = \sum_{ik} \psi_i^{(\mu)}\phi_k^{(\mu)}[D_{ij}^{(\mu)}(R)D_{kl}^{(\mu)}(R) - D_{il}^{(\mu)}(R)D_{kj}^{(\mu)}(R)]$$
$$= \tfrac{1}{2}\sum_{ik} (\psi_i^{(\mu)}\phi_k^{(\mu)} - \psi_k^{(\mu)}\phi_i^{(\mu)})[D_{ij}^{(\mu)}(R)D_{kl}^{(\mu)}(R) - D_{il}^{(\mu)}(R)D_{kj}^{(\mu)}(R)]. \tag{5–27}$$

The product $D^{(\mu)} \times D^{(\mu)}$ of the representation $D^{(\mu)}$ with itself is always (aside from the trivial case of $n_\mu = 1$) reducible into the sum of the symmetric product representation $[D^{(\mu)} \times D^{(\mu)}]$ of Eq. (5–26) and the antisymmetric product representation $\{D^{(\mu)} \times D^{(\mu)}\}$ of Eq. (5–27):

$$D^{(\mu)} \times D^{(\mu)} = [D^{(\mu)} \times D^{(\mu)}] + \{D^{(\mu)} \times D^{(\mu)}\}. \tag{5–28}$$

These representations may in turn be reducible. The matrices of the symmetric and antisymmetric product representations are

$$[D^{(\mu)} \times D^{(\mu)}(R)]_{kl,ij} = \tfrac{1}{2}[D^{(\mu)}_{ki}(R)D^{(\mu)}_{lj}(R) + D^{(\mu)}_{li}(R)D^{(\mu)}_{kj}(R)], \quad (5\text{-}29)$$

$$\{D^{(\mu)} \times D^{(\mu)}(R)\}_{kl,ij} = \tfrac{1}{2}[D^{(\mu)}_{ki}(R)D^{(\mu)}_{lj}(R) - D^{(\mu)}_{li}(R)D^{(\mu)}_{kj}(R)]. \quad (5\text{-}30)$$

The dimensions of these two representations are $\tfrac{1}{2}n_\mu(n_\mu + 1)$ and $\tfrac{1}{2}n_\mu(n_\mu - 1)$, respectively. The characters of the symmetric and antisymmetric product representations are denoted by $[\chi^{(\mu)} \times \chi^{(\mu)}(R)]$ and $\{\chi^{(\mu)} \times \chi^{(\mu)}(R)\}$, respectively. For our special case, we have from Eqs. (5–20) through (5–23):

$$[\chi \times \chi(R)] = [D_{11}(R)]^2 + D_{11}(R)D_{22}(R) + D_{12}(R)D_{21}(R) + [D_{22}(R)]^2$$

$$= \tfrac{1}{2}[D_{11}(R)$$

$$+ D_{22}(R)]^2 + \tfrac{1}{2}[[D_{11}(R)]^2 + D_{12}(R)D_{21}(R)$$

$$+ D_{21}(R)D_{12}(R) + [D_{22}(R)]^2]$$

$$= \tfrac{1}{2}[D_{11}(R) + D_{22}(R)]^2 + \tfrac{1}{2}[D_{11}(R^2) + D_{22}(R^2)]$$

$$= \tfrac{1}{2}[(\chi(R))^2 + \chi(R^2)] \quad (5\text{-}31)$$

and

$$\{\chi \times \chi(R)\} = \tfrac{1}{2}[(\chi(R))^2 - \chi(R^2)]. \quad (5\text{-}31a)$$

Similarly, in the general case, from Eqs. (5–26) and (5–27):

$$[\chi \times \chi(R)] = \tfrac{1}{2}\sum_{ij}[D_{ii}(R)D_{jj}(R) + D_{ij}(R)D_{ji}(R)]$$

$$= \tfrac{1}{2}\Big[\sum_{ij}D_{ii}(R)D_{jj}(R) + \sum_i D_{ii}(R^2)\Big]$$

$$= \tfrac{1}{2}[(\chi(R))^2 + \chi(R^2)] \quad (5\text{-}32)$$

and

$$\{\chi \times \chi(R)\} = \tfrac{1}{2}[(\chi(R))^2 - \chi(R^2)]. \quad (5\text{-}33)$$

Finally, we note that if $\mu = \nu$ and $\psi_i^{(\mu)} = \phi_i^{(\mu)}$, the antisymmetric products (5–27) are identically zero, and we obtain only the symmetric representation (5–29).

5–3 The adjoint representation.   The complex conjugate representation.   In this section we consider some other methods for deriving new representations.   Suppose that $D(R)$ is an irreducible representation of the group G.   If we take the inverse transpose of each of the matrices, we again obtain a representation since

$$\widetilde{D}^{-1}(RS) = [\widetilde{D}(RS)]^{-1} = [\widetilde{D}(S)\widetilde{D}(R)]^{-1} = \widetilde{D}^{-1}(R)\widetilde{D}^{-1}(S). \quad (5\text{–}34)$$

This representation is called the *adjoint representation* $\overline{D}$:

$$\overline{D}(R) \equiv \widetilde{D}^{-1}(R). \quad (5\text{–}35)$$

If we take the complex conjugate of the representation $D(R)$ [i.e., the set of matrices whose elements are the complex conjugates of the elements of $D(R)$], we get the *complex conjugate representation* $D^*(R)$:

$$D^*(RS) = D^*(R)D^*(S). \quad (5\text{–}36)$$

Problem.   Prove that D, $\overline{D}$, and D^* are either all reducible or all irreducible.

The characters of the adjoint representation are

$$\bar{\chi}(R) = \chi(R^{-1}) \quad (5\text{–}37)$$

or

$$\bar{\chi}_i = \chi_{i'}, \quad (5\text{–}38)$$

where $K_{i'}$ is the class of the elements inverse to those in class K_i.   The characters of the complex conjugate representation D^* are the complex conjugates of the characters of D.   In terms of the adjoint representation, the orthogonality theorems for characters (3–146) and (3–148a) become

$$\sum_R \chi^{(\mu)}(R)\bar{\chi}^{(\nu)}(R) = g\,\delta_{\mu\nu}, \quad (5\text{–}39)$$

$$\sum_i g_i \chi_i^{(\mu)}\bar{\chi}_i^{(\nu)} = g\,\delta_{\mu\nu}; \quad (5\text{–}40)$$

the completeness relation (3–177) is

$$\sum_\nu \chi_i^{(\nu)}\bar{\chi}_j^{(\nu)} = \left(\frac{g}{g_i}\right)\delta_{ij}, \quad (5\text{–}41)$$

and Eqs. (3–150a), (3–151a), and (3–155a), which are used to analyze a representation into its irreducible components, become

$$a_\mu = \left(\frac{1}{g}\right) \sum_i g_i \chi_i \tilde{\chi}_i^{(\mu)}, \tag{5-42}$$

$$\sum_i g_i \chi_i \bar{\chi}_i = g \sum_\mu a_\mu^2, \tag{5-43}$$

$$\sum_i \left(\frac{g_i}{g}\right) \chi_i \bar{\phi}_i = \sum_\mu a_\mu b_\mu. \tag{5-44}$$

5–4 Conditions for existence of invariants. If the irreducible representation $D(R)$ is *unitary*, the scalar product of vectors in the Hilbert space of the representation is invariant since

$$(D\mathbf{y}, D\mathbf{x}) = (\mathbf{y}, D^\dagger D\mathbf{x}) = (\mathbf{y}, \mathbf{x}). \tag{5-45}$$

Moreover, in this case the scalar product provides us with a Hermitian invariant. For finite groups [or for any group in which (3–101) holds], we can always make the representations unitary.

If the adjoint representation $\bar{D}$ and the complex conjugate representation D^* are equivalent, that is,

$$\tilde{D}^{-1}(R) \approx D^*(R) \tag{5-46}$$

or

$$D(R) \approx D^{\dagger-1}(R), \tag{5-46a}$$

there exists a matrix F such that

$$D^*(R) = \tilde{F}^{-1}\tilde{D}^{-1}(R)\tilde{F}$$

or

$$D^\dagger(R) = FD^{-1}(R)F^{-1}, \tag{5-47}$$

$$D^\dagger(R)FD(R) = F. \tag{5-48}$$

Then

$$(\mathbf{y}, F\mathbf{x}) \tag{5-49}$$

is invariant under all transformations of the group G:

$$(D\mathbf{y}, FD\mathbf{x}) = (\mathbf{y}, D^\dagger FD\mathbf{x}) = (\mathbf{y}, F\mathbf{x}). \tag{5-50}$$

Taking the adjoint of Eq. (5–48), we get

$$D^\dagger(R)F^\dagger D(R) = F^\dagger, \tag{5-48a}$$

so that

$$(\mathbf{y}, F^\dagger \mathbf{x}) \tag{5-49a}$$

is also an invariant. Combining (5–49) and (5–49a), we find two Hermitian invariants:

$$\left(\mathbf{y}, \frac{F + F^\dagger}{2}\, \mathbf{x}\right), \quad \left(\mathbf{y}, \frac{F - F^\dagger}{2i}\, \mathbf{x}\right). \tag{5-51}$$

These invariants are not both identically zero, since this would imply that $F = 0$.

Conversely, if there exists a nonsingular matrix F such that $(\mathbf{y}, F\mathbf{x})$ is invariant,

$$(\mathbf{y}, F\mathbf{x}) = (D\mathbf{y}, FD\mathbf{x}) = (\mathbf{y}, D^\dagger FD\mathbf{x}),$$

then

$$D^\dagger FD = F, \quad D^* = \widetilde{F}^{-1}\widetilde{D}^{-1}\widetilde{F},$$

and the adjoint representation $\overline{D}$ and the complex conjugate representation D^* are equivalent. Thus, the *necessary* and *sufficient* condition for the existence of a *Hermitian invariant* is the equivalence of $\overline{D}$ and D^*.

For an *irreducible* representation, there can be *no more than one* invariant of the form (5–49). If in addition to (5–49), $(\mathbf{y}, H\mathbf{x})$ is invariant, then

$$D^\dagger HD = H, \quad D^{-1}H^{-1}D^{\dagger-1} = H^{-1},$$

$$FH^{-1} = (D^\dagger FD)(D^{-1}H^{-1}D^{\dagger-1})$$

$$= D^\dagger FH^{-1}D^{\dagger-1},$$

$$(FH^{-1})D^\dagger = D^\dagger(FH^{-1}). \tag{5-52}$$

The matrix FH^{-1} commutes with all the matrices of the irreducible representation $D^\dagger$, and hence it follows from Schur's lemma that F is a multiple of H.

For an irreducible representation, the two invariants in (5–51) cannot be independent, and (except for a constant factor) the matrix F must be Hermitian.

We have shown that if $\overline{D}$ and D^* are not equivalent (i.e., $D \approx D^{\dagger-1}$), we cannot construct a Hermitian invariant. However, we shall show that this can be done by taking the direct sum of D and $D^{\dagger-1}$. Let the irreducible representation D act in the space x and the irreducible representation $D^{\dagger-1}$ act in the space y. (Both spaces clearly have the same dimension n.) Then,

$$x' = Dx, \quad y' = D^{\dagger-1}y, \tag{5-53}$$

and the quantity $y^\dagger x$ is invariant under all the transformations of the group:

$$y'^\dagger x' = (D^{\dagger-1}y)^\dagger(Dx) = y^\dagger D^{-1}Dx = y^\dagger x. \qquad (5\text{--}54)$$

Similarly $x^\dagger y$ is invariant, and hence we have found two Hermitian invariants:

$$A = \tfrac{1}{2}(y^\dagger x + x^\dagger y),$$

$$B = \frac{1}{2i}(y^\dagger x - x^\dagger y). \qquad (5\text{--}55)$$

We take the direct sum of the two spaces and form vectors

$$\psi = \begin{pmatrix} x \\ y \end{pmatrix} \qquad (5\text{--}56)$$

having $2n$ components. Under a transformation of the group G, ψ goes over into

$$\psi' = \begin{pmatrix} Dx \\ D^{\dagger-1}y \end{pmatrix}. \qquad (5\text{--}57)$$

The two Hermitian matrices

$$f_1 = \begin{bmatrix} 0 & E \\ E & 0 \end{bmatrix}, \qquad f_2 = \begin{bmatrix} 0 & -iE \\ iE & 0 \end{bmatrix}, \qquad (5\text{--}58)$$

in which E is the n-by-n unit matrix, provide us with two Hermitian invariants:

$$\phi^\dagger f_1 \psi = (\phi, f_1\psi) \qquad \text{and} \qquad \phi^\dagger f_2 \psi. \qquad (5\text{--}59)$$

This result does not contradict our earlier theorem since the representation space (5–56) is reducible.

Problem. Show that for the representation found in (5–57) the adjoint and complex conjugate representations are equivalent. Show that either f_1 or f_2 in Eq. (5–58) accomplishes the transformation.

5–5 Real representations. In this section, we wish to investigate the conditions under which the irreducible representations of a group G can be brought to real form, so that

$$D_{ij}(R) = D_{ij}^*(R). \qquad (5\text{--}60)$$

We shall assume that all representations are unitary (as is certainly the case for finite groups), so that

$$D^\dagger(R)D(R) = 1 \tag{5-61}$$

or

$$\tilde{D}^{-1}(R) \equiv \overline{D}(R) = D^*(R), \tag{5-61a}$$

i.e., the adjoint and complex conjugate representations coincide.

If the representation D is *real*, $D(R) = D^*(R)$, and $\chi(R)$ is real. Conversely, if $\chi(R)$ is real, then, since

$$\text{tr}\,[D^*(R)] = [\text{tr}\,D(R)]^* = [\chi(R)]^*, \tag{5-62}$$

the characters of D^* are identical with those of D, and the representation D is equivalent to its complex conjugate D^*. If $\chi(R)$ is complex, D and D^* are not equivalent, and this is the only case in which non-equivalence occurs.

The irreducible representations of the group G can therefore be divided into three types:

(1) D is real (i.e., D can be brought to real form).

(2) D is equivalent to D^*, but cannot be brought to real form.

(3) D is not equivalent to D^*.

If $\chi(R)$ is real, so that we are dealing with cases 1 and 2, D is equivalent to D^*, and consequently, we have from (5–61a)

$$D(R) \approx \tilde{D}^{-1}(R). \tag{5-63}$$

Taking traces, we find

$$\chi(R) = \chi(R^{-1}). \tag{5-64}$$

Moreover, Eq. (5–63) implies the existence of a nonsingular matrix S such that

$$SD(R)S^{-1} = \tilde{D}^{-1}(R) = D^*(R), \tag{5-65}$$

$$\tilde{D}(R)SD(R) = S. \tag{5-66}$$

Since the representation was assumed to be unitary, the matrix S is also unitary. Thus there exists a bilinear form

$$\tilde{x}Sy \tag{5-67}$$

which is invariant under all the transformations

$$x' = D(R)x, \tag{5-68}$$

since

$$\widetilde{D(R)x}SD(R)y = \tilde{x}\widetilde{D}(R)SD(R)y = \tilde{x}Sy.$$

Conversely, if such a bilinear form exists, then

$$SD(R) = \widetilde{D}^{-1}(R)S$$

for all R. Since the representation D is irreducible and S is not identically zero, it follows from Schur's lemma that S is nonsingular, so that Eq. (5-65) is satisfied and D is equivalent to D^*. We conclude that in case 3, for which $\chi(R)$ is complex, there is *no* bilinear form which is invariant under the transformations (5-68).

In cases 1 and 2, we have Eq. (5-66). The transposed equation is

$$\widetilde{D}(R)\widetilde{S}D(R) = \widetilde{S},$$

and the inverse is

$$D^{-1}(R)S^{-1}\widetilde{D}^{-1}(R) = S^{-1}.$$

Multiplying, we find

$$S^{-1}\widetilde{S} = D^{-1}(R)S^{-1}\widetilde{S}D(R)$$

or

$$D(R)S^{-1}\widetilde{S} = S^{-1}\widetilde{S}D(R). \tag{5-69}$$

Since $D(R)$ is irreducible, $S^{-1}\widetilde{S}$ must be a multiple of the unit matrix E:

$$S^{-1}\widetilde{S} = cE, \tag{5-70}$$

$$\widetilde{S} = cS. \tag{5-70a}$$

The transpose of (5-70a) is

$$S = c\widetilde{S}, \tag{5-70b}$$

so that

$$S = c\widetilde{S} = c(cS) = c^2S,$$

and

$$c^2 = 1, \quad c = \pm 1.$$

We conclude that if $\chi(R)$ is real [cases (1) and (2)], there exists an invariant bilinear form whose unitary matrix S is symmetric or skew-symmetric:

$$\widetilde{S} = S \quad \text{or} \quad \widetilde{S} = -S. \tag{5-71}$$

Since

$$\det \widetilde{S} = \det S, \qquad \det (-S) = (-1)^n \det S,$$

where n is the dimension of the representation, we see that the minus sign in Eq. (5–71) can occur *only* for representations of *even* dimension.

If the plus sign in Eq. (5–71) holds, we are dealing with case 1: The matrix S is unitary and symmetric. Then there exists a unitary symmetric matrix B such that

$$B^2 = S. \tag{5–72}$$

Since $B^\dagger B = 1$ and $\widetilde{B} = B$,

$$B^* = B^{-1}, \qquad B = B^{*-1}. \tag{5–73}$$

Substituting (5–72) in (5–65) and using (5–73), we obtain

$$B^2 D(R) B^{-2} = D^*(R),$$

$$BD(R)B^{-1} = B^{-1}D^*(R)B = B^*D^*(R)B^{*-1} = [BD(R)B^{-1}]^*. \tag{5–74}$$

The last equation shows that the representation $D(R)$ is transformed to *real* form by the matrix B. Since B and $D(R)$ were unitary, the matrices $D'(R) = BD(R)B^{-1}$ are also unitary. They form therefore a *real, orthogonal* representation of the group G:

$$\widetilde{D}'(R)D'(R) = E. \tag{5–75}$$

We shall omit the more complicated treatment of case 2. Instead we proceed to find a simple criterion for the three types of representations. This criterion is expressed in terms of sums over characters, so we shall assume that such sums are meaningful. (This is certainly the case for finite groups.)

The matrix

$$S = \sum_G \widetilde{D}(R)XD(R), \tag{5–76}$$

where X is an arbitrary matrix, satisfies Eq. (5–66) and provides us with an invariant bilinear form (5–67). In case 3, for which no invariant bilinear form can exist, the matrix S must be the null matrix for any choice of X:

$$\sum_G D_{\alpha\beta}(R)D_{\gamma\delta}(R) = 0 \qquad \text{for all} \quad \alpha, \beta, \gamma, \delta. \tag{5–77}$$

Setting $\beta = \gamma$ and summing over β, we have

$$0 = \sum_G \sum_\beta D_{\alpha\beta}(R)D_{\beta\delta}(R) = \sum_G D_{\alpha\delta}(R^2),$$

so that in case 3,

$$\sum_G D(R^2) = 0. \tag{5-78}$$

We can combine the results for all three cases in the single statement:

$$S = c\tilde{S}, \qquad c = \begin{cases} +1 & \text{for case} & 1 \\ -1 & \text{for case} & 2 \\ 0 & \text{for case} & 3 \end{cases}. \tag{5-79}$$

From Eqs. (5–76) and (5–79),

$$\sum_G \sum_{\beta\gamma} D_{\beta\alpha}(R)X_{\beta\gamma}D_{\gamma\delta}(R) = c \sum_G \sum_{\beta\gamma} D_{\beta\alpha}(R)X_{\gamma\beta}D_{\gamma\delta}(R),$$

so that

$$\sum_G D_{\beta\alpha}(R)D_{\gamma\delta}(R) = c \sum_G D_{\gamma\alpha}(R)D_{\beta\delta}(R). \tag{5-80}$$

In particular, we set $\alpha = \gamma$, $\beta = \delta$ and sum over α and β:

$$\sum_G \sum_{\alpha,\beta} D_{\beta\alpha}(R)D_{\alpha\beta}(R) = c \sum_G \sum_{\alpha,\beta} D_{\alpha\alpha}(R)D_{\beta\beta}(R) \tag{5-81}$$

or

$$\sum_G \chi(R^2) = c \sum_G \chi(R)\chi(R). \tag{5-82}$$

In cases 1 and 2, we have Eq. (5–64), $\chi(R) = \chi(R^{-1})$, and the orthogonality relation (3–146) gives

$$\sum_G \chi(R)\chi(R) = g, \tag{5-83}$$

while in case 3, D is not equivalent to $\overline{D}$, so that

$$\sum_G \chi(R)\chi(R) = 0. \tag{5-83a}$$

Combining these results with Eq. (5–82) yields

$$\sum_G \chi^{(\mu)}(R^2) = c^{(\mu)} \cdot g, \qquad \text{where} \quad c^{(\mu)} = \begin{cases} +1 & \text{for case} & 1 \\ -1 & \text{for case} & 2 \\ 0 & \text{for case} & 3 \end{cases}. \tag{5-84}$$

Wigner calls the representations of type 1, which can be transformed to real form, *integer representations*. For such representations, $c^{(\mu)} = 1$, while $c^{(\mu)} = -1$ for the *half-integer representations* of type 2, which cannot be transformed to real form but are equivalent to their complex conjugate. Finally, $c^{(\mu)} = 0$ for the representations of type 3, which are not equivalent to their complex conjugate.

Equation (5–84) provides us with a simple criterion for the three types of representations. In particular, a representation can be brought to real form only if the sum of the characters of the squares of the group elements is equal to $+g$.

Let $\zeta(S)$ be the number of solutions of the equation

$$R^2 = S. \tag{5–85}$$

Combining those terms in Eq. (5–84) for which $R^2 = S$, we have

$$\sum_G \zeta(S)\chi^{(\mu)}(S) = c^{(\mu)} \cdot g. \tag{5–86}$$

Using the completeness relation (3–177) and noting that $\zeta(S^{-1}) = \zeta(S)$, we solve (5–86) to give

$$\zeta(S) = \sum_\mu c^{(\mu)}\chi^{(\mu)}(S). \tag{5–87}$$

Equation (5–87) states a remarkable theorem:

THEOREM. The number of solutions R of the equation $R^2 = S$ (i.e., the number of square roots of the elements S) is given by (5–87) in which $c^{(\mu)} = 0$ if $\chi^{(\mu)}(S)$ is complex. If $\chi^{(\mu)}(S)$ is real, $c^{(\mu)} = +1$ or -1, depending on whether the representation $D^{(\mu)}$ is or is not equivalent to a real representation.

In particular, setting S equal to the identity E, we find that the number of solutions of the equation $R^2 = E$ is

$$\zeta(E) = \sum_\mu c^{(\mu)}n_\mu; \tag{5–88}$$

i.e., the number of solutions is obtained by taking the sum of the degrees of all irreducible representations of type 1 and subtracting the sum of the degrees of all those of type 2.

The matrix

$$\sum_G D(R^2)$$

commutes with all the matrices of the irreducible representation $D(R)$ since

$$D^{-1}(A) \sum_G D(R^2)D(A) = \sum_G D^{-1}(A)D(R^2)D(A)$$

$$= \sum_G [D^{-1}(A)D(R)D(A)][D^{-1}(A)D(R)D(A)]$$

$$= \sum_G D(R)D(R)$$

$$= \sum_G D(R^2). \tag{5-89}$$

[As $D(R)$ runs through G, so does $D^{-1}(A)D(R)D(A)$.]   From Schur's lemma, $\sum_G D(R^2)$ must be a multiple of the unit matrix:

$$\sum_G D(R^2) = \lambda E. \tag{5-90}$$

Taking traces and using (5–84), we obtain

$$c^{(\mu)}g = \lambda n_\mu,$$

where n_μ is the dimension of the representation, so that (5–90) becomes

$$\sum_G D^{(\mu)}(R^2) = \frac{c^{(\mu)}g}{n_\mu} E. \tag{5-91}$$

We multiply Eq. (5–91) by $D^{(\mu)}(A)$:

$$\sum_G D^{(\mu)}(AR^2) = \frac{c^{(\mu)}g}{n_\mu} D^{(\mu)}(A), \tag{5-92}$$

and take the trace:

$$\sum_G \chi^{(\mu)}(AR^2) = \frac{c^{(\mu)}g}{n_\mu} \chi^{(\mu)}(A). \tag{5-93}$$

Let $A = S^2$ and sum over S:

$$\sum_R \sum_S \chi^{(\mu)}(S^2 R^2) = \frac{c^{(\mu)}g}{n_\mu} \sum_S \chi^{(\mu)}(S^2).$$

Using Eq. (5–84) once again, we get

$$\sum_R \sum_S \chi^{(\mu)}(S^2 R^2) = \frac{(c^{(\mu)}g)^2}{n_\mu}. \tag{5-94}$$

If $\eta(T)$ is the number of solutions of the equation

$$S^2 R^2 = T, \tag{5-95}$$

then

$$\sum_T \eta(T) \chi^{(\mu)}(T) = \frac{(c^{(\mu)}g)^2}{n_\mu}. \tag{5-96}$$

Again using the completeness relation, we solve to obtain

$$\eta(T) = \sum_\mu \frac{(c^{(\mu)}g)^2}{g n_\mu} \chi^{(\mu)}(T). \tag{5-97}$$

In particular, for $T = E$,

$$\eta(E) = g \sum_\mu [c^{(\mu)}]^2; \tag{5-98}$$

i.e., the number of solutions of the equation $S^2 R^2 = E$ (or $S^2 = R^2$) is equal to g times the number of irreducible representations with real characters.

This process can be repeated. Multiplying equations of type (5–91) for the s elements $R_1, R_2, \ldots, R_s$, we find

$$\sum_{R_1, \ldots, R_s} D^{(\mu)}(R_1^2 R_2^2 \cdots R_s^2) = \left(\frac{c^{(\mu)}g}{n_\mu}\right)^s E. \tag{5-99}$$

Taking the trace yields

$$\sum_{R_1, \ldots, R_s} \chi^{(\mu)}(R_1^2 R_2^2 \cdots R_s^2) = n_\mu \left(\frac{c^{(\mu)}g}{n_\mu}\right)^s. \tag{5-100}$$

By the same procedure as before, we now find that the number of solutions of the equation

$$R_1^2 R_2^2 \cdots R_s^2 = E \tag{5-101}$$

is

$$g^{s-1} \sum_\mu \frac{[c^{(\mu)}]^s}{[n_\mu]^{s-2}}.$$

Problems. (1) Construct the character table for the quaternion group (see the problem on p. 28). Show that the two-dimensional representation is of type 2. Check the various theorems of this section for the quaternion group.

(2) Assign the representations of the crystal point groups to types 1, 2, 3.

If we take the square of Eq. (5–87) for $\zeta(S)$ and $\zeta(S^{-1}) = \zeta(S)$, and sum over S, we get

$$\sum_G [\zeta(S)]^2 = \sum_{\mu,\nu} c^{(\mu)} c^{(\nu)} \sum_G \chi^{(\mu)}(S) \chi^{(\nu)}(S^{-1})$$

$$= \sum_{\mu,\nu} c^{(\mu)} c^{(\nu)} g\, \delta_{\mu\nu} \qquad (5\text{–}102)$$

$$= g \sum_\mu [c^{(\mu)}]^2,$$

which is the same result as in Eq. (5–98). Equation (5–102) states that the sum of the squares of the numbers of square roots of all elements of the group G is equal to g times the number of real primitive characters.

This last result can be changed to a more useful form. We know that the inverses R^{-1} of all the elements R in the class K_i form a class $K_{i'}$. The completeness relation (3–177) for $i = j$ was

$$\sum_\mu g_i \chi_i^{(\mu)} \chi_i^{(\mu)} = g\, \delta_{ii'}. \qquad (5\text{–}103)$$

Summing over i, we have

$$\sum_\mu \left(\sum_i g_i \chi_i^{(\mu)} \chi_i^{(\mu)} \right) = g \sum_i \delta_{ii'}. \qquad (5\text{–}104)$$

According to Eqs. (5–83) and (5–83a), the quantity in parentheses on the left side is g if the character $\chi^{(\mu)}$ is real, and zero if $\chi^{(\mu)}$ is complex; thus the left-hand side is equal to g times the number of real characters. The sum on the right is the number of classes i which coincide with i'. Such classes are called *ambivalent* classes. Thus Eq. (5–104) states that the number of real characters is equal to the number of ambivalent classes. Combining this with Eq. (5–102), we obtain the following theorem:

THEOREM. The sum of the squares of the numbers of square roots of all elements of a group is equal to the order of the group multiplied by the number of ambivalent classes.

We also note that if all the classes of the group G are ambivalent, all the characters must be real. This is the case for the symmetric group S_n, since a permutation and its inverse have the same cycle structure. Thus all the characters of the symmetric group are real.

Problem. Find the number of ambivalent classes in each of the crystal point groups, and check this against the number of real characters in the character table. Enumerate those point groups for which all classes are ambivalent.

5-6 The reduction of Kronecker products.   The Clebsch-Gordan series. In Section 5-1 we defined the Kronecker product of two representations. These products play a fundamental role in the theory of coupled systems and in the derivation of selection rules. In this section, we obtain the mathematical basis for such applications.

The product of two irreducible representations of the group G will, in general, be reducible. To analyze the product representation, we use Eq. (5-15),

$$\chi^{(\mu \times \nu)}(R) = \chi^{(\mu)}(R) \cdot \chi^{(\nu)}(R), \tag{5-15}$$

and Eq. (5-42),

$$a_\sigma = \frac{1}{g} \sum_G \chi^{(\mu \times \nu)}(R)\bar{\chi}^{(\sigma)}(R), \tag{5-42}$$

which gives the number of times, a_σ, that the irreducible representation $D^{(\sigma)}$ is contained in the representation $D^{(\mu \times \nu)}$. Combining these two equations, we find

$$a_\sigma = \frac{1}{g} \sum_G \chi^{(\mu)}(R)\chi^{(\nu)}(R)\bar{\chi}^{(\sigma)}(R). \tag{5-105}$$

This equation can be put into various useful special forms.

First we take the Kronecker product of $D^{(\mu)}$ and $\overline{D}^{(\nu)}$ and ask for the number of times the identity representation is contained in $D^{(\mu)} \times \overline{D}^{(\nu)}$. Since the characters of the identity representation $D^{(1)}$ are all equal to unity, this number is

$$a_1 = \frac{1}{g} \sum_G \chi^{(\mu)}(R)\bar{\chi}^{(\nu)}(R) = \delta_{\mu\nu}, \tag{5-106}$$

where the last equality follows from the orthogonality theorem. Thus the product of $D^{(\mu)}$ and $\overline{D}^{(\nu)}$ contains the identity representation (once) if and only if $\mu = \nu$. This result can be restated in terms of basis functions. The (single) basis function of the identity representation is invariant under all transformations of the group G. Equation (5-106) states that we can construct an invariant linear combination of product wave functions if and only if the representations $D^{(\mu)}$ and $\overline{D}^{(\nu)}$ are adjoint ($\mu = \nu$). If the representations are unitary, $\overline{D} = D^*$. In this case we can say that the product contains the identity if and only if the factor representations are complex conjugates.

Again, if we replace $D^{(\sigma)}$ by $\overline{D}^{(\sigma)}$ in Eq. (5-105), we obtain

$$a_{\bar{\sigma}} = \frac{1}{g} \sum_G \chi^{(\mu)}(R)\chi^{(\nu)}(R)\chi^{(\sigma)}(R), \tag{5-107}$$

which gives the number of times that $\overline{D}^{(\sigma)}$ is contained in $D^{(\mu)} \times D^{(\nu)}$.

The numbers a_σ in Eq. (5–105) are the coefficients in the expansion of $D^{(\mu)} \times D^{(\nu)}$ in irreducible representations:

$$D^{(\mu)} \times D^{(\nu)} = \sum_\sigma a_\sigma D^{(\sigma)}. \tag{5–108}$$

The expansion (5–108) is called the *Clebsch-Gordan series.*

If the representations are unitary, $\bar{\chi} = \chi^*$, and Eq. (5–105) becomes

$$a_\sigma = \frac{1}{g} \sum_G \chi^{(\mu)}(R)\chi^{(\nu)}(R)\chi^{*(\sigma)}(R). \tag{5–105a}$$

A more suggestive notation than that in Eq. (5–105) or (5–108) is

$$D^{(\mu)} \times D^{(\nu)} = \sum_\sigma (\mu\nu\sigma)D^{(\sigma)}; \tag{5–108a}$$

i.e., $(\mu\nu\sigma)$ is the number of times that $D^{(\sigma)}$ occurs in the Kronecker product of $D^{(\mu)}$ and $D^{(\nu)}$. Clearly, $(\mu\nu\sigma) = (\nu\mu\sigma)$.

Problems. (1) Prove that the numbers of times that

$\overline{D}^{(\sigma)}$ is contained in $D^{(\mu)} \times D^{(\nu)}$,

$\overline{D}^{(\mu)}$ is contained in $D^{(\nu)} \times D^{(\sigma)}$,

$\overline{D}^{(\nu)}$ is contained in $D^{(\sigma)} \times D^{(\mu)}$

are all equal. Show that if all the characters of the group G are real, the symbol $(\mu\nu\sigma)$ in (5–108a) is completely symmetric.

(2) Find the conditions under which the Kronecker product $D^{(\mu)} \times D^{(\nu)}$ is irreducible.

(3) Prove that the Kronecker product of two irreducible representations of dimension n_1, n_2 ($n_1 \geq n_2$) cannot contain representations of dimension less than n_1/n_2.

(4) Find the coefficients in the Clebsch-Gordan series for the product of the two-dimensional representation of the quaternion group with itself.

(5) If the characters of $D^{(\mu)}$ and $D^{(\nu)}$ are real and the character of $D^{(\sigma)}$ is complex, show that $D^{(\mu)} \times D^{(\nu)}$ must contain $D^{(\sigma)}$ and $D^{*(\sigma)}$ the *same* number of times.

5–7 Clebsch-Gordan coefficients. In the preceding section we discussed the problem of finding those irreducible representations which are contained in the Kronecker product of two irreducible representations. In the next chapter we shall show how the Clebsch-Gordan series enables us to determine selection rules.

Of even more importance for physical applications is the problem of finding the basis functions for the representations which are contained in the Kronecker product. We are given n_μ basis functions $\psi_j^{(\mu)}$ ($j = 1, \ldots, n_\mu$) for the irreducible representation $D^{(\mu)}(G)$, and n_ν basis functions $\phi_l^{(\nu)}$ ($l = 1, \ldots, n_\nu$) for the irreducible representation $D^{(\nu)}(G)$. We want to find n_λ functions $\Psi_s^{(\lambda)}$ ($s = 1, \ldots, n_\lambda$) which are linear combinations of products $\psi_j^{(\mu)}\phi_l^{(\nu)}$, and which form the basis for the representation $D^{(\lambda)}(G)$. The functions $\Psi_s^{(\lambda)}$ are a set of *partners* belonging to $D^{(\lambda)}(G)$. From the last section we know that such a set exists *only* if $D^{(\lambda)}$ is contained in $D^{(\mu)} \times D^{(\nu)}$, that is, only if $(\mu\nu\lambda) \neq 0$. On the other hand, if $(\mu\nu\lambda) > 1$, we can form several independent sets of partners $\Psi_s^{(\lambda)}$; there will in fact be $(\mu\nu\lambda)$ "correct linear combinations" of product functions. To distinguish these different sets of partners, we use the notation $\Psi_s^{(\lambda\tau_\lambda)}$, $s = 1, \ldots, n_\lambda$, $\tau_\lambda = 1, \ldots, (\mu\nu\lambda)$. The functions $\Psi_s^{(\lambda\tau_\lambda)}$ will be linear combinations of products $\psi_j^{(\mu)}\phi_l^{(\nu)}$:

$$\Psi_s^{(\lambda\tau_\lambda)} = \psi_j^{(\mu)}\phi_l^{(\nu)}(\mu j, \nu l|\lambda\tau_\lambda s). \tag{5–109}$$

(We shall use the convention of summing over repeated Latin letters; sums over Greek letters will always be shown explicitly.) The coefficients $(\mu j, \nu l|\lambda\tau_\lambda s)$ in Eq. (5–109) are called *Clebsch-Gordan coefficients*. (Other names are *Wigner coefficients* or *vector-addition coefficients*. Since these names are used for particular types of groups, we prefer the generic name.) The total number of functions $\Psi_s^{(\lambda\tau_\lambda)}$ must equal the total number of product functions $\psi_j^{(\mu)}\phi_l^{(\nu)}$:

$$\sum_\lambda (\mu\nu\lambda)n_\lambda = n_\mu n_\nu, \tag{5–110}$$

so the CG-coefficients $(\mu j, \nu l|\lambda\tau_\lambda s)$ form a matrix of degree $n_\mu n_\nu$. Equation (5–109) is merely a description of the relation between the two different bases $\Psi_s^{(\lambda\tau_\lambda)}$ and $\psi_j^{(\mu)}\phi_l^{(\nu)}$ for the representation space of the Kronecker product. Another description of this relation would be to express the products $\psi_j^{(\mu)}\phi_l^{(\nu)}$ as linear combinations of the $\Psi_s^{(\lambda\tau_\lambda)}$:

$$\psi_j^{(\mu)}\phi_l^{(\nu)} = \sum_{\lambda\tau_\lambda} \Psi_s^{(\lambda\tau_\lambda)}(\lambda\tau_\lambda s|\mu j, \nu l), \qquad \begin{matrix} j = 1, \ldots, n_\mu, \\ l = 1, \ldots, n_\nu, \end{matrix} \tag{5–109a}$$

which is the inverse of Eq. (5–109), so that

$$(\lambda'\tau_{\lambda'}s'|\mu j, \nu l)(\mu j, \nu l|\lambda\tau_\lambda s) = \delta_{\lambda\lambda'}\,\delta_{\tau_\lambda\tau_{\lambda'}}\,\delta_{ss'}, \tag{5–111}$$

$$\sum_{\lambda\tau_\lambda} (\mu j', \nu l'|\lambda\tau_\lambda s)(\lambda\tau_\lambda s|\mu j, \nu l) = \delta_{jj'}\,\delta_{ll'}. \tag{5–111a}$$

Problem. Using Eqs. (5–109), (5–109a), and the linear independence of the basis functions, derive Eqs. (5–111) and (5–111a).

To simplify our problem we shall assume that we deal throughout with *unitary* representations. In this case, Eq. (5–109) is the shift from one orthonormal system $\psi_j^{(\mu)}\phi_l^{(\nu)}$ to another, $\Psi_s^{(\lambda\tau_\lambda)}$, by means of the *unitary* matrix $(\mu j, \nu l | \lambda\tau_\lambda s)$, so that

$$(\lambda\tau_\lambda s | \mu j, \nu l) = (\mu j, \nu l | \lambda\tau_\lambda s)^*, \qquad (5\text{–}112)$$

$$(\mu j, \nu l | \lambda'\tau'_{\lambda'}s')^*(\mu j, \nu l | \lambda\tau_\lambda s) = \delta_{\lambda\lambda'}\,\delta_{\tau_\lambda\tau'_{\lambda'}}\,\delta_{ss'}, \qquad (5\text{–}111b)$$

$$\sum_{\lambda\tau_\lambda}(\mu j', \nu l' | \lambda\tau_\lambda s)(\mu j, \nu l | \lambda\tau_\lambda s)^* = \delta_{jj'}\,\delta_{ll'}. \qquad (5\text{–}111c)$$

We now act on Eq. (5–109) with the operator O_R corresponding to the element R of the group G:

$$O_R\Psi_s^{(\lambda\tau_\lambda)} = \Psi_{s'}^{(\lambda\tau_\lambda)}D_{s's}^{(\lambda\tau_\lambda)}(R),$$

where the matrices $D^{(\lambda\tau_\lambda)}(R)$ are to be selected according to some fixed scheme; in particular, we can (and will) choose them to be the same matrices for all τ_λ, so that $D_{s's}^{(\lambda\tau_\lambda)}(R) = D_{s's}^{(\lambda)}(R)$. We now use Eqs. (5–7) and (5–7a):

$$O_R\Psi_s^{(\lambda\tau_\lambda)} = \Psi_{s'}^{(\lambda\tau_\lambda)}D_{s's}^{(\lambda\tau_\lambda)}(R) = \psi_i^{(\mu)}\phi_k^{(\nu)}(\mu i, \nu k | \lambda\tau_\lambda s')D_{s's}^{(\lambda\tau_\lambda)}(R); \quad (5\text{–}113)$$

$$\begin{aligned}O_R\Psi_s^{(\lambda\tau_\lambda)} &= O_R[\psi_j^{(\mu)}\phi_l^{(\nu)}](\mu j, \nu l | \lambda\tau_\lambda s)\\ &= O_R\psi_j^{(\mu)}O_R\phi_l^{(\nu)}(\mu j, \nu l | \lambda\tau_\lambda s)\\ &= \psi_i^{(\mu)}\phi_k^{(\nu)}D_{ij}^{(\mu)}(R)D_{kl}^{(\nu)}(R)(\mu j, \nu l | \lambda\tau_\lambda s). \qquad (5\text{–}113a)\end{aligned}$$

Since the $\psi_i^{(\mu)}\phi_k^{(\nu)}$ are a set of linearly independent functions,

$$D_{ij}^{(\mu)}(R)D_{kl}^{(\nu)}(R)(\mu j, \nu l | \lambda\tau_\lambda s) = (\mu i, \nu k | \lambda\tau_\lambda s')D_{s's}^{(\lambda\tau_\lambda)}(R). \qquad (5\text{–}114)$$

Equation (5–114) can be rewritten in a variety of useful forms. The CG-coefficients can be moved to the left side by applying Eq. (5–111), giving

$$(\lambda'\tau'_{\lambda'}t | \mu i, \nu k)D_{ij}^{(\mu)}(R)D_{kl}^{(\nu)}(R)(\mu j, \nu l | \lambda\tau_\lambda s) = D_{s's}^{(\lambda\tau_\lambda)}(R)\,\delta_{\lambda\lambda'}\,\delta_{\tau_\lambda\tau'_{\lambda'}}\,\delta_{ts'}.$$

$$(5\text{–}115)$$

Equation (5–115) states that the matrices $D^{(\mu)}(R) \times D^{(\nu)}(R)$ of the

Kronecker product are brought to reduced form by transforming with the matrix $(\mu j, \nu l | \lambda \tau_\lambda s)$ of the CG-coefficients. The first factor on the left can also be written as $(\mu i, \nu k | \lambda' \tau'_\lambda, t)^*$. We can also shift the CG-coefficients to the right side of Eq. (5–114) by using Eq. (5–111a) to give

$$D_{ij}^{(\mu)}(R) D_{kl}^{(\nu)}(R) = \sum_{\lambda \tau_\lambda} (\mu i, \nu k | \lambda \tau_\lambda s') D_{s's}^{(\lambda \tau_\lambda)}(R)(\lambda \tau_\lambda s | \mu j, \nu l), \quad (5\text{–}116)$$

which shows the inverse transformation from the reduced form to the Kronecker product.

We could also shift the matrix representative from the right side of (5–114) to the left by multiplying by $D_{is}^{*(\lambda \tau_\lambda)}(R)$. Similarly, in (5–116) we could use the orthogonality relations to move all matrices to one side of the equation. However, we see that this yields equations which are not symmetric in μ, ν, λ.

5–8 Simply reducible groups. We can obtain addition coefficients which have a higher degree of symmetry if we choose particular types of groups. We shall assume first that the characters of all the irreducible representations of the group G are *real*. This implies that each representation is equivalent to its complex conjugate, and no representations of type 3 (cf. Section 5–5) occur. All irreducible representations of G are integral or half-integral. A very large class of groups satisfies this condition: the rotation group, the quaternion group, most of the crystal point groups, and all the symmetric groups S_n. In fact, we shall see later that all irreducible representations of the symmetric group S_n can be chosen to be real, so that we obtain only integer representations.

A second difficulty in the general case arises because $(\mu \nu \lambda)$ can be greater than unity. This fact makes the definition of the CG-coefficients ambiguous, since we can take arbitrary linear combinations:

$$\sum_{\tau_\lambda} c_{\lambda \tau_\lambda} \Psi_s^{(\lambda \tau_\lambda)} \qquad (5\text{–}117)$$

and still keep the representation in reduced form. If each representation $D^{(\lambda)}$ appeared at most once in the Kronecker product, there would be less ambiguity in the CG-coefficients. From Eq. (5–117), we see that in this case we could replace only $\Psi_s^{(\lambda)}$ by $c_\lambda \Psi_s^{(\lambda)}$. Since we are dealing with unitary representations, this would mean that $|c_\lambda| = 1$; thus the only arbitrariness in the CG-coefficients would be a phase factor which is the same for all coefficients with the same μ, ν, λ.

When we discuss the symmetric group, we shall show that one can proceed despite this second difficulty. In this chapter we shall make both of the restrictions discussed above.

We say that a group is *simply reducible* (SR *group*) if

(a) Every element of G is equivalent to its reciprocal;
(b) the Kronecker product of two irreducible representations of G contains each irreducible representation no more than once.

Condition (a) means that all classes of G are ambivalent, that all characters are real, that all irreducible representations of G are integral or half-integral ($c^{(\mu)} = +1$ or -1).

The symmetric groups S_3 and S_4, the quaternion group, the two-dimensional unimodular unitary group, and the three-dimensional rotation group are SR groups.

Condition (b) is important for physical applications. It implies that the "correct linear combinations" of products of basis functions are determined to within a phase factor, and that the solution of the physical problem is uniquely determined from symmetry arguments.

We first prove the following lemma:

LEMMA. The Kronecker product of two integer or of two half-integer representations of an SR group contains *only* integer representations; the Kronecker product of an integer and a half-integer representation contains *only* half-integer representations.

The unitary matrix S_μ, which transforms $D^{(\mu)}$ to its complex conjugate,

$$S_\mu D^{(\mu)} S_\mu^{-1} = D^{*(\mu)}, \tag{5-65}$$

satisfies the equation

$$\widetilde{S}_\mu = c^{(\mu)} S_\mu, \tag{5-79}$$

where $c^{(\mu)} = +1$ if $D^{(\mu)}$ is integral, and $c^{(\mu)} = -1$ if $D^{(\mu)}$ is half-integral [condition (a)]. Similarly,

$$S_\nu D^{(\nu)} S_\nu^{-1} = D^{*(\nu)}, \qquad \widetilde{S}_\nu = c^{(\nu)} S_\nu.$$

Combining these equations and using Eq. (5-6b), we have

$$(S_\mu \times S_\nu)(D^{(\mu)} \times D^{(\nu)})(S_\mu \times S_\nu)^{-1} = (D^{*(\mu)} \times D^{*(\nu)}) = (D^{(\mu)} \times D^{(\nu)})^*; \tag{5-118}$$

$$\widetilde{(S_\mu \times S_\nu)} = c^{(\mu)} c^{(\nu)} (S_\mu \times S_\nu). \tag{5-119}$$

Thus the unitary matrix $S = S_\mu \times S_\nu$ which transforms the Kronecker product $M = D^{(\mu)} \times D^{(\nu)}$ to its complex conjugate is symmetric if $D^{(\mu)}$ and $D^{(\nu)}$ are both integral or both half-integral, and is antisymmetric if one of the representations is integral and the other half-integral:

$$SMS^{-1} = M^*, \qquad \tilde{S} = \pm S. \tag{5-120}$$

We denote the unitary matrix of the CG-coefficients by U. Then

$$UMU^{-1} = M_r, \qquad U^*M^*U^{-1*} = M_r^*, \tag{5-121}$$

where M_r is the reduced form of the Kronecker product M. From condition (b), M_r has the form

$$M_r = \begin{bmatrix} D_1 & & & \\ & D_2 & & \\ & & \ddots & \\ & & & D_s \end{bmatrix}, \tag{5-122}$$

where $D_1, D_2, \ldots, D_s$ are *nonequivalent* irreducible representations.

From (5–121) we have

$$M = U^{-1}M_rU, \tag{5-123}$$

and combining with Eq. (5–120), we obtain

$$SU^{-1}M_rUS^{-1} = M^*,$$
$$U^*SU^{-1}M_rUS^{-1}U^{-1*} = M_r^* \tag{5-124}$$

or

$$S_rM_rS_r^{-1} = M_r^*, \qquad S_rM_r = M_r^*S_r, \tag{5-125}$$

where

$$S_r = U^*SU^{-1}. \tag{5-126}$$

Using Eqs. (5–120), (5–126), and the unitarity of U yields

$$\tilde{S}_r = \tilde{U}^{-1}\tilde{S}\tilde{U}^* = U^*\tilde{S}U^{-1} = \pm U^*SU^{-1} = \pm S_r. \tag{5-127}$$

We divide the matrix S_r in the same way as M_r (Eq. 5–122) and denote the submatrices by S_{ij} $(i, j = 1, \ldots, s)$. Equation (5–127) states that

$$S_{ji} = \pm \tilde{S}_{ij}, \tag{5-128}$$

and Eq. (5–125) gives

$$S_{ij}D_j = D_i^*S_{ij}. \tag{5-129}$$

For $i \neq j$, D_i^* and D_j are not equivalent, so from Eq. (5–129) and Schur's lemmas $S_{ij} = 0$ for $i \neq j$. Thus the matrix S_r has the same step form (5–122) as M_r. From Eq. (5–129) for $i = j$,

$$S_{ii}D_iS_{ii}^{-1} = D_i^*, \tag{5-130}$$

and from (5-128),

$$S_{ii} = \pm \widetilde{S}_{ii}. \qquad (5\text{-}131)$$

Thus each of the submatrices D_i of M_r is transformed into its complex conjugate by a matrix S_{ii} which is symmetric if the representations are both integral or both half-integral, and which is antisymmetric if one representation is integral and the other half-integral.

The coefficients $(\mu\nu\lambda)$ in the Clebsch-Gordan series are necessarily ≥ 0. For any group, the numbers $c^{(\mu)}$ are ± 1 or 0. Thus,

$$(\mu\nu\lambda)^2 \geq c^{(\mu)}c^{(\nu)}c^{(\lambda)}(\mu\nu\lambda). \qquad (5\text{-}132)$$

The equality sign can hold in (5-132) only if $(\mu\nu\lambda) = 0$, or $(\mu\nu\lambda) = 1$ and $c^{(\mu)}c^{(\nu)}c^{(\lambda)} = 1$. Summing over all irreducible representations, we obtain

$$\sum_{\mu\nu\lambda} (\mu\nu\lambda)^2 \geq \sum_{\mu\nu\lambda} c^{(\mu)}c^{(\nu)}c^{(\lambda)}(\mu\nu\lambda), \qquad (5\text{-}133)$$

where the equality sign holds if, for all μ, ν, λ, either $(\mu\nu\lambda) = 0$, or $(\mu\nu\lambda) = 1$ and $c^{(\mu)}c^{(\nu)}c^{(\lambda)} = 1$. But from our lemma, this is precisely the case for SR groups. So the equality sign in Eq. (5-133) is a necessary and sufficient condition for the group to be an SR group.

For any group G, the right side of Eq. (5-133) can be rewritten by means of Eqs. (5-105a) and (5-87) as

$$\sum_{\mu\nu\lambda} c^{(\mu)}c^{(\nu)}c^{(\lambda)}(\mu\nu\lambda) = \frac{1}{g} \sum_{\mu\nu\lambda} \sum_{R} c^{(\mu)}c^{(\nu)}c^{(\lambda)}\chi^{(\mu)}(R)\chi^{(\nu)}(R)\chi^{*(\lambda)}(R)$$

$$= \frac{1}{g} \sum_{R} [\zeta(R)]^3. \qquad (5\text{-}134)$$

Again using Eq. (5-105a), we find that the left side of (5-133) becomes

$$\sum_{\mu\nu\lambda} (\mu\nu\lambda)^2 = \frac{1}{g^2} \sum_{\mu\nu\lambda} \sum_{RR'} \chi^{(\mu)}(R)\chi^{*(\mu)}(R')\chi^{(\nu)}(R)\chi^{*(\nu)}(R')\chi^{*(\lambda)}(R)\chi^{(\lambda)}(R').$$

The sums over μ, ν, λ can be evaluated separately by means of the completeness relation (5-41), to give

$$\sum_{\mu\nu\lambda} (\mu\nu\lambda)^2 = \frac{1}{g^2} \sum_{R} g_R \left(\frac{g}{g_R}\right)^3 = g \sum_{R} \frac{1}{(g_R)^2}, \qquad (5\text{-}135)$$

where g_R is the number of elements in the class of R. Substituting (5-135) and (5–134) in Eq. (5–133), we find

$$\sum_R \left[\zeta(R)\right]^3 \le \sum_R \left(\frac{g}{g_R}\right)^2 ; \qquad (5\text{–}136)$$

thus we have shown the following:

The inequality (5–136) holds for every finite group. The equality sign holds if and only if the group is simply reducible.

According to our lemma, for SR groups the Kronecker product $D^{(\mu)} \times D^{(\mu)}$ of a representation with *itself* contains only *integer* representations. This product can always be decomposed into a symmetrized and an antisymmetrized part:

$$D^{(\mu)} \times D^{(\mu)} = [D^{(\mu)} \times D^{(\mu)}] + \{D^{(\mu)} \times D^{(\mu)}\}. \qquad (5\text{–}28)$$

If $D^{(\mu)}$ is an integer representation, the irreducible representations contained in $[D^{(\mu)} \times D^{(\mu)}]$ are called *even representations*, and the irreducible representations contained in $\{D^{(\mu)} \times D^{(\mu)}\}$ are called *odd representations*. Conversely, if $D^{(\mu)}$ is half-integral, the representations contained in the *symmetrized* square are *odd*, and those in the *antisymmetrized* square are *even*. We now prove the following theorem.

THEOREM. No representation of an SR group can be both even and odd. To prove the theorem, we use Eqs. (5–32) and (5–33):

$$[\chi^{(\mu)} \times \chi^{(\mu)}(R)] = \tfrac{1}{2}[(\chi^{(\mu)}(R))^2 + \chi^{(\mu)}(R^2)], \qquad (5\text{–}32)$$

$$\{\chi^{(\mu)} \times \chi^{(\mu)}(R)\} = \tfrac{1}{2}[(\chi^{(\mu)}(R))^2 - \chi^{(\mu)}(R^2)]. \qquad (5\text{–}33)$$

The condition for two representations to have no irreducible parts in common requires that

$$\sum_R \chi^{(\alpha)}(R)\bar{\chi}^{(\beta)}(R) = 0, \qquad (5\text{–}44)$$

where, moreover, the sum can never be less than zero. Using Eqs. (5–32), (5–33), (5–44), and the definitions of even and odd representations, we find that our theorem is equivalent to the statement that

$$\sum_R [(\chi^{(\mu)}(R))^2 + c^{(\mu)}\chi^{(\mu)}(R^2)][(\chi^{(\nu)}(R))^2 - c^{(\nu)}\chi^{(\nu)}(R^2)] = 0 \quad (5\text{–}137)$$

for all μ, ν, or

$$\sum_{\mu\nu} \sum_R [(\chi^{(\mu)}(R))^2 + c^{(\mu)}\chi^{(\mu)}(R^2)][(\chi^{(\nu)}(R))^2 - c^{(\nu)}\chi^{(\nu)}(R^2)] = 0.$$
$$(5\text{–}137\text{a})$$

We now use Eq. (5–87) with $S = R^2$ for the second terms in the factors in (5–137a),

$$\zeta(R^2) = \sum_{\mu} c^{(\mu)}\chi^{(\mu)}(R^2),$$

and apply the completeness relation (5–41) to the first terms,

$$\sum_{\mu} \chi^{(\mu)}(R)\chi^{(\mu)}(R) = \frac{g}{g_R},$$

so that the left side of (5–137a) becomes

$$\sum_{R} \left(\frac{g}{g_R} + \zeta(R^2)\right)\left(\frac{g}{g_R} - \zeta(R^2)\right) = \sum_{R}\left[\left(\frac{g}{g_R}\right)^2 - \{\zeta(R^2)\}^2\right]. \quad (5\text{–}138)$$

But from the definition of $\zeta(S)$ it follows that

$$\sum_{S} [\zeta(S)]^3 = \sum_{S} [\zeta(S)]^2\zeta(S)$$

$$= \sum_{S,R} [\zeta(S)]^2\, \delta_{S,R^2} = \sum_{R} [\zeta(R^2)]^2. \quad (5\text{–}139)$$

Our previous theorem [Eq. (5–136) with the equality sign, since the group is SR] then shows that (5–138) vanishes, and proves the theorem.

We should emphasize that the theorem does *not* state that every integer representation is either even or odd. It may happen that a particular integer representation is not contained in the square of any representation. Such a representation is neither even nor odd.

Problem. Consider the group G consisting of the eight elements 1, -1, x, $-x$, y, $-y$, z, $-z$, with the multiplication table:

$$x^2 = y^2 = 1, \quad z^2 = -1, \quad xy = -yx = z,$$

$$xz = -zx = y, \quad zy = -yz = x.$$

What is the relation between the group G and the quaternion group? Construct the character table of the group G. Show that the two-dimensional representation is integral, but is not contained in the square of any representation.

5–9 Three-j symbols. In this section we shall use the notation introduced by Wigner, which is based on that for the three-dimensional rotation group. The irreducible representations will be labeled by j_1, j_2, etc., and the rows and columns of matrices by Greek letters. The summation convention will be used for all repeated Greek indices, while other sums will be indicated explicitly. The dimension of the representation $D^{(j)}$ will be denoted by $[j]$.

The analysis of the Kronecker product in Section 5–7 can be performed just as well in terms of the complex conjugates of the irreducible representations as in terms of the representations themselves. We define a unitary matrix U with matrix elements

$$U_{j_3\kappa_3,\kappa_1\kappa_2} = [j_3]^{1/2} \begin{pmatrix} j_1 j_2 j_3 \\ \kappa_1 \kappa_2 \kappa_3 \end{pmatrix} \tag{5–140}$$

which reduces the Kronecker product of $D^{(j_1)}$ and $D^{(j_2)}$ to a sum of representations $D^{*(j_3)}$. The quantity

$$\begin{pmatrix} j_1 j_2 j_3 \\ \kappa_1 \kappa_2 \kappa_3 \end{pmatrix} \tag{5–141}$$

is called a 3-j *symbol*.

All equations of Section 5–7 can be rewritten in this notation. In place of (5–111b) and (5–111c) we now have the unitary conditions:

$$[j] \begin{pmatrix} j_1 j_2 j \\ \kappa_1 \kappa_2 \kappa \end{pmatrix} \begin{pmatrix} j_1 j_2 j' \\ \kappa_1 \kappa_2 \kappa' \end{pmatrix}^* = (j_1 j_2 j) \, \delta_{jj'} \, \delta_{\kappa\kappa'}, \tag{5–142}$$

$$\sum_{j_3} [j_3] \begin{pmatrix} j_1 j_2 j_3 \\ \kappa_1 \kappa_2 \kappa_3 \end{pmatrix} \begin{pmatrix} j_1 j_2 j_3 \\ \lambda_1 \lambda_2 \kappa_3 \end{pmatrix}^* = \delta_{\kappa_1\lambda_1} \, \delta_{\kappa_2\lambda_2}. \tag{5–142a}$$

Equations (5–115) and (5–116) are replaced by

$$[j_3] \begin{pmatrix} j_1 j_2 j_3 \\ \kappa_1 \kappa_2 \kappa_3 \end{pmatrix} D^{(j_1)}_{\kappa_1\lambda_1}(R) D^{(j_2)}_{\kappa_2\lambda_2}(R) \begin{pmatrix} j_1 j_2 j_3' \\ \lambda_1 \lambda_2 \lambda_3 \end{pmatrix} = \delta_{j_3 j_3'} D^{*(j_3)}_{\kappa_3\lambda_3}(R), \tag{5–143}$$

$$D^{(j_1)}_{\kappa_1\lambda_1}(R) D^{(j_2)}_{\kappa_2\lambda_2}(R) = \sum_{j_3} [j_3] \begin{pmatrix} j_1 j_2 j_3 \\ \kappa_1 \kappa_2 \kappa_3 \end{pmatrix}^* D^{*(j_3)}_{\kappa_3\lambda_3} \begin{pmatrix} j_1 j_2 j_3 \\ \lambda_1 \lambda_2 \lambda_3 \end{pmatrix}, \tag{5–144}$$

and Eq. (5–114) by

$$\begin{pmatrix} j_1 j_2 j_3 \\ \kappa_1 \kappa_2 \kappa_3 \end{pmatrix} D^{(j_1)}_{\kappa_1\lambda_1}(R) D^{(j_2)}_{\kappa_2\lambda_2}(R) = D^{*(j_3)}_{\kappa_3\lambda_3}(R) \begin{pmatrix} j_1 j_2 j_3 \\ \lambda_1 \lambda_2 \lambda_3 \end{pmatrix}. \tag{5–145}$$

Since the representations are unitary, we can shift the matrix representative to the left side of (5–145) by multiplying by $D^{(j_3)}_{\kappa_3\lambda_3'}(R)$, and obtain

$$D^{(j_1)}_{\kappa_1\lambda_1}(R) D^{(j_2)}_{\kappa_2\lambda_2}(R) D^{(j_3)}_{\kappa_3\lambda_3}(R) \begin{pmatrix} j_1 j_2 j_3 \\ \kappa_1 \kappa_2 \kappa_3 \end{pmatrix} = \begin{pmatrix} j_1 j_2 j_3 \\ \lambda_1 \lambda_2 \lambda_3 \end{pmatrix}. \tag{5–146}$$

We take the conjugate of (5–146), use the unitarity of the representations, and replace R^{-1} by R to give

$$D_{\lambda_1\kappa_1}^{(j_1)}(R)D_{\lambda_2\kappa_2}^{(j_2)}(R)D_{\lambda_3\kappa_3}^{(j_3)}(R)\begin{pmatrix}j_1 j_2 j_3\\\kappa_1\kappa_2\kappa_3\end{pmatrix}^* = \begin{pmatrix}j_1 j_2 j_3\\\lambda_1\lambda_2\lambda_3\end{pmatrix}^*. \quad (5\text{–}146a)$$

The 3-j symbol is set equal to zero if $(j_1 j_2 j_3) = 0$, so we need not indicate limits on summations over j's. Setting $j' = j = j_3$, $\kappa' = \kappa = \kappa_3$ in (5–142), and summing over κ_3, we have

$$\begin{pmatrix}j_1 j_2 j_3\\\kappa_1\kappa_2\kappa_3\end{pmatrix}\begin{pmatrix}j_1 j_2 j_3\\\kappa_1\kappa_2\kappa_3\end{pmatrix}^* = (j_1 j_2 j_3). \quad (5\text{–}147)$$

We adopt the notation appropriate to the three-dimensional rotation group and set $(-1)^{2j} = 1$ if j is an integer representation, $(-1)^{2j} = -1$ if j is half-integral. We can then write

$$\begin{pmatrix}j_1 j_2 j_3\\\kappa_1\kappa_2\kappa_3\end{pmatrix} = (-1)^{2j_1+2j_2+2j_3}\begin{pmatrix}j_1 j_2 j_3\\\kappa_1\kappa_2\kappa_3\end{pmatrix} \quad (5\text{–}148)$$

since, according to the lemma of Section 5–8, the 3-j symbol vanishes unless $(-1)^{2j_1+2j_2+2j_3} = 1$. One can also use the convention that $(-1)^j = 1$ if j is an even representation, and $(-1)^j = -1$ if j is an odd representation. If j is an integer representation which is neither even nor odd, one can arbitrarily set $(-1)^j$ equal to 1. For half-integral j, we can arbitrarily set $(-1)^j$ equal to i or $-i$, but must keep its value fixed throughout.

The reduced Kronecker product commutes with any diagonal matrix in which all the diagonal elements corresponding to a given irreducible representation are the same. We can therefore multiply the 3-j symbol by a factor $\omega(j_1, j_2, j_3)$ depending on j_1, j_2, j_3, but independent of $\kappa_1, \kappa_2, \kappa_3$. If the representation is to be unitary, $|\omega|^2 = 1$.

We use the orthogonality relations (3–137) to change Eq. (5–144) to

$$\sum_R D_{\kappa_1\lambda_1}^{(j_1)}(R)D_{\kappa_2\lambda_2}^{(j_2)}(R)D_{\kappa_3\lambda_3}^{(j_3)}(R) = g\begin{pmatrix}j_1 j_2 j_3\\\kappa_1\kappa_2\kappa_3\end{pmatrix}^*\begin{pmatrix}j_1 j_2 j_3\\\lambda_1\lambda_2\lambda_3\end{pmatrix}. \quad (5\text{–}149)$$

The replacement of the CG-coefficients by the 3-j symbols with the factor $[j_3]^{1/2}$ enabled us to make this equation symmetric in the matrix representatives. Setting $\kappa_i = \lambda_i$ in Eq. (5–149), we find that

$$\left|\begin{pmatrix}j_1 j_2 j_3\\\kappa_1\kappa_2\kappa_3\end{pmatrix}\right|^2 = \left|\begin{pmatrix}j_2 j_1 j_3\\\kappa_2\kappa_1\kappa_3\end{pmatrix}\right|^2 = \left|\begin{pmatrix}j_3 j_1 j_2\\\kappa_3\kappa_1\kappa_2\end{pmatrix}\right|^2, \quad \text{etc.}, \quad (5\text{–}150)$$

i.e., the absolute value of the 3-j symbol is unaltered if we interchange columns.

Next we prove that it is always possible to choose the phase factors $\omega(j_1 j_2 j_3)$ so that

$$\begin{pmatrix} j_1 j_2 j_3 \\ \kappa_1 \kappa_2 \kappa_3 \end{pmatrix} = (-1)^{j_1+j_2+j_3} \begin{pmatrix} j_2 j_1 j_3 \\ \kappa_2 \kappa_1 \kappa_3 \end{pmatrix} = (-1)^{j_1+j_2+j_3} \begin{pmatrix} j_1 j_3 j_2 \\ \kappa_1 \kappa_3 \kappa_2 \end{pmatrix}, \qquad (5\text{--}151)$$

whence, using (5–148), we find

$$\begin{pmatrix} j_1 j_2 j_3 \\ \kappa_1 \kappa_2 \kappa_3 \end{pmatrix} = \begin{pmatrix} j_2 j_3 j_1 \\ \kappa_2 \kappa_3 \kappa_1 \end{pmatrix} = \begin{pmatrix} j_3 j_1 j_2 \\ \kappa_3 \kappa_1 \kappa_2 \end{pmatrix}. \qquad (5\text{--}152)$$

In other words, the phase factors can be chosen so that the 3-j symbols are unchanged by an even permutation of the columns, and are multiplied by $(-1)^{j_1+j_2+j_3}$ (which is equal to ± 1) for an odd permutation.

To obtain Eqs. (5–151) and (5–152), we choose some particular set $\kappa_{10}, \kappa_{20}, \kappa_{30}$ for which the 3-j symbol $\begin{pmatrix} j_1 & j_2 & j_3 \\ \kappa_{10} \kappa_{20} \kappa_{30} \end{pmatrix}$ does not vanish. From Eq. (5–150) we see that the 3-j symbols with permuted columns also do not vanish for this triple of κ's. We now choose the phase factor $\omega(j_1 j_2 j_3)$ to make $\begin{pmatrix} j_1 & j_2 & j_3 \\ \kappa_{10} \kappa_{20} \kappa_{30} \end{pmatrix}$ real and positive. Next we choose the phase factor $\omega(j_2 j_1 j_3)$, so that the first part of Eq. (5–151) is valid. This is always possible if $j_1 \neq j_2$, since in this case $\omega(j_2 j_1 j_3)$ is independent of $\omega(j_1 j_2 j_3)$. If $j_1 = j_2$, the first part of Eq. (5–151) merely states that

$$\begin{pmatrix} j_1 j_1 j_3 \\ \kappa_1 \kappa_2 \kappa_3 \end{pmatrix} = (-1)^{2j_1+j_3} \begin{pmatrix} j_1 j_1 j_3 \\ \kappa_2 \kappa_1 \kappa_3 \end{pmatrix},$$

i.e., j_3 is in the symmetric square of j_1 if $j_3 + 2j_1$ is even, and in the antisymmetric square of j_1 if $j_3 + 2j_1$ is odd. Similarly, we can make the second part of (5–151) [and consequently, also, (5–152)] hold for κ_{10}, κ_{20}, κ_{30}. But then, if we set κ_i in κ_{i0} in Eq. (5–149), we see from the symmetry of the left-hand side that Eqs. (5–151) and (5–152) hold for *all* $\lambda_1, \lambda_2, \lambda_3$. These equations will remain valid if we multiply all 3-j symbols by phase factors $\omega(j_1 j_2 j_3)$ which are completely symmetric in j_1, j_2, j_3.

Next we consider Eq. (5–144) for the special case where j_2 is the identity representation. We label this representation by $j_2 = 0$ and denote its single row (and column) by the index $\kappa_2 = 0$. The Kronecker product of $D^{(j_1)}$ and $D^{(0)}$ contains only $D^{(j_1)}$, so for this case Eq. (5–144) gives

$$D^{(j_1)}_{\kappa_1 \lambda_1}(R) = [j_1] \begin{pmatrix} j_1 0 j_1 \\ \kappa_1 0 \kappa_3 \end{pmatrix}^* D^{*(j_1)}_{\kappa_3 \lambda_3}(R) \begin{pmatrix} j_1 0 j_1 \\ \lambda_1 0 \lambda_3 \end{pmatrix}; \qquad (5\text{--}144a)$$

that is, the unitary matrix

$$\begin{pmatrix} j \\ \kappa \lambda \end{pmatrix} = [j]^{1/2} \begin{pmatrix} j 0 j \\ \kappa 0 \lambda \end{pmatrix}, \qquad (5\text{--}153)$$

which we shall call a *1-j symbol*, transforms $D^{(j)}$ into its complex conjugate:

$$D^{(j)}_{\kappa\lambda}(R) = \begin{pmatrix} j \\ \kappa\mu \end{pmatrix}^* D^{*(j)}_{\mu\nu}(R) \begin{pmatrix} j \\ \lambda\nu \end{pmatrix}, \qquad (5\text{-}154)$$

or, taking the complex conjugate of (5-154), we have

$$D^{*(j)}_{\kappa\lambda}(R) = \begin{pmatrix} j \\ \kappa\mu \end{pmatrix} D^{(j)}_{\mu\nu}(R) \begin{pmatrix} j \\ \nu\lambda \end{pmatrix}^\dagger. \qquad (5\text{-}154\text{a})$$

For this special case, the unitarity conditions (5-142) and (5-142a) reduce to

$$\begin{pmatrix} j \\ \kappa\lambda \end{pmatrix}^* \begin{pmatrix} j \\ \kappa\lambda' \end{pmatrix} = \delta_{\lambda\lambda'} \qquad (5\text{-}155)$$

and

$$\begin{pmatrix} j \\ \kappa\lambda \end{pmatrix}^* \begin{pmatrix} j \\ \kappa'\lambda \end{pmatrix} = \delta_{\kappa\kappa'}. \qquad (5\text{-}155\text{a})$$

Equation (5-151) reduces to

$$\begin{pmatrix} j \\ \kappa\lambda \end{pmatrix} = (-1)^{2j} \begin{pmatrix} j \\ \lambda\kappa \end{pmatrix}, \qquad (5\text{-}156)$$

which is a restatement of Eq. (5-79): the matrix which transforms the representation to its complex conjugate is symmetric or antisymmetric, depending on whether the representation is integral or half-integral.

The actual evaluation of the 3-*j* symbols (or the CG-coefficients) depends on the particular structure of the group and will be discussed when we treat the representations of individual types of groups.

CHAPTER 6

PHYSICAL APPLICATIONS

6–1 Classification of spectral terms. On several occasions we have noted the relation between group representations and quantum-mechanical problems. If we are studying an atomic system, we must first find the symmetry group of the Hamiltonian, i.e., the set of transformations which leave the Hamiltonian invariant. The existence of a symmetry group for the system raises the possibility of degeneracy. If ψ is an eigenfunction belonging to the energy ϵ, then $O_R\psi$ is degenerate with ψ (R is any element of the symmetry group G). Unless $O_R\psi = C\psi$ for all R, the level is degenerate. The eigenfunctions belonging to a given energy ϵ form the basis for a representation of the group G. In most cases this representation will be irreducible. Only in rare cases, for very special choices of parameters, will we have "accidental" degeneracy, so that sets of functions belonging to different irreducible representations coincide in energy. It is clear that the partners which form the basis for one of the irreducible representations of G *must* be degenerate, since they are transformed into one another by operations of the symmetry group. But two distinct sets of partners, $\psi_i^{(\mu)}$ and $\phi_j^{(\nu)}$, even if they form bases for the *same* irreducible representation of $G(\mu = \nu)$, transform only among themselves, and are not compelled by symmetry considerations to be degenerate with one another.

So we may assume, in general, that the set of eigenfunctions belonging to a given energy ϵ are a set of partners, and form the basis for one of the irreducible representations of the symmetry group. This already tells us a great deal about the degree of degeneracy to be expected. For example, if we consider a system having the symmetry group O, we can see from the character table (Table 4–19) that the energy levels of the system can only be single, or doubly or triply degenerate. The single levels will be of two types, depending on whether they belong to the representations A_1 or A_2. The eigenfunctions of these two types of simple levels differ in their behavior under the operations C_4 and C_2. The doubly degenerate levels will all be of the same type, belonging to the two-dimensional representation E. Finally, there will be two different types of triply degenerate levels belonging to the representations F_1 and F_2. If we disregard possible accidental degeneracy, these are the only possible level types. Though the labels which we use may appear strange, we are actually doing exactly what is done in standard quantum-mechanical treatments—we are assigning two quantum numbers, μ and i, to each eigenfunction $\psi_i^{(\mu)}$ to describe its behavior under the operations of the point-symmetry group. In the

161

same way, as we shall later see, when the symmetry group is the full rotation group, we assign quantum numbers l and m to $\psi_m^{(l)}$ to characterize its behavior under rotation and inversion (by assigning it to the mth row of the lth irreducible representation).

Thus the following level scheme might be typical for a system with symmetry O:

$$\overline{}\quad E;\qquad \phi_1^{(E)},\ \phi_2^{(E)}.$$

$$\overline{\overline{\overline{}}}\quad F_1;\qquad \psi_1^{(F_1)},\ \psi_2^{(F_1)},\ \psi_3^{(F_1)}.$$

$$\overline{}\quad A_2;\qquad \psi_1^{(A_2)}.$$

$$\overline{}\quad A_1;\qquad \phi_1^{(A_1)}.$$

$$\overline{\overline{\overline{}}}\quad F_2;\qquad \psi_1^{(F_2)},\ \psi_2^{(F_2)},\ \psi_3^{(F_2)}.$$

$$\overline{\overline{}}\quad E;\qquad \psi_1^{(E)},\ \psi_2^{(E)}.$$

$$\overline{}\quad A_1;\qquad \psi_1^{(A_1)}.$$

In this diagram we have drawn two levels which belong to the A_1-representation. The fact that they are pictured as having different energies implies that the functions $\psi_1^{(A_1)}$ and $\phi_1^{(A_1)}$ are linearly independent; if they were linearly related, they would necessarily have the same energy. Similarly, for the two levels labeled E, $\psi_1^{(E)}$ and $\psi_2^{(E)}$ are partners which transform according to E and are thus necessarily degenerate; $\phi_1^{(E)}$ and $\phi_2^{(E)}$ are also partners, but the ψ's and ϕ's are linearly independent of one another.

6–2 Perturbation theory. The unperturbed Hamiltonian H_0 was invariant under its symmetry group G. Suppose now that the system is subjected to a perturbation V. The perturbed Hamiltonian $H = H_0 + V$ will have a symmetry group which is necessarily a subgroup of G. If the perturbation V has symmetry *at least* as great as H_0, then G will *still* be the symmetry group of the total Hamiltonian H. In this case the possible types of levels will be unchanged by the perturbation. In fact, no splitting of the degenerate levels can occur. For example, consider a level ϵ and its eigenfunctions $\psi_i^{(\mu)}$. Since V is invariant under all operations of G, $\phi_i^{(\mu)} = V\psi_i^{(\mu)}$ belongs to the ith row of the μth irreducible representation. But we claim that the nondiagonal elements $(\psi_i^{(\mu)},\ V\psi_j^{(\mu)})$ of the perturbation matrix are zero, and that all the diagonal elements are equal, so that no splitting of the level occurs. We prove this in a more general case for later use. We claim that the scalar product of two functions which do not belong to the *same* row of the *same* irreducible representation is zero. Call the functions $\psi_i^{(\mu)}$ and $\phi_j^{(\nu)}$. The unitary operators of the symmetry group do not change the scalar product, so

$$(\psi_i^{(\mu)}, \phi_j^{(\nu)}) = (O_R\psi_i^{(\mu)}, O_R\phi_j^{(\nu)}) = \frac{1}{g}\sum_R (O_R\psi_i^{(\mu)}, O_R\phi_j^{(\nu)})$$

$$= \frac{1}{g}\sum_{kl}\sum_R (\psi_k^{(\mu)}D_{ki}^{(\mu)}(R), \phi_l^{(\nu)}D_{lj}^{(\nu)}(R))$$

$$= \frac{1}{g}\sum_{kl} (\psi_k^{(\mu)}, \phi_l^{(\nu)}) \sum_R D_{ki}^{*(\mu)}(R)D_{lj}^{(\nu)}(R)$$

$$= \frac{1}{g}\sum_{kl} (\psi_k^{(\mu)}, \phi_l^{(\nu)}) \frac{g}{n_\mu}\delta_{\mu\nu}\,\delta_{kl}\,\delta_{ij} = \frac{1}{n_\mu}\sum_k (\psi_k^{(\mu)}, \phi_k^{(\nu)})\,\delta_{\mu\nu}\,\delta_{ij}.$$

$$(6\text{--}1)$$

Setting $\mu = \nu$, $i = j$, we find that

$$(\psi_i^{(\mu)}, \phi_i^{(\mu)}) \qquad \text{is independent of } i. \qquad (6\text{--}2)$$

For the special case $\phi_i^{(\mu)} = V\psi_i^{(\mu)}$, we obtain the above result. We should point out that the level cannot split in any approximation. For if it did, this would mean that the original representation was reducible, in contradiction to our assumption.

If there is an accidental degeneracy, we shall assume that the rth set of partners (which form a basis for the rth irreducible representation) have the energy E_r, and that the energies E_r accidentally coincide with one another. If a symmetric perturbation V is applied to the system, it can at most remove this accidental degeneracy. If the representation D of the level is a sum of irreducible representations, each of which appears just once, then the perturbation theory is simple, for according to (6–1) only diagonal elements appear, and according to (6–2) they are the same for all the partners: $(\psi_i^{(\mu)}, V\psi_i^{(\mu)}) = V_\mu$. If the same irreducible representation is contained several times in D, there will be nondiagonal elements like $(\psi_i^{(\mu)}, V\phi_i^{(\mu)}) = V_\mu'$. To obtain the proper zero-order functions we must then solve the secular equation for all functions belonging to the same row of the same irreducible representation. This is not much work; for example, if some irreducible representation appears twice, we need solve only one 2×2 determinant, since we get the same secular equation for all rows i of the representation.

If the perturbation V has lower symmetry than H_0, the total Hamiltonian H will have a symmetry group G' which is a subgroup G' of G. Suppose that we are given a representation $D(G)$ of the group G. We immediately obtain a representation of the subgroup G' by selecting from among the matrices $D(G)$ those which correspond to elements in G'. Even if $D(G)$ is an irreducible representation of G, the representation of G'

which we derive in this way [let us call it $D'(G')$] may be reducible. In other words, even though we cannot find a subset of basis vectors of $D(G)$ which is invariant under all transformations of the group G, we may be able to find a subspace which is invariant under all transformations of the subgroup G'. Restated in physical terms, this means that even though the eigenfunctions belonging to the energy ϵ form the basis of an irreducible representation of G, this representation may be reducible for the subgroup G'. The perturbation V will then split the level. We illustrate the method by several examples.

Suppose that the symmetry group of the unperturbed Hamiltonian is $\mathcal{C}_4$. As shown in Chapter 4, all irreducible representations are one-dimensional, but the complex-conjugate pair of representations E are degenerate with one another. If we apply a perturbation having the symmetry group $\mathcal{C}_2$, then a degenerate E-level will split into a pair of levels belonging to the representation B of $\mathcal{C}_2$.

Next suppose that the symmetry group G is $\mathcal{C}_{3v}$. The terms are classified according to the representations A_1, A_2, E of $\mathcal{C}_{3v}$. The last type is doubly degenerate. If the symmetry group G' of the total Hamiltonian is $\mathcal{C}_s$, the degeneracy is removed. To find the representations of $\mathcal{C}_s$ which are contained in the representation E of $\mathcal{C}_{3v}$, we write out the part of the character table of $\mathcal{C}_{3v}$ which refers to the operations of the subgroup $\mathcal{C}_s$:

	E	σ_v
$E; x, y$	2	0

and use Eq. (3–150) to find which irreducible representations of $\mathcal{C}_s$ are contained in E. For A' of $\mathcal{C}_s$ we find

$$a_{A'} = \tfrac{1}{2}[2(1) + 0(1)] = 1,$$

and for A'',

$$a_{A''} = \tfrac{1}{2}[2(1) + 0(-1)] = 1,$$

so an E-level of $\mathcal{C}_{3v}$ splits into levels of type A', A'' of $\mathcal{C}_s$.

As a further example, suppose the symmetry group G is the tetrahedral group T. Consider a level belonging to the F-representation. It will be 3-fold degenerate, with characters

	E	$C_2(3)$	$C_3(4)$	$C_3^2(4)$	
F	3	-1	0	0	.

Suppose that the perturbation is such that the total Hamiltonian has the symmetry group $D_2 \equiv V$. From the table given above, we select the characters for the elements of the subgroup V:

	E	C_{2x}	C_{2y}	C_{2z}
F	3	-1	-1	-1

From Eq. (3–150) and the character table of V (Table 4–11), we find

$$a_{A_1} = \tfrac{1}{4}[3(1) - 1(1) - 1(1) - 1(1)] = 0,$$

$$a_{B_3} = \tfrac{1}{4}[3(1) - 1(-1) - 1(-1) - 1(1)] = 1,$$

$$a_{B_2} = a_{B_1} = 1.$$

So the representation F of the group T contains the representations B_1, B_2, B_3 of the group V.

Again for the F-representation of T, suppose that the total Hamiltonian has the symmetry group $\mathcal{C}_3$. We write that part of the character table which refers to the subgroup $\mathcal{C}_3$:

	E	C_3	C_3^2
F	3	0	0

Using Eq. (3–150) and the character table for $\mathcal{C}_3$, we find

$$a_A = \tfrac{1}{3}[3(1) + 0(1) + 0(1)] = 1.$$

Thus an F-level of T splits into an A- and an E-level of $\mathcal{C}_3$.

Finally, suppose the total Hamiltonian has symmetry $\mathcal{C}_2$,

	E	C_2
F	3	-1

From (3–150),

$$a_A = \tfrac{1}{2}[3(1) - 1(1)] = 1,$$

$$A_B = \tfrac{1}{2}[3(1) - 1(-1)] = 2.$$

Thus an F-level splits into three nondegenerate levels, one belonging to A, the other two to the B-representation of $\mathcal{C}_2$.

Problem. Consider a system having the symmetry O. Suppose a perturbation is applied which reduces the symmetry to: (a) T. (b) D_3. (c) V, with the three 2-fold axes joining midpoints of opposite faces in Fig. 2–35. (d) V, with two of the three 2-fold axes joining midpoints of opposite edges in Fig. 2–35. (e) $\mathcal{C}_4$.

In each case, find how levels belonging to E, F_1, and F_2 of O are split by the perturbation.

6–3 Selection rules. In Section 6–1 we discussed the classification of spectral terms according to irreducible representations of the symmetry group of the unperturbed Hamiltonian. Once the term scheme is obtained, we must ask what the selection rules are for optical transitions between various (degenerate) levels. Or, more generally, we may ask which elements of the perturbation matrix of a perturbation V vanish.

In the absence of accidental degeneracy, the eigenfunctions of each level form the basis for one of the irreducible representations of the symmetry group. The matrix elements of a quantity f, between states belonging to the μ- and ν-representations of the symmetry group, will be of the form

$$f_{ij}^{\mu\nu} = \int \psi_i^{(\mu)*} f \phi_j^{(\nu)} \, d\tau = (\psi_i^{(\mu)}, f\phi_j^{(\nu)}), \qquad (6\text{--}3)$$

where $\psi_i^{(\mu)}$ belongs to the ith row of the μ-representation, and $\phi_j^{(\nu)}$ belongs to the jth row of the ν-representation. Equation (6–1) stated that

$$\int \psi_i^{(\mu)*} \phi_j^{(\nu)} \, d\tau = (\psi_i^{(\mu)}, \phi_j^{(\nu)}) = 0 \qquad \begin{array}{l} \text{if} \quad \mu \neq \nu, \\ \text{or} \quad i \neq j. \end{array} \qquad (6\text{--}4)$$

If in Eq. (6–4) we choose the μ-representation to be the identity representation (we shall take $\mu = 1$ for the identity representation), then $\psi_i^{(\mu)}$ can be taken equal to a constant (a constant is invariant under all transformations of the point group and therefore belongs to the identity representation). Then (6–4) becomes

$$\int \phi_j^{(\nu)} \, d\tau = 0 \qquad \text{unless} \quad \nu = 1. \qquad (6\text{--}5)$$

In terms of the expansion theorem, Eq. (3–193), we may say that, for any function,

$$\int \psi \, d\tau = \int \psi^{(1)} \, d\tau = \frac{1}{g} \int \sum_R O_R \psi \, d\tau, \qquad (6\text{--}6)$$

where $\psi^{(1)}$ is the part of ψ which belongs to the identity representation.

Equation (6–5) is easy to derive independently for the case where the ν-representation is one-dimensional. The integral of ψ over all space is not changed if we transform ψ into $O_R\psi$, where R is any rotation or reflection; thus

$$\int \psi \, d\tau = \int O_R \psi \, d\tau$$

for all R in G. For a one-dimensional representation, $O_R\psi = r\psi$, where r is a numerical factor, so $\int \psi \, d\tau = r \int \psi \, d\tau = 0$ unless $r = 1$ for all R. Hence $\int \psi^{(\nu)} d\tau = 0$ unless $\nu = 1$. The extension of this proof to representations of higher degree is tedious, while the proof given above is quite general.

We may now restate the result of Eq. (6-1) as follows: The product $\psi_i^{(\mu)*}\phi_j^{(\nu)}$ contains a part which is invariant under all operations of G if and only if $\mu = \nu$, $i = j$. For the integral (6-3) we may say that $f_{ij}^{\mu\nu} = 0$ unless $f\phi_j^{(\nu)}$ contains a function which belongs to the ith row of the μ-representation. If the function f is a scalar, it belongs to the identity representation. If we are dealing with the dipole moment of the system, then f has three components, etc. In any case, the function (or functions) f which appear in (6-3) are basis functions for one or several irreducible representations of G. Suppose that f runs through a set of functions $\theta_k^{(\rho)}$ which form the basis of the ρ-representation of G. Then the transition between the μ- and ν-levels will be forbidden only if none of the products $\theta_k^{(\rho)}\phi_j^{(\nu)}$ or linear combinations of these products belongs to the μ-representation. If the μ-representation is not contained in the product representation formed from $D^{(\rho)}$ and $D^{(\nu)}$, then the transition is forbidden.

We carry out the reduction of the product representation by using (5-15) to compute the character of each element in the product representation, and then using Eq. (3-150) to find the number of times each irreducible representation is contained in the product representation. [These two steps give us the coefficients in the Clebsch-Gordan series (5-108).] For the complete tabulation of the selection rules, it is more convenient to follow this procedure than to combine the steps and use Eq. (5-105). We illustrate the procedure for the group $\mathcal{C}_{3v}$. In this first example, we shall copy the character table (Table 4-14), but will omit it in later examples. We append to the character table the characters of all product representations:

$\mathcal{C}_{3v}$	E	$C_3(2)$	$\sigma_v(3)$	
$A_1; z$	1	1	1	
$R_z; A_2$	1	1	-1	
$R_x, R_y; E; x, y$	2	-1	0	
$A_1 \times A_1$	1	1	$1 = A_1$	
$A_1 \times A_2$	1	1	$-1 = A_2$	(6-7)
$A_1 \times E$	2	-1	$0 = E$	
$A_2 \times A_2$	1	1	$1 = A_1$	
$A_2 \times E$	2	-1	$0 = E$	
$E \times E$	4	1	$0 = A_1 + A_2 + E$	

It is clear that most of the results can be obtained by inspection. First the product of the identity representation with any other representation merely gives this other representation (since $\psi_1^{(1)}\phi_j^{(\nu)} = \phi_j^{(\nu)}$). Secondly, the product of two one-dimensional representations is a one-dimensional representation and must be irreducible. The only reducible product representation in the table is $E \times E$. This is a four-dimensional representation, and the sum of the absolute squares of its characters is $(4)^2 + 2(1)^2 = 18$. From Eq. (3–151)

$$18 = g \sum_\mu a_\mu^2 \quad \text{or} \quad \sum_\mu a_\mu^2 = 3,$$

with the unique solution $a_{A_1} = a_{A_2} = a_E = 1$, so

$$E \times E = A_1 + A_2 + E. \tag{6-8}$$

Or, using Eq. (3–150), we have

$$a_{A_1} = \tfrac{1}{6}[(1)(4)(1) + (2)(1)(1) + (3)(0)(1)] = 1,$$
$$a_{A_2} = \tfrac{1}{6}[(1)(4)(1) + (2)(1)(1) + (3)(0)(1)] = 1, \tag{6-8a}$$
$$a_E = \tfrac{1}{6}[(1)(4)(2) + (2)(1)(-1) + (3)(0)(0)] = 1,$$

which gives the same result.

Equation (6–8) states that if we take products of one set of partners with another set of partners belonging to the representation E, we obtain four functions which form the basis for the representation $E \times E$; we can find a linear combination of these product functions which belongs to A_1, another which belongs to A_2, and a pair of linear combinations which form a set of partners belonging to E. Note that in the character table, z belongs to A_1, while x and y are a set of partners belonging to E. A triplet of functions $\mathbf{P} = (P_x, P_y, P_z)$ is said to be a (polar) vector if the functions transform like the components of the position vector $\mathbf{r} = (x, y, z)$. Thus, for any vector, P_z belongs to A_1, while P_x and P_y are a set of partners belonging to E. Given any two (polar) vectors $\mathbf{P}$ and $\mathbf{Q}$, we have two sets of partners belonging to E: $P_x, P_y; Q_x, Q_y$. The product representation $E \times E$ has the basis functions $P_x Q_x, P_x Q_y, P_y Q_x, P_y Q_y$. In this simple example, the linear combinations which belong to A_1, A_2, E can be found by inspection. Nevertheless, we shall find the result by methods of brute force in order to become more familiar with product representations. In Section 4–2, we gave the matrices of $\mathfrak{C}_{3v}$ in the E-representation:

$$
\begin{array}{cccccc}
E & C_3 & C_3^2 & \sigma_v & \sigma_{v''} & \sigma_{v'} \\
\begin{bmatrix} 1 & 0 \\ 0 & 1 \end{bmatrix}, &
\begin{bmatrix} \epsilon & 0 \\ 0 & \epsilon^2 \end{bmatrix}, &
\begin{bmatrix} \epsilon^2 & 0 \\ 0 & \epsilon \end{bmatrix}, &
\begin{bmatrix} 0 & 1 \\ 1 & 0 \end{bmatrix}, &
\begin{bmatrix} 0 & \epsilon \\ \epsilon^2 & 0 \end{bmatrix}, &
\begin{bmatrix} 0 & \epsilon^2 \\ \epsilon & 0 \end{bmatrix},
\end{array} \tag{6-9}
$$

where the basis functions were $e^{i\phi}$ and $e^{-i\phi}$, or $x + iy$ and $x - iy$, or

$P_x + iP_y$ and $P_x - iP_y$. Now we use Eq. (5–4) to find the matrices of the product representation:

$$
\begin{array}{ccc}
E & C_3 & C_3^2
\end{array}
$$

$$
\begin{bmatrix}
1 & 0 & 0 & 0 \\
0 & 1 & 0 & 0 \\
0 & 0 & 1 & 0 \\
0 & 0 & 0 & 1
\end{bmatrix},
\quad
\begin{bmatrix}
\epsilon^2 & 0 & 0 & 0 \\
0 & 1 & 0 & 0 \\
0 & 0 & 1 & 0 \\
0 & 0 & 0 & \epsilon
\end{bmatrix},
\quad
\begin{bmatrix}
\epsilon & 0 & 0 & 0 \\
0 & 1 & 0 & 0 \\
0 & 0 & 1 & 0 \\
0 & 0 & 0 & \epsilon^2
\end{bmatrix},
$$

$$
\begin{array}{ccc}
\sigma_v & \sigma_{v''} & \sigma_{v'}
\end{array}
$$
(6–10)

$$
\begin{bmatrix}
0 & 0 & 0 & 1 \\
0 & 0 & 1 & 0 \\
0 & 1 & 0 & 0 \\
1 & 0 & 0 & 0
\end{bmatrix},
\quad
\begin{bmatrix}
0 & 0 & 0 & \epsilon \\
0 & 0 & 1 & 0 \\
0 & 1 & 0 & 0 \\
\epsilon^2 & 0 & 0 & 0
\end{bmatrix},
\quad
\begin{bmatrix}
0 & 0 & 0 & \epsilon^2 \\
0 & 0 & 1 & 0 \\
0 & 1 & 0 & 0 \\
\epsilon & 0 & 0 & 0
\end{bmatrix},
$$

where the rows and columns are arranged in dictionary order. We note that, in all the matrices, only the components 11 and 22, and the components 12 and 21 are coupled. The matrices are therefore reducible. The matrices coupling 11 and 22 are just the matrices of E, and hence $(P_x + iP_y) \times (Q_x + iQ_y)$ and $(P_x - iP_y)(Q_x - iQ_y)$ belong to E. Taking linear combinations, we may also state that $P_x Q_x - P_y Q_y$ and $P_x Q_y + P_y Q_x$ belong to E. If we look at the matrices coupling 12 and 21 [basis functions $(P_x + iP_y)(Q_x - iQ_y)$ and $(P_x - iP_y)(Q_x + iQ_y)$], we see that they have the form $\begin{pmatrix} 1 & 0 \\ 0 & 1 \end{pmatrix}$ for E, C_3, C_3^2, and are $\begin{pmatrix} 0 & 1 \\ 1 & 0 \end{pmatrix}$ for $\sigma_v, \sigma_{v'}, \sigma_{v''}$. They can be reduced by taking as new basis functions the linear combinations

$$(P_x + iP_y)(Q_x - iQ_y) + (P_x - iP_y)(Q_x + iQ_y) = 2(P_x Q_x + P_y Q_y)$$

and

$$(P_x + iP_y)(Q_x - iQ_y) - (P_x - iP_y)(Q_x + iQ_y) = 2i(P_y Q_x - P_x Q_y).$$

Thus $P_x Q_x + P_y Q_y$ belongs to A_1, and $P_x Q_y - P_y Q_x$ belongs to A_2; $P_x Q_x + P_y Q_y$ is the scalar product of the polar vectors $\mathbf{P}$ and $\mathbf{Q}$ (in $\mathcal{C}_{3v}$ the z-component is not affected by any of the operations) and is therefore an invariant (a scalar). The quantity $P_x Q_y - P_y Q_x$ is the z-component of the vector product $\mathbf{P} \times \mathbf{Q}$. The vector product transforms like the coordinates under rotations, but does not change sign under inversion. Quantities of this type are called *axial vectors*. So we may say that the z-component of any axial vector belongs to the representation A_2 of $\mathcal{C}_{3v}$.

In the character table we note that z (and therefore P_z) belongs to A_1. The product representation $A_1 \times E = E$; typical partners belonging to $A_1 \times E$ are therefore $\pm z$ (or $\pm P_z$, or $\pm Q_z$) multiplied by x, y (or P_x, P_y, or Q_x, Q_y). Thus we find that xz, yz belong to E; so do $P_x P_z, P_y P_z$; so do

$P_x Q_z$, $P_y Q_z$; and so do $P_z Q_x - P_x Q_z$, $P_y Q_z - P_z Q_y$. Hence the x- and y-components of the vector product $\mathbf{P} \times \mathbf{Q}$ belong to E, as do the x- and y-components of any axial vector. Frequently the symbol R is used for an axial vector (R for rotation, which is a typical axial vector). In the character table we would assign R_z to A_2, while R_x and R_y belong to E.

Now we consider the selection rules in this same case of symmetry under $\mathcal{C}_{3v}$. The matrix elements are given by (6–3). If $\mathbf{f} = (x, y, z)$, we obtain the selection rules for electric dipole radiation. Consider the integrals

$$\int \psi_i^{(\mu)*} \mathbf{f} \phi_j^{(\nu)} \, d\tau \tag{6–11}$$

as i goes through 1 to n_μ and j goes through 1 to n_ν. The transition will be forbidden (for electric dipole radiation) only if all these integrals vanish. We proceed as follows: From the table, z belongs to the representation A_1 of $\mathcal{C}_{3v}$. Therefore the set of functions $z\phi_j^{(\nu)}(j = 1, \ldots, n_\nu)$ forms a basis for $A_1 \times D^{(\nu)}$. Unless $A_1 \times D^{(\nu)}$ contains $D^{(\mu)}$, all the integrals (6–11) will vanish. Similarly, x and y belong to the representation E, so the sets of functions $x\phi_j^{(\nu)}$, $y\phi_j^{(\nu)}$ ($j = 1, \ldots, n_\nu$) form the basis for the representation $E \times D^{(\nu)}$. Unless $E \times D^{(\nu)}$ contains $D^{(\mu)}$, all these integrals (matrix elements of x and y) vanish. Hence a transition between a level belonging to $D^{(\nu)}$ and one belonging to $D^{(\mu)}$ will be forbidden (for electric dipole radiation) if neither $A_1 \times D^{(\nu)}$ nor $E \times D^{(\nu)}$ contains $D^{(\mu)}$. From the table [Eq. (6–7)] we find

$$A_1 \times A_1 = A_1, \quad A_1 \times A_2 = A_2, \quad A_1 \times E = E,$$
$$E \times A_1 = E, \quad E \times A_2 = E, \quad E \times E = A_1 + A_2 + E. \tag{6–12}$$

The transitions $A_1 \leftrightarrow A_2$ are forbidden.

The magnetic dipole moment is an axial vector, so magnetic dipole transitions will be forbidden if neither $A_2 \times D^{(\nu)}$ nor $E \times D^{(\nu)}$ contain $D^{(\mu)}$. From (6–7)

$$A_2 \times A_1 = A_2, \quad A_2 \times A_2 = A_1, \quad A_2 \times E = E,$$
$$E \times A_1 = E, \quad E \times A_2 = E, \quad E \times E = A_1 + A_2 + E. \tag{6–13}$$

Since the pairs $A_1 \leftrightarrow A_1$; $A_2 \leftrightarrow A_2$ do not appear, magnetic dipole transitions between two states belonging to A_1 (or A_2) are forbidden.

One point to note is that if the symmetry group is a pure rotation group, there is no distinction between axial and polar vectors, and the selection rules for electric and magnetic dipole radiation are identical.

As a second example, we find the selection rules for electric (and magnetic) dipole radiation for a system with symmetry T. All the components of a vector belong to the representation F (cf. Table 4–18). So we form the product representations $F \times A$, $F \times E$, $F \times F$, and compute the characters by means of (5–15):

T	E	$C_2(3)$	$C_3(4)$	$C_3^2(4)$
$F \times A$	3	-1	0	0
$F \times E$	$\begin{cases} 3 \\ 3 \end{cases}$	$\begin{matrix} -1 \\ -1 \end{matrix}$	$\begin{matrix} 0 \\ 0 \end{matrix}$	$\begin{matrix} 0 \\ 0 \end{matrix}$
$F \times F$	9	-1	0	0

Now using the above character table and Eq. (3–150), we find

$$F \times A = F, \qquad F \times E = 2F, \qquad F \times F = A + E + 2F. \quad (6\text{–}14)$$

Hence the forbidden transitions are $A \leftrightarrow E$, $A \leftrightarrow A$, $E \leftrightarrow E$.

As a last example, we consider the symmetry group D_{2d}. From the character table, we see that P_x and P_y are partners belonging to the E-representation, while P_z belongs to B_2. Products of components of polar vectors like $P_x Q_z$, $P_y Q_z$, or $P_z Q_x$, $P_z Q_y$, or $P_z Q_x - P_x Q_z$, $P_y Q_z - P_z Q_y$ form pairs of partners belonging to $E \times B_2 = E$. Thus for an axial vector, R_x and R_y form a pair of partners belonging to E. On the other hand, R_z belongs to A_2. (This can be seen geometrically by the methods of Chapter 2. Consider a rotation about the z-axis. What is the effect on this rotation if we turn through 180° about a horizontal axis or reflect in a vertical plane? Another method which the reader should try is the reduction of the matrices of $E \times E$ as illustrated for $\mathcal{C}_{3v}$.)

To find the selection rules for electric dipole radiation, we must determine which representations are contained in the products of B_2 and E with all the representations of D_{2d}:

$$
\begin{aligned}
B_2 \times A_1 &= B_2, & B_2 \times A_2 &= B_1, & B_2 \times B_1 &= A_2, \\
B_2 \times B_2 &= A_1, & B_2 \times E &= E, & E \times A_1 &= E, \\
E \times A_2 &= E, & E \times B_1 &= E, & E \times B_2 &= E, \\
& & E \times E &= A_1 + A_2 + B_1 + B_2. &
\end{aligned}
\qquad (6\text{–}15)
$$

Note that in groups like D_{2d}, where there is a preferred axis (the z-direction), the matrix elements of P_z differ from those of P_x, P_y. This distinction is similar to that between π- and σ-components in ordinary spectroscopy. For P_z, the allowed transitions are

$$A_1 \leftrightarrow B_2, \qquad A_2 \leftrightarrow B_1, \qquad E \leftrightarrow E,$$

while for P_x, P_y, the allowed transitions are

$$A_1, A_2, B_1, B_2 \leftrightarrow E.$$

For magnetic dipole radiation, R_z belongs to A_2 and R_x, R_y belong to E. For R_z, $A_2 \times A_1 = A_2$, $A_2 \times A_2 = A_1$, $A_2 \times B_1 = B_2$, $A_2 \times B_2 = B_1$, $A_2 \times E = E$, and the allowed transitions therefore are $A_1 \leftrightarrow A_2$, $B_1 \leftrightarrow B_2$, $E \leftrightarrow E$. For R_x, R_y, as for P_x, P_y, the allowed transitions are A_1, A_2, B_1, $B_2 \leftrightarrow E$.

Problem. Find the selection rules for electric and magnetic dipole transitions if the symmetry group is (a) D_{3d}; (b) O.

Using the results for (b), indicate the allowed transitions in the level scheme of Section 6–1.

In evaluating matrix elements like (6–3), we have really been dealing with a product representation formed from three factors. In (6–3), the function f was allowed to run through the basis functions $\theta_k^{(\rho)}$ of some irreducible representation (say the ρ-representation). The set of functions $\psi_i^{(\mu)*}$ coincides with the set $\psi_i^{(\mu)}$ in all the cases under consideration, so if we take all the triple products $\psi_i^{(\mu)*} \theta_k^{(\rho)} \phi_j^{(\nu)}$, we obtain a representation which should be denoted as $D^{(\mu)} \times D^{(\rho)} \times D^{(\nu)}$. Our procedure up to now has been to see whether the product $D^{(\rho)} \times D^{(\nu)}$ contained $D^{(\mu)}$. But the triple-product representation can be reduced in any order we wish to choose, and we should always obtain the same final result. In other words, we could also take $D^{(\mu)} \times D^{(\nu)}$ first and see whether it contains $D^{(\rho)}$. (When the integrand consists of any number of factors, the general procedure is to take the product of the representations of the factors, $D^{(1)} \times D^{(2)} \times D^{(3)} \times \cdots$, and see whether it contains the identity representation.) For example, in the group D_{2d}, we form the products given in Table 6–1.

To find the selection rules for electric dipole radiation, we look for B_2 (matrix elements of P_z) or E (matrix elements of P_x, P_y) in the expansions of the products. In Table 6–1, we have circled the first type of transition

TABLE 6–1

$A_1 \times A_1 = A_1$	$A_1 \times A_2 = A_2$	$A_1 \times B_1 = B_1$	$\boxed{A_1 \times B_2 = B_2}$	$\boxed{A_1 \times E = E}$
	$A_2 \times A_2 = A_1$	$\boxed{A_2 \times B_1 = B_2}$	$A_2 \times B_2 = B_1$	$\boxed{A_2 \times E = E}$
		$B_1 \times B_1 = A_1$	$B_1 \times B_2 = A_2$	$\boxed{B_1 \times E = E}$
			$B_2 \times B_2 = A_1$	$\boxed{B_2 \times E = E}$
				$\boxed{E \times E = A_1 + A_2 + B_1 + B_2}$

and put a rectangle around the second type. The results are, of course, the same as before.

Problem. Apply this method to find the selection rules for dipole radiation under symmetry T.

We have so far very carefully chosen all our functions to be independent. For example, when we wrote the matrix elements

$$\int \psi_i^{(\mu)^*} f \phi_j^{(\nu)} \, d\tau,$$

it was understood that either $\nu \neq \mu$ or, if the μ- and ν-representations are the same, then the ψ's and ϕ's are two independent sets of partners. The case where the ψ's and ϕ's are the same will occur if we ask for the diagonal matrix elements of an operator f. We must then consider the symmetrized Kronecker products which we discussed in Section 5–2. To illustrate the difference between the various types of products, we take as an example the group $\mathcal{C}_{3v}$. The character table for the various Kronecker products is given in Table 6–2. To illustrate how we obtain this table, consider the element C_3. We find that $\chi(C_3) = -1$, $\chi(C_3^2) = -1$, so $[\chi \times \chi(C_3)] = \frac{1}{2}[(-1)^2 - 1] = 0$. For the element σ_v, $\chi(\sigma_v) = 0$, but $\chi(\sigma_v^2) = \chi(E) = 2$, and hence $[\chi \times \chi(\sigma_v)] = \frac{1}{2}[0^2 + 2] = 1$. Resolving the products into irreducible representations, we have

$$E \times E = A_1 + A_2 + E, \quad [E \times E] = A_1 + E, \quad \{E \times E\} = A_2.$$

$$(6\text{--}16)$$

Finally, if we let μ equal ν and let the ψ's and ϕ's be the same functions (which is the case for diagonal matrix elements), the antisymmetric products are all identically zero, and we obtain only the symmetric product representation. For example, in the group $\mathcal{C}_{3v}$, x and y are partners belonging to E. Taking products of x and y among themselves, we cannot form a basis for $E \times E$, but instead get a basis for $[E \times E]$. We obtain

TABLE 6–2

$\mathcal{C}_{3v}$	E	$C_3(2)$	$\sigma_v(3)$
E	2	-1	0
$E \times E$	4	1	0
$[E \times E]$	3	0	1
$\{E \times E\}$	1	1	-1

only three independent products: x^2, xy, y^2. Earlier, when we were discussing selection rules for magnetic dipole radiation [Eq. (6-13)], we found that two different states belonging to the E-representation were coupled through R_z (which belongs to A_2). This is not the case for diagonal matrix elements. In other words, the integrals $\int \psi_i^{(E)*} f \psi_j^{(E)} \, d\tau$ for a function f belonging to A_2 are all zero. For if we first form the products $\psi_i^{(E)*} \psi_j^{(E)}$, we obtain the symmetric product representation $[E \times E]$ which, according to Eq. (6-17), does not contain A_2. Another way of stating this result is to say that the identity representation is not contained in $[E \times E] \times A_2$, although it is contained in $E \times E \times A_2$. When the ψ's and ϕ's are the same functions, we are forced to use the $[E \times E]$. Earlier we proceeded in a different order; taking first the products $f \psi_j^{(E)}$, we obtained the representation $E \times A_2 = E$. (Since A_2 and E are different representations, this was certainly valid.) One is tempted now to combine $E \times A_2 = E$ with the representation of the functions $\psi_i^{(E)*}$, yielding $E \times E = A_1 + A_2 + E$. But this step is not valid, because the functions obtained by taking $E \times A_2$ are related to the basis functions of E, and the results must be symmetrized. We illustrate this with $\mathcal{C}_{3v}$.

We take $\psi_1^{(E)} = e^{i\phi}$ and $\psi_2^{(E)} = e^{-i\phi}$ for the first level, $\phi_1^{(E)} = e^{-2i\phi}$, $\phi_2^{(E)} = e^{2i\phi}$ for the second level, and choose $f = R_z$ belonging to A_2. If we proceed as before, i.e., consider transitions between the two levels, the matrix elements have integrands $R_z e^{\pm 3i\phi}$, $R_z e^{\pm i\phi}$. To obtain a nonvanishing integral, we must be able to find a linear combination of these four product functions which belongs to the identity representation A_1. This is achieved by taking $R_z \sin 3\phi$, the *antisymmetric* combination. But if we take diagonal elements for the first level, the integrands are R_z, $R_z e^{\pm 2i\phi}$, and we cannot form a function belonging to the identity representation.

The general procedure for finding selection rules for the diagonal matrix elements is to form $[D^{(\mu)} \times D^{(\mu)}]$ for each irreducible representation of the symmetry group, and see which ones contain the representation to which f belongs.

For $\mathcal{C}_{3v}$, we find the symmetric products:

$$[A_1 \times A_1] = A_1, \quad [A_2 \times A_2] = A_1, \quad [E \times E] = A_1 + E.$$

$$(6\text{-}17)$$

The electric dipole moment belongs to the representation $A_1 + E$. Since A_1 and/or E appear on the right for each case in (6-17), we conclude that none of the diagonal elements of the electric dipole moment vanish. The magnetic dipole moment belongs to $A_2 + E$. Since A_2 and/or E appear only in $[E \times E]$, we conclude that the diagonal elements of the magnetic dipole moment vanish for states of type A_1 and A_2.

For the group D_{2d}, we find, using Table 4–16:

D_{2d}	E	C_2	$S_4(2)$	$C_2(2)$	$\sigma_d(2)$
$[E \times E]$	3	3	-1	1	1

$$[A_1 \times A_1] = [A_2 \times A_2] = [B_1 \times B_1] = [B_2 \times B_2] = A_1;$$
$$E \times E = A_1 + B_1 + B_2. \tag{6-18}$$

The electric dipole moment belongs to $B_2 + E$; the only nonvanishing diagonal element is that for a state of type E. The magnetic dipole moment belongs to $A_2 + E$; all of its diagonal elements vanish.

Problem. Find the selection rules for diagonal elements of the electric and magnetic dipole moments if the symmetry group is (a) D_{3d}; (b) O.

Next we consider the selection rules for the electric quadrupole moment. A *tensor* of the second rank in three dimensions is defined to be any set of 9 quantities $A_{ij}(i, j = 1, 2, 3)$ which transform under rotations and reflections like the products of the components of two independent vectors. Thus, if the transformation of coordinates is

$$x_i' = \sum_k a_{ik} x_k, \tag{6-19}$$

then

$$x_i' y_j' = \sum_{k,l} a_{ik} a_{jl} x_k y_l, \tag{6-20}$$

and the components of a tensor of second rank must transform as follows:

$$A_{ij}' = \sum_{k,l} a_{ik} a_{jl} A_{kl}. \tag{6-21}$$

A tensor is said to be symmetric if

$$A_{ij} = A_{ji}. \tag{6-22}$$

A symmetric tensor of the second rank in three dimensions has six independent components. A typical symmetric tensor is obtained from the products of x, y, z with one another:

$$\begin{bmatrix} x^2 & xy & xz \\ xy & y^2 & yz \\ xz & yz & z^2 \end{bmatrix}.$$

The quadrupole-moment tensor is symmetric, but has the additional property that its trace is zero, that is, $A_{xx} + A_{yy} + A_{zz} = 0$, so that it has only five independent components (spherical harmonics for $l = 2$). To find the selection rules for electric quadrupole transitions, we must first assign the components of the symmetric tensor to the various irreducible representations of the symmetry group.

As our first example, we again choose $\mathfrak{C}_{3v}$. Since z belongs to A_1, z^2 will also belong to A_1. The products zx, zy are a pair of partners belonging to $A_1 \times E = E$. The products x^2, xy, y^2 form a basis for the symmetric product representation $[E \times E] = A_1 + E$. It is easily seen that $x^2 + y^2$ belongs to A_1, while $x^2 - y^2$ and xy are a pair of partners belonging to E. So the components of a symmetric tensor are assigned as follows:

$$A_1: \begin{Bmatrix} A_{zz} \\ A_{xx} + A_{yy} \end{Bmatrix}; \qquad A_2: \text{none}; \qquad E: A_{yz}, A_{xz}; A_{xx} - A_{yy}, A_{xy}.$$

The trace of the quadrupole moment tensor is zero, so $A_{xx} + A_{yy} + A_{zz} = 0$. But since A_{zz} and $A_{xx} + A_{yy}$ independently belong to A_1, the results are not affected. To find the selection rules for quadrupole transitions, we must see whether A_1 and/or E are contained in $D^{(\mu)} \times D^{(\nu)}$; for diagonal elements we do the same with respect to $[D^{(\mu)} \times D^{(\mu)}]$. In $\mathfrak{C}_{3v}$,

$$A_1 \times A_1 = A_1, \qquad A_1 \times A_2 = A_2, \qquad A_1 \times E = E,$$
$$A_2 \times A_2 = A_1, \qquad A_2 \times E = E,$$
$$E \times E = A_1 + A_2 + E.$$
$$\text{(6-23)}$$

We look for A_1 and/or E on the right side and find that electric quadrupole transitions are forbidden for $A_1 \leftrightarrow A_2$. Similarly, for diagonal elements, we form

$$[A_1 \times A_1] = A_1, \qquad [A_2 \times A_2] = A_1, \qquad [E \times E] = A_1 + E,$$
$$\text{(6-24)}$$

and find that none of the diagonal elements vanish.

Next consider the group D_{2d}. Since z belongs to B_2, z^2 belongs to $B_2 \times B_2 = A_1$. The functions zx, zy are partners belonging to $B_2 \times E = E$. The functions x^2, xy, y^2 belong to the symmetric product $[E \times E] = A_1 + B_1 + B_2$. The assignment is easily made: $x^2 + y^2$ belongs to A_1, xy to B_2, and $x^2 - y^2$ to B_1. Thus the tensor components are assigned as follows:

$$A_1: \begin{Bmatrix} A_{zz} \\ A_{xx} + A_{yy} \end{Bmatrix}; \qquad A_2: \text{none}; \qquad B_1: A_{xx} - A_{yy};$$

$$B_2: A_{xy}; \qquad E: A_{xz}, A_{yz}.$$

Again the zero-trace condition does not affect our results. For a transition to occur between levels belonging to the μ- and ν-representations, $D^{(\mu)} \times D^{(\nu)}$ must contain at least one of A_1, B_1, B_2, E. From Table 6–1 we see that the transitions $A_1 \leftrightarrow A_2$, $B_1 \leftrightarrow B_2$ are forbidden. For the diagonal elements, we see from Eq. (6–18) that none are forced to vanish because of the symmetry.

A more interesting case is that of the group T. Here x, y, z are partners belonging to F. $[F \times F]$ has the characters 6, 2, 0, 0, so $[F \times F] = A + E + F$. We could find the functions belonging to the various irreducible representations by writing out the matrices of $[F \times F]$, but this can be easily done by inspection. We see that $x^2 + y^2 + z^2$ belongs to A, $x^2 + \epsilon y^2 + \epsilon^2 z^2$ and $x^2 + \epsilon^2 y^2 + \epsilon z^2$ belong to E, and xy, xz, yz belong to F. So the components of the quadrupole-moment tensor have the assignments:

$$A: A_{xx} + A_{yy} + A_{zz}; \quad E: \begin{Bmatrix} A_{xx} + \epsilon A_{yy} + \epsilon^2 A_{zz} \\ A_{xx} + \epsilon^2 A_{yy} + \epsilon A_{zz} \end{Bmatrix}; \quad F: A_{xy}, A_{yz}, A_{zx}.$$

Now the fact that the trace of the quadrupole-moment tensor is zero does have an effect—the function which we assigned to A is actually equal to zero. Thus, in determining the selection rules, we must look for E and/or F (and *not* for A) in the product representations. We find:

$$A \times A = A, \quad A \times E = E, \quad A \times F = F,$$
$$E \times E = E + 2A, \quad E \times F = 2F,$$
$$F \times F = A + E + 2F.$$

$$(6\text{–}25)$$

The forbidden transitions are $A \leftrightarrow A$.

For the diagonal elements, we have

$$[A \times A] = A, \quad [E \times E] = A + E, \quad [F \times F] = A + E + F,$$

$$(6\text{–}26)$$

so the diagonal element is zero for a state of type A.

Problem. Find the selection rules for quadrupole transitions if the symmetry group is (a) D_{3d}; (b) O.

The procedure we have used can be extended to tensors of higher rank. We shall return to this problem later, after we have considered the rotation group.

6–4 Coupled systems. Our discussion of product representations can also be applied to the problem of coupled systems. We start with two independent systems, with coordinates **r** and **r̄**, respectively. The Hamiltonians for the two systems have the same form, and are invariant under the same group. If we are considering some operator of the symmetry group and the operator acts on the coordinates of the first system, we shall denote it by O_R (in the group G), while we use $O_{\overline{R}}$ (in the group $\overline{G}$) when it acts on the second system. R and $\overline{R}$ are the same geometrical transformation applied to system 1 or system 2. If we consider system 1 alone, we can classify its states according to the symmetry group, and will denote its wave functions by $\psi_i^{(\mu)}$. Similarly for system 2, we denote the wave functions by $\overline{\psi}_i^{(\mu)}$. As long as the systems are uncoupled, the total energy is the sum of the energies of the separate systems, and the Hamiltonian $H = H_1 + H_2$ is invariant under all operations $O_R O_{\overline{S}}$. (These products of an element of G and an element of $\overline{G}$ form a group—the direct product $G \times \overline{G}$. Note that R acts only on the coordinates of the first system, while $\overline{S}$ acts only on the second system.) If now the systems are coupled to each other, we add to the Hamiltonian terms involving the distance between the two systems. Operators like $O_R O_{\overline{S}}$ will not leave this interaction term unchanged, *unless* $S = R$; i.e., unless both systems are subjected to the same symmetry operation. So in the presence of the interaction, the symmetry group of the total Hamiltonian consists of products of the form $O_R O_{\overline{R}}$, where R and $\overline{R}$ are the same geometrical operation applied to the coordinates 1 and 2, respectively. (This group is isomorphic to the group G consisting of the elements R.) Suppose the uncoupled systems are in states with eigenfunctions $\psi_i^{(\mu)}$, $\overline{\psi}_j^{(\nu)}$. The introduction of the coupling reduces the symmetry group from the whole set of operations $O_R O_{\overline{S}}$ to the subgroup of operations $O_R O_{\overline{R}}$. Thus our problem is similar to the perturbation problems of Section 6–2, where the perturbation reduced the symmetry group to one of its subgroups. The product wave functions $\psi_i^{(\mu)} \overline{\psi}_j^{(\nu)}$ which formed the basis for an irreducible representation of the group of all elements $R\overline{S}$ (and were therefore necessarily degenerate with one another) will also provide a representation for the subgroup of elements $R\overline{R}$. But this representation will be reducible, and the degenerate states will be split by the coupling. We choose the matrices of the representation in the same way for both G and $\overline{G}$, so that $D_{ik}^{(\mu)}(R) = D_{ik}^{(\mu)}(\overline{R})$. Then

$$O_R O_{\overline{R}} \psi_i^{(\mu)} \overline{\psi}_j^{(\nu)} = (O_R \psi_i^{(\mu)})(O_{\overline{R}} \overline{\psi}_j^{(\nu)}) = \sum_{k,l} \psi_k^{(\mu)} \overline{\psi}_l^{(\nu)} D_{ki}^{(\mu)}(R) D_{lj}^{(\nu)}(R). \quad (6\text{–}27)$$

We see that Eq. (6–28) is formally identical with the product representation in which the ψ's refer to the same system. To find the types of states of the total system which are contained among the products $\psi_i^{(\mu)} \overline{\psi}_j^{(\nu)}$, we must (just as before) reduce the product representation $D^{(\mu)} \times D^{(\nu)}$ to its irreducible components.

We are here applying to the finite symmetry groups the same procedure that is usual for the rotation group. For the rotation group (i.e., electrons in the central field of an atom, where the symmetry group is the full rotation group), the individual electrons are assigned to angular momenta (irreducible representations) l_1, l_2, etc. When the coupling terms are included, we are required to find how the product functions break up into linear combinations belonging to different values of the total angular momentum L (i.e., to different irreducible representations of the total Hamiltonian).

For example, consider an electron in an atom located in a crystal. If we neglect the interaction between the electrons in the atom and assume that the field produced by the ions in the lattice is large, the states of the individual electrons will be classified according to the representations of the symmetry group of the crystalline field. We shall denote the representations for the individual electrons by small letters, a, e, f, and reserve the capital letters for representations of the total system.

Suppose that the symmetry group is $\mathcal{C}_{3v}$. Let the first electron be in a state belonging to the representation a, and the second in a state belonging to the representation e. Then the states of the total system will belong to the representation $a \times e = E$. (This result is somewhat analogous to the case, for the rotation group, where $l_1 = 0$, so that $L = l_2$. However, for the finite symmetry groups, it is clear that all levels belonging to one-dimensional representations behave in this way.) If the electrons are both in states belonging to e, the total system belongs to the representation $e \times e = A_1 + A_2 + E$. The energy diagram might look like Table 6–3.

TABLE 6–3

STRONG CRYSTALLINE FIELD

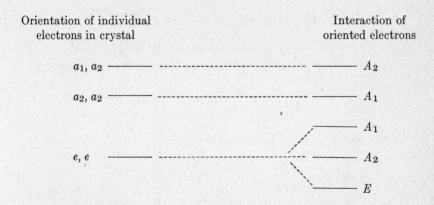

Orientation of individual electrons in crystal	Interaction of oriented electrons
a_1, a_2	A_2
a_2, a_2	A_1
	A_1
e, e	A_2
	E

Here we are not taking account of the identity of the particles (Pauli principle). Our results are correct provided that the one-particle states from which we start are not the same for both systems.

We might also try to find the wave functions, for the coupled systems, which belong to a given row of a given irreducible representation. In other words, we may ask what linear combination of the products $\psi_i^{(\mu)}\bar{\psi}_j^{(\nu)}$ yields $\Psi_k^{(\rho)}$. (This is the problem of finding the Clebsch-Gordan coefficients for the crystal point groups.) For the finite groups this calculation is simple; as a matter of fact, we have already done it in this chapter. Remember that we found the product representation formed from any two representations of the symmetry group. Then we reduced it and obtained those linear combinations of product functions which belonged to the various irreducible representations. Thus for the group $\mathcal{C}_{3v}$ we found $A_1 \times A_1 = A_1$. In our present notation, we write this as $a_1 \times a_1 = A_1$. In terms of the basis functions we have:

$$\Psi_1^{(A_1)} = \psi_1^{(a_1)}\bar{\psi}_1^{(a_1)}. \tag{6-28}$$

Similarly, $a_1 \times e = E$, so

$$\Psi_1^{(E)} = \psi_1^{(a_1)}\bar{\psi}_1^{(e)} \quad \text{and} \quad \Psi_2^{(E)} = \psi_1^{(a_1)}\bar{\psi}_2^{(e)}. \tag{6-29}$$

Finally we had $e \times e = A_1 + A_2 + E$ for which we found the product representation in Eq. (6-10) and then performed the reduction. Translating those results into our present notation, we have:

$$\Psi_1^{(A_1)} = \frac{1}{\sqrt{2}} (\psi_1^{(e)}\bar{\psi}_2^{(e)} + \psi_2^{(e)}\bar{\psi}_1^{(e)}); \tag{6-30}$$

$$\Psi_1^{(A_2)} = \frac{1}{\sqrt{2}} (\psi_1^{(e)}\bar{\psi}_2^{(e)} - \psi_2^{(e)}\bar{\psi}_1^{(e)}); \tag{6-31}$$

$$\Psi_1^{(E)} = \psi_1^{(e)}\bar{\psi}_1^{(e)}, \quad \Psi_2^{(E)} = \psi_2^{(e)}\bar{\psi}_2^{(e)}. \tag{6-32}$$

Similarly, for the group T, we determined the proper linear combinations of products of functions belonging to the F-representation (p. 177). Thus we found that

$$\psi_1^{(f)}\bar{\psi}_1^{(f)} + \psi_2^{(f)}\bar{\psi}_2^{(f)} + \psi_3^{(f)}\bar{\psi}_3^{(f)}$$

belongs to the A_1-representation, and hence

$$\Psi_1^{(A_1)} = \frac{1}{\sqrt{3}} (\psi_1^{(f)}\bar{\psi}_1^{(f)} + \psi_2^{(f)}\bar{\psi}_2^{(f)} + \psi_3^{(f)}\bar{\psi}_3^{(f)}). \tag{6-33}$$

We also have

$$\Psi_1^{(E)} = \frac{1}{\sqrt{3}} \left(\psi_1^{(f)} \bar{\psi}_1^{(f)} + \epsilon \psi_2^{(f)} \bar{\psi}_2^{(f)} + \epsilon^2 \psi_3^{(f)} \bar{\psi}_3^{(f)} \right) \qquad (6\text{--}34)$$

and

$$\Psi_2^{(E)} = \frac{1}{\sqrt{3}} \left(\psi_1^{(f)} \bar{\psi}_1^{(f)} + \epsilon^2 \psi_2^{(f)} \bar{\psi}_2^{(f)} + \epsilon \psi_3^{(f)} \bar{\psi}_3^{(f)} \right). \qquad (6\text{--}35)$$

Since $f \times f = A + E + 2F$, there are two sets of functions, for the coupled system, which belong to the F-representation. (This situation does not occur for two particles in the case of a central field; each L appears only once for given l_1, l_2. But if we have more than two particles, the same problem arises.) All we can do is to give two sets of partners belonging to F. Then the proper zero-order combinations must be found by solving the secular equation, as we have already pointed out in Section 6–2. Thus we obtain

$$\Psi_1^{(F)} = \psi_2^{(f)} \bar{\psi}_3^{(f)} \quad \text{or} \quad \psi_3^{(f)} \bar{\psi}_2^{(f)},$$
$$\Psi_2^{(F)} = \psi_3^{(f)} \bar{\psi}_1^{(f)} \quad \text{or} \quad \psi_1^{(f)} \bar{\psi}_3^{(f)}, \qquad (6\text{--}36)$$
$$\Psi_3^{(F)} = \psi_1^{(f)} \bar{\psi}_2^{(f)} \quad \text{or} \quad \psi_2^{(f)} \bar{\psi}_1^{(f)}.$$

Problem. For the symmetry group D_{2d}, consider all combinations of individual states of two particles. Express the eigenfunctions of states of the coupled system in terms of the single-particle wave functions.

CHAPTER 7

THE SYMMETRIC GROUP

The symmetric group S_n, the group of all permutations on n symbols, is of central importance for both mathematics and physics. For a mathematician, the solution of the problem of finding the irreducible representations of S_n is one of the classic examples of algebraic technique. Despite all the effort that has been expended on the permutation groups, one can still use them as a tool to find new and remarkable combinatorial formulas. The classification of tensors into irreducible sets with respect to any group of linear transformations in n dimensions is easily done once the representations of the symmetric groups are known. In physics, whenever we deal with systems of n identical particles, the symmetry group of the Hamiltonian will contain the group S_n. The classification of atomic and nuclear states depends essentially on the properties of S_n.

In the present chapter we shall determine the characters and the matrices of the irreducible representations of S_n. The problem will be approached in several different ways, some very powerful (but highbrow!), others the physicist's substitute for well-known mathematical techniques.

7-1 The deduction of the characters of a group from those of a subgroup. In Section 3–5 we pointed out that if we can find the simple characters for an invariant subgroup of G, then we can immediately derive some of the simple characters of the group G itself. This result is not of much use. For $n > 4$, the symmetric group S_n has only one invariant subgroup—the alternating group $\mathcal{C}_n$, of index two. From the two one-dimensional representations of the factor group $S_n/\mathcal{C}_n$ we derive two obvious one-dimensional representations of S_n: the identity (symmetric) representation with character (matrix) unity for all elements of S_n; and the antisymmetric representation with character $+1$ for even and -1 for odd permutations. All other irreducible representations are faithful. What we need is some connection between the simple characters of G and those of *any* of its subgroups. This can be achieved through a theorem of Frobenius which we now derive.

Let the group G, of order g, have a subgroup H, of order h. Suppose that we are given a representation $D(G)$ of the group G. We immediately obtain a representation of the subgroup H by selecting from among the matrices $D(G)$ those which correspond to elements in H. Even if $D(G)$ is an irreducible representation of G, the representation of H which we derive in this way [let us call it $D'(H)$] may be reducible. In other words, even though we cannot find a subset of the basis vectors of $D(G)$ which is invariant under all the transformations $D(G)$, we may be able to find one which is invariant under the subset of transformations $D'(H)$. (This breaking up of an irreducible representation when we reduce the symmetry, i.e., go to a subgroup of the original symmetry group, was the basis for the application of group theory to perturbation problems in Chapter 6.)

Now we reverse the question of the previous paragraph. It may be easier to find irreducible representations for the subgroup H than for G (since H is of lower order than G). Can we deduce the characters of G from those of H?

We separate the elements of G into classes K_i; the number of elements in K_i is g_i. We denote the simple characters of G by $\chi_i^{(\mu)}$ and the corresponding representations by $D^{(\mu)}(G)$. Now two elements of the subgroup H which were originally in the same class K_i in G may not be in the same class in H since the element which performed the transformation may not be in H; also some elements in the class K_i in G are not *in* H. On the other hand, two elements which are in the same class in H were *certainly* in the same class in G. In the process of forming the subgroup H from the group G, only h_i elements of the total number g_i were selected; of these elements, h_{i_1} are in the class K_{i_1}, h_{i_2} are in the class K_{i_2}, etc., in H. We denote the simple characters of H by $\phi_{i_\tau}^{(\mu)}$, and the corresponding representations by $\Delta^{(\mu)}(H)$. We saw above that, starting from $D^{(\mu)}(G)$ and considering only elements R in the subgroup H, we obtain a representation (reducible, in general) of H:

$$D'^{(\mu)}(R) = \sum_\nu a_{\mu\nu} \Delta^{(\nu)}(R); \qquad R \text{ in } H, \qquad (7\text{-}1)$$

where the $a_{\mu\nu}$ are positive integers.

If *at least* one element of the class K_i of G is in the subgroup H, then from (7-1) we have

$$\chi_i^{(\mu)} = \sum_\nu a_{\mu\nu}\phi_{i_1}^{(\nu)} = \sum_\nu a_{\mu\nu}\phi_{i_2}^{(\nu)} = \cdots . \qquad (7\text{-}2)$$

Suppose that the class K_j of G has *none* of its elements in the subgroup H. Multiplying (7-2) by $\chi_j^{(\mu)*}$ and summing over μ, we obtain

$$\sum_\mu \chi_j^{(\mu)*}\chi_i^{(\mu)} = 0 = \sum_{\mu,\nu} a_{\mu\nu}\chi_j^{(\mu)*}\phi_{i_\tau}^{(\nu)} \qquad (7\text{-}3)$$

for all classes K_i of G which have at least one element in H and for all classes K_{i_τ} into which those elements are put in H. But this means that for all classes of H

$$\sum_\nu \left(\sum_\mu a_{\mu\nu}\chi_j^{(\mu)*}\right)\phi_k^{(\nu)} = 0. \qquad (7\text{-}3\text{a})$$

Now multiply by $h_k \phi_k^{(\rho)*}/h$ and sum over k (i.e., over all classes of H). From the orthogonality theorem, we obtain

$$\sum_\mu a_{\mu\rho}\chi_j^{(\mu)*} = 0 \qquad \text{for all } \rho, \tag{7-4}$$

or

$$\sum_\mu a_{\mu\rho}\chi_j^{(\mu)} = 0 \qquad \text{for all } \rho. \tag{7-4a}$$

Equation (7–4a) applies to *every* class in G which has *none* of its elements in H. Now we return to consider those classes in G which do have at least one element in H. Multiply (7–2) by $h_{i_\tau}\phi_{i_\tau}^{(\sigma)*}$ and sum over *all* classes of H:

$$h a_{\mu\sigma} = \sum h_{i_\tau}\phi_{i_\tau}^{(\sigma)*}\chi_i^{(\mu)}, \tag{7-5}$$

where the summation extends over all classes of H. Now multiply by $g_l \chi_l^{(\mu)*}$ and sum over μ:

$$hg_l \sum_\mu a_{\mu\sigma}\chi_l^{(\mu)*} = \sum h_{i_\tau}\phi_{i_\tau}^{(\sigma)*} \sum_\mu g_l \chi_i^{(\mu)}\chi_l^{(\mu)*}$$

$$= \sum h_{i_\tau}\phi_{i_\tau}^{(\sigma)*} g\, \delta_{il} = g \sum_{K_{l_\tau}} h_{l_\tau}\phi_{l_\tau}^{(\sigma)*},$$

where the last sum is over all classes K_{l_τ} of H which originated from the class K_l of G. Taking the complex conjugate, we have

$$\psi_l^{(\sigma)} = \sum_\mu a_{\mu\sigma}\chi_l^{(\mu)} = \sum_{K_{l_\tau}} \frac{gh_{l_\tau}}{g_l h}\phi_{l_\tau}^{(\sigma)}. \tag{7-6}$$

Equations (7–2), (7–4a), and (7–6) are the full statement of our theorem. The $a_{\mu\nu}$ in (7–2) were positive integers, so the left side of (7–6) is a compound character $\psi_l^{(\sigma)}$ of G. If we know any simple character of H (i.e., the $\phi_{l_\tau}^{(\sigma)}$), then (7–6) states that the sum on the right-hand side is necessarily a compound character of G. Furthermore, (7–4a) states that the compound character thus obtained will be zero for all classes of G which have no elements in H.

A very important special case of (7–6) is the following: There is one simple character which we know for any group, namely the character of the identity representation, $\chi_i^{(0)} = 1$ for all i. Setting $\phi_{l_\tau}^{(\sigma)} = 1$ for all l_τ in (7–6), we get

$$\frac{gh_l}{g_l h} = \sum_\mu a_{\mu 0}\chi_l^{(\mu)}; \tag{7-7}$$

i.e., the quantities $gh_l/g_l h$ form a compound character of G.

Equation (7–7) can be applied to any subgroup H of G and involves only counting to obtain h_l and g_l. Later in this chapter we shall show how (7–7) can be used to give the Frobenius formula, which expresses in closed form all the simple characters of the symmetric group. In this section we shall confine ourselves to a simple ladder procedure based on (7–6). We shall find the simple characters of S_n by applying (7–6) to the characters of S_{n-1}. As in Section 1–5, we designate by $(1^\alpha 2^\beta 3^\gamma \ldots)$ the class of permutations having α 1-cycles, β 2-cycles, γ 3-cycles, etc. For any S_n we know two simple characters: the symmetric one, with $+1$ for all permutations; and the antisymmetric one, with $+1$ for even and -1 for odd permutations (only if $n > 1$).

For $n = 1, 2$, the trivial results are:

S_1:

	1
	(1)
$\chi^{(1)}$	1

S_2:

	1	1
	(1^2)	(2)
$\chi^{(1)}$	1	1
$\chi^{(2)}$	1	-1

where the number of elements in the class is shown next to the partition symbol. For S_3 we know only the part of the table shown:

S_3:

	1	3	2
	(1^3)	(2, 1)	(3)
$\chi^{(1)}$	1	1	1
$\chi^{(2)}$	1	-1	1
$\chi^{(3)}$			

The last character is, of course, easily written by means of the orthogonality theorems, but let us use (7–6). Since the class (3) of S_3 has no elements in S_2, its compound character will be 0. Start with $\phi^{(1)}$ (i.e., $\chi^{(1)}$ in S_2). Then (7–6) gives

$$\psi_{(1^3)} = \frac{6}{2} \cdot \frac{1}{1} = 3, \qquad \psi_{(2,1)} = \frac{6}{2} \cdot \frac{1}{3} = 1.$$

We obtain the compound character 3, 1, 0. Now

$$\frac{1}{g} \sum_l g_l \psi_l \chi_l^{(1)} = \frac{1}{6}(1 \cdot 3 + 3 \cdot 1 + 2 \cdot 0) = 1,$$

so χ contains $\chi^{(1)}$ once. Subtracting, we obtain the character 2, 0, -1.

This is a simple character, since $\frac{1}{6}[1 \cdot 2^2 + 3 \cdot 0^2 + 2 \cdot (-1)^2] = 1$. Hence our table for S_3 is

$$S_3: \begin{array}{c|ccc} & 1 & 3 & 2 \\ & (1^3) & (2, 1) & (3) \\ \hline \chi^{(1)} & 1 & 1 & 1 \\ \chi^{(2)} & 1 & -1 & 1 \\ \chi^{(3)} & 2 & 0 & -1 \end{array} \ .$$

Now we consider S_3 as a subgroup of S_4. We know the following portion of the character table of S_4:

$$S_4: \begin{array}{c|ccccc} & 1 & 6 & 3 & 8 & 6 \\ & (1^4) & (2, 1^2) & (2^2) & (3, 1) & (4) \\ \hline \chi^{(1)} & 1 & 1 & 1 & 1 & 1 \\ \chi^{(2)} & 1 & -1 & 1 & 1 & -1 \end{array}$$

The factors gh_l/g_lh of Eq. (7–6) are 4, 2, 0, 1, 0. First we take $\phi^{(1)}$ (i.e., $\chi^{(1)}$ for S_3) and get the compound character ψ: 4, 2, 0, 1, 0. Then

$$\sum \frac{g_l}{g} \psi_l^2 = \frac{1}{24}[1 \cdot 4^2 + 6 \cdot 2^2 + 3 \cdot 0^2 + 8 \cdot 1^2 + 6 \cdot 0^2] = 2,$$

and hence ψ contains two simple characters, each of which appears once. ψ contains $\chi^{(1)}$ once since

$$\sum \frac{g_l}{g} \psi_l \chi_l^{(1)} = \frac{1}{24}[1 \cdot 4 + 6 \cdot 2 + 3 \cdot 0 + 8 \cdot 1 + 6 \cdot 0] = 1.$$

Subtracting $\chi^{(1)}$, we obtain the character $\chi^{(4)}$: 3, 1, -1, 0, -1. This character is simple since

$$\sum \frac{g_l}{g} (\chi_l^{(4)})^2 = \frac{1}{24}[1 \cdot 3^2 + 6 \cdot 1^2 + 3 (-1)^2 + 8 \cdot 0^2 + 6 (-1)^2] = 1.$$

Next we use $\phi^{(2)}$ (i.e., $\chi^{(2)}$ for S_3) and get ψ: 4, -2, 0, 1, 0. This compound character contains $\chi^{(2)}$ once since

$$\sum \frac{g_l}{g} \psi_l \chi_l^{(2)}$$

$$= \frac{1}{24}[1 \cdot 4 \cdot 1 + 6 (-2)(-1) + 3 \cdot 0 \cdot 1 + 8 \cdot 1 \cdot 1 + 6 \cdot 0 \cdot (-1)]$$

$$= 1.$$

Subtracting $\chi^{(2)}$, we are left with the simple character $\chi^{(5)} = 3, -1, -1, 0, 1$. Finally we start with $\phi^{(3)}$ (i.e., $\chi^{(3)}$ for S_3) and obtain ψ: 8, 0, 0, -1, 0.

We find

$$\sum \frac{g_l}{g} \psi_l^2 = \frac{1}{24} [1 \cdot 8^2 + 8 (-1)^2] = 3,$$

so ψ is a sum of three distinct simple characters. ψ contains the previously derived $\chi^{(4)}$ once since

$$\sum \frac{g_l}{g} \psi_l \chi_l^{(4)} = \frac{1}{24} [1 \cdot 8 \cdot 3 + 0 + 0 + 0] = 1.$$

Similarly, ψ contains $\chi^{(5)}$ once. Subtracting $\chi^{(4)}$ and $\chi^{(5)}$ from ψ, we obtain the character $\chi^{(3)}$: 2, 0, 2, -1, 0 which is simple since

$$\sum \frac{g_l}{g} (\chi_l^{(3)})^2 = \frac{1}{24} [1 \cdot 2^2 + 3(2)^2 + 8 (-1)^2] = 1.$$

We have obtained the complete character table of S_4 (Table 7–1).

TABLE 7–1

S_4:

	1	6	3	8	6
	(1^4)	$(2, 1^2)$	(2^2)	$(3, 1)$	(4)
$\chi^{(1)}$	1	1	1	1	1
$\chi^{(2)}$	1	-1	1	1	-1
$\chi^{(3)}$	2	0	2	-1	0
$\chi^{(4)}$	3	1	-1	0	-1
$\chi^{(5)}$	3	-1	-1	0	1

Now we apply the same procedure to S_5, considering S_4 as a subgroup. We have the following information about S_5:

S_5:

	1	10	15	20	20	30	24
	(1^5)	$(2, 1^3)$	$(2^2, 1)$	$(3, 1^2)$	$(3, 2)$	$(4, 1)$	(5)
$\chi^{(1)}$	1	1	1	1	1	1	1
$\chi^{(2)}$	1	-1	1	1	-1	-1	1

The factors $gh_l/g_l h$ in Eq. (7–6) are 5, 3, 1, 2, 0, 1, 0. Using $\phi^{(1)}$ from S_4 yields this same set of numbers as a compound character ψ of S_5, that is,

$$\sum \frac{g_l}{g} \psi_l^2 = \frac{1}{120} [1 \cdot 5^2 + 10 \cdot 3^2 + 15 \cdot 1^2 + 20 \cdot 2^2 + 30 \cdot 1^2] = 2,$$

and ψ therefore contains two different simple characters. It contains $\chi^{(1)}$ since $\sum (g_l/g) \psi_l = 1$; subtracting $\chi^{(1)}$, we obtain the simple character $\chi^{(3)}$: 4, 2, 0, 1, -1, 0, -1. In similar fashion, we have from $\phi^{(2)}$ the simple character $\chi^{(4)}$: 4, -2, 0, 1, 1, 0, -1.

TABLE 7–2

S_5:	1 (1^5)	10 $(2, 1^3)$	15 $(2^2, 1)$	20 $(3, 1^2)$	20 $(3, 2)$	30 $(4, 1)$	24 (5)
$\chi^{(1)}$	1	1	1	1	1	1	1
$\chi^{(2)}$	1	−1	1	1	−1	−1	1
$\chi^{(3)}$	4	2	0	1	−1	0	−1
$\chi^{(4)}$	4	−2	0	1	1	0	−1
$\chi^{(5)}$	5	1	1	−1	1	−1	0
$\chi^{(6)}$	5	−1	1	−1	−1	1	0
$\chi^{(7)}$	6	0	−2	0	0	0	1

Starting from $\phi^{(3)}$, $\phi^{(4)}$, $\phi^{(5)}$, respectively, we form the compound characters

$$10, 0, 2, -2, 0, 0, 0: \qquad \sum \frac{g_l}{g} \psi_l^2 = 2,$$

$$15, 3, -1, 0, 0, -1, 0: \qquad \sum \frac{g_l}{g} \psi_l^2 = 3,$$

$$15, -3, -1, 0, 0, 1, 0: \qquad \sum \frac{g_l}{g} \psi_l^2 = 3.$$

The second character contains $\chi^{(3)}$ once, the third contains $\chi^{(4)}$ once; subtracting these, we have the compound characters

$$\chi^{(5)} + \chi^{(6)}: 10, 0, 2, -2, 0, 0, 0,$$
$$\chi^{(5)} + \chi^{(7)}: 11, 1, -1, -1, 1, -1, 1,$$
$$\chi^{(6)} + \chi^{(7)}: 11, -1, -1, -1, -1, 1, 1,$$

each of which consists of two simple characters. From the last two we find

$$\chi^{(5)} - \chi^{(6)}: 0, 2, 0, 0, 2, -2, 0,$$

and combining with the first, we obtain

$$\chi^{(5)}: 5, 1, 1, -1, 1, -1, 0,$$
$$\chi^{(6)}: 5, -1, 1, -1, -1, 1, 0.$$

Then we find $\chi^{(7)}: 6, 0, -2, 0, 0, 0, 1$. The complete character table is given in Table 7–2.

The task for S_6 is already formidable. Our difficulty really stems from the fact that we have used only the subgroup S_{n-1} of S_n. In the following sections we shall show how the general problem can be reduced to a routine.

Problem. Use the method of this section to find the character table of S_6.

7–2 Frobenius' formula for the characters of the symmetric group.

In Section 7–1, we derived the following result:

Given a group G of order g, having g_l elements in the class K_l. If a subgroup H of G has order h and contains h_l elements from K_l, then the set of numbers

$$\chi_l^{(H)} = \frac{gh_l}{g_l h} \tag{7–7}$$

forms a compound character of G.

We shall show in this section how, starting from (7–7), we can solve completely the problem of finding the irreducible representations of the symmetric group. The idea (again due to Frobenius) is the following: Equation (7–7) supplies us with a character $\chi^{(H)}$ of $G(\equiv S_n)$ for each subgroup H chosen. Consider any partition $(\lambda) = (\lambda_1, \lambda_2, \ldots, \lambda_n)$ of n,

$$\lambda_1 + \lambda_2 + \cdots + \lambda_n = n, \tag{7–8}$$

$$\lambda_1 \geq \lambda_2 \geq \cdots \geq \lambda_n \geq 0. \tag{7–8a}$$

Corresponding to the partition (λ) we can construct a subgroup of G: Distribute the symbols $1, \ldots, n$ over the partition; group λ_1 symbols in one bunch, λ_2 others in a second bunch, etc. (the choice of symbols in each bunch is irrelevant). Now construct the symmetric group on the first bunch of symbols; call it G_{λ_1}. Do the same thing for each of the other bunches. Now take the direct product of all these groups. (Remember they have no symbols in common.) The direct product is

$$G_{(\lambda)} = G_{\lambda_1} \times G_{\lambda_2} \times \cdots \times G_{\lambda_n}, \tag{7–9}$$

and is a subgroup of S_n; using (7–7), we can obtain a compound character of S_n which we label $\phi^{(\lambda)}$. Each partition (λ) of n can be handled in this way. Since the number of partitions of n is equal to the number of classes in S_n, we thus obtain a number of compound characters of S_n which is equal to the number of classes in S_n. We shall prove that the $\phi^{(\lambda)}$ are linearly independent vectors; they must therefore yield all the simple characters of S_n when proper linear combinations are taken. We shall also give the final result obtained by Frobenius, a closed formula giving all the simple characters of S_n.

In order to use Eq. (7–7) we must calculate the quantities h_l, h, and g_l. Suppose we wish to construct the character corresponding to the partition (λ). From (7–9),

$$h = \lambda_1! \cdot \lambda_2! \cdots \lambda_n!. \tag{7–10}$$

Next consider the class K_l of S_n. We can characterize K_l by giving its cycle structure $(1^\alpha, 2^\beta, 3^\gamma, \ldots)$, the symbol stating that the permutations in K_l contain α 1-cycles, β 2-cycles, γ 3-cycles, etc., where

$$\alpha + 2\beta + 3\gamma + \cdots = n. \tag{7-11}$$

The number of permutations in K_l was found in Chapter 1, Eq. (1-27):

$$g_l = \frac{n!}{1^\alpha \cdot \alpha! 2^\beta \cdot \beta! 3^\gamma \cdot \gamma! \cdots}. \tag{7-12}$$

The quantity h_l is the number of elements in $G_{(\lambda)}$ which have the cycle structure $(1^\alpha, 2^\beta, 3^\gamma \ldots)$. In order that an element of $G_{(\lambda)}$ have this structure it must contain

$\qquad\qquad \alpha$ 1-cycles, $\qquad \beta$ 2-cycles, $\qquad \gamma$ 3-cycles, etc.

Such an element can be constructed according to (7-9) if the direct factor from G_{λ_1} has

$\qquad\qquad \alpha_1$ 1-cycles, $\qquad \beta_1$ 2-cycles, $\qquad \gamma_1$ 3-cycles, etc.;

the factor from G_{λ_2} has

$\qquad\qquad \alpha_2$ 1-cycles, $\qquad \beta_2$ 2-cycles, $\qquad \gamma_2$ 3-cycles, etc.;

and finally the factor from G_{λ_n} has

$\qquad\qquad \alpha_n$ 1-cycles, $\qquad \beta_n$ 2-cycles, $\qquad \gamma_n$ 3-cycles, etc.,

provided that

$$\sum_{i=1}^{n} \alpha_i = \alpha, \qquad \sum_{i=1}^{n} \beta_i = \beta, \qquad \sum_{i=1}^{n} \gamma_i = \gamma, \quad \text{etc.} \tag{7-13}$$

Since G_{λ_i} consists of permutations on λ_i symbols, we have

$$\alpha_i + 2\beta_i + 3\gamma_i + \cdots = \lambda_i. \tag{7-14}$$

The number of permutations of G_{λ_i} having the cycle structure

$$(1^{\alpha_i}, 2^{\beta_i}, 3^{\gamma_i}, \ldots)$$

for fixed α_i, β_i, γ_i, $\ldots$ is

$$\frac{\lambda_i!}{1^{\alpha_i} \cdot \alpha_i! 2^{\beta_i} \cdot \beta_i! \cdots}.$$

If we select any solution of (7-13) and (7-14), we obtain

$$\prod_{i=1}^{n} \frac{\lambda_i!}{1^{\alpha_i} \cdot \alpha_i! 2^{\beta_i} \cdot \beta_i! \cdots}$$

members of the class $(1^\alpha, 2^\beta, 3^\gamma \ldots)$. Summing over all solutions, we have

$$h_l = \sum_{\substack{\alpha_i, \beta_i, \ldots \\ \alpha_i + 2\beta_i + \cdots = \lambda_i}} \prod_i \frac{\lambda_i!}{1^{\alpha_i} \cdot \alpha_i! 2^{\beta_i} \cdot \beta_i! \cdots}. \tag{7-15}$$

Substituting (7–10), (7–12), and (7–15) in Eq. (7–7), we find

$$\phi_l^{(\lambda)} = \frac{gh_l}{g_l h} = \sum_{\substack{\alpha_i, \beta_i, \ldots \\ \Sigma\alpha_i = \alpha, \Sigma\beta_i = \beta, \ldots \\ \alpha_i + 2\beta_i + \cdots = \lambda_i}} \frac{\alpha!}{\alpha_1! \alpha_2! \cdots} \cdot \frac{\beta!}{\beta_1! \beta_2! \cdots} \cdots. \tag{7-16}$$

Before proceeding, we wish to illustrate the method in a particular case. We consider S_4 and construct the subgroup corresponding to the partition (2^2). Then $G_{(2^2)} = G_2 \times G_2$ where, say, we form the symmetric group on elements 1 and 2, and multiply by the symmetric group on elements 3 and 4: e, (12), (34), (12)(34). The order of the subgroup is $h = 2! \cdot 2! = 4$. Consider the various classes of S_4. Classes (4) and (31) have no members in $G_{(2^2)}$, so (7–7) will give zero for them. The class (2^2) of S_4 appears once in $G_{(2^2)}$, the class (21^2) appears twice, the class (1^4) once, so we get the compound character

$$(1^4), (21^2), (2^2), (31), (4),$$
$$6 , 2 , 2 , 0 , 0 .$$

The quantities $\phi_l^{(\lambda)}$ are the coefficients in a certain multinomial which we now construct. Suppose we consider the class $(1^\alpha 2^\beta 3^\gamma \ldots)$ in S_n. The polynomial

$$(x_1 + x_2 + \cdots + x_n)^\alpha (x_1^2 + x_2^2 + \cdots + x_n^2)^\beta (x_1^3 + x_2^3 + \cdots + x_n^3)^\gamma \cdots$$

in the variables $x_1, \ldots, x_n$ can be expanded to give

$$\left(\sum_{\substack{\alpha_1, \ldots, \alpha_n \\ \Sigma\alpha_i = \alpha}} \frac{\alpha!}{\alpha_1! \cdots \alpha_n!} x_1^{\alpha_1} \cdots x_n^{\alpha_n} \right) \left(\sum_{\substack{\beta_1, \ldots, \beta_n \\ \Sigma\beta_i = \beta}} \frac{\beta!}{\beta_1! \cdots \beta_n!} x_1^{2\beta_1} \cdots x_n^{2\beta_n} \right) \cdots$$

$$= \sum_{\substack{\alpha_i, \beta_i, \ldots \\ \Sigma\alpha_i = \alpha, \Sigma\beta_i = \beta, \ldots}} \frac{\alpha!}{\alpha_1! \cdots \alpha_n!} \times \frac{\beta!}{\beta_1! \cdots \beta_n!} \times \cdots \times x_1^{\alpha_1 + 2\beta_1 + 3\gamma_1 + \cdots}$$

$$\times \cdots \times x_n^{\alpha_n + 2\beta_n + 3\gamma_n + \cdots}.$$

We now collect all the coefficients of a given monomial $x_1^{\mu_1} x_2^{\mu_2} \cdots x_n^{\mu_n}$. The expression can then be written as

$$\sum_{\Sigma\mu_i = n} x_1^{\mu_1} \cdots x_n^{\mu_n} \sum_{\substack{\alpha_i, \beta_i, \ldots \\ \Sigma\alpha_i = \alpha, \Sigma\beta_i = \beta, \ldots \\ \alpha_i + 2\beta_i + \cdots = \mu_i}} \frac{\alpha!}{\alpha_1! \cdots \alpha_n!} \times \frac{\beta!}{\beta_1! \cdots \beta_n!} \times \cdots.$$

All monomials which are obtained from $x_1^{\mu_1} \ldots x_n^{\mu_n}$ by permuting the variables have the same coefficient. We can, therefore, order the μ_i in decreasing sequence and identify the μ_i with the parts of the partition (λ). Then making use of Eq. (7–16), we find

$$\left(\sum x_i\right)^\alpha \left(\sum x_i^2\right)^\beta \left(\sum x_i^3\right)^\gamma \cdots$$

$$= \sum_{\substack{(\lambda) \\ \text{perm}}} x_1^{\lambda_1} \cdots x_n^{\lambda_n} \sum_{\substack{\alpha_i, \beta_i \ldots \\ \Sigma\alpha_i=\alpha, \Sigma\beta_i=\beta, \ldots \\ \alpha_i+2\beta_i+\cdots=\lambda_i}} \frac{\alpha!}{\alpha_1! \cdots \alpha_n!} \times \frac{\beta!}{\beta_1! \cdots \beta_n!} \times \cdots$$

$$= \sum_{(\lambda)} \phi_l^{(\lambda)} \sum_{\text{perm}} x_1^{\lambda_1} \cdots x_n^{\lambda_n}. \tag{7–17}$$

The sum runs through all partitions (λ) and all distinct monomials obtained by permuting $x_1 \ldots x_n$.

We define new variables s_r:

$$s_r = \sum_i x_i^r, \qquad r = 1, \ldots, n. \tag{7–18}$$

For each class l with cycle structure $(1^\alpha, 2^\beta, 3^\gamma, \ldots)$, we define still a third set of variables, namely, the left-hand side of Eq. (7–17):

$$s_{(l)} = s_1^\alpha s_2^\beta \cdots, \tag{7–19}$$

where (l) is the class $(1^\alpha, 2^\beta, 3^\gamma, \ldots)$. Then Eq. (7–17) can be written as

$$s_{(l)} = \sum_{(\lambda)} \phi_{(l)}^{(\lambda)} \sum_{\text{perm}} x_1^{\lambda_1} \cdots x_n^{\lambda_n}. \tag{7–20}$$

The s_r defined by Eq. (7–18) are n functions of the n independent variables x_i. The s_r are functionally independent; their Jacobian is the determinant

$$J = \begin{vmatrix} 1 & 1 & 1 & \cdots \\ 2x_1 & 2x_2 & 2x_3 & \cdots \\ 3x_1^2 & 3x_2^2 & 3x_3^2 & \cdots \\ \vdots & & & \vdots \\ nx_1^{n-1} & nx_2^{n-1} & nx_3^{n-1} & \cdots \end{vmatrix} = n! \begin{vmatrix} 1 & 1 & 1 & \cdots \\ x_1 & x_2 & x_3 & \cdots \\ x_1^2 & x_2^2 & x_3^2 & \cdots \\ \vdots & & & \vdots \\ x_1^{n-1} & x_2^{n-1} & x_3^{n-1} & \cdots \end{vmatrix}$$

$$= n! \prod_{i<j} (x_i - x_j). \tag{7–21}$$

This last result is obtained by noting that the determinant is a multinomial of total degree $0 + 1 + 2 + \cdots + (n - 1) = n(n - 1)/2$. The determinant vanishes if any two variables are made equal, and hence it contains all possible factors $(x_i - x_j)$ with $i \neq j$; since there are $n(n - 1)/2$ such factors, we obtain Eq. (7–21) (the sign being identified by finding the coefficient of $x_1^{n-1} x_2^{n-2} \cdots x_n^0$). Since J is not identically zero, the s_r can be introduced as new independent variables. Furthermore, the $s_{(l)}$ defined by Eq. (7–19) are linearly independent (for if they were linearly dependent, this would imply a functional dependence among the s_r, which we have just proved to be independent variables). By the same argument, the quantities

$$\sum_{\text{perm}} x_1^{\lambda_1} \cdots x_n^{\lambda_n}$$

corresponding to the various partitions (λ) are also linearly independent. Thus as (l) runs through the c classes of S_n, we obtain c equations like (7–20) expressing the c linearly independent quantities $s_{(l)}$ in terms of the c linearly independent quantities $\sum_{\text{perm}} x_1^{\lambda_1} \cdots x_n^{\lambda_n}$. But, in a transformation from one linearly independent set to another, the matrix of the transformation is nonsingular. The columns of this matrix, which are just the characters $\phi_{(l)}^{(\lambda)}$ for fixed (λ), must therefore be linearly independent. So the c simple characters must be expressible as linear combinations of the $\phi_i^{(\lambda)}$.

For the derivation of general statements, we have used n variables x_i. In practical calculations, the only requirement is that the number of x's should not be less than the number of parts in the partition (λ).

We apply Eq. (7–20) to S_4. Then

$$(l) = (1^4): \qquad s_{(1^4)} = (s_1)^4 = \left(\sum x_i\right)^4 = \sum x_1^4 + 4\sum x_1^3 x_2$$
$$+ 6\sum x_1^2 x_2^2 + 12\sum x_1^2 x_2 x_3 + 24\sum x_1 x_2 x_3 x_4,$$

where $\sum x_1^4$, etc., implies a sum over all permutations of indices which give distinct monomials. Similarly,

$$(l) = (21^2): \qquad s_{(21^2)} = s_2(s_1)^2 = (x_1^2 + \cdots + x_n^2)(x_1 + \cdots + x_n)^2$$
$$= [\sum x_1^2 + 2\sum x_1 x_2][\sum x_3^2]$$
$$= \sum x_1^4 + 2\sum x_1^3 x_2 + 2\sum x_1^2 x_2^2$$
$$+ 2\sum x_1^2 x_2 x_3;$$

$$(l) = (2^2): \qquad s_{(2^2)} = (s_2)^2 = \left(\sum x_1^2\right)^2 = \sum x_1^4 + 2\sum x_1^2 x_2^2;$$

$$(l) = (3, 1): \qquad s_{(3,1)} = s_3 s_1 = \left(\sum x_1^3\right)\left(\sum x_2\right) = \sum x_1^4 + \sum x_1^3 x_2;$$

$$(l) = (4): \qquad s_{(4)} = s_4 = \sum x_1^4.$$

From these equations we form the matrix of the $\phi_{(l)}^{(\lambda)}$; we transpose the matrix so that (λ) designates rows and (l) columns:

$$
\begin{array}{c}
(l) = (1^4)^1 \quad (21^2)^6 \quad (2^2)^3 \quad (31)^8 \quad (4)^6 \\
(\lambda) = \begin{array}{c} (4) \\ (31) \\ (2^2) \\ (21^2) \\ (1^4) \end{array}
\left[
\begin{array}{ccccc}
1 & 1 & 1 & 1 & 1 \\
4 & 2 & 0 & 1 & 0 \\
6 & 2 & 2 & 0 & 0 \\
12 & 2 & 0 & 0 & 0 \\
24 & 0 & 0 & 0 & 0
\end{array}
\right] . \quad (7\text{-}22)
\end{array}
$$

From this matrix we could now determine the simple characters $\chi_{(l)}^{(\lambda)}$ in the same way as in Section 7–1.

Problem. Apply this method to S_5 and obtain the $\phi^{(\lambda)}$; use Table 7–2 to express the $\phi^{(\lambda)}$ as linear combinations of the simple characters of S_5.

Formula (7–20) enables us to find a complete set of compound characters $\phi^{(\lambda)}$ for S_n. Going beyond this, Frobenius has given a similar formula which yields directly all the simple characters $\chi^{(\lambda)}$. For this purpose we use a determinant of the type already encountered in Eq. (7–21), that is,

$$
D(x_i) \equiv D(x_1, \ldots, x_m) = \prod_{i<j} (x_i - x_j)
$$

$$
= \sum_P \delta_P P x_1^{m-1} x_2^{m-2} \cdots x_{m-1} x_m^0, \quad (7\text{-}23)
$$

where P is any permutation of the variables x_i, and $\delta_P = \pm 1$, depending on whether P is an even or odd permutation. Note that $D(x_i)$ is an alternating function of the x_i, changing sign under any interchange of two variables. The quantities $s_{(l)}$ in Eq. (7–20) are symmetric multinomials in the x_i [cf. Eqs. (7–18) and (7–19)], so the product $s_{(l)}D(x_i)$ changes sign under any interchange of two variables. If we write the expansion of $s_{(l)}D(x_i)$ as a sum of monomials $x_1^{\nu_1}, \ldots, x_m^{\nu_m}$, no terms having equal powers for two of the variables can occur, since the interchange of those variables in $s_{(l)}D(x_i)$ would change the sign of all terms in the expansion, while the interchange acting on the monomial leaves the sign unchanged; thus such monomials must have zero coefficient. We can, therefore, rearrange the variables in the monomials so that the highest power comes first, the next lower power second, etc. Hence the quantity $s_{(l)}D(x_i)$ can be written as

$$
s_{(l)}D(x_i) = \sum_{(\lambda)} \chi_{(l)}^{(\lambda)} \sum_P \delta_P P x_1^{\lambda_1+m-1} x_2^{\lambda_2+m-2} \cdots x_m^{\lambda_m}. \quad (7\text{-}24)
$$

The first sum goes over all partitions (λ) of n, the second sum over all permutations of the variables $x_1, \ldots, x_m$. Clearly the coefficients $\chi_{(l)}^{(\lambda)}$ are certain linear combinations of the $\phi_{(l)}^{(\lambda)}$. Frobenius' theorem states that the quantities $\chi_{(l)}^{(\lambda)}$ in Eq. (7–24) are precisely the simple characters of S_n.

We shall give many examples to make the meaning of (7–24) clear, but first we present the proof of the theorem.

The $\chi_{(l)}^{(\lambda)}$ will be simple characters if [cf. Eq. (3–152)]

$$\sum_{(l)} g_{(l)} |\chi_{(l)}^{(\lambda)}|^2 = g, \tag{7–25}$$

where g is the order of S_n and $g_{(l)}$ the number of elements in the class (l).

Just as we introduced the variables x_i, we now introduce a second set y_i and define

$$t_r = \sum_i y_i^r, \tag{7–18a}$$

$$t_{(l)} = t_1^\alpha t_2^\beta t_3^\gamma \cdots \quad \text{for} \quad (l) = (1^\alpha, 2^\beta, 3^\gamma, \ldots). \tag{7–19a}$$

Construct the sum $\sum_{(l)} (g_{(l)}/g) s_{(l)} t_{(l)}$. Using Eqs. (7–12), (7–19) and (7–19a), we have

$$\sum_{(l)} \frac{g_{(l)}}{g} s_{(l)} t_{(l)} = \sum_{\substack{\alpha, \beta, \ldots \\ \alpha + 2\beta + \cdots = n}} \frac{1}{1^\alpha \alpha!} (s_1 t_1)^\alpha \frac{1}{2^\beta \beta!} (s_2 t_2)^\beta \cdots. \tag{7–26}$$

We now sum over n from 0 to ∞:

$$\sum_{\substack{\text{all partitions} \\ \text{of all } n}} \frac{g_{(l)}}{g} s_{(l)} t_{(l)} = \exp\left[s_1 t_1 + \frac{1}{2} s_2 t_2 + \frac{1}{3} s_3 t_3 + \cdots \right]. \tag{7–27}$$

Now substitute for s_r, t_r from Eqs. (7–18) and (7–18a):

$$\sum_{r=1}^\infty \frac{s_r t_r}{r} = \sum_{r=1}^\infty \frac{\sum_1^m x_i^r \sum_1^m y_j^r}{r} = \sum_{i,j=1}^m \sum_{r=1}^\infty \frac{(x_i y_j)^r}{r}$$

$$= - \sum_{i,j=1}^m \ln(1 - x_i y_j) = -\ln \prod_{i,j=1}^m (1 - x_i y_j). \tag{7–28}$$

Thus the right side of (7–27) is

$$\frac{1}{\prod_{i,j=1}^m (1 - x_i y_j)}.$$

Now we prove that

$$\frac{D(x_i) D(y_j)}{\prod_{i,j=1}^m (1 - x_i y_j)} = \left| \frac{1}{1 - x_r y_s} \right|, \tag{7–29}$$

where D is the determinant defined by Eq. (7–23) and $|1/(1 - x_r y_s)|$ is a determinant whose rs-element is $1/(1 - x_r y_s)$. This determinant has the form

$$\begin{vmatrix} (1 - x_1 y_1)^{-1} & (1 - x_1 y_2)^{-1} & (1 - x_1 y_3)^{-1} & \cdots \\ (1 - x_2 y_1)^{-1} & (1 - x_2 y_2)^{-1} & (1 - x_2 y_3)^{-1} & \cdots \\ (1 - x_3 y_1)^{-1} & \cdots\cdots\cdots\cdots\cdots\cdots\cdots\cdots\cdots\cdots \\ \vdots & & & \vdots \end{vmatrix}. \qquad (7\text{–}30)$$

Subtract the first from the ith row:

$$(1 - x_i y_j)^{-1} - (1 - x_1 y_j)^{-1} = \frac{x_i - x_1}{1 - x_1 y_j} \cdot \frac{y_j}{1 - x_i y_j}.$$

The factor $(x_i - x_1)$ is common to all the elements of the ith row, and the factor $(1 - x_1 y_j)$ is common to all the elements of the jth column; factoring these, we obtain

$$\frac{(x_2 - x_1)(x_3 - x_1)\ldots(x_m - x_1)}{\prod_{j=1}^m (1 - x_1 y_j)} \begin{vmatrix} 1 & 1 & 1 & \cdot\cdot \\ \dfrac{y_1}{1 - x_2 y_1} & \dfrac{y_2}{1 - x_2 y_2} & \cdots \\ \dfrac{y_1}{1 - x_3 y_1} & \dfrac{y_2}{1 - x_3 y_2} & \cdots \\ \vdots & \vdots & \end{vmatrix}.$$

Now subtract the first column from the others:

$$\frac{y_j}{1 - x_i y_j} - \frac{y_1}{1 - x_i y_1} = \frac{y_j - y_1}{1 - x_i y_1} \cdot \frac{1}{1 - x_i y_j}.$$

Again the factor $(y_j - y_1)$ is common to all elements of the jth column and $(1 - x_i y_1)$ is common to all elements of the ith row, so the determinant becomes

$$\frac{(x_2 - x_1)\ldots(x_m - x_1)(y_2 - y_1)\ldots(y_m - y_1)}{\prod_{j=1}^m (1 - x_1 y_j)\prod_{i=2}^m (1 - x_i y_1)}$$

$$\times \begin{vmatrix} 1 & 0 & 0 & \cdots \\ \dfrac{y_1}{1 - x_2 y_1} & \dfrac{1}{1 - x_2 y_2} & \dfrac{1}{1 - x_2 y_3} & \cdots \\ \dfrac{y_1}{1 - x_3 y_1} & \dfrac{1}{1 - x_3 y_2} & \dfrac{1}{1 - x_3 y_3} & \cdots \\ \vdots & & & \vdots \end{vmatrix},$$

and

$$\left|\frac{1}{1 - x_iy_j}\right|_{i,j=1,\ldots,m}$$

$$= \frac{(x_2 - x_1)\ldots(x_m - x_1)(y_2 - y_1)\ldots(y_m - y_1)}{\prod_{j=1}^{m}(1 - x_1y_j)\prod_{i=2}^{m}(1 - x_iy_1)}\left|\frac{1}{1 - x_iy_j}\right|_{i,j=2,\ldots,m}.$$

$$(7\text{–}31)$$

Repeating the induction, we arrive at (7–29). Now multiplying both sides of (7–27) by $D(x_i)D(y_j)$ and using (7–29), we obtain

$$\sum_{(l),n=1}^{\infty}\frac{g_{(l)}}{g}s_{(l)}t_{(l)}D(x_i)D(y_j) = \left|\frac{1}{1 - x_iy_j}\right|.$$

But

$$(1 - x_iy_j)^{-1} = \sum_{\nu=0}^{\infty}x_i^{\nu}y_j^{\nu},$$

so

$$\left|\frac{1}{1 - x_iy_j}\right| = \sum_{\nu_1,\ldots,\nu_m=1}^{\infty}\sum_{P}\delta_P x_{P_1}^{\nu_1}y_1^{\nu_1}x_{P_2}^{\nu_2}y_2^{\nu_2}\ldots x_{P_m}^{\nu_m}y_m^{\nu_m}.\quad(7\text{–}32)$$

All the ν's must be different since the function alternates under any interchange. The indices can be arranged in descending order; the $\sum_P$ then goes over all permutations of the x's and y's, and δ_P is the sign of the permutation of one set relative to the other. Since the ν's are now in descending order, we can set $\nu_i = \lambda_i + m - i$ and get

$$\sum_{(l)}\frac{g_{(l)}}{g}s_{(l)}D(x_i)t_{(l)}D(y_j) = \sum_{(\lambda)}\sum_{P}\delta_P x_{P_1}^{\lambda_1+m-1}y_1^{\lambda_1+m-1}\cdots x_{P_m}^{\lambda_m}y_m^{\lambda_m}.$$

$$(7\text{–}33)$$

Substituting (7–24) and the similar expression in y, and comparing coefficients of monomials in $x_1, \ldots, x_m, y_1, \ldots, y_m$, we find

$$\sum_{(l)}\frac{g_{(l)}}{g}\chi_{(l)}^{(\lambda)}\chi_{(l)}^{(\mu)} = 0 \quad\text{for}\quad (\lambda) \neq (\mu),$$

$$\sum_{(l)}\frac{g_{(l)}}{g}[\chi_{(l)}^{(\lambda)}]^2 = 1,$$

$$(7\text{–}25)$$

which proves that formula (7–24) gives us independent simple characters. Equation (7–25) still leaves the possibility that the $\chi_{(l)}^{(\lambda)}$ are $\pm$ a simple character, but we shall see in the process of working out the results that the sign is given correctly.

Formula (7–24) looks (and is!) formidable, and we shall therefore deal with it to devise methods which are tractable.

7–3 Graphical methods. Lattice permutations. Young patterns. Young tableaux. First let us consider the identity element of S_n, i.e., the class $(l) = (1^n)$. The left side of (7–24) is just $D(x_i)(\sum x_i)^n$. We start with $D(x_i)$ and multiply successively n times by $\sum x_i$. Since the product at each stage is an alternating function, any monomial having two equal powers must have zero coefficient. Thus $\chi_{(1^n)}^{(\lambda)}$ will be the coefficient of

$$x_1^{\lambda_1+m-1} x_2^{\lambda_2+m-2} \cdots x_m^{\lambda_m}$$

in (7–24). We start from

$$D(x_i) = \sum_P \delta_P P x_1^{m-1} x_2^{m-2} \cdots x_{m-1} x_m^0.$$

If we multiply by $\sum x_i$, we increase one of the powers by unity. But if at any stage two of the exponents become equal, the term must vanish. So we see that, as we raise the degree of the polynomial by one at each stage, we must always raise the power of x_1 at least as fast as x_2, etc. Our final goal is to raise the power of x_1 by λ_1, that of x_i by λ_i. While doing this, one factor at a time, we must be sure at each stage that the number of multiplications by x_1 is greater than or equal to the number of multiplications by x_2, etc. The total number of ways by which we can reach our goal will be $\chi_{(1^n)}^{(\lambda)}$, i.e., the degree of the representation (λ).

For example, suppose that we wish to obtain the dimension of the irreducible representation of S_4 corresponding to $(\lambda) = (2, 1^2)$. Then we must raise the power of x_1 by 2, and the powers of x_2 and x_3 by 1. In other words we wish to arrive at $x_1^2 x_2 x_3$ by applying one x-factor at a time, always keeping more x_1's than x_2's than x_3's, etc. The possible ways of doing this are

$$x_1^2 x_2 x_3 = x_1 x_1 x_2 x_3, \qquad x_1 x_2 x_1 x_3, \qquad x_1 x_2 x_3 x_1. \qquad (7\text{–}34)$$

These arrangements are called the *lattice permutations* of $x_1^2 x_2 x_3$. There are three such permutations, so the dimension of the representation associated with $(2, 1^2)$ is 3. Now replace the words x_1 by "dot on the first line," x_2 by "dot on the second line," etc. We wish to arrive at a *graph* with two dots on the first line (x_1^2) and one dot on the second and third lines $(x_2 x_3)$, by applying dots, one at a time, so that at every stage each line has no fewer dots than the succeeding lines. (Such graphs are said to be *regular graphs*.) Pictorially, we wish to arrive at the graph

$$\begin{matrix} \bullet \;\; \bullet \\ \bullet \\ \bullet \end{matrix} \qquad \text{or} \qquad \begin{matrix} \square\square \\ \square \\ \square \end{matrix} \qquad\qquad (7\text{–}35)$$

Partition $(2, 1^2)$

(where the second diagram uses squares instead of dots) by placing objects, *one at a time*, on the diagram, making sure that at each stage the number of objects in the first line is greater than or equal to the number of objects in the second line, etc. (i.e., the graph shall be regular at each stage), and ending with all the dots (or squares) of the graph filled. Such a procedure is called a *regular application* (of one *node*, or *dot*, or *square* at a time). In our example, if we label the objects in order of application, we have the following possibilities:

$$
\begin{array}{ccc}
\begin{matrix} 1 & 2 \\ 3 \\ 4 \end{matrix} &
\begin{matrix} 1 & 3 \\ 2 \\ 4 \end{matrix} &
\begin{matrix} 1 & 4 \\ 2 \\ 3 \end{matrix}
\end{array}
\qquad (7\text{–}36)
$$

Equation (7–36) really says the same thing as Eq. (7–34): The first graph says "first line, first line, second line, third line," i.e., $x_1 x_1 x_2 x_3$, etc.

By the same procedure we can find the dimensions of all irreducible representations of S_4. To each partition there corresponds a graph:

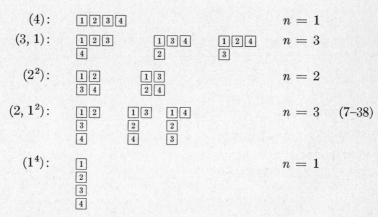

$$ (7\text{–}37) $$

The possible ways of building these graphs by regular applications of single nodes are:

(4):	$\boxed{1}\boxed{2}\boxed{3}\boxed{4}$			$n = 1$	
(3, 1):	$\begin{matrix}\boxed{1}\boxed{2}\boxed{3}\\\boxed{4}\end{matrix}$	$\begin{matrix}\boxed{1}\boxed{3}\boxed{4}\\\boxed{2}\end{matrix}$	$\begin{matrix}\boxed{1}\boxed{2}\boxed{4}\\\boxed{3}\end{matrix}$	$n = 3$	
(2^2):	$\begin{matrix}\boxed{1}\boxed{2}\\\boxed{3}\boxed{4}\end{matrix}$	$\begin{matrix}\boxed{1}\boxed{3}\\\boxed{2}\boxed{4}\end{matrix}$		$n = 2$	
(2, 1^2):	$\begin{matrix}\boxed{1}\boxed{2}\\\boxed{3}\\\boxed{4}\end{matrix}$	$\begin{matrix}\boxed{1}\boxed{3}\\\boxed{2}\\\boxed{4}\end{matrix}$	$\begin{matrix}\boxed{1}\boxed{4}\\\boxed{2}\\\boxed{3}\end{matrix}$	$n = 3$	(7–38)
(1^4):	$\begin{matrix}\boxed{1}\\\boxed{2}\\\boxed{3}\\\boxed{4}\end{matrix}$			$n = 1$	

The results given in (7–38) call attention to the intimate connection between *conjugate* or *associated partitions*, i.e., two partitions which are obtained from each other by interchange of rows and columns: for example, (4) and (1^4), (3, 1) and $(2, 1^2)$. On the other hand, the partition (2^2) is *self-conjugate;* it is transformed into itself by interchange of rows and columns. We see from (7–38) that the dimensions of the representations associated with conjugate partitions are equal.

As another example, we find the dimension of the representation of S_7 associated with the partition (4, 2, 1):

(1) 1234 56 7	(3) 1235 46 7	(5) 1236 45 7	(7) 1237 45 6	(6) 1236 47 5	(8) 1237 46 5
(2) 1234 57 6	(4) 1235 47 6	(9) 1245 36 7	(10) 1245 37 6	(11) 1246 35 7	(13) 1247 35 6
(12) 1246 37 5	(14) 1247 36 5	(15) 1256 34 7	(17) 1257 34 6	(19) 1267 34 5	(16) 1256 37 4
(18) 1257 36 4	(20) 1267 35 4	(21) 1345 26 7	(22) 1345 27 6	(23) 1346 25 7	(25) 1347 25 6
(24) 1346 27 5	(26) 1347 26 5	(27) 1356 24 7	(29) 1357 24 6	(31) 1367 24 5	
(28) 1356 27 4	(30) 1357 26 4	(32) 1367 25 4	(33) 1456 27 3	(34) 1457 26 3	
(35) 1467 25 3		$n = 35$			(7–39)

Problem. Construct diagrams similar to (7–38) and (7–39) for all representations of S_5.

The quantities shown in (7–39) are examples of *Young tableaux*—a graph of a given shape (determined by the partition) in which the numbers $1, \ldots, n$ are placed in the various boxes. When the boxes are filled according to our rule of regular application, the tableau is said to be a *standard tableau.* The number of standard tableaux is just the dimension of the representation. The standard tableaux can be put in a certain order by writing all the numbers on one line, i.e., write the first line, follow with the second, etc; then arrange them in lowest page order of the resulting numbers: for example,

$$1367$$
$$25 \quad \rightarrow 1367254$$
$$4$$

comes after

$$1367$$
$$24 \quad \rightarrow 1367245.$$
$$5$$

We shall return to standard tableaux later and show how they enable us to construct basis functions for the irreducible representations.

7–4 Graphical method for determining characters. The same graphical method which we have used to find the dimensions of the irreducible representations can be applied to calculate the characters. Again we start from Eq. (7–24). To find the character $\chi_{(l)}^{(\lambda)}$ we must find the coefficient of $x_1^{\lambda_1+m-1} x_2^{\lambda_2+m-2} \cdots x_m^{\lambda_m}$ in $D(x_i) s_{(l)}$. We write (l) in cycle form $(l_1, l_2, \ldots)$, and multiply $D(x_i)$ successively by $s_{l_1} = \sum_i x_i^{l_1}, \ldots, s_{l_j} = \sum_i x_i^{l_j}$. We start from $D(x_i) = \sum_P \delta_P P x_1^{m-1} x_2^{m-2} \cdots x_{m-1} x_m^0$ and are trying to determine the coefficient of $x_1^{\lambda_1+m-1} x_2^{\lambda_2+m-2} \cdots x_m^{\lambda_m}$. At each stage of the multiplication the coefficient will be zero if two of the exponents are equal. Suppose that l_1 is a one-cycle (our previous case). Then multiplying $D(x_i)$ by $\sum x_i$, we find that the only term which will not give zero is multiplication by x_1; we put a dot in the first row. Suppose $l_1 = 2$; then we multiply by $\sum x_i^2$. There are now two nonzero cases, arising from x_1^2 and x_2^2. The first leaves $x_1^{m+1} x_2^{m-2} \cdots x_{m-1}$ with the exponents still in decreasing order. The second gives $x_1^{m-1} x_2^m x_3^{m-3} \cdots x_{m-1}$ with the exponents out of order. Among the permutations $\sum_P \delta_P P$ there will be one which reorders the exponents in decreasing order, namely, the transposition (12), which changes the sign, giving $-x_1^m x_2^{m-1} x_3^{m-3} \cdots x_{m-1}$. We can express the result as follows: We added a dot to the second line, then added one to the first line and changed the sign. For $l_1 = 3$ we would obtain

$$x_1^{m+2} x_2^{m-2} \cdots x_{m-1}, \qquad x_1^{m-1} x_2^{m+1} x_3^{m-3} \cdots x_{m-1},$$
$$x_1^{m-1} x_2^{m-2} x_3^m x_4^{m-4} \cdots x_{m-1}.$$

Selecting the terms from $\sum_P \delta_P P$ whose exponents are in decreasing order, we have:

$$x_1^{m+2} x_2^{m-2} \cdots x_{m-1}, \qquad -x_1^{m+1} x_2^{m-1} x_3^{m-3} \cdots x_{m-1},$$
$$x_1^m x_2^{m-1} x_3^{m-2} x_4^{m-4} \cdots x_{m-1}.$$

The three results correspond to the following possibilities:

(a) Put three dots on the first line.

(b) Put a dot on the second line and two dots on the first line.

(c) Put one dot on each of the first three lines.

The same procedure applies for $l_1 > 3$. We see that if the set of l_1 dots is placed on an odd number of lines, the sign is unchanged (*even application*), but if the dots are put on an even number of lines, the sign is changed (*odd application*). The result is even more pictorial if we omit the common factor $x_1^{m-1} \cdots x_{m-1}$, in which case our three possibilities above would be x_1^3, $-x_1^2 x_2$, $x_1 x_2 x_3$. We may say that we have applied the dots as follows: Add dots to any line until the number of dots exceeds the number in the preceding line by one; then go to the preceding line and repeat the procedure, etc. When all dots have been added, the result must be a legitimate graph. The sign is $+(-)$, depending on whether the number of lines involved is odd (even).

Now consider any later stage in the multiplication process. Suppose that after adding dots for $l_1, l_2, \ldots$, we arrive at a typical (permissible) result:

$$\pm (x_1^{m-1} x_2^{m-2} \cdots x_{m-1}) x_1^{\nu_1} x_2^{\nu_2} \cdots x_q^{\nu_q} \cdots x_m^{\nu_m}.$$

If this is a permissible result, we must have $\nu_1 \geq \nu_2 \geq \cdots \geq \nu_m$. If we now multiply by $s_r = \sum x_i^r$ for a cycle on r letters, we consider the effect of the factor x_q^r. If two exponents become equal, the result is zero. Suppose that $\nu_{p-1} + (m - p + 1) > \nu_q + (m - q) + r > \nu_p + (m - p)$. We obtain the result

$$\pm x_1^{\nu_1 + m - 1} \cdots x_{p-1}^{\nu_{p-1} + m - p + 1} x_p^{\nu_p + m - p} \cdots x_q^{\nu_q + m - q + r} \cdots x_m^{\nu_m}$$

$$= \pm x_1^{\nu_1 + m - 1} \cdots x_{p-1}^{\nu_{p-1} + m - p + 1} x_q^{\nu_q + m - q + r} x_p^{\nu_p + m - p} \cdots$$

$$\qquad\qquad x_{q-1}^{\nu_{q-1} + m - q + 1} x_{q+1}^{\nu_{q+1} + m - q - 1} \cdots x_m^{\nu_m}$$

$$= \pm (x_1^{m-1} \cdots x_{p-1}^{m-p+1} x_q^{m-p} x_p^{m-p-1} \cdots x_{m-1}) x_1^{\nu_1} \cdots$$

$$\qquad\qquad x_{p-1}^{\nu_{p-1}} x_q^{\nu_q + p - q + r} x_p^{\nu_p + 1} x_{p+1}^{\nu_{p+1} + 1} \cdots x_{q-1}^{\nu_{q-1} + 1} x_{q+1}^{\nu_{q+1}} \cdots x_m^{\nu_m}.$$

$$(7\text{--}40)$$

There will exist a corresponding term with $x_1, \ldots, x_m$ in natural order, the sign being determined by the number of variables $x_p, \ldots, x_q$ involved in the shift: $+(-)$ if $(q - p)$ is odd (even). This reordered term will be

$$(\pm 1)^{q-p+1} (x_1^{m-1} \cdots x_{p-1}^{m-p+1} x_p^{m-p} \cdots x_{m-1}) x_1^{\nu_1} \cdots$$

$$\qquad\qquad x_{p-1}^{\nu_{p-1}} x_p^{\nu_q + p - q + r} x_{p+1}^{\nu_p + 1} x_{p+2}^{\nu_{p+1} + 1} \cdots x_q^{\nu_{q-1} + 1} x_{q+1}^{\nu_{q+1}} \cdots x_m^{\nu_m}.$$

$$(7\text{--}41)$$

In terms of the graphs, we may say that we added dots to the qth line until the number exceeded that of the previous line by one (that is, $x_q^{\nu_q}$ has become $x_q^{\nu_{q-1}+1}$), and then repeated the procedure in the preceding line, etc., making sure that the final result of applying all r dots left us with a regular graph. This process is called a *regular application of r dots*. The general rule is:

To find the character in the representation (λ) of the class (l) having cycles of lengths $l_1, l_2, \ldots$, build up the graph of the partition (λ) by successive regular applications of l_1, l_2, etc., dots. The character $\chi_{(l)}^{(\lambda)}$ equals the number of ways of building the graph which contain an even number of negative applications minus the number of ways which contain an odd number of negative applications.

As a first example, we find $\chi_{(2,1^2)}^{(3,1)}$. First draw the graph of $(3, 1)$:

$$\square\square\square \atop \square \tag{7–42}$$

We make regular applications of 2, 1, and 1 dots. This can be done in any order. For example, first apply a pair of dots (we label them 1), then a single dot (2) and a single dot (3). The two dots can be entered in two ways:

$$\text{(a)} \quad \boxed{1}\boxed{1}\square \atop \square \qquad \text{or} \qquad \text{(b)} \quad \boxed{1}\square\square \atop \boxed{1} \tag{7–42a}$$

Now make a regular application of dot 2:

$$\text{(a}_1\text{)} \quad \boxed{1}\boxed{1}\square \atop \boxed{2} \qquad \text{(a}_2\text{)} \quad \boxed{1}\boxed{1}\boxed{2} \atop \square \qquad \text{(b)} \quad \boxed{1}\boxed{2}\square \atop \boxed{1} \tag{7–42b}$$

Clearly dot 3 is put in the remaining empty space:

$$\text{(a}_1\text{)} \quad \boxed{1}\boxed{1}\boxed{3} \atop \boxed{2} \qquad \text{(a}_2\text{)} \quad \boxed{1}\boxed{1}\boxed{2} \atop \boxed{3} \qquad \text{(b)} \quad \boxed{1}\boxed{2}\boxed{3} \atop \boxed{1} \tag{7–42c}$$

In (a$_1$) and (a$_2$) the two 1's are on the same line, so these constructions contain no odd applications. In (b) the two 1's are on two lines, and hence (b) contains one negative application. Thus we obtain $+1$ from (a$_1$), $+1$ from (a$_2$), and -1 from (b), and

$$\chi_{(2,1^2)}^{(3,1)} = +1 + 1 - 1 = 1. \tag{7–43}$$

The order in which the applications are made is irrelevant (though wise choice may reduce the labor considerably). For example, suppose that we first apply one dot (1), then two dots (2), then one dot (3). The first dot (1) must go in the upper left corner:

Now we add the two 2's. We cannot place the first 2 in the second line since our rule requires that we continue until all dots are used or the number of dots exceeds by one the number of dots in the preceding line. Hence the only regular application is $\boxed{\begin{smallmatrix}1&2&2\\\hline\end{smallmatrix}}$, and placing the third dot, we obtain $\boxed{\begin{smallmatrix}1&2&2\\3&&\end{smallmatrix}}$. We have one build-up, with no negative applications, so $\chi^{(3,1)}_{(2,1^2)} = +1$.

As a next example, we find $\chi^{(2,1^5)}_{(4,1^3)}$. The graph of $(2, 1^5)$ is

One order is one dot (1), four dots (2), one dot (3), one dot (4). The 1 must go in the upper left box. Then the only regular application of the four 2's gives

Now the 3 and 4 can be inserted in two ways. In each case the four 2's occupy an even number of lines (negative application), so $\chi^{(2,1^5)}_{(4,1^3)} = -2$.

Next we try $\chi^{(8,2)}_{(5^2)}$. The graph of $(8, 2)$ is

If we enter five 1's in the first line, we have

When we start to apply the five 2's, we violate our rule since we must start in the second line and continue there until we exceed the number of occupants of the first line; hence, this method does not build the graph by

regular applications. The alternative is to put the first 1 in the second line and the other four 1's in the first:

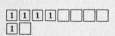

but we still cannot make a regular application of five 2's to build the graph. Thus there are no ways of building the graph by regular applications, so

$$\chi_{(5^2)}^{(8,2)} = 0.$$

As a last example, we find $\chi_{(14,8,1)}^{(20,2,1)}$. The graph of $(20, 2, 1)$ is:

We make the applications in the order 1, 8, 14. The 1 goes in the upper left box. Then, if we put the eight 2's in the first line, we cannot apply the 3's regularly. Similarly, if we start the 2's in the second line and enter two there and six in the first line, the dangling box in the third line prevents regular application of the 3's. Hence we must place the first 2 in the third line, then two 2's in the second, and five in the first. The fourteen 3's then fill the first line, and we have one regular buildup:

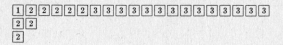

It contains only positive applications, so $\chi_{(14,8,1)}^{(20,2,1)} = +1$.

––––––––––

Problems. (1) Calculate the following characters:

(a) $\chi_{(2^3)}^{(4,2)}$,

(b) $\chi_{(8)}^{(3^2,1^2)}$,

(c) $\chi_{(3,2,1^3)}^{(5,2,1)}$,

(d) $\chi_{(3,2^2,1^2)}^{(4,3,1^2)}$,

(e) $\chi_{(5,4,1)}^{(6,2,1^2)}$,

(f) $\chi_{(3,2^3,1)}^{(8,1^2)}$.

(2) Prove the following theorem:

THEOREM. The character of the class (n) in the irreducible representations of S_n corresponding to the partitions $(p, 1^q)$, where $p + q = n$, is $(-1)^q$ for $q = 0, 1, \ldots, (n-1)$. The character of this class is zero in the other irreducible representations.

[*Hint:* Try to make a regular application of n dots to the various partitions; check the conditions for the application to be even or odd.]

(3) Derive the general formulas:

(a) $\chi^{(n-1,1)}_{(1^\alpha,2^\beta,3^\gamma\ldots)} = \alpha - 1;$

(b) $\chi^{(n-2,1^2)}_{(1^\alpha,2^\beta,3^\gamma\ldots)} = \dfrac{(\alpha-1)(\alpha-2)}{2} - \beta;$

(c) $\chi^{(n-2,2)}_{(1^\alpha,2^\beta,3^\gamma\ldots)} = \dfrac{(\alpha-1)(\alpha-2)}{2} + \beta - 1.$

Try to generalize the formulas.

The irreducible representation corresponding to the partition (n) is the identity representation, i.e., $\chi^{(n)}_{(l)} = 1$ for every class (l). Since the graph consists of a single row, there is only one method for building the graph by regular applications.

The irreducible representation corresponding to the partition (1^n) is the alternating representation: $\chi^{(1^n)}_{(l)} = +1$ if (l) is a class of even permutations, $\chi^{(1^n)}_{(l)} = -1$ if (l) is a class of odd permutations. The proof is similar to that of the above theorem. Now there is only a single column, and again there is only one way of building the graph regularly. Each cycle on an even number of letters (odd permutation) contributes a factor -1. Since even (odd) permutations contain an even (odd) number of cycles of even order, the theorem follows.

We earlier called attention to conjugate or associated partitions. Two partitions (λ), $(\tilde{\lambda})$ are conjugate to each other if the diagram of one is obtained from that of the other by interchanging rows and columns. The representations corresponding to conjugate partitions are also said to be *conjugate representations*. We now prove that the characters of a class in conjugate representations are equal if the class is even, and opposite in sign if the class is odd, i.e.,

$$\chi^{(\lambda)}_{(l)} = \chi^{(\tilde{\lambda})}_{(l)}\chi^{(1^n)}_{(l)}. \tag{7–44}$$

There are many ways of proving (7–44). The method we choose is valuable in getting accustomed to thinking in terms of graphs and patterns. Suppose that we are building a graph by regular applications and arrive at:

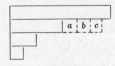

A possible regular application of three dots to the second line is shown;

they would be applied in the order a, b, c. In the conjugate graph, the same application, in the order c, b, a, would again be regular.

Again, suppose that we arrive at the stage

$$(7\text{–}45)$$

and make a regular application of 8 dots as shown:

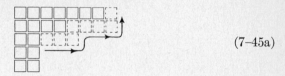

$$(7\text{–}45a)$$

In the conjugate diagram, the same application, but in reverse order, is again regular:

$$(7\text{–}45b)$$

Note the shape of the piece which has been added to our jigsaw puzzle. It is always of the type

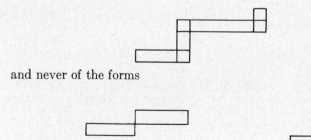

and never of the forms

In other words, each dot is either in the same row or the same column as its predecessor. The total number of different rows and columns occupied by the r dots in a regular application is $r + 1$, for the first dot starts both a row and a column, while the later dots are (as we just pointed out) either in the same row or the same column as their predecessors. But we see from

(7–45a) and (7–45b) that the applications to conjugate partitions mean an interchange of rows and columns. Thus if r is odd ($r + 1$ even), the total number of rows and columns involved is even, and the applications to the conjugate graphs are therefore both positive or both negative. If r is even ($r + 1$ odd), then if the application to (λ) is positive, the application to $(\tilde{\lambda})$ is negative, and vice versa. So to each way of building the graph (λ) for the class (l) there corresponds a method of building the graph $(\tilde{\lambda})$. For an even class the number of cycles of even r is even, and hence the methods contribute the same number $+1$ or -1 in each case. For an odd class, the number of cycles with even r is odd, so we obtain $+1$ in one case and -1 in the other.

7–5 Recursion formulas for characters. Branching laws. For the detailed task of constructing character tables of the symmetric groups, one can also use recursion formulas, so that the characters of S_n can be determined from the already known answers for symmetric groups of lower order.

The recursion formulas are derived by starting from Eq. (7–24):

$$s_{(l)}D(x_i) = \sum_{(\lambda)} \chi_{(l)}^{(\lambda)} \sum_P \delta_P P x_1^{\lambda_1+m-1} x_2^{\lambda_2+m-2} \cdots x_m^{\lambda_m}, \qquad (7\text{--}24)$$

where (l) is a partition of n.

Now suppose that we consider the class of S_{n+r} which is obtained by adding one cycle on r letters to the partition (l): $(k) = ((l), r)$. Then

$$s_{(k)}D(x_i) = \sum_{(\mu)} \chi_{(k)}^{(\mu)} \sum_P \delta_P P x_1^{\mu_1+m-1} x_2^{\mu_2+m-2} \cdots x_m^{\mu_m}. \qquad (7\text{--}24a)$$

But

$$s_{(k)} = s_{(l)} s_r, \qquad (7\text{--}46)$$

so

$$\sum_{(\mu)} \chi_{(k)}^{(\mu)} \sum_P \delta_P P x_1^{\mu_1+m-1} \cdots x_m^{\mu_m}$$
$$= (x_1^r + \cdots + x_m^r) \sum_{(\lambda)} \chi_{(l)}^{(\lambda)} \sum_P \delta_P P x_1^{\lambda_1+m-1} \cdots x_m^{\lambda_m}. \qquad (7\text{--}47)$$

We now equate coefficients of the same monomial on both sides of Eq. (7–47) to get

$$\chi_{(k)}^{(\mu)} = {\sum_{(\lambda)}}' \pm \chi_{(l)}^{(\lambda)} \qquad \text{with} \qquad (k) = ((l), r), \qquad (7\text{--}48)$$

where the sum $\sum'$ extends over the partitions (λ) which we shall now characterize. From (7–47) it is clear that we take the set of exponents

$$\lambda_1 + m - 1, \lambda_2 + m - 2, \ldots, \lambda_m, \qquad (7\text{--}49)$$

and, in turn, increase one of the numbers by r. We want the resulting sequence to be a permutation of the sequence

$$\mu_1 + m - 1, \mu_2 + m - 2, \ldots, \mu_m. \qquad (7\text{-}50)$$

If it is, then $\chi_{(l)}^{(\lambda)}$ will appear in (7–48); its sign will be determined by whether the sequences (7–49) and (7–50) are obtained from each other by an even ($+$) or an odd ($-$) permutation. Since, in practice, we start with (μ) and the sequence (7–50), we *subtract* r in turn from each number in the sequence (7–50) and compare with (7–49). To standardize our procedure, we shall always choose m equal to the number of parts in the partition (μ), although any choice of m is permissible.

For example, consider the character $\chi_{(k)}^{(4,3^2,1^2)}$. We try to express it in terms of $\chi_{(l)}^{(\lambda)}$ of S_{10}, where the class (l) is obtained from (k) by removing a 2-cycle ($r = 2$). We choose $m = 5$ and write the sequence (7–50):

$$8, 6, 5, 2, 1.$$

We now subtract 2, in turn, from each number and write the resulting sequences. To save labor we note that, with our choice of m, any sequence containing negative numbers should be discarded. Also, if two of the numbers in the resulting sequence are equal, it cannot yield (7–49) and should again be dropped. The sequences are:

$$6, 6, 5, 2, 1;$$
$$8, 4, 5, 2, 1;$$
$$8, 6, 3, 2, 1;$$
$$8, 6, 5, 0, 1;$$
$$8, 6, 5, 2, -1;$$

so we should discard the first and last sequences. We reorder each remaining sequence in descending order and mark it $+$ or $-$, depending on whether the required permutation is even or odd:

$$-8, 5, 4, 2, 1;$$
$$+8, 6, 3, 2, 1;$$
$$-8, 6, 5, 1, 0.$$

Since these are now the sequences (7–49), we obtain the λ's by subtracting $m - 1 = 4$ from the first number, $m - 2 = 3$ from the second number, etc.:

$$-4, 2, 2, 1, 1;$$
$$+4, 3, 1, 1, 1;$$
$$-4, 3, 3, 0, 0.$$

Our final recursion formula is:

$$\chi_{((l),2)}^{(4,3^2,1^2)} = -\chi_{(l)}^{(4,2^2,1^2)} + \chi_{(l)}^{(4,3,1^3)} - \chi_{(l)}^{(4,3^2)}. \tag{7-51}$$

As a second example, we find the recursion formula for $\chi_{((l),4)}^{(7,6,5)}$ in terms of $\chi_{(l)}^{(\lambda)}$. The μ's are 7, 6, 5, and $m = 3$, so the sequence (7–50) is 9, 7, 5. Now subtract 4 from each number in turn:

$$\underline{5, 7, 5;} \qquad 9, 3, 5; \qquad 9, 7, 1.$$

Rearrange: $-9, 5, 3; +9, 7, 1$; and now subtract 2, 1, 0, giving $-7, 4, 3;$ $+7, 6, 1$, so $\chi_{((l),4)}^{(7,6,5)} = \chi_{(l)}^{(7,6,1)} - \chi_{(l)}^{(7,4,3)}$.

A very important case is that in which we strip off a 1-cycle ($r = 1$). In this case we must subtract 1 successively from each term in (7–50) and identify with (7–49). Here the terms cannot get out of order (the only untoward result might be that two terms become equal) and we can therefore work with the λ's and μ's directly. The result of the comparison is the *branching law*,

$$\chi_{((l),1)}^{(\mu)} = \sum_r \chi_{(l)}^{(\mu_1,\dots\mu_r-1,\dots)} \tag{7-52}$$

summed over all resulting partitions which are regular (i.e., for which $\mu_r - 1 \geq \mu_{r+1}$). Suppose we strip off a second 1-cycle. Repeating the argument, we have

$$\chi_{((l),1^2)}^{(\mu)} = \sum_{\substack{r \\ \mu_r-2 \geq \mu_{r+1}}} \chi_{(l)}^{(\mu_1,\dots\mu_r-2,\dots)} + \sum_{\substack{r,s \\ \mu_r-1 \geq \mu_{r+1} \\ \mu_s-1 \geq \mu_{s+1}}} a_{rs}\chi_{(l)}^{(\mu_1,\dots\mu_r-1,\dots\mu_s-1,\dots)},$$

$$\tag{7-53}$$

where $a_{rs} = 2$ unless r and s are consecutive and $\mu_r = \mu_s$, in which case $a_{rs} = 1$.

The last statement is quite awkward and is much better expressed graphically. To understand the recursion formula in terms of graphs, we first consider the case of (7–52). In building the graph of the partition (μ) by regular application of the nodes corresponding to the class $((l), 1)$, imagine that we save the 1-cycle for the last regular application (of one node). Just before this last step, we have built up partitions π of $n - 1$ by regular applications, and we now apply the last node to build (μ). This last application is positive. The number of ways of building π is the character of (l) for that partition of $n - 1$. So the character of $((l), 1)$ in (μ) is the sum of the characters of (l) in all regular graphs obtained from

(μ) by a (regular) removal of one dot. For example, from the diagram

$$(7\text{-}54)$$

we can remove any one of the dots marked with a cross, and hence

$$\chi_{((l),1)}^{(6,4^2,3,2)} = \chi_{(l)}^{(5,4^2,3,2)} + \chi_{(l)}^{(6,4,3^2,2)} + \chi_{(l)}^{(6,4^2,2^2)} + \chi_{(l)}^{(6,4^2,3,1)}. \qquad (7\text{-}54a)$$

The stripping-off of two 1-cycles which yielded (7–53) can be described graphically as two successive regular removals of a node. From our diagram we can see why (7–53) is complicated. Our removals can be first from line 1, then from line 3, or vice versa, so the resulting partition appears twice in the sum. But for lines 2 and 3, we must carry out the removal first from line 3 and then from line 2, so this partition appears only once in the sum.

The stripping-off of a 2-cycle means a regular removal of 2 dots. In order that the strip removed be regular, it must be either

$$\square\square \qquad \text{or} \qquad \begin{array}{c}\square\\\square\end{array} \qquad\qquad (7\text{-}55)$$

where the first strip is positive and the second negative (even number of lines). Thus in the partition

only the crosshatched strips of 2 can be removed, so that

$$\chi_{((l),2)}^{(5,4^2,2)} = \chi_{(l)}^{(5,4^2)} + \chi_{(l)}^{(5,4,2^2)} - \chi_{(l)}^{(5,3^2,2)}. \qquad (7\text{-}56)$$

If we remove a 3-cycle, the regular strips are

$$\underset{+}{\square\square\square} \qquad \underset{-}{\square\square}^{\displaystyle\square} \qquad \underset{-}{{}^{\displaystyle\square\square}\square} \qquad \underset{+}{\begin{array}{c}\square\\\square\\\square\end{array}} \qquad (7\text{-}57)$$

In general, there are 2^{r-1} different strips for an r-cycle.

Problem. Use the recursion methods to evaluate

(a) $\chi_{((l),2^2)}^{(6,4,2)}$ in terms of $\chi_{(l)}^{(\lambda)}$;   (b) $\chi_{((l),3,1)}^{(4,2,1)}$ in terms of $\chi_{(l)}^{(\lambda)}$.

7–6 Calculation of characters by means of the Frobenius formula.
The last method for calculating characters is the direct use of the Frobenius formula (7–24). We illustrate the method for some simple cases which we need later on. According to (7–24), $\chi_0 \equiv \chi^{(\lambda)}_{(1^n)}$ and $\chi \equiv \chi^{(\lambda)}_{(2,1^{n-2})}$ are, respectively, the coefficients of $x_1^{h_1} x_2^{h_2} \ldots x_m^{h_m}$ in

$$D(x_1, \cdots, x_m)(x_1 + \cdots + x_m)^n \qquad (7\text{–}58)$$

and

$$D(x_1, \cdots, x_m)(x_1 + \cdots + x_m)^{n-2}(x_1^2 + \cdots + x_m^2), \qquad (7\text{–}59)$$

where

$$h_i = \lambda_i + m - i. \qquad (7\text{–}60)$$

First we deal with (7–58). Each term in the expansion of $D(x_i)$ is of the form

$$\pm x_1^{k_1} \ldots x_m^{k_m}, \qquad (7\text{–}61)$$

where the k's are a permutation of the integers $m - 1, m - 2, \ldots, 0$, and the sign is determined by whether the permutation is even or odd. To obtain the desired multinomial from (7–61), we must multiply by the term

$$\frac{n!}{(h_1 - k_1)! \cdots (h_m - k_m)!} x_1^{h_1 - k_1} \ldots x_m^{h_m - k_m}, \qquad (7\text{–}62)$$

so the required coefficient is

$$\chi_0 = n! \sum_{(k)} \pm \frac{1}{(h_1 - k_1)! \cdots (h_m - k_m)!}, \qquad (7\text{–}63)$$

where the sum runs over all permutations of $m - 1, \ldots, 0$. This sum is the expansion of the determinant

$$\chi_0 = n! \begin{vmatrix} \dfrac{1}{[h_1 - (m-1)]!} & \dfrac{1}{[h_1 - (m-2)]!} & \cdots & \dfrac{1}{h_1!} \\[2mm] \dfrac{1}{[h_2 - (m-1)]!} & \dfrac{1}{[h_2 - (m-2)]!} & \cdots & \dfrac{1}{h_2!} \\ \vdots & \vdots & & \vdots \\ \dfrac{1}{[h_m - (m-1)]!} & \dfrac{1}{[h_m - (m-2)]!} & \cdots & \dfrac{1}{h_m!} \end{vmatrix}. \qquad (7\text{–}64)$$

Removing the factor $1/(h_1! \ldots h_m!)$, we have

$$\chi_0 = \frac{n!}{h_1! \cdots h_m!} \begin{vmatrix} h_1(h_1 - 1) \cdots (h_1 - m + 2) & h_1(h_1 - 1) \cdots (h_1 - m + 3) & \cdots & h_1 & 1 \\ h_2(h_2 - 1) \cdots (h_2 - m + 2) & h_2(h_2 - 1) \cdots (h_2 - m + 3) & \cdots & h_2 & 1 \\ \vdots & & & \vdots & \vdots \\ h_m(h_m - 1) \cdots (h_m - m + 2) & h_m(h_m - 1) \cdots (h_m - m + 3) & \cdots & h_m & 1 \end{vmatrix}. \qquad (7\text{–}65)$$

The columns of (7-65) are polynomials of degree $m-1, m-2, \ldots, 0$, with leading coefficient equal to one. Hence, by taking differences, we can reduce (7-65) to

$$\chi_0 = \frac{n!}{h_1! \cdots h_m!} \begin{vmatrix} h_1^{m-1} & h_1^{m-2} & \cdots & h_1 & 1 \\ h_2^{m-1} & h_2^{m-2} & \cdots & h_2 & 1 \\ \vdots & & & & \vdots \\ h_m^{m-1} & h_m^{m-2} & \cdots & h_m & 1 \end{vmatrix}$$

or

$$\chi_0 = \frac{n!}{h_1! \cdots h_m!} D(h_1, \ldots, h_m). \tag{7-66}$$

To evaluate $\chi \equiv \chi_{(2,1^{n-2})}^{(\lambda)}$, we apply (7-66) to (7-59):

$$\chi = (n-2)! \sum_{i=1}^{m} \frac{D(h_1, h_2, \ldots, h_i - 2, \ldots, h_m)}{h_1! h_2! \cdots (h_i - 2)! \cdots h_m!}. \tag{7-67}$$

We shall need the quantity

$$\xi = \frac{n(n-1)}{2} \frac{\chi}{\chi_0}. \tag{7-68}$$

From (7-66) and (7-67),

$$\xi = \frac{1}{2} \sum_{i=1}^{m} \frac{h_i(h_i - 1) D(h_1, h_2, \ldots, h_i - 2, \ldots, h_m)}{D(h_1, \ldots, h_m)}. \tag{7-69}$$

The denominator of (7-69) is an alternating function of the h_i [cf. the remarks about Eq. (7-23)]. The sum in the numerator is also an alternating function of the h_i. [Reader, show that it contains all factors $(h_i - h_j)$.] Hence the ratio must be a symmetric polynomial of second degree in the h_i:

$$\sum_i h_i(h_i - 1) D(h_1, h_2, \ldots, h_i - 2, \ldots, h_m)$$
$$= D(h_1, \ldots, h_m)[A\textstyle\sum h_i^2 + B\sum h_i h_j + C\sum h_i + F]. \tag{7-70}$$

We can determine A, B, C, F, by comparing coefficients on both sides of Eq. (7-70). First, we look at terms in which the power of h_1 is $m+1$ or m, and the power of h_2 is $\geq m-2$. On the left, such terms occur only in the summand $i = 1$:

$$h_1(h_1 - 1) \begin{vmatrix} (h_1 - 2)^{m-1} & (h_1 - 2)^{m-2} \\ h_2^{m-1} & h_2^{m-2} \end{vmatrix} \cdot |h_3^{m-3} \cdots h_{m-1}|, \tag{7-71}$$

where $|h_3^{m-3} \ldots h_{m-1}|$ is the determinant remaining after crossing out the first two rows and columns in $D(h)$. Expanding (7–71), we obtain

$$h_1(h_1 - 1)[h_2^{m-2}\{h_1^{m-1} - 2(m - 1)h_1^{m-2} \cdots\}$$
$$- h_2^{m-1}\{h_1^{m-2} - 2(m - 2)h_1^{m-3}\}] \cdot |h_3^{m-3} \cdots h_{m-1}|$$
$$= [h_1^{m+1}h_2^{m-2} - h_1^m h_2^{m-2} - 2(m - 1)h_1^m h_2^{m-2} - h_1^m h_2^{m-1} \cdots]$$
$$\times |h_3^{m-3} \cdots h_{m-1}|. \quad (7\text{–}72)$$

The right side of Eq. (7–70) gives

$$[h_1^{m-1}h_2^{m-2} - h_1^{m-2}h_2^{m-1}][Ah_1^2 + Bh_1 h_2 + Ch_1 + \cdots] \cdot |h_3^{m-3} \cdots h_{m-1}|.$$
$$(7\text{–}72\text{a})$$

Comparing coefficients, we find

$$A = 1, \quad B = 0, \quad C = -(2m - 1). \quad (7\text{–}73)$$

The constant F is the coefficient of $h_1^{m-1}h_2^{m-2} \cdots h_{m-1}$ on the right side of (7–70). On the left side of (7–70) we get this monomial only from the main diagonal of the determinant. In the ith summand the diagonal term is $h_1^{m-1}h_2^{m-2} \cdots h_i(h_i - 1)(h_i - 2)^{m-i} \cdots h_{m-1}$, so the monomial $h_1^{m-1}h_2^{m-2} \cdots h_{m-1}$ has the coefficient $2(m - i)(m - i - 1) + 2(m - i) = 2(m - i)^2$, and

$$F = 2 \sum_{i=1}^{m} (m - i)^2 = \frac{1}{3} m(m - 1)(2m - 1). \quad (7\text{–}74)$$

Using (7–69) and (7–70), we find

$$\xi = \frac{1}{2} \Sigma h_i^2 - \frac{2m - 1}{2} \Sigma h_i + \frac{m(m - 1)(2m - 1)}{6}. \quad (7\text{–}75)$$

Now we substitute for h_i from (7–60) and obtain

$$\xi = \frac{1}{2} \sum_{i=1}^{m} \lambda_i(\lambda_i + 1) - \sum_{i=1}^{m} i\lambda_i. \quad (7\text{–}76)$$

Problem. Use Eq. (7–66) to find the dimension of the irreducible representation of S_{10} corresponding to the partition $(4, 3^2)$.

7–7 The matrices of the irreducible representations of S_n. Yamanouchi symbols.

Now we turn to the problem of constructing the matrices of the irreducible representations of S_n. We shall follow a paper by Yamanouchi (*Phys. Math. Soc. Japan* **19**, 436, 1937).

We consider an irreducible representation $D(R)$ characterized by the partition (λ). Let us assume that we know the irreducible representations of S_{n-1}, the permutation group on the symbols $1, 2, \ldots, n-1$. All the permutations of S_n which are contained in this subgroup S_{n-1} have the property that they leave the last symbol, n, unchanged. Thus every element R of S_n which is contained in S_{n-1} has the form $((l), 1)$, and hence we can apply the branching law (7–52). Equation (7–52) can be expressed in terms of representations: For elements R in S_{n-1}, the matrix representative $D(R)$ is the sum of matrix representatives $D_r(R)$, where D_r is the irreducible representation of S_{n-1} corresponding to the partition $(\lambda_1, \lambda_2, \ldots, \lambda_r - 1, \ldots, \lambda_m)$. Symbolically,

$$D(R) = \sum_{\substack{r \\ \lambda_r - 1 \geq \lambda_{r+1}}} D_r(R) \tag{7-77}$$

for any permutation R which leaves the last symbol n unchanged. For example, in the partition shown in Eq. (7–54), any permutation on the first 18 letters has a representative,

$$D(R) = D_1(R) + D_3(R) + D_4(R) + D_5(R), \tag{7-78}$$

where, say, D_1 is the representative in the representation characterized by $(5, 4^2, 3, 1)$. We can choose the representation so that $D(R)$ is in explicitly reduced form, with the matrices D_r along the main diagonal. The subscript r labels the collection of rows (columns) in which D_r stands:

$$\tag{7-79}$$

For those elements R which leave the last *two* symbols, $n-1$ and n, unchanged, and hence are in S_{n-2}, D_r splits once more [cf. Eq. (7–53)] into a sum of representations D_{rs} corresponding to regular partitions

$$(\lambda_1, \ldots, \lambda_r - 2, \ldots, \lambda_m), \qquad (\lambda_1, \ldots, \lambda_r - 1, \ldots, \lambda_s - 1, \ldots, \lambda_m). \tag{7-80}$$

The irreducible representations which are reached in this manner are of three types: (1) $D_{rr}(\lambda_r \geq \lambda_{r+1} + 2)$ occurs once. (2) $D_{rs} = D_{sr}$ occurs twice if $\lambda_r - 1 \geq \lambda_{r+1}$ and $\lambda_s - 1 \geq \lambda_{s+1}$. (3) D_{rs} occurs, but D_{sr} does not; this happens if $s = r - 1$ and $\lambda_{r-1} = \lambda_r$.

For example, consider again the partition shown in (7–54). For permutations R which leave the last two symbols unchanged, we obtain

$$\begin{aligned}
D = \; &D_{11} + D_{13} + D_{14} + D_{15} \\
&+ D_{31} + D_{32} + D_{34} + D_{35} \\
&+ D_{41} + D_{43} + D_{45} \\
&+ D_{51} + D_{53} + D_{54} + D_{55}.
\end{aligned} \qquad (7\text{–}81)$$

(Note that D_{32} is of Type 2.) For such permutations we get a subdivision of r into sets rs, and the matrix has the form

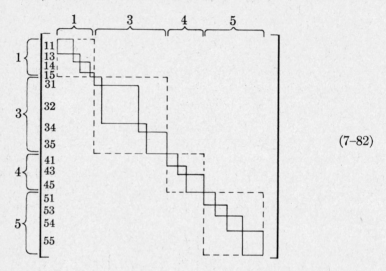

$$(7\text{–}82)$$

Equation (7–82) shows the matrix for any permutation in S_{n-2}; the dotted lines show the matrix (7–79) of elements in S_{n-1}. Again in (7–82) we have chosen the basis vectors to exhibit the matrix in reduced form.

Now we wish to construct the matrix representative of elements R in S_n which are *not* in S_{n-1}; i.e., elements which permute the last symbol n. For this purpose we need only the matrix of the transposition $(n-1, n)$, which we denote by U, because

$$(i, n) = (n-1, n)(i, n-1)(n-1, n). \qquad (7\text{–}83)$$

The transposition $(i, n-1)$ is in S_{n-1}, so its matrix is known; knowing U, we can get the matrix for any transposition (i, n) and, consequently, for any permutation. The matrix V for any permutation in S_{n-2} has the form shown in (7–82):

$$V_{rs,pq} = V_{rs,rs}\, \delta_{rs,pq}. \qquad (7\text{–}84)$$

Since the transposition $(n - 1, n)$ commutes with all the elements of S_{n-2},

$$VU = UV \tag{7-85}$$

or

$$V_{rs,rs}U_{rs,pq} = U_{rs,pq}V_{pq,pq} \tag{7-86}$$

for all elements in S_{n-2}. The matrices $V_{rs,rs}$ and $V_{pq,pq}$ are irreducible representations of S_{n-2} which are equivalent only for $pq = rs$ or $pq = sr$. Hence by Schur's lemmas, $U_{rs,pq} = 0$ unless $pq = rs$ or $pq = sr$, and

$$U_{rs,rs} = \sigma_{rs,rs}E_{rs,rs}, \qquad U_{rs,sr} = \sigma_{rs,sr}E_{rs,sr}, \tag{7-87}$$

where $E_{lm,pq}$ is a unit matrix at the lm-rows, pq-columns in (7–82), and the σ's are constants.

If we require our representation to be unitary and look at the form of U, we see that if rs is of type 1 or type 3, then

$$|\sigma_{rs,rs}|^2 = 1, \tag{7-88}$$

and if rs is of type 2, then

$$|\sigma_{rs,rs}|^2 + |\sigma_{rs,sr}|^2 = 1. \tag{7-89}$$

In order to find the matrix U of the transposition $(n - 1, n)$, we need only the constants σ. They can be found as follows:

We showed earlier [Eqs. (3–168) through (3–170)] that if we sum the matrices of an irreducible representation for all the elements of a class K, we obtain a multiple of the unit matrix:

$$\sum_{R \subset K} U(R) = \frac{n_K \chi_K E}{\chi_0}, \tag{7-90}$$

where E is the unit matrix, n_K the number of elements in the class K, χ_K its character, and χ_0 the dimension of the representation. We apply this to the class of transpositions. For S_n,

$$\sum_{i<j}^{n} U(ij) = \frac{n(n - 1)}{2} \frac{\chi}{\chi_0} E = \xi E, \tag{7-91}$$

where ξ is the quantity defined in Eq. (7–68) and evaluated in Eq. (7–76). Similarly, if we consider the irreducible representations D_r of S_{n-1} and D_{rs} of S_{n-2}, we have

$$\sum_{i<j}^{n-1} U^{(r)}(ij) = \xi_r E_r, \qquad \sum_{i<j}^{n-2} U^{(rs)}(ij) = \xi_{rs}E_{rs}, \tag{7-92}$$

where the E_r, E_{rs} are the unit matrices of D_r and D_{rs}, and ξ_r, ξ_{rs} are defined similarly to ξ. From (7–91), (7–92) we find that

$$\sum_{i=1}^{n-1} U(in) = \sum_{i<j}^{n} U(ij) - \sum_{i<j}^{n-1} U(ij) = A_n \qquad (7\text{–}93)$$

is a diagonal matrix, with submatrices

$$(\xi - \xi_r)E_{rr} \qquad (7\text{–}93a)$$

along the diagonal in (7–79). Repeating the procedure, using both expressions in (7–92), we find that

$$\sum_{i=1}^{n-2} U(i, n-1) = A_{n-1} \qquad (7\text{–}94)$$

is also a diagonal matrix, with submatrices

$$(\xi_r - \xi_{rs})E_{rs,rs} \qquad (7\text{–}94a)$$

along the diagonal in (7–82).

Multiplying (7–93) by $U(n-1, n)$, we obtain

$$A_n U(n-1, n) = \sum_{i=1}^{n-1} U(i, n)U(n-1, n)$$

$$= \sum_{i=1}^{n-2} U(i, n)U(n-1, n) + E. \qquad (7\text{–}95)$$

But

$$\sum_{i=1}^{n-2} U(i, n)U(n-1, n)$$

$$= U(n-1, n)\left[U(n-1, n) \sum_{i=1}^{n-2} U(i, n)U(n-1, n) \right]$$

$$= U(n-1, n) \sum_{i=1}^{n-2} U(i, n-1) = U(n-1, n)A_{n-1}, \qquad (7\text{–}95a)$$

so

$$A_n U(n-1, n) - U(n-1, n)A_{n-1} = E. \qquad (7\text{–}96)$$

Now we take the matrix element pq, rs of this equation and make use of (7–93a) and (7–94a). Taking the rs, rs element, we have

$$(\xi - \xi_r)U_{rs,rs} - U_{rs,rs}(\xi_r - \xi_{rs}) = E_{rs,rs}; \qquad (7\text{–}97)$$

and from the rs, sr element,

$$(\xi - \xi_r)U_{rs,sr} - U_{rs,sr}(\xi_s - \xi_{sr}) = 0. \qquad (7\text{–}98)$$

Substituting from (7–87), we find

$$\sigma_{rs,rs} = \frac{1}{\xi - 2\xi_r + \xi_{rs}}, \qquad (7\text{–}99)$$

$$\sigma_{rs,sr} = 0 \qquad \text{unless} \qquad \xi - \xi_r - \xi_s + \xi_{sr} = 0. \qquad (7\text{–}100)$$

Now we use (7–76). In D_r we decrease λ_r by 1, so

$$\xi - \xi_r = \lambda_r - r. \qquad (7\text{–}101)$$

For type 1($r = s$), we need $\xi_r - \xi_{rr}$. In D_{rr}, λ_r is decreased by 2, so that $\xi_r - \xi_{rr} = \lambda_r - 1 - r$, and

$$\sigma_{rr,rr} = 1. \qquad (7\text{–}102)$$

For $r \neq s$,

$$\xi_r - \xi_{rs} = \lambda_s - s, \qquad (7\text{–}103)$$

so that

$$\sigma_{rs} \equiv \sigma_{rs,rs} = \frac{1}{\lambda_r - \lambda_s + s - r} = -\sigma_{sr,sr} \qquad (7\text{–}104)$$

for type 2. We also note that $\xi - \xi_r = \xi_s - \xi_{sr}$, so the condition (7–100) is automatically fulfilled. Finally, for type 3, where $s = r - 1$, $\lambda_{r-1} = \lambda_r$, we find from (7–104):

$$\sigma_{rs,rs} = -1. \qquad (7\text{–}105)$$

The matrix $U(n - 1, n)$ is very simple. From Eq. (7–102), it has $+1$'s at rr, rr; from (7–105), it has -1 at diagonal positions rs, rs of type 3. From Eq. (7–104), we have the reciprocal of an integer at (rs, rs)-positions of type 2, and the negative of this number at sr, sr. All of these appear along the diagonal in (7–82). From the unitary condition (7–89) we see that for type 2 we get an off-diagonal submatrix which is a multiple of the unit matrix,

$$U_{rs,sr} = \pm\sqrt{1 - \sigma_{rs,rs}^2}\, E_{rs,sr} = \sigma_{rs,sr} E_{rs,sr}; \qquad (7\text{–}106)$$

$\sigma_{rs,sr}$ is real, since $|\sigma_{rs,rs}| < 1$. The choice of sign in (7–86) is arbitrary; it can be changed by transforming with a diagonal matrix having -1's in the (rs, rs)-box and $+1$'s elsewhere. We adopt the $+$ sign. We define the integer

$$\tau_{rs,rs} = \frac{1}{\sigma_{rs,rs}} = \lambda_r - \lambda_s + s - r, \qquad (7\text{–}107)$$

so that

$$\sigma_{rs,rs} = \frac{\sqrt{\tau_{rs,rs}^2 - 1}}{\tau_{rs,rs}}. \qquad (7\text{–}108)$$

Thus the matrix U contains only rational numbers along the main diagonal; at off-diagonal positions, square roots $\sqrt{\tau_{rs,rs}^2 - 1}$ appear. Since U is unitary and real, $\widetilde{U}U = E$; but for a transposition, $U^2 = E$, so $U = \widetilde{U}$, and the matrix is symmetric,

$$\sigma_{rs,sr} = \sigma_{sr,rs}. \tag{7-109}$$

The integer $\tau_{rs,rs} \equiv \tau_{rs} = \lambda_r - \lambda_s + s - r$ has a simple geometrical meaning. In the pattern, we move from the square r by horizontal and vertical steps until we reach the square s; then τ_{rs} is the total number of steps made. The steps are counted as positive if they are to the left or down, as negative if they are to the right or up. The quantity τ_{rs} is called the *axial distance* from r to s. For a given n, the largest τ_{rs} will occur if $\lambda_1 = n - 1, \lambda_2 = 1$ or $\lambda_1 = 2, \lambda_2 = \cdots = \lambda_{n-1} = 1$; then $\tau_{rs} = n - 1$.

Using $U(n - 1, n)$ and the matrices (7-82) for the elements in S_{n-1}, we can determine the matrices for all elements in S_n by multiplication. We can now repeat the entire process for the transposition $(n - 2, n - 1)$ and the elements of S_{n-2}, and arrive by induction at the following result:

All the matrices of irreducible representations of S_n can be constructed by means of *real numbers*. The diagonal elements are rational; the off-diagonal elements contain square roots of integers no greater than $n^2 - 2n$.

In this process of successive reduction, we start from the given partition and remove one node. We repeat this procedure until the entire pattern is exhausted. Consider the pattern

$$\tag{7-110}$$

Suppose we remove the node in the third line, so that $r = 3$; we then get D_r corresponding to the partition . If we now remove a node from the first line, so that $s = 1$, we obtain D_{rs} corresponding to the pattern . Continuing with the removal of a node from the second line, $t = 2$, we get D_{rst} with the pattern . If we now remove nodes in turn from line 1 and line 2, we obtain the patterns , . Clearly, the submatrix D_{rstuv} which we get at the last stage has one row and one column. By virtue of our method of construction, the basis function corresponding to this row (column) has the property of being a basis vector of the irreducible representation (321) for S_6, of $(3, 2)$ for S_5, of (2^2) for S_4, of $(2, 1)$ for S_3, of (1^2) for S_2, and of (1) for S_1. This basis function ϕ can be labeled by indicating the rows from which nodes are dropped in turn; in our case, it would be labeled as $\phi[3, 1, 2, 1, 2, 1]$. The sequence

$$[3, 1, 2, 1, 2, 1]$$

is called a *Yamanouchi symbol* (Y-symbol). We should note that there is a simple connection between Yamanouchi symbols and standard Young tableaux. In our example, we first removed a node from line 3. The symbol removed was 6, leaving us with S_5. At the second step, we removed a symbol from line 1 and arrived at S_4; clearly this was 5. So our original tableau was

$$\begin{array}{|c|c|c|} \hline 1 & 3 & 5 \\ \hline \end{array}$$
$$\begin{array}{|c|c|} \hline 2 & 4 \\ \hline \end{array}$$
$$\begin{array}{|c|} \hline 6 \\ \hline \end{array}$$

We should also note that our Yamanouchi symbol consists of three 1's, two 2's, and one 3, in agreement with $\lambda_1 = 3$, $\lambda_2 = 2$, $\lambda_3 = 1$. Moreover, if we read the numbers in the symbol from right to left, we see that they are a lattice permutation of 1, 2, 3. Thus, there is a one-to-one correspondence between Yamanouchi symbols, lattice permutations, and standard tableaux. For example,

$$\begin{array}{lll} 1\ 2\ 3 & & 1\ 2\ 6 \\ 4\ 5 & \rightarrow [322111]; & 3\ 5 & \rightarrow [123211]. \\ 6 & & 4 \end{array}$$

For each partition (λ) we get a set of Yamanouchi symbols equal in number to the dimension of the representation $D^{(\lambda)}$. If $(\lambda) = (\lambda_1, \ldots, \lambda_m)$, with $\lambda_m > 0$, the possible Y-symbols are

$$[r_n, r_{n-1}, \ldots, r_2, 1],$$

where the r_i are integers from 1 to m, the integer k appears λ_k times, and if we read from right to left, we obtain a lattice permutation of

$$1^{\lambda_1}, 2^{\lambda_2}, \ldots, m^{\lambda_m}.$$

We can now express the effect of the permutation $(n-1, n)$ in terms of the ϕ's and find that

$$(n-1, n)\phi[r, r, r_{n-2}, \ldots, 1] = +\phi[r, r, r_{n-2}, \ldots, 1],$$
$$(n-1, n)\phi[r, r-1, r_{n-2}, \ldots, 1] = -\phi[r, r-1, r_{n-2}, \ldots, 1]$$
$$(7\text{–}111)$$

if the symbol $[r-1, r, r_{n-2}, \ldots, 1]$ is not permissible; and

$$(n-1, n)\phi[r, s, r_{n-2}, \ldots, 1] = \sigma_{rs}\phi[r, s, r_{n-2}, \ldots, 1]$$
$$+ \sqrt{1 - (\sigma_{rs})^2}\ \phi[s, r, r_{n-2}, \ldots, 1]$$

if both symbols are permissible and $s \neq r$.

The most convenient method for ordering the Y-symbols is to arrange them in decreasing page order. Thus for the diagram of (7–110), the permissible symbols are:

$$\begin{array}{llll}
322111, & 321211, & 321121, & 312211, \\
312121, & 232111, & 231211, & 231121, \\
213211, & 213121, & 211321, & 132211, \\
132121, & 123211, & 123121, & 121321.
\end{array}$$

The advantage of this ordering is that it helps us to picture the successive buildup of the matrices as we go from S_n to S_{n+1}. The first five symbols are the basis functions for the irreducible representation $(\lambda) \equiv (3, 2)$ of S_5 corresponding to the pattern ⊞⊞ obtained after we drop the symbol 6 from line 3. The matrices for the permutations of S_5 in this representation will appear as submatrices in our representation of S_6. Again, if we go past the letters 32 in the first three symbols and the letters 23 in symbols 6 through 8, we arrive at identical sequences corresponding to the pattern ⊞⊞ which refers to S_4. Let us construct the matrix for the permutation (56) in this representation, $(\lambda) \equiv (321)$. Using (7–102), (7–104), (7–105) or (7–111), we obtain the matrix

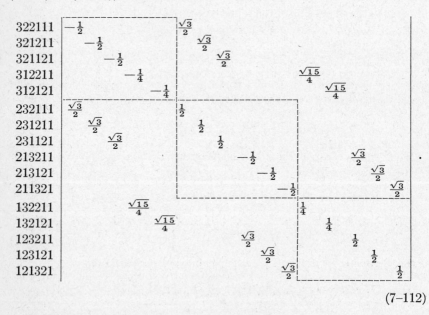

$$(7\text{–}112)$$

Permutations in S_5 (i.e., permutations leaving the letter 6 fixed) will have matrix elements only within the dotted squares drawn along the main diagonal. Their matrices are already known from S_5. The matrices for permutations $(i\,6)$ with $i < 5$ can be obtained by transforming the matrix $(i\,5)$ which is in S_5 by the matrix (7–112) for (56).

Starting with S_1, we can successively build up the real, orthogonal matrix representations of the permutation groups. We present tables for these up to S_5 (see Table 7–3). Because of the symmetry, we show only the elements on and above the main diagonal. Also, we give only the matrices for transpositions; the rest of the representation can be obtained by multiplications.

Problem. Use the tables (Table 7–3) to build up the matrices for the permutations (23), (45), and (56) in the representations $(\lambda) = (4, 2)$ and (3^2) of S_6.

The mere fact that all irreducible representations of the symmetric group can be brought to real form enables us to prove a surprising theorem. Since all irreducible representations of S_n are real, the quantities $c^{(\mu)}$ in Eq. (5–88) are all equal to $+1$, so that

$$\zeta(E) = \sum_\mu n_\mu, \qquad (7\text{–}113)$$

where $\zeta(E)$ is the number of solutions R of the equation

$$R^2 = E. \qquad (7\text{–}113a)$$

In other words, for S_n the sum of the dimensions of all irreducible representations is equal to the number of square roots of the identity element. In order to satisfy Eq. (7–113a), the element R must contain only 1-cycles and 2-cycles, i.e., the cycle structure of R is $(1^\alpha 2^\beta)$, where $\alpha + 2\beta = n$. The number of elements in the class $(1^\alpha 2^\beta)$ is $g_{(1^\alpha 2^\beta)} = n!/1^\alpha \alpha!\,2^\beta \beta!$, so the total number of solutions of Eq. (7–113a) is

$$\sum_{\substack{\alpha,\beta \\ \alpha+2\beta=n}} g_{(1^\alpha 2^\beta)}.$$

Substituting in Eq. (7–113), we get the relation

$$\sum_\mu n_\mu = \sum_{\substack{\alpha,\beta \\ \alpha+2\beta=n}} \frac{n!}{\alpha!\,2^\beta \beta!} = \sum_{\beta=0}^{[n/2]} \frac{n!}{2^\beta \beta!(n-2\beta)!}, \qquad (7\text{–}114)$$

where $[n/2]$ is the largest integer in $n/2$. The sum on the right of Eq. (7–114) is very easily evaluated for any n. Comparing it with Eq. (3–159)

TABLE 7-3

ORTHOGONAL MATRICES OF IRREDUCIBLE REPRESENTATIONS OF THE PERMUTATION GROUPS

S_1:   □   1   [1]   for all elements

S_2: $\begin{matrix}□\\□\end{matrix}$ 11   [1]   for all elements

    □□   21
$\begin{array}{cc} e & (ij) \\ [1] & [-1] \end{array}$

S_3: $\begin{matrix}□\\□\\□\end{matrix}$ 111   [1]   for all elements

    □□□   321
$\begin{array}{cc} e & (ij) \\ [1] & [-1] \end{array}$

    $\begin{matrix}□□\\□\end{matrix}$
$\begin{array}{c} 211 \\ 121 \end{array}$

$$
\begin{array}{cccc}
e & (12) & (23) & (13) \\[4pt]
\begin{bmatrix} 1 & 0 \\ 1 & 1 \end{bmatrix} &
\begin{bmatrix} 1 & 0 \\ 1 & -1 \end{bmatrix} &
\begin{bmatrix} -\dfrac{1}{2} & \dfrac{\sqrt{3}}{2} \\[4pt] \dfrac{1}{2} & \dfrac{1}{2} \end{bmatrix} &
\begin{bmatrix} -\dfrac{1}{2} & -\dfrac{\sqrt{3}}{2} \\[4pt] \dfrac{1}{2} & \dfrac{1}{2} \end{bmatrix}
\end{array}
$$

S_4: $\begin{matrix}□\\□\\□\\□\end{matrix}$ 1111   [1]   for all elements

    □□□□   4321
$\begin{array}{cc} e & (ij) \\ [1] & [-1] \end{array}$

First tableau (partition $(2,1,1)$), basis $\{2111, 1211, 1121\}$:

e and (12), (13) (rows 1211, 1121 shown):

$$
\begin{bmatrix} 1 & 0 \\ & 1 \end{bmatrix}\qquad
\begin{bmatrix} 1 & 0 \\ & -1 \end{bmatrix}\qquad
\begin{bmatrix} -\tfrac12 & -\tfrac{\sqrt3}{2} \\ & \tfrac12 \end{bmatrix}
$$

(34):
$$
\begin{bmatrix}
-\tfrac13 & \tfrac{\sqrt8}{3} & 0 \\
 & \tfrac13 & 0 \\
 & & 1
\end{bmatrix}
$$

(14):
$$
\begin{bmatrix}
-\tfrac13 & \tfrac{\sqrt2}{3} & \tfrac{\sqrt6}{3} \\
 & \tfrac56 & -\tfrac{\sqrt3}{6} \\
 & & -\tfrac12
\end{bmatrix}
$$

(24):
$$
\begin{bmatrix}
-\tfrac13 & -\tfrac{\sqrt2}{3} & \tfrac{\sqrt6}{3} \\
 & \tfrac56 & \tfrac{\sqrt3}{6} \\
 & & -\tfrac12
\end{bmatrix}
$$

Second tableau (partition $(3,1)$), basis $\{3211, 3121, 1321\}$:

e:
$$
\begin{bmatrix}
1 & 0 & 0 \\
1 & 0 & 0 \\
 & 1 & 1
\end{bmatrix}
$$

(12):
$$
\begin{bmatrix}
1 & 0 & 0 \\
 & 1 & 0 \\
 & -1 & 1
\end{bmatrix}
$$

(13):
$$
\begin{bmatrix}
-\tfrac12 & -\tfrac{\sqrt3}{2} & 0 \\
 & \tfrac12 & 0 \\
 & & -1
\end{bmatrix}
$$

(24):
$$
\begin{bmatrix}
-\tfrac13 & \tfrac{\sqrt3}{6} & \tfrac{\sqrt6}{3} \\
-\tfrac12 & -\tfrac56 & -\tfrac{\sqrt2}{3} \\
 & \tfrac{\sqrt3}{6} & \tfrac13
\end{bmatrix}
$$

(14):
$$
\begin{bmatrix}
-\tfrac12 & \tfrac{\sqrt3}{6} & \tfrac{\sqrt6}{3} \\
 & -\tfrac56 & -\tfrac{\sqrt2}{3} \\
 & \tfrac{\sqrt3}{6} & \tfrac13
\end{bmatrix}
$$

(34):
$$
\begin{bmatrix}
-1 & 0 & 0 \\
 & -\tfrac13 & \tfrac{2\sqrt2}{3} \\
 & -\tfrac13 & \tfrac13
\end{bmatrix}
$$

(continued)

TABLE 7–3 (continued)

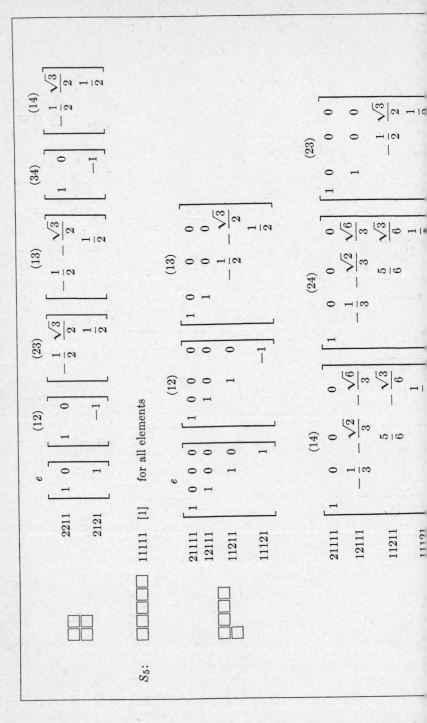

S_5:

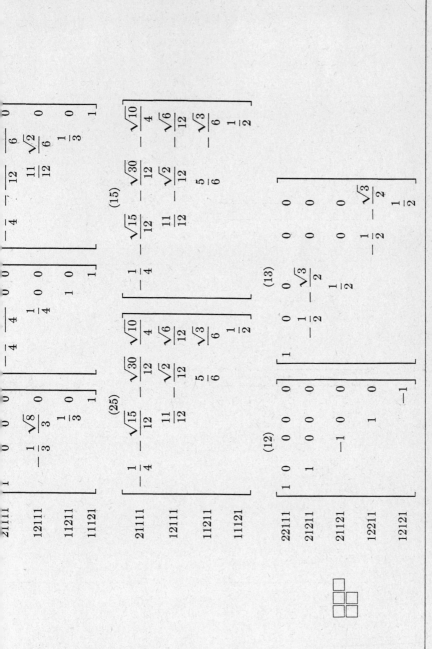

(continued)

TABLE 7–3 (continued)

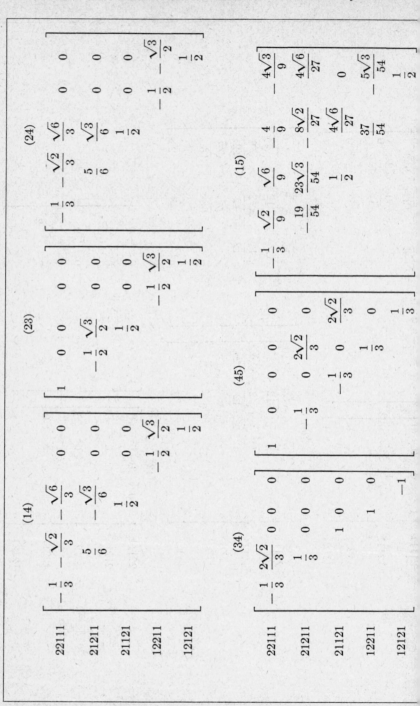

Basis order (columns): 22111, 21211, 21121, 12211, 12121

(14):
$$\begin{bmatrix}
-\frac{1}{3} & -\frac{\sqrt{2}}{3} & -\frac{\sqrt{6}}{3} & 0 & 0\\[4pt]
-\frac{\sqrt{2}}{3} & \frac{5}{6} & -\frac{\sqrt{3}}{6} & 0 & 0\\[4pt]
-\frac{\sqrt{6}}{3} & -\frac{\sqrt{3}}{6} & \frac{1}{2} & 0 & 0\\[4pt]
0 & 0 & 0 & -\frac{1}{2} & \frac{\sqrt{3}}{2}\\[4pt]
0 & 0 & 0 & \frac{\sqrt{3}}{2} & \frac{1}{2}
\end{bmatrix}$$

(23):
$$\begin{bmatrix}
1 & 0 & 0 & 0 & 0\\[4pt]
0 & -\frac{1}{2} & \frac{\sqrt{3}}{2} & 0 & 0\\[4pt]
0 & \frac{\sqrt{3}}{2} & \frac{1}{2} & 0 & 0\\[4pt]
0 & 0 & 0 & -\frac{1}{2} & \frac{\sqrt{3}}{2}\\[4pt]
0 & 0 & 0 & \frac{\sqrt{3}}{2} & \frac{1}{2}
\end{bmatrix}$$

(24):
$$\begin{bmatrix}
-\frac{1}{3} & -\frac{\sqrt{2}}{3} & \frac{\sqrt{6}}{3} & 0 & 0\\[4pt]
-\frac{\sqrt{2}}{3} & \frac{5}{6} & \frac{\sqrt{3}}{6} & 0 & 0\\[4pt]
\frac{\sqrt{6}}{3} & \frac{\sqrt{3}}{6} & \frac{1}{2} & 0 & 0\\[4pt]
0 & 0 & 0 & -\frac{1}{2} & \frac{\sqrt{3}}{2}\\[4pt]
0 & 0 & 0 & \frac{\sqrt{3}}{2} & \frac{1}{2}
\end{bmatrix}$$

(34):
$$\begin{bmatrix}
-\frac{1}{3} & \frac{2\sqrt{2}}{3} & 0 & 0 & 0\\[4pt]
\frac{2\sqrt{2}}{3} & \frac{1}{3} & 0 & 0 & 0\\[4pt]
0 & 0 & 1 & 0 & 0\\[4pt]
0 & 0 & 0 & 1 & 0\\[4pt]
0 & 0 & 0 & 0 & -1
\end{bmatrix}$$

(45):
$$\begin{bmatrix}
1 & 0 & 0 & 0 & 0\\[4pt]
0 & 1 & 0 & 0 & 0\\[4pt]
0 & 0 & -\frac{1}{3} & 0 & \frac{2\sqrt{2}}{3}\\[4pt]
0 & 0 & 0 & 1 & 0\\[4pt]
0 & 0 & \frac{2\sqrt{2}}{3} & 0 & \frac{1}{3}
\end{bmatrix}$$

(15):
$$\begin{bmatrix}
-\frac{1}{3} & \frac{\sqrt{2}}{9} & \frac{\sqrt{6}}{9} & -\frac{4}{9} & -\frac{4\sqrt{3}}{9}\\[4pt]
\frac{\sqrt{2}}{9} & -\frac{19}{54} & \frac{23\sqrt{3}}{54} & -\frac{8\sqrt{2}}{27} & \frac{4\sqrt{6}}{27}\\[4pt]
\frac{\sqrt{6}}{9} & \frac{23\sqrt{3}}{54} & \frac{1}{2} & \frac{4\sqrt{6}}{27} & 0\\[4pt]
-\frac{4}{9} & -\frac{8\sqrt{2}}{27} & \frac{4\sqrt{6}}{27} & \frac{37}{54} & -\frac{5\sqrt{3}}{54}\\[4pt]
-\frac{4\sqrt{3}}{9} & \frac{4\sqrt{6}}{27} & 0 & -\frac{5\sqrt{3}}{54} & \frac{1}{2}
\end{bmatrix}$$

(35)

$$
\begin{array}{c}
22111\\21211\\21121\\12211\\11211
\end{array}
\left[
\begin{array}{ccccc}
-\dfrac{5}{27} & -\dfrac{8\sqrt{2}}{27} & \dfrac{8}{9} & 0 & 0\\[2mm]
-\dfrac{8\sqrt{2}}{27} & \dfrac{23}{27} & \dfrac{2\sqrt{2}}{9} & 0 & 0\\[2mm]
\dfrac{8}{9} & \dfrac{2\sqrt{2}}{9} & \dfrac{1}{3} & 0 & 0\\[2mm]
0 & 0 & 0 & -\dfrac{1}{3} & -\dfrac{2\sqrt{2}}{3}\\[2mm]
0 & 0 & 0 & -\dfrac{2\sqrt{2}}{3} & \dfrac{1}{3}
\end{array}
\right]
$$

(25)

$$
\left[
\begin{array}{ccccc}
-\dfrac{1}{3} & \dfrac{\sqrt{2}}{9} & \dfrac{\sqrt{6}}{9} & -\dfrac{4}{9} & \dfrac{4\sqrt{3}}{9}\\[2mm]
\dfrac{\sqrt{2}}{9} & -\dfrac{19}{54} & -\dfrac{23\sqrt{3}}{54} & -\dfrac{8\sqrt{2}}{27} & -\dfrac{4\sqrt{6}}{27}\\[2mm]
\dfrac{\sqrt{6}}{9} & -\dfrac{23\sqrt{3}}{54} & \dfrac{1}{2} & -\dfrac{4\sqrt{6}}{27} & 0\\[2mm]
-\dfrac{4}{9} & -\dfrac{8\sqrt{2}}{27} & -\dfrac{4\sqrt{6}}{27} & \dfrac{37}{54} & \dfrac{5\sqrt{3}}{54}\\[2mm]
\dfrac{4\sqrt{3}}{9} & -\dfrac{4\sqrt{6}}{27} & 0 & \dfrac{5\sqrt{3}}{54} & \dfrac{1}{2}
\end{array}
\right]
$$

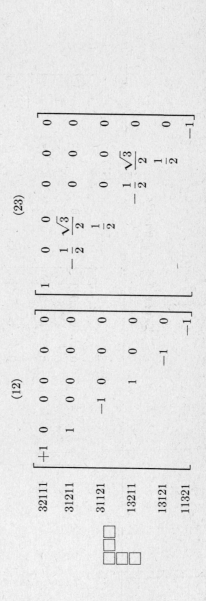

(12)

$$
\begin{array}{c}
32111\\31211\\31121\\13211\\13121\\11321
\end{array}
\left[
\begin{array}{cccccc}
+1 & 0 & 0 & 0 & 0 & 0\\
0 & 1 & 0 & 0 & 0 & 0\\
0 & 0 & -1 & 0 & 0 & 0\\
0 & 0 & 0 & 1 & 0 & 0\\
0 & 0 & 0 & 0 & -1 & 0\\
0 & 0 & 0 & 0 & 0 & -1
\end{array}
\right]
$$

(23)

$$
\left[
\begin{array}{cccccc}
1 & 0 & 0 & 0 & 0 & 0\\[1mm]
0 & -\dfrac{1}{2} & \dfrac{\sqrt{3}}{2} & 0 & 0 & 0\\[2mm]
0 & \dfrac{\sqrt{3}}{2} & \dfrac{1}{2} & 0 & 0 & 0\\[2mm]
0 & 0 & 0 & -\dfrac{1}{2} & \dfrac{\sqrt{3}}{2} & 0\\[2mm]
0 & 0 & 0 & \dfrac{\sqrt{3}}{2} & \dfrac{1}{2} & 0\\[2mm]
0 & 0 & 0 & 0 & 0 & -1
\end{array}
\right]
$$

(continued)

TABLE 7–3 (continued)

(34)

	32111	31211	31121	13211	13121	11321
32111	$-\frac{1}{3}$	$\frac{\sqrt{8}}{3}$	0	0	0	0
31211	$\frac{\sqrt{8}}{3}$	$\frac{1}{3}$	0	0	0	0
31121	0	0	1	0	0	0
13211	0	0	0	-1	0	0
13121	0	0	0	0	$-\frac{1}{3}$	$\frac{\sqrt{8}}{3}$
11321	0	0	0	0	$\frac{\sqrt{8}}{3}$	$\frac{1}{3}$

(45)

	32111	31211	31121	13211	13121	11321
32111	-1	0	0	0	0	0
31211	0	$-\frac{1}{4}$	0	$\frac{\sqrt{15}}{4}$	0	0
31121	0	0	$-\frac{1}{4}$	0	$\frac{\sqrt{15}}{4}$	0
13211	0	$\frac{\sqrt{15}}{4}$	0	$\frac{1}{4}$	0	0
13121	0	0	$\frac{\sqrt{15}}{4}$	0	$\frac{1}{4}$	0
11321	0	0	0	0	0	1

which, for the symmetric group S_n, becomes

$$\sum_\mu n_\mu^2 = n!, \tag{7-115}$$

we obtain the inequality

$$\sum_{\beta=0}^{[n/2]} \frac{1}{2^\beta \beta!(n-2\beta)!} > \sqrt{\frac{1}{n!}}. \tag{7-116}$$

The quantities n_μ are given by Eqs. (7–66) and (7–60). Substituting in Eq. (7–114), we get the remarkable sum formula

$$\sum_{(\lambda)} \frac{D(\lambda_1+m-1, \lambda_2+m-2, \cdots, \lambda_m)}{(\lambda_1+m-1)!(\lambda_2+m-2)!\cdots\lambda_m!} = \sum_{\beta=0}^{[n/2]} \frac{1}{2^\beta \beta!(n-2\beta)!}. \tag{7-117}$$

7–8 Hund's method. The method of treating the symmetric group which was developed by F. Hund (*Z. Physik* **43**, 788, 1927) is a physicist's version of the mathematical method based on group algebras.

Suppose that we are dealing with a system of equivalent particles. (By the term "equivalent" we mean that the Hamiltonian of the problem, which may be an approximation to the full problem, is invariant under the interchange of all coordinates of such particles. The term "identical" implies that we believe, in the present state of our knowledge, that the exact Hamiltonian is invariant under interchange of the particles.) If we are given any eigenfunction $\psi(1, \ldots, n)$ belonging to a given eigenvalue of the Hamiltonian, then any one of the $n!$ permutations of the particle coordinates will yield an eigenfunction belonging to the same energy. (In ψ, the symbol 1 means *all* the coordinates of the first particle, etc.) Thus, if we permute the first two particles in ψ, we can obtain the linear combinations

$$\begin{aligned}
\psi(1, 2, \ldots, n) + \psi(2, 1, \ldots, n) &= (e + (12))\psi(1, 2, \ldots, n), \\
\psi(1, 2, \ldots, n) - \psi(2, 1, \ldots, n) &= (e - (12))\psi(1, 2, \ldots, n)
\end{aligned} \tag{7-118}$$

which are also eigenfunctions, and are respectively symmetric and antisymmetric in particles 1 and 2. The permutation operator $e + (12)$ is called the *symmetrizer*, and $e - (12)$ the *antisymmetrizer* of the particles 1 and 2. By continuing this process we can attempt to construct, from the initial eigenfunction ψ, functions which are symmetric (antisymmetric) in larger sets of particles. The operator which symmetrizes ψ in a certain set of particles is

$$\sum_P P, \tag{7-119}$$

where we sum over all permutations P of the particles in the set; similarly, the antisymmetrizer for the set is

$$\sum_P \delta_P P. \tag{7–119a}$$

By applying operators like (7–119) and (7–119a) to ψ we can attempt to construct degenerate eigenfunctions having different symmetries. To indicate the symmetry properties of a function we shall put braces around particles in which the function is antisymmetric, and brackets around particles in which the function is symmetric; if particle numbers appear outside of these enclosures, nothing is implied about symmetry in their coordinates. For example,

$$\psi(\{123\}[45]67) \tag{7–120}$$

is antisymmetric in particles 1, 2, 3; it is symmetric in 4 and 5; no claim is made as to symmetry or antisymmetry for particles 6 and 7. Furthermore, we claim no *a priori* knowledge as to whether the function is symmetric or antisymmetric in any particles which do not appear inside the same enclosure.

In any case, starting from our initial eigenfunction, we can perform permutations and linear combinations to construct equivalent eigenfunctions having certain symmetries. For example, starting from $\psi(1, \ldots, n)$, we can try to symmetrize ψ in all the particles by applying the symmetrizer (7–119) for all n particles; this gives an equivalent eigenfunction

$$\psi([123 \ldots n]). \tag{7–121}$$

Now it may happen that the original function $\psi(1, \ldots, n)$, from which we started, is such that the function in (7–121) is identically zero. This will occur if the initial ψ is antisymmetric in any one pair of particles. If $\psi([12 \ldots n]) \equiv 0$, we try to symmetrize in any $n - 1$ particles; e.g., we apply to our initial ψ the symmetrizer in all particles except the nth, then in all except the $(n - 1)$st, etc. If all resulting functions vanish identically, we try to symmetrize in $n - 2$ particles, etc. If, at the end of our travail, we find that we cannot symmetrize even in one pair of particles, this must mean that our original function was completely antisymmetric in all n particles. In completely analogous fashion we might take our original function $\psi(1, \ldots, n)$ and try to antisymmetrize in all particles by applying the operator (7–119a) for n particles. If the resulting function $\psi(\{1, \ldots, n\})$ vanishes identically, we try to antisymmetrize in any $n - 1$ particles. If we find by continuing this process that we cannot antisymmetrize ψ even in one pair of particles, then our original function was totally symmetric.

As an example we consider the case of three particles. We start from an eigenfunction $\psi(1, 2, 3)$. We apply the symmetrizer and obtain $\psi([123])$.

If this function is not identically zero, we say that ψ has the *symmetry type* $S(3)$. If $\psi([123]) \equiv 0$, then we try to symmetrize in a pair of particles and obtain, say,

$$\psi([12]3) \quad \text{and} \quad \psi([13]2).$$

If these are not both identically zero, then we say that ψ has the symmetry type $S(2 + 1)$; if both vanish identically, then ψ is antisymmetric in the three particles, and has the symmetry type $S(1 + 1 + 1)$. In precisely the same fashion, we could try to antisymmetrize, and the following alternatives would result:

$$\psi(\{123\}) - \text{antisymmetry type } A(3),$$
$$\psi(\{12\}3) - \text{antisymmetry type } A(2 + 1),$$
$$\psi(123) \quad - \text{antisymmetry type } A(1 + 1 + 1).$$

We proceed in precisely the same way in the general case. We try to symmetrize in as many particles as possible. If the maximum number of particles in which we can symmetrize ψ (without getting a result which is identically zero) is λ_1, we can relabel the coordinates so that these particles have the labels $1, \ldots, \lambda_1$, and obtain an eigenfunction $\psi([12 \ldots \lambda_1] \ldots n)$. We now leave these particles alone and try to symmetrize in as large a group of the remaining particles as possible. Proceeding in this fashion, we finally arrive at an equivalent eigenfunction which is in *normal form* (*S-form*),

$$\psi\left([12 \ldots \lambda_1][\lambda_1 + 1, \ldots, \lambda_1 + \lambda_2] \ldots \left[\sum_{\nu=1}^{m-1} \lambda_\nu + 1, \ldots, \sum_{\nu=1}^{m} \lambda_\nu\right]\right),$$

$$(7\text{--}122)$$

where $\lambda_1 \geq \lambda_2 \geq \cdots \geq \lambda_m$ and $\sum_{\nu=1}^{m} \lambda_\nu = n$. By the normal form we mean that we cannot use permutations and linear combinations to construct from ψ a function symmetric in *more than* λ_1 particles; if we permute the particles $\lambda_1 + 1, \ldots, n$, we cannot construct a function symmetric in *more than* λ_2 particles, etc. If in (7–122) several symbols appear separately at the end,

$$\psi([\quad], [\quad] \ldots n - 2, n - 1, n),$$

then the function must be antisymmetric in all the separate symbols. The function in (7–122) is said to belong to the *symmetry type*

$$S(\lambda_1 + \lambda_2 + \cdots + \lambda_m). \tag{7--123}$$

Note that there is a complete correspondence between the symmetry types and the Young patterns which we used earlier.

In similar fashion we can bring ψ to a *normal antisymmetric form* (*A-form*),

$$\psi(\{1 \ldots \mu_1\}\{\mu_1 + 1, \ldots, \mu_1 + \mu_2\} \ldots), \qquad (7\text{-}122a)$$

belonging to the *antisymmetry type*

$$A(\mu_1 + \mu_2 + \cdots + \mu_l). \qquad (7\text{-}123a)$$

It should be emphasized that we can apply to a given ψ either of the two processes; in other words, a given eigenfunction belongs to a definite symmetry type $S(\lambda_1 + \cdots + \lambda_m)$ *and* to a definite antisymmetry type $A(\mu_1 + \cdots + \mu_l)$. If we make the assumption that there are no other degeneracies than those due to equivalence of particles, then the degenerate (equivalent) eigenfunctions must have the same S-type or A-type. An eigenfunction belonging to a given eigenvalue can be written as a linear combination of basis functions of a definite S-type or A-type. (This is essentially our earlier statement that the degenerate functions form the basis for an irreducible representation of the symmetry group.)

We now prove some simple theorems about symmetrization.

LEMMA. If an eigenfunction

$$\psi(\{12 \ldots \lambda\}[\lambda + 1, \ldots, \lambda + \mu] \ldots) \qquad (7\text{-}124)$$

is antisymmetric in the particles 1, 2, $\ldots \lambda$, and symmetric in the particles $\lambda + 1, \lambda + 2, \ldots, \lambda + \mu$ (regardless of its dependence on the remaining particles), then by permuting 1, 2, $\ldots, \lambda + \mu$ and taking linear combinations we can form one or the other (*but not both*) of

$$\begin{aligned}
&\phi(\{12 \ldots \lambda, \lambda + 1\}[\lambda + 2, \ldots, \lambda + \mu] \ldots), \\
&\chi(\{12 \ldots \lambda - 1\}[\lambda, \lambda + 1, \ldots, \lambda + \mu] \ldots).
\end{aligned} \qquad (7\text{-}125)$$

To construct ϕ, we need only permute 1, 2, $\ldots, \lambda, \lambda + 1$; to construct χ, we need only permute $\lambda, \lambda + 1, \ldots, \lambda + \mu$.

Proof: We use $\alpha, \beta, \ldots$ to denote letters 1, 2, $\ldots, \lambda$, and $x, y, \ldots$ to denote letters $\lambda + 1, \ldots, \lambda + \mu$. The operator

$$e - \sum_{\alpha=1}^{\lambda} (\alpha x) \qquad (7\text{-}126)$$

applied to ψ gives a function which is antisymmetric in 1, 2, $\ldots, \lambda, x$. To prove this we apply all transpositions of these letters to

$$\left[e - \sum_{\alpha=1}^{\lambda} (\alpha x) \right] \psi(\{12 \ldots \lambda\}[\lambda + 1, \ldots, \lambda + \mu] \ldots);$$

then

$$(\beta\gamma)\left[e - \sum_{\alpha=1}^{\lambda} (\alpha x)\right]\psi = (\beta\gamma)\left[e - (\beta x) - (\gamma x) - \sum_{\alpha \neq \beta, \gamma} (\alpha x)\right]\psi.$$

Since $(\beta\gamma)(\beta x) = (\gamma x)(\beta\gamma)$ and $(\beta\gamma)\psi = -\psi$, we have

$$(\beta\gamma)\left[e - (\beta x) - (\gamma x) - \sum_{\alpha \neq \beta, \gamma} (\alpha x)\right]\psi = \left[e - \sum_{\alpha} (\alpha x)\right](\beta\gamma)\psi$$

$$= -\left[e - \sum_{\alpha} (\alpha x)\right]\psi.$$

Similarly, since $(\beta x)(\alpha x) = (\alpha x)(\alpha\beta)$,

$$(\beta x)\left[e - \sum_{\alpha=1}^{\lambda} (\alpha x)\right]\psi = (\beta x)\left[e - (\beta x) - \sum_{\alpha \neq \beta} (\alpha x)\right]\psi$$

$$= \left[(\beta x) - e - \sum_{\alpha \neq \beta} (\alpha x)(\alpha\beta)\right]\psi,$$

Thus

$$(\beta x)\left[e - \sum_{\alpha} (\alpha x)\right]\psi = [-e + (\beta x)]\psi - \left(\sum_{\alpha \neq \beta} (\alpha x)\right)(\alpha\beta)\psi$$

$$= -\left[e - \sum_{\alpha} (\alpha x)\right]\psi.$$

Now it may happen with a given choice of x that the function

$$\left[e - \sum_{\alpha} (\alpha x)\right]\psi$$

is identically zero. We now show that, as we select different choices for x (that is, $x = \lambda + 1, \ldots, \lambda + \mu$), either *all* or *none* of the resultant functions vanish. For if $[e - \sum_{\alpha} (\alpha x)]\psi = 0$ for x, then applying the transposition (xy) yields

$$0 = (xy)\left[e - \sum_{\alpha} (\alpha x)\right]\psi$$

$$= (xy)\left[e - \sum_{\alpha} (\alpha x)\right](xy)(xy)\psi$$

$$= (xy)\left[e - \sum_{\alpha} (\alpha x)\right](xy)\psi = \left[e - \sum_{\alpha} (\alpha y)\right]\psi$$

for any y.

Similarly, by applying the operator

$$\left[e + \sum_{x=\lambda+1}^{\lambda+\mu} (\alpha x) \right] \tag{7-127}$$

to ψ, we get a function which is symmetric in α, $\lambda + 1, \ldots, \lambda + \mu$. By a procedure similar to that used above, we show that either *all* or *none* of the functions

$$\left[e + \sum_{x=\lambda+1}^{\lambda+\mu} (\alpha x) \right] \psi$$

are identically zero ($\alpha = 1, \ldots, \lambda$).

Next we show that the two types of functions obtained by operating on ψ with (7–126) and (7–127) cannot *both* be identically zero. For if

$$\left[e - \sum_{\alpha} (\alpha x) \right] \psi = 0$$

for all x and

$$\left[e + \sum_{x} (\alpha x) \right] \psi = 0$$

for all α, then, summing the first over x, the second over α, and adding, we would obtain $(\lambda + \mu)\psi = 0$, which would imply that the original function ψ was identically zero.

Finally we show that both forms are not possible—one or the other must be identically zero. For, if this is not the case, then

$$\chi(\{12 \ldots \lambda - 1\}[\lambda, \lambda + 1, \ldots, \lambda + \mu] \ldots)$$

is expressible as a linear combination of functions obtained from

$$\phi(\{12 \ldots \lambda, \lambda + 1\}[\lambda + 2, \ldots, \lambda + \mu] \ldots)$$

by applying permutations P (on the symbols $1, \ldots, \lambda + \mu$) and taking linear combinations, i.e.,

$$\chi(\{12 \ldots \lambda - 1\}[\lambda, \lambda + 1, \ldots, \lambda + \mu] \ldots)$$
$$= \sum_{P} a_P P \phi(\{12 \ldots \lambda, \lambda + 1\}[\lambda + 2, \ldots, \lambda + \mu] \ldots). \tag{7-128}$$

But then it is clear that at least two of the symbols λ, $\lambda + 1, \ldots, \lambda + \mu$ which appear in the brackets on the left side of Eq. (7–128) must be contained in the braces on the right, after P is applied to ϕ. In other words, each term in $\sum_P$ on the right is antisymmetric in at least two of λ, $\lambda + 1$,

$\ldots, \lambda + \mu$. We now apply the symmetrizer on λ, $\lambda + 1, \ldots, \lambda + \mu$ to both sides of (7–128). On the left we get back $n!\chi$, while the right yields zero.

The use of this lemma enables us to shift from S-forms to A-forms, and conversely. For example, suppose that we have a normal S-form $\psi([123][45])$. Since this is a normal S-form, it must be antisymmetric under interchange of 4 (or 5) and at least one of the particles 1, 2, 3. We can therefore antisymmetrize in, say 5 and 3, the result being $\psi([12]4\{35\})$. Since our original ψ was symmetric in 1, 2, 3 and in 4, 5, we cannot antisymmetrize in more than two particles; i.e., one from each pair of brackets in the S-form. According to our lemma, we can now form either (1) $\psi([124]\{35\})$ or (2) $\psi(1\{24\}\{35\})$. If the first alternative were possible, then, using the lemma again, we should be able to form either (a) $\psi([1234]5)$ or (b) $\psi([12]\{345\})$: (a) is impossible since the original S-form implies that we cannot symmetrize in more than three particles, and (b) is impossible because the original S-form implies that we cannot antisymmetrize in more than two particles. Thus from the symmetry type $S(3 + 2)$ we can form, by permutation and linear combination, only the antisymmetry type $A(2 + 2 + 1)$. Note that these are conjugate partitions. Before applying this result, we derive the following theorem:

THEOREM. A normal S-form of the type $\psi([12 \ldots r][r + 1, \ldots, 2r])$ can be brought to the A-form $\psi(\{1a\}\{2b\} \ldots \{rs\})$, where $a, b, \ldots, s$ are taken from the set $r + 1, \ldots, 2r$.

Since we start from an S-form, the function must be antisymmetric under interchange of 1 and some symbol from the second pair of brackets; so we can form

$$\psi(\{1a\}[2 \ldots r][r + 1, \ldots, 2r]).$$

We now try to continue this process of antisymmetrizing on pairs, taking one symbol from the first and one from the second bracket. Suppose that after pulling out p pairs, we find that we can no longer continue to do so. This means that the function is symmetric in the $2r - 2p$ symbols remaining in the brackets. At this stage we have

$$\psi(\{1a\}\{2b\} \ldots \{p\rho\}[\ldots]).$$

Applying our lemma to the term in brackets and to each term in braces in turn, we see that we can draw back one symbol from each pair of braces into the brackets. But then we have a ψ which is symmetric in $2r - 2p + p = 2r - p$ symbols. This will contradict our assumption that we started from an S-form unless $p = r$, i.e., unless ψ is converted to

$$\psi(\{1a\}\{2b\} \ldots \{rs\}).$$

Now we note that if we interchange the two groups of particles, ψ is multiplied by the factor $(-1)^r$.

Our previous result concerning conversion from S-form to A-form can be generalized: A function of the type $S(\lambda_1 + \lambda_2 + \cdots) = S(\lambda)$ can be converted by permutations and linear combination to a function of the type $A(\mu_1 + \mu_2 + \cdots) = A(\mu)$, where the partitions (λ) and (μ) are conjugate. The proof is essentially that given in our previous example. To make it clear we give a second example. We start from the normal form $\psi([12345][678]9)$. Beginning at the right, we find that our lemma allows us to form only $\psi([12345][67]\{89\})$. Now we apply the lemma to [12345] and {89}. We cannot form [123458]9 since this would exceed our original normal form, so we pull one symbol out of the brackets and obtain

$$\psi([1234][67]\{589\}).$$

We see that we have antisymmetrized on a set consisting of one symbol from each pair of brackets of our original normal S-form, so we cannot bring any more symbols into {589}. Leaving the term in braces alone, we repeat the process on [1234][67]. Up to now we have permuted only 5, 8, 9. If we could form [12346]7, we could apply the inverse of our previous operation on 5, 8, 9 and get a result exceeding our normal form; hence we again take one symbol from each pair of brackets and get

$$\psi([123]6\{47\}\{589\}).$$

We now repeat the process on [123]6 and finally obtain

$$\psi(12\{36\}\{47\}\{589\}),$$

so that

$$S(5 + 3 + 1) \rightarrow A(3 + 2 + 2 + 1 + 1).$$

The normal forms $S(\lambda)$ clearly are just the basis functions for the irreducible representation (λ). The symmetry type of a given function is determined by giving either its S-normal form $S(\lambda)$ or its A-normal form $A(\mu)$, where (λ) and (μ) are conjugate partitions. Two functions ψ and ϕ are said to have *reciprocal* or *conjugate symmetry type* if their S-forms (or A-forms) correspond to conjugate partitions.

Problem. Prove that terms of different symmetry type do not combine; i.e., if ψ and ϕ have different symmetry types and the function f is symmetric in all particles, then

$$\int \psi f \phi \, d\Omega = 0.$$

7–9 Group algebra. Though Hund's method is simple, it does not provide very effective techniques for constructing functions of a particular symmetry type. A very powerful method for treating the symmetric group which, as we shall see in a later chapter, also enables us to find the representations of linear transformation groups in n dimensions was developed by Young and Frobenius. This method is based on the general procedure for analyzing the structure of a finite algebra. We used the notion of a group algebra in Section 3–17, but we derived most of the theorems concerning representations by other methods. We steered clear of the algebraic method in Chapter 3 because most physicists are unnecessarily wary of the jargon and formalism. In this section we shall give a sketch of this method. We shall appeal to the results of Chapter 3 to make some of our statements plausible, and use the symmetric group as an example to prevent the presentation from becoming too abstract.

We start with the symmetric group, and consider the permutations as operators acting on a function of n variables, $\psi(1, \ldots, n)$. The effect of the transposition (ij) is to give a new function

$$(ij)\psi(1, \ldots, i, \ldots, j, \ldots, n) = \psi(1, \ldots, j, \ldots, i, \ldots, n). \quad (7\text{–}129)$$

Since the transpositions generate S_n, Eq. (7–129) tells us the effect of any permutation operator on ψ. If $\phi_1, \ldots, \phi_n$ are independent functions of a single set of variables, we may choose for ψ the product function

$$\phi_1(1)\phi_2(2) \cdots \phi_n(n),$$

where $1, \ldots, n$ denote n different sets of variables. Then

$$(ij)\phi_1(1) \cdots \phi_i(i) \cdots \phi_j(j) \cdots \phi_n(n)$$
$$= \phi_1(1) \cdots \phi_i(j) \cdots \phi_j(i) \cdots \phi_n(n). \quad (7\text{–}129a)$$

By applying the permutation operators R to a general function ψ we obtain a set of $n!$ linearly independent functions $R\psi$. If we now take linear combinations of these functions, we obtain vectors

$$x = \sum_R x_R(R\psi) \quad (7\text{–}130)$$

in a vector space having $n!$ dimensions. The coefficients x_R are the coordinates of the function x in the basis provided by the functions $R\psi$.

If we apply a permutation operator S to the function x in Eq. (7–130), we get a new function

$$x' = Sx = S\sum_R x_R(R\psi) = \sum_R x_R S(R\psi) = \sum_R x_R SR\psi.$$

Setting $SR = T$, $R = S^{-1}T$, we have

$$x' = \sum_T x_{S^{-1}T}(T\psi),$$

and comparing with $x' = \sum_T x'_T(T\psi)$, we find

$$x'_T = x_{S^{-1}T}. \tag{7–131}$$

Equation (7–131) associates with the permutation S a linear transformation in the $n!$-dimensional vector space. (In Section 3–17 we considered only those vectors x which have a single nonzero component, and obtained the regular representation of the group.) Clearly the operator $\alpha_S S$ will be represented by the linear transformation

$$x'_T = \alpha_S x_{S^{-1}T}, \tag{7–131a}$$

and the sum $S + U$ of two permutation operators by the transformation

$$x'_T = x_{S^{-1}T} + x_{U^{-1}T}; \tag{7–131b}$$

hence the general operator $s = \sum_S \alpha_S S$ will be represented by the transformation

$$x'_T = \sum_S \alpha_S x_{S^{-1}T} = \sum_R \alpha_{TR^{-1}} x_R. \tag{7–132}$$

If we examine the preceding equations, we notice that the presence of the function ψ which provided an object on which the operator could act was irrelevant to the results. From the elements R of any group of order g we can construct a g-dimensional linear vector space of vectors

$$x = \sum_R x_R R, \tag{7–130a}$$

where the x_R are the coordinates of the vector x in the particular basis which is obtained by taking the group elements themselves as basis vectors. The vector $z = \alpha x + \beta y$ has components

$$z_R = \alpha x_R + \beta y_R, \tag{7–133}$$

and the product of two vectors is contained in the space:

$$x = \sum_R x_R R, \qquad y = \sum_S y_S S,$$

$$z = (xy) = \sum_{R,S} x_R y_S RS = \sum_T \left(\sum_S x_{TS^{-1}} y_S \right) T = \sum_T \left(\sum_S x_S y_{S^{-1}T} \right) T,$$

so that

$$(xy)_T = \sum_S x_{TS^{-1}} y_S = \sum_S x_S y_{S^{-1}T}. \tag{7–134}$$

A linear vector space which is closed under some multiplication law is called an *algebra*. In our case, since the multiplication law is taken over from the multiplication law of the group, which is associative, the *group algebra A* is an *associative algebra*.

Equation (7–132) provides us with a representation of the group algebra A. It is in fact the regular representation.

Any representation of the group G automatically gives a representation of the algebra A: If $s = \sum_S \alpha_S S$,

$$D(s) = \sum_S \alpha_S D(S). \tag{7–135}$$

Conversely, any representation of the algebra A gives a representation of G. Moreover, if one of these representations is reducible (or irreducible), then so is the other.

We know from Chapter 3 that the regular representation (Eq. 7–132) is fully reducible and contains each irreducible representation a number of times equal to the dimension of the representation.

By a *subalgebra B* of the algebra A we mean a linear vector space which is contained in A and which is closed under the law of multiplication of the algebra A.

If a subalgebra B has the property that, for u in B, su is also in B for *any* element s of the *whole* algebra A, then B is called a *left ideal*.

In the regular representation (7–132), a left ideal $\mathcal{L}_1$ is an invariant subspace since $s\mathcal{L}_1 = \mathcal{L}_1$ for any element s of the algebra A. Since the regular representation is *fully reducible*, the space A must be a direct sum of left ideals; that is,

$$A = \mathcal{L}_1 + \mathcal{L}_2, \tag{7–136}$$

where

$$s\mathcal{L}_1 = \mathcal{L}_1, \qquad s\mathcal{L}_2 = \mathcal{L}_2. \tag{7–137}$$

Every element of the algebra A is uniquely expressible as the sum of an element in $\mathcal{L}_1$ and an element in $\mathcal{L}_2$; only the element 0 is common to $\mathcal{L}_1$ and $\mathcal{L}_2$. The matrices $D(s)$ of the regular representation are reducible to

$$D(s) = D_1(s) + D_2(s), \tag{7–138}$$

where $D_i(s)$ is the matrix of the linear transformation induced in $\mathcal{L}_i$ by left multiplication with s.

The unit element e of the group G is in the group algebra A and has the property that

$$es = se = s \tag{7–139}$$

for all elements s in A. If A is the direct sum of two left ideals as in Eq.

(7–136), the unit element e is uniquely expressible as a sum

$$e = e_1 + e_2, \tag{7–140}$$

where e_1 is in $\mathcal{L}_1$ and e_2 in $\mathcal{L}_2$. Similarly, any element s of A is uniquely expressible as

$$s = s_1 + s_2. \tag{7–141}$$

Substituting Eqs. (7–140) and (7–141) in (7–139), we find

$$s = s_1 + s_2 = se = s(e_1 + e_2) = se_1 + se_2. \tag{7–142}$$

Since $\mathcal{L}_1$ and $\mathcal{L}_2$ are left ideals, se_1 is in $\mathcal{L}_1$ and se_2 in $\mathcal{L}_2$, so

$$s_1 = se_1, \qquad s_2 = se_2. \tag{7–143}$$

If s is in $\mathcal{L}_1$, then $s = s_1$, $s_2 = 0$, and Eq.(7–143) states that

$$s = se_1, \qquad se_2 = 0 \qquad \text{for } s \text{ in } \mathcal{L}_1, \tag{7–144}$$

and in particular for $s = e_1$,

$$e_1 = e_1^2, \qquad e_1 e_2 = 0. \tag{7–144a}$$

Similarly, for s in $\mathcal{L}_2$ we get

$$s = se_2, \qquad se_1 = 0, \qquad e_2 = e_2^2, \qquad e_2 e_1 = 0. \tag{7–144b}$$

The element e_1 is *idempotent*, that is, $e_1^2 = e_1$; furthermore, it is a *generator* of the ideal $\mathcal{L}_1$ since se_1 is in $\mathcal{L}_1$ for all s in A. If s is in $\mathcal{L}_1$ then $se_1 = s$. The same remarks apply for e_2 in $\mathcal{L}_2$. Moreover, $e_1 e_2 = e_2 e_1 = 0$.

The resolution of the unit element in Eq. (7–140) into its parts in $\mathcal{L}_1$ and $\mathcal{L}_2$ gave us the generators of the left ideals $\mathcal{L}_1$ and $\mathcal{L}_2$ (Peirce resolution).

The left ideals $\mathcal{L}_1$ and $\mathcal{L}_2$ may in turn contain subalgebras which are left ideals. If an ideal $\mathcal{L}$ contains no proper subideal, it provides us with an irreducible representation of the algebra A. Such an ideal is said to be *minimal*. Continuing this process, we end up with the algebra A expressed as a direct sum of minimal left ideals:

$$A = \mathcal{L}_1 + \mathcal{L}_2 + \cdots + \mathcal{L}_k. \tag{7–145}$$

The left ideal $\mathcal{L}_i$ is generated by the idempotent e_i, and

$$e_i^2 = e_i; \qquad e_i e_j = 0 \qquad \text{for } i \neq j. \tag{7–146}$$

From our previous argument it is clear that we find the generators e_i by resolving the unit element e into components in the spaces $\mathcal{L}_1, \ldots, \mathcal{L}_k$:

$$e = e_1 + e_2 + \cdots + e_k. \tag{7-147}$$

An idempotent ϵ is said to be *primitive* if it cannot be resolved into a sum of idempotents satisfying Eq. (7-146). We leave to the reader the simple proof of the following theorem:

THEOREM. If ϵ is a primitive idempotent, the left ideal $\mathcal{L} = A\epsilon$ is minimal. Conversely, if $\mathcal{L}$ is a minimal left ideal, any generating unit of $\mathcal{L}$ is primitive.

One can continue this argument and derive all the theorems of Chapter 3, but instead we shall now turn to the problem for the particular case of the symmetric group.

7-10 Young operators. From the discussion of the last section, we know that any idempotent e_i will generate a left ideal $\mathcal{L}_i$ giving a representation which is contained in the regular representation. Furthermore, since (nonzero) numerical factors are irrelevant, an element e_i satisfying $e_i^2 = \alpha e_i$ will serve our purpose since then e_i/α is idempotent.

In particular, we consider the element

$$P = \sum_R R, \tag{7-148}$$

where the sum runs over all the permutations of S_n. Since for any permutation S, $SP = \sum_R SR = P$, $P^2 = n!P$, it follows that P is a generator. If we multiply P on the left by the quantity $s = \sum_S \alpha_S S$, we get $sP = \sum_S \alpha_S SP = (\sum \alpha_S)P$; thus the left ideal AP generated by P consists of the multiples of P. This is a one-dimensional vector space. Left-multiplying by a permutation S does not change αP, so our representation assigns the number 1 to every group element, and is the identity representation.

Similarly $Q = \sum_R \delta_R R$ is idempotent except for a factor, since $SQ = \sum_R \delta_R SR = \delta_S Q$, so that $Q^2 = n!Q$. Q generates the one-dimensional ideal consisting of multiples of Q. Multiplying on the left by the group element S, we see that we obtain the alternating representation.

To find the other irreducible representations, we use the Young tableaux which we discussed earlier in this chapter.

For any partition (λ) of n, draw the Young pattern. Insert the numbers $1, 2, \ldots, n$ into the pattern in any order to give a Young tableau. Once the tableau has been fixed, we consider two types of permutations. *Horizontal permutations* p are permutations which interchange only symbols in the same row. *Vertical permutations* q interchange only symbols in the same column. We construct the quantities

$$P = \sum_p p \quad \text{``symmetrizer,''} \tag{7-149}$$

$$Q = \sum_q \delta_q q \quad \text{``antisymmetrizer,''} \tag{7-150}$$

where δ_q is the parity of the permutation q; the sum is over all horizontal permutations of the given tableau in (7–149) and over all vertical permutations in (7–150).

Then the Young operator

$$Y = QP \tag{7-151}$$

is essentially idempotent, and generates a left ideal which provides an irreducible representation of S_n. The representations obtained by this process for different patterns are inequivalent. Different tableaux with the same pattern give equivalent irreducible representations.

If we always choose the standard tableaux of Section 7–3, we will obtain in this way the complete reduction of the regular representation of S_n.

The proof of these statements is elegant and straightforward, but has been given in identical form in so many texts that we shall omit it here. Instead we shall illustrate its use for some simple cases.

For $n = 2$, the problem is trivial. The quantities $e + (12)$ corresponding to the pattern ▢1▢2, and $e - (12)$ corresponding to the pattern $\frac{\boxed{1}}{\boxed{2}}$ are essentially idempotent:

$$[e \pm (12)][e \pm (12)] = e^2 \pm 2e(12) + (12)(12) = 2[e \pm (12)].$$

The resolution of the unit element into idempotents is

$$e = \frac{e + (12)}{2} + \frac{e - (12)}{2}.$$

The first generator gives the identity representation, the second the alternating representation. Each is contained once in the regular representation.

For $n = 3$:

Tableau ▢1▢2▢3 with generator $= \dfrac{1}{6} \sum_R R$ gives the identity representation.

Tableau $\dfrac{\boxed{1}}{\boxed{2}}_{\boxed{3}}$ with generator $= \dfrac{1}{6} \sum_R \delta_R R$ gives the alternating representation.

The element $e + (12)$ is again essentially idempotent. Let us see what left ideal it generates. We multiply $e + (12)$ on the left by each of the permutations

$$e[e + (12)] = e + (12) = (12)[e + (12)],$$
$$(13)[e + (12)] = (13) + (123) = (123)[e + (12)],$$
$$(23)[e + (12)] = (23) + (132) = (132)[e + (12)].$$

We see that $e + (12)$ generates the three-dimensional left ideal $\mathcal{L}_1$ with basis vectors $e + (12)$, $(13) + (123)$, $(23) + (132)$.

Since $e = \frac{1}{2}[e + (12)] + \frac{1}{2}[e - (12)]$, the quantity $[e - (12)]$ generates a left ideal $\mathcal{L}_2$ such that $A = \mathcal{L}_1 + \mathcal{L}_2$. The left ideal $\mathcal{L}_2$ has basis vectors $e - (12)$, $(13) - (123)$, $(23) - (132)$. In agreement with Eq. (7–144a), $[e \pm (12)][e \mp (12)] = 0$.

The ideals $\mathcal{L}_1$ and $\mathcal{L}_2$ are not minimal; $\mathcal{L}_1$ contains the idempotent $\frac{1}{6}\sum R$ and $\mathcal{L}_2$ the idempotent $\frac{1}{6}\sum \delta_R R$. So $\mathcal{L}_1$ gives the identity representation and a two-dimensional representation, while $\mathcal{L}_2$ gives the alternating representation and a two-dimensional representation. The resolution of the unit element obtained in this way is

$$e = \frac{1}{6}\sum R + \frac{1}{6}\sum \delta_R R + \left[\frac{e + (12)}{2} - \frac{1}{6}\sum R\right]$$
$$+ \left[\frac{e - (12)}{2} - \frac{1}{6}\sum \delta_R R\right].$$

If instead we continue with our recipe, we construct the standard tableau $\boxed{\begin{smallmatrix}1&2\\3&\end{smallmatrix}}$, for which $P = e + (12)$, $Q = e - (13)$, $Y = QP = e + (12) - (13) - (123)$. The other standard tableau $\boxed{\begin{smallmatrix}1&3\\2&\end{smallmatrix}}$ gives $P' = e + (13)$, $Q' = e - (12)$, and $Y' = Q'P' = e - (12) + (13) - (132)$. We obtain the following resolution of the unit element:

$$e = \frac{1}{6}\sum R + \frac{1}{6}\sum \delta_R R + \frac{1}{3}Y + \frac{1}{3}Y'.$$

Problem. Verify that $Y/3$ and $Y'/3$ are idempotent and that $YY' = Y'Y = 0$.

For $n = 4$:

　　　Tableau $\boxed{1}\boxed{2}\boxed{3}\boxed{4}$ gives the identity representation.

　　　Tableau $\begin{smallmatrix}\boxed{1}\\\boxed{2}\\\boxed{3}\\\boxed{4}\end{smallmatrix}$ gives the alternating representation.

For the pattern $\begin{array}{|c|c|c|}\hline &&\\\hline\end{array}\,\begin{array}{|c|}\hline\\\hline\end{array}$, the standard tableau $\begin{array}{|c|c|c|}\hline 1&2&3\\\hline 4&&\\\hline\end{array}$ has $P_1 = e +$ (12) + (23) + (13) + (123) + (132), $Q_1 = e - (14)$, and $Y_1 = Q_1 P_1$. The standard tableau $\begin{array}{|c|c|c|}\hline 1&2&4\\\hline 3&&\\\hline\end{array}$ has $P_2 = e + (12) + (14) + (24) + (124)$ + (142), $Q_2 = e - (13)$, and $Y_2 = Q_2 P_2$. The standard tableau $\begin{array}{|c|c|c|}\hline 1&3&4\\\hline 2&&\\\hline\end{array}$ has $P_3 = e + (13) + (14) + (34) + (134) + (143)$, $Q_3 = e - (12)$, and $Y_3 = Q_3 P_3$.

For the pattern $\begin{array}{|c|c|}\hline &\\\hline &\\\hline\end{array}$, the standard tableau $\begin{array}{|c|c|}\hline 1&2\\\hline 3&4\\\hline\end{array}$ has $P_4 = [e + (12)]$ $[e + (34)]$, $Q_4 = [e - (13)][e - (24)]$, and $Y_4 = Q_4 P_4$. The standard tableau $\begin{array}{|c|c|}\hline 1&3\\\hline 2&4\\\hline\end{array}$ has $P_5 = [e + (13)][e + (24)]$, $Q_5 = [e - (12)][e - (34)]$, and $Y_5 = Q_5 P_5$.

Problem. For $n = 4$, find the resolution of the identity in terms of the Young operators.

7–11 The construction of product wave functions of a given symmetry. Fock's cyclic symmetry conditions. Let us assume that we have a system containing n equivalent particles and that a typical eigenfunction belonging to the energy E is $\psi = u(1)v(2)w(3)\ldots z(n)$. We assume that the single-particle functions in the product are all different and orthonormal (since they are solutions of the same Schrödinger equation); later we shall examine the case where the number of single-particle states is limited. Since the particles are equivalent, we can construct degenerate eigenfunctions by permutation and linear combination. We want to construct basis functions for the various irreducible representations of S_n. The solution to the problem is given by Section 7–10: We apply to the function the Young operators corresponding to all the standard tableaux for a given pattern to obtain the basis functions for the corresponding irreducible representation.

For $n = 2$, the normalized function for $\begin{array}{|c|c|}\hline 1&2\\\hline\end{array}$ is

$$\frac{1}{\sqrt{2}}\,[u(1)v(2) + u(2)v(1)],$$

and the basis function for $\begin{array}{|c|}\hline 1\\\hline 2\\\hline\end{array}$ is

$$\frac{1}{\sqrt{2}}\,[u(1)v(2) - u(2)v(1)] = \frac{1}{\sqrt{2}}\begin{vmatrix} u(1) & u(2) \\ v(1) & v(2) \end{vmatrix}.$$

For $n = 3$, the completely symmetric function is

$$f_1 = \frac{1}{\sqrt{6}}\left[\begin{array}{c} u(1)v(2)w(3) + u(1)v(3)w(2) + u(2)v(1)w(3) \\ + u(3)v(2)w(1) + u(2)v(3)w(1) + u(3)v(1)w(2) \end{array}\right],$$

and the completely antisymmetric function is the determinant

$$f_2 = \frac{1}{\sqrt{6}}\begin{vmatrix} u(1) & u(2) & u(3) \\ v(1) & v(2) & v(3) \\ w(1) & w(2) & w(3) \end{vmatrix}.$$

To find the basis functions for the two-dimensional representation, we use the operators Y and Y' of Section 7-10 and get

$$f_3 = \tfrac{1}{2}[u(1)v(2)w(3) + u(2)v(1)w(3) - u(3)v(2)w(1) - u(2)v(3)w(1)]$$

$$= \tfrac{1}{2}[v(2)\{u(1)w(3) - u(3)w(1)\} + u(2)\{v(1)w(3) - v(3)w(1)\}];$$

$$f_4 = \tfrac{1}{2}[u(1)v(2)w(3) - u(2)v(1)w(3) + u(3)v(2)w(1) - u(3)v(1)w(2)].$$

Problem. Use the Young operators of Section 7-10 to construct linear combinations of product wave functions of four particles and tabulate the basis functions for the irreducible representations of S_4.

We have assumed that the single-particle wave functions $u, v, \ldots$ were all different. We were then able to obtain all possible symmetry types by applying suitable operators. In many physical problems the number of different single-particle states is limited; e.g., if we consider the spin function of a many-electron system to be made up of products of single-particle functions, only two spin states are available. Similar restrictions would apply to aggregates of identical particles with spin other than one-half. For nucleons, only four single-particle states (of spin and I-spin) are available. When the number of single-particle states is restricted, not all irreducible representations can occur. We shall give a brief discussion here, and return to this problem when we take up physical applications.

Let us assume that we have a system of n equivalent particles. If only one single-particle state is possible, then the product wave function from which we start our symmetrization is necessarily

$$u(1)u(2) \ldots u(n). \tag{7-152}$$

This wave function is already completely symmetric, so that the only representation which can occur is that corresponding to the partition (n).

Suppose two single-particle states u and v are possible. We start from a product function

$$u(1) \ldots u(m)v(m + 1) \ldots v(n) \qquad (7\text{-}153)$$

or, in terms of Hund's method,

$$\psi([1 \ldots m][m + 1, \ldots, n]). \qquad (7\text{-}153a)$$

The possible irreducible representations can be obtained in two ways. We see that we cannot obtain patterns which have more than two rows, since any antisymmetrizer on more than two particles annihilates (7-153). The same argument essentially applies to (7-153a) in Hund's method.

Similarly, if the number of different single-particle states is k, we can have no partitions with more than k rows. (In the nucleon problem, at most four-row partitions occur for the spin-I spin function.)

In connection with the problem of constructing coordinate wave functions of a definite symmetry for the n-electron problem, Fock has used the procedure of imposing on the function ψ the following conditions:

(1) Antisymmetry in the first k arguments ($k \geq n/2$).
(2) Antisymmetry in the remaining $n - k$ arguments.
(3) Cyclic symmetry:

$$\left[e - \sum_{i=1}^{k} (ni) \right] \psi = 0. \qquad (7\text{-}154)$$

It is easy to show that a function ψ with Hund antisymmetry type $A(k + \{n - k\})$ satisfies these three conditions. The tableau in question is

1	$k + 1$
.	.
.	.
.	.
.	.
.	n
.	
.	
k	

$$(7\text{-}154a)$$

The function ψ with normal form $A(k + \{n - k\})$ clearly satisfies conditions 1 and 2. The operator (7-154) is identical with the operator we used in Eq. (7-126). As we showed there, this operator applied to ψ gives a function which is antisymmetric in the $k + 1$ arguments $1, 2, \ldots k, n$. Since ψ has normal form $A(k + \{n - k\})$ with $k \geq n/2$, the resulting function must be identically zero.

Fock's conditions can also be stated for the general case. The function ψ has a definite symmetry (i.e., it is one of the basis functions for an irreducible representation of S_n) if

(1) It is antisymmetric in a set of λ_1 arguments, antisymmetric in another set of λ_2 arguments, etc.

(2) In the partition (λ), $\lambda_1 \geq \lambda_2 \geq \cdots \geq \lambda_i \cdots$ etc., the function ψ satisfies the cyclic symmetry conditions

$$\left[e - \sum_{\alpha \text{ in } \lambda_i} (\alpha x) \right] \psi = 0, \qquad (7\text{--}155)$$

where x is in λ_j for all combinations of λ_i and λ_j with $\lambda_i \geq \lambda_j$.

Again the argument following Eq. (7–126) shows that a function having the normal form $A(\lambda_1 + \lambda_2 + \cdots)$ satisfies the Fock conditions. The operators in Eq. (7–155), when applied to ψ, give functions which are of the form $A(\lambda_1 + \cdots + (\lambda_i + 1) + \cdots + (\lambda_j - 1) + \cdots)$. Since ψ was in normal form, all these functions are identically zero.

The general conditions given above are very useful for testing the symmetry of a given function.

Problem. Prove that the Young operator Y for the tableau (7–154a) is annihilated by the cyclic symmetry operator in (7–154), i.e.,

$$\left[e - \sum_{i=1}^{k} (ni) \right] Y = 0.$$

Similarly, prove that the Young operator for a general tableau is annihilated by the cyclic symmetry operators in Eq. (7–155).

7–12 Outer products of representations of the symmetric group.

For the symmetric group, the Kronecker products of representations are usually called *inner products* since they refer to products of two representations each of which refers to the *same* particles. Thus for three particles we earlier learned to construct inner products like $D^\alpha \times D^\beta$ where, say, α and β are both the partition $(2, 1)$ for particles labeled 1, 2, 3. We also learned how to reduce the inner product to its irreducible constituents; e.g., for the case just mentioned,

$$(2, 1) \times (2, 1) = (3) + (2, 1) + (1^3).$$

We shall discuss this inner product (Clebsch-Gordan series) in the next section.

In the case of the symmetric group, another type of product is important for many physical problems. Suppose we have two separate systems, the first consisting of particles 1, 2, 3, and the second containing particles 4, 5, 6. All particles are considered to be equivalent. First we assume that the systems are not interacting with each other. In this case, we would classify the states of the first system according to the irreducible representations of S_3 on particles 1, 2, 3; similarly, the states of the second system would be classified according to representations of the symmetric group on particles 4, 5, 6. Thus we may be dealing with a two-fold degenerate level of system 1 corresponding to the pattern ⊞ with basis functions $\boxed{1}\boxed{2}\ \boxed{3}$, $\boxed{1}\boxed{3}\ \boxed{2}$ and a 2-fold degenerate level of system 2 with pattern ⊞ and basis functions $\boxed{4}\boxed{5}\ \boxed{6}$, $\boxed{4}\boxed{6}\ \boxed{5}$. When the two systems interact with each other, we should classify the states of the amalgamated systems according to representations of the symmetric group on all six particles. We then say that we are dealing with the *outer product* (symbol ⊗) of the two representations,

$$(2, 1) \otimes (2, 1),$$

which is to be resolved into irreducible representations of S_6. We should note that if we restrict ourselves to permuting 1, 2, 3 among themselves and 4, 5, 6 among themselves, the outer product is an irreducible representation of this particular subgroup of S_6. But when we permute all six particles, it becomes reducible in terms of S_6. Thus to find the total number of basis functions of the outer product we may argue as follows: Three particles are selected from the six to form system one, and the remaining three are assigned to system two. This can be done in ${}^6C_3 = 20$ ways. For each such choice, we obtain two basis functions for each representation, so the total number of product functions is $2 \times 2 \times 20 = 80$. The result of the resolution, for which we shall give the rule in a moment, is

$$(2, 1) \otimes (2, 1) = (4, 2) + (4, 1^2) + (3^2) + 2(3, 2, 1)$$
$$+ (3, 1^3) + (2^3) + (2^2, 1^2). \quad (7\text{--}156)$$

Counting the dimensionalities of the representations on the right, we find

$$9 + 10 + 5 + 2(16) + 10 + 5 + 9 = 80.$$

This counting process is a useful check on such expansions. Now for the general rule, which we give without proof: To find the components in the outer product, draw the pattern for one of the factors. In the pattern for the other factor, assign the same symbol, say a, to *all* boxes in the first row, the same symbol b to *all* boxes in the second row, etc. Now apply the symbols a to the first pattern, and enlarge it in all possible ways subject to the rule that no two a's appear in the same column and that the resultant graph be regular; repeat with the b's, etc. One further restriction: If, after all symbols have been added to the pattern, we read the added symbols from right to left in the first row, then the second row, etc., they must form a lattice permutation of the a's, b's, ...

To illustrate the procedure, we shall work out the expansion (7–156). We start with the first pattern

$$\begin{matrix} \bullet & \bullet \\ \bullet & \end{matrix}$$

and label the second pattern

$$\begin{matrix} a & a \\ b & \end{matrix}$$

First we enlarge the first pattern by adjoining the two a's. The possible results are:

$$
\begin{array}{llll}
\bullet\ \bullet\ a\ a & \bullet\ \bullet\ a & \bullet\ \bullet\ a & \bullet\ \bullet \\
(1)\ \bullet & (2)\ \bullet\ a & (3)\ \bullet & (4)\ \bullet\ a \\
& & \quad\ a & \quad\ a
\end{array}
$$

(Note that

$$\begin{matrix} \bullet & \bullet \\ \bullet & \\ a & \\ a & \end{matrix}$$

is not allowed.) Now we adjoin the symbol b.

From (1) we get

$$
\begin{matrix}
\bullet\ \bullet\ a\ a & \qquad & \bullet\ \bullet\ a\ a \\
\bullet\ b & \qquad & \bullet \\
& & \quad\quad b
\end{matrix}
$$

(Note that

$$\begin{matrix} \bullet & \bullet & a & a & b \\ \bullet & & & & \end{matrix}$$

is not allowed, since baa is not a lattice permutation.)

From (2), we obtain

$$
\begin{array}{ccc}
\bullet & \bullet & a \\
\bullet & a & b \\
\end{array}
\qquad
\begin{array}{ccc}
\bullet & \bullet & a \\
\bullet & a & \\
b & & \\
\end{array}
$$

From (3),

$$
\begin{array}{ccc}
\bullet & \bullet & a \\
\bullet & b & \\
a & & \\
\end{array}
\qquad
\begin{array}{ccc}
\bullet & \bullet & a \\
\bullet & & \\
a & & \\
b & & \\
\end{array}
$$

From (4),

$$
\begin{array}{ccc}
\bullet & \bullet & \\
\bullet & a & \\
a & b & \\
\end{array}
\qquad
\begin{array}{ccc}
\bullet & \bullet & \\
\bullet & a & \\
a & & \\
b & & \\
\end{array}
$$

which checks (7–156).

Frequently, equations like (7–156) are written with patterns instead of partitions:

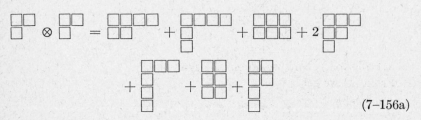

$$(7\text{–}156a)$$

We have given the general rule without proof because for physical applications the outer products are usually so simple that the results can be obtained by elementary methods.

The outer product of any representation (λ) with a representation corresponding to a horizontal strip is easily found. Consider first the outer product $\square \otimes \square$. No symmetry conditions have been imposed on the functions $\psi(1)$ and $\phi(2)$, so by permuting the particles we can obtain both the symmetric and antisymmetric combinations $\psi(1)\phi(2) \pm \psi(2)\phi(1)$. Thus

$$\square \otimes \square = \square\square + \begin{array}{c}\square\\\square\end{array} \qquad (7\text{–}157)$$

Next consider the outer product of (λ) and (1):

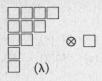

Since the wave function for the single added particle is not subject to any symmetry conditions, we can construct functions which are antisymmetric in all the particles in any column of (λ) plus the added particle, or we can insert it in the first row of (λ). Thus

$$(\lambda) \otimes (1) = \sum_{\lambda_i} (\lambda_1 \cdots \lambda_i + 1 \cdots \lambda_n). \qquad (7\text{-}158)$$

In the outer product $(\lambda) \otimes (2)$,

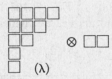

the wave function for the two particles in ☐☐ is symmetric in these two particles, but is not subjected to any symmetry conditions under permutations of these two particles with the particles in (λ). We may therefore add the two dots to the pattern (λ) in all possible ways, provided we do not put them in the same column.

Similarly we see that the outer product of (λ) and a horizontal strip (m) is obtained by applying the nodes of (m) to (λ) in all possible ways subject to the restriction that no two of the nodes be placed in the same column. If we interchange the role of rows and columns and go to the conjugate representations (p. 206), we immediately obtain the rule for evaluating $(\lambda) \otimes (1^m)$: We apply the m nodes to (λ) in all possible ways in which no two nodes are inserted in the same row.

In this manner we evaluate

$$\square\square \otimes \square \;=\; \square\square\square + {\square\square \atop \square} \qquad (7\text{-}159)$$

$$\square\square \otimes \square\square = \square\square\square\square + {\square\square\square \atop \square} + {\square\square \atop \square\square} \qquad (7\text{-}160)$$

$$\square\square \otimes {\square \atop \square} \;=\; {\square\square\square \atop \square} + {\square\square \atop \square \atop \square} \qquad (7\text{-}161)$$

$$ {\square \atop \square} \otimes {\square \atop \square} \;=\; {\square \atop \square \atop \square} + {\square\square \atop \square \atop \square} + {\square\square \atop \square\square} \qquad (7\text{-}162)$$

We can now evaluate more complicated outer products by a ladder process. Successive outer products can be evaluated in any order,

$$(\lambda) \otimes (\mu) \otimes (\nu) = [(\lambda) \otimes (\mu)] \otimes (\nu)$$
$$= [(\mu) \otimes (\nu)] \otimes (\lambda) = [(\lambda) \otimes (\nu)] \otimes (\mu), \quad (7\text{-}163)$$

and the distributive law,

$$(\lambda) \otimes [(\mu) + (\nu)] = (\lambda) \otimes (\mu) + (\lambda) \otimes (\nu), \quad (7\text{-}164)$$

is valid. So from Eq. (7-159),

$$(\lambda) \otimes (2, 1) = (\lambda) \otimes [(2) \otimes (1)] - (\lambda) \otimes (3)$$
$$= [(\lambda) \otimes (2)] \otimes (1) - (\lambda) \otimes (3). \quad (7\text{-}165)$$

All expressions on the right side of Eq. (7-165) involve the application of horizontal strips, which we did above, and hence we can evaluate

$$(\lambda) \otimes (2, 1).$$

Problems. (1) Evaluate the outer products:

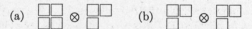

(a)   (b)

(2) Extend the ladder method to evaluate $(\lambda) \otimes (2, 2)$ and $(\lambda) \otimes (3, 1)$. Try to derive the general rule for outer products.

7–13 Inner products. Clebsch-Gordan series for the symmetric group.

In Chapter 6, we discussed the reduction of products of irreducible representations of a given group G. In the case of the symmetric group, such Kronecker products are called inner products to distinguish them from the outer products which we discussed in the preceding section. In Chapter 6, we showed how the reduction can be done by a sort of Fourier resolution, using the character table for the group. Though this method is straightforward, it does not give general formulas. We know that for the rotation group the Clebsch-Gordan series (vector model) gives a completely general result:

$$D^l \times D^{l'} = \sum_{L=|l-l'|}^{l+l'} D^L, \quad (7\text{-}166)$$

and it is of interest to ask whether there are similar general formulas for the symmetric group. We must expect the formulas to be more complicated than (7-166); the symmetric group offers a more complex structure than the rotation group, despite (or because of!) the fact that the latter is a continuous group. We should also note that, from the point of view of Example 4, Section 1–1, S_n can be regarded as a subgroup of the group of linear transformations in n dimensions.

A straightforward procedure for the reduction of inner products (Clebsch-Gordan series) has been given by Murnaghan and by Gamba. We shall not describe their method, but merely give some of the general formulas as well as tables (Table 7–4) for all nontrivial inner products up through $n = 5$:

$$(n - 1, 1) \times (n - 1, 1) \ = (n) + (n - 1, 1) + (n - 2, 2)$$
$$+ (n - 2, 1^2).$$
$$(7\text{–}167)$$

$$(n - 1, 1) \times (n - 2, 2) \ = (n - 1, 1) + (n - 2, 2)$$
$$+ (n - 2, 1^2) + (n - 3, 3)$$
$$+ (n - 3, 2, 1). \quad (n > 4)$$
$$(7\text{–}168)$$

$$(n - 1, 1) \times (n - 2, 1^2) = (n - 1, 1) + (n - 2, 2) + (n - 2, 1^2)$$
$$+ (n - 3, 2, 1) + (n - 3, 1^3).$$
$$(7\text{–}169)$$

$$(n - 2, 2) \times (n - 2, 2) \ = (n) + (n - 1, 1) + 2(n - 2, 2)$$
$$+ (n - 2, 1^2) + (n - 3, 3)$$
$$+ 2(n - 3, 2, 1) + (n - 3, 1^3)$$
$$+ (n - 4, 4) + (n - 4, 3, 1)$$
$$+ (n - 4, 2^2).$$
$$(7\text{–}170)$$

The tables give the coefficients in the expansion of the inner products. For products at the left, the coefficients refer to the partitions at the top of the columns; for products on the right, the coefficients refer to the partitions at the bottom of the columns. Not all possibilities are shown; those omitted can be obtained from simple theorems about conjugate partitions, which we have used earlier, but collect here for convenience. We know from Eq. (7–44) that

$$(\lambda) \times (1^n) = (\tilde{\lambda}), \qquad (7\text{–}171)$$

where $(\tilde{\lambda})$ is the partition conjugate to (λ). Also

$$\{(\lambda) \times (\mu)\} \times (\nu) = (\lambda) \times \{(\mu) \times (\nu)\} = \{(\lambda) \times (\nu)\} \times (\mu). \quad (7\text{–}172)$$

TABLE 7–4

CLEBSCH-GORDAN SERIES FOR $n = 4,\ 5$

(1) $n = 4$

	(4)	(3, 1)	(2^2)	$(2, 1^2)$	(1^4)	
$(3, 1) \times (3, 1)$	1	1	1	1		$(3, 1) \times (2^2)$
$(3, 1) \times (2^2)$		1		1		$(3, 1) \times (2^2)$
$(2^2) \times (2^2)$	1		1		1	$(2^2) \times (2^2)$
	(1^4)	$(2, 1^2)$	(2^2)	$(3, 1)$	(4)	

(2) $n = 5$

	(5)	(4, 1)	(3, 2)	$(3, 1^2)$	$(2^2, 1)$	$(2, 1^3)$	(1^5)	
$(4, 1) \times (4, 1)$	1	1	1	1				$(4, 1) \times (2, 1^3)$
$(4, 1) \times (3, 2)$		1	1	1	1			$(4, 1) \times (2^2, 1)$
$(4, 1) \times (3, 1^2)$		1	1	1	1	1		$(4, 1) \times (3, 1^2)$
$(3, 2) \times (3, 2)$	1	1	1	1	1	1		$(3, 2) \times (2^2, 1)$
$(3, 2) \times (3, 1^2)$		1	1	2	1	1		$(3, 2) \times (3, 1^2)$
$(3, 1^2) \times (3, 1^2)$	1	1	2	1	2	1	1	$(3, 1^2) \times (3, 1^2)$
	(1^5)	$(2, 1^3)$	$(2^2, 1)$	$(3, 1^2)$	$(3, 2)$	$(4, 1)$	(5)	

Letting $(\nu) = (1^n)$, we have

$$\widetilde{(\lambda) \times (\mu)} = (\lambda) \times (\tilde{\mu}) = (\tilde{\lambda}) \times (\mu). \qquad (7\text{–}173)$$

From the second equality, replacing (λ) by $(\tilde{\lambda})$, we obtain

$$(\tilde{\lambda}) \times (\tilde{\mu}) = (\lambda) \times (\mu). \qquad (7\text{–}174)$$

We also showed in Chapter 6 that if $(\lambda) \times (\mu)$ contains (ν), then $(\lambda) \times (\nu)$ contains (μ), and $(\mu) \times (\nu)$ contains (λ); in fact, the expansion coefficient in all three cases is just the coefficient of the identity representation (n) in $(\lambda) \times (\mu) \times (\nu)$. Thus we see that $(\lambda) \times (\mu)$ contains the identity representation ($once$) if and only if $(\lambda) \equiv (\mu)$; and $(\lambda) \times (\mu)$ contains the alternating representation (1^n) $once$ if and only if $(\lambda) = (\tilde{\mu})$.

The table for $n = 5$ already shows that the symmetric group presents problems beyond that of the rotation group; some representations appear more than once in the expansion of the Kronecker product of two representations. As we shall see later, this phenomenon appears in the rotation group only for triple products (cf. also Section 5–8).

Inner products can also be evaluated by graphical methods. Gamba and Radicati have given a complicated derivation of a graphical procedure for the special case of $(\lambda) \times (n-1, 1)$. We shall derive this in a much simpler way and show how the method can be generalized.

From Problem 3 of Section 7–4, we know that

$$\chi^{(n-1,1)}_{(1^\alpha 2^\beta \dots)} = \alpha - 1. \tag{7-175}$$

To find the number of times, $a_{(\mu)}$, that the representation (μ) is contained in $(\lambda) \times (n-1, 1)$, we use Eqs. (5–107), (7–12), and (7–175):

$$a_{(\mu)} = \sum_{\substack{\alpha,\beta,\gamma,\dots \\ \alpha+2\beta+\dots=n}} \frac{1}{\alpha! 2^\beta \beta! \cdots} \chi^{(n-1,1)}_{(1^\alpha 2^\beta \dots)} \chi^{(\lambda)}_{(1^\alpha 2^\beta \dots)} \chi^{(\mu)}_{(1^\alpha 2^\beta \dots)}$$

$$= \sum_{\substack{\alpha,\beta,\gamma,\dots \\ \alpha+2\beta+\dots=n}} \frac{1}{\alpha! 2^\beta \beta! \cdots} (\alpha - 1) \chi^{(\lambda)}_{(1^\alpha 2^\beta \dots)} \chi^{(\mu)}_{(1^\alpha 2^\beta \dots)}. \tag{7-176}$$

The sum with the factor (-1) is immediately evaluated by means of the orthogonality theorem, and gives $-\delta_{\lambda\mu}$. In the sum with the factor α we apply the branching theorem (7–52) to $\chi^{(\lambda)}$ and $\chi^{(\mu)}$:

$$\sum_{\substack{\alpha,\beta,\gamma,\dots \\ \alpha+2\beta+\dots=n}} \frac{\alpha}{\alpha! 2^\beta \beta! \cdots} \chi^{(\lambda)}_{(1^\alpha 2^\beta \dots)} \chi^{(\mu)}_{(1^\alpha 2^\beta \dots)}$$

$$= \sum_{\substack{\alpha\neq 0,\beta,\gamma,\dots \\ \alpha+2\beta+\dots=n}} \frac{1}{(\alpha-1)! 2^\beta \beta! \cdots} \chi^{(\lambda)}_{(1^\alpha 2^\beta \dots)} \chi^{(\mu)}_{(1^\alpha 2^\beta \dots)}$$

$$= \sum_{\substack{\alpha\neq 0,\beta,\gamma,\dots \\ \alpha+2\beta+\dots=n}} \frac{1}{(\alpha-1)! 2^\beta \beta! \cdots} \sum_{(\lambda')} \chi^{(\lambda')}_{(1^{\alpha-1} 2^\beta \dots)} \sum_{(\mu')} \chi^{(\mu')}_{(1^{\alpha-1} 2^\beta \dots)}$$

$$= \sum_{\substack{\alpha',\beta,\gamma,\dots \\ \alpha'+2\beta+\dots=n-1}} \frac{1}{\alpha'! 2^\beta \beta! \cdots} \sum_{(\lambda')} \chi^{(\lambda')}_{(1^{\alpha'} 2^\beta \dots)} \sum_{(\mu')} \chi^{(\mu')}_{(1^{\alpha'} 2^\beta \dots)}, \tag{7-177}$$

where we have set $\alpha - 1 = \alpha'$. The partitions (λ') and (μ') are obtained from (λ) and (μ) by regular removal of one node. But now we see that each product in $\sum_{(\lambda')} \sum_{(\mu')}$ is in precisely the form for applying the orthogonality theorem for $n-1$ particles, and gives $\delta_{\lambda'\mu'}$. Thus the Kronecker product $(\lambda) \times (n-1, 1)$ will contain (μ) if the removal of one node from (λ) and one node from (μ) leads to the same partition of $n-1$. Or we can make the simpler statement: The Kronecker product $(\lambda) \times (n-1, 1)$ contains (μ) if the removal of one node from (λ) and the addition of one node to the resulting partition gives (μ). Combining this with the result for the factor -1, we can make the statement: $(\lambda) \times (n - 1, 1)$ contains $(\lambda)N - 1$ times, where N is the number of different λ_i's. All those representations (μ) whose graphs are obtained from that of (λ) by shifting one node appear once.

For example,

$$\boxed{\begin{array}{ccc}\square\square\square\\\square\end{array}} \times \boxed{\begin{array}{cc}\square\square\\\square\square\end{array}} = \boxed{\begin{array}{ccc}\square\square\square\\\square\end{array}} + \boxed{\begin{array}{cc}\square\square\\\square\\\square\end{array}}$$

The graphical result contains Eqs. (7–167) through (7–169) as special cases.

The method used above can be extended to any product $(\lambda) \times (\mu)$ for which we know the characters $\chi^{(\lambda)}_{(1^\alpha 2^\beta \dots)}$ of one of the factors as analytic expressions (polynomials) in $\alpha, \beta, \dots$. For example, let us consider the Kronecker products $(n - 2, 1^2) \times (\lambda)$ and $(n - 2, 2) \times (\lambda)$. In Problem 3 of Section 7–4 we found

$$\chi^{(n-2,1^2)}_{(1^\alpha 2^\beta \dots)} = \frac{(\alpha - 1)(\alpha - 2)}{2} - \beta, \tag{7–178}$$

$$\chi^{(n-2,2)}_{(1^\alpha 2^\beta \dots)} = \frac{(\alpha - 1)(\alpha - 2)}{2} + \beta - 1. \tag{7–179}$$

We rewrite the polynomials in terms of the set:

$$1, \; \alpha, \; \alpha(\alpha - 1), \quad \alpha(\alpha - 1)(\alpha - 2), \dots$$
$$2\beta, \; 2^2\beta(\beta - 1), \; 2^3\beta(\beta - 1)(\beta - 2), \dots$$
$$3\gamma, \; 3^2\gamma(\gamma - 1), \; 3^3\gamma(\gamma - 1)(\gamma - 2), \dots \text{ etc.}$$

The two polynomials above contain $1, \alpha, \alpha(\alpha - 1), 2\beta$. When we evaluate Eq. (7–176) (with $\alpha - 1$ replaced by the appropriate polynomial), we obtain terms with the factors 1 or α (which we have just evaluated), one term with the factor $\alpha(\alpha - 1)$, and another with the factor 2β. These are evaluated separately.

For the factor $\alpha(\alpha - 1)$, the contribution to $a_{(\mu)}$ is

$$a' = \sum_{\substack{\alpha, \beta, \gamma, \dots \\ \alpha + 2\beta + \dots = n}} \frac{1}{\alpha! 2^\beta \beta! \cdots} \alpha(\alpha - 1) \chi^{(\lambda)}_{(1^\alpha 2^\beta \dots)} \chi^{(\mu)}_{(1^\alpha 2^\beta \dots)}$$

$$= \sum_{\substack{\alpha > 1, \beta, \gamma, \dots \\ \alpha + 2\beta + \dots = n}} \frac{1}{(\alpha - 2)! 2^\beta \beta! \cdots} \chi^{(\lambda)}_{(1^\alpha 2^\beta \dots)} \chi^{(\mu)}_{(1^\alpha 2^\beta \dots)}. \tag{7–180}$$

Now we apply to $\chi^{(\lambda)}$ and $\chi^{(\mu)}$ the branching law for two successive removals of a single node (Eq. 7–53):

$$a' = \sum_{\substack{\alpha > 1, \beta, \gamma, \ldots \\ \alpha + 2\beta + \cdots = n}} \frac{1}{(\alpha - 2)! 2^\beta \beta! \cdots} \sum_{(\lambda')} \chi^{(\lambda')}_{(1^{\alpha-2}2^\beta \ldots)} \sum_{(\mu')} \chi^{(\mu')}_{(1^{\alpha-2}2^\beta \ldots)}$$

$$= \sum_{\substack{\alpha', \beta, \gamma, \ldots \\ \alpha' + 2\beta + \cdots = n-2}} \frac{1}{\alpha'! 2^\beta \beta! \cdots} \sum_{(\lambda')} \chi^{(\lambda')}_{(1^{\alpha'}2^\beta \ldots)} \sum_{(\mu')} \chi^{(\mu')}_{(1^{\alpha'}2^\beta \ldots)}, \qquad (7\text{–}181)$$

where (λ') and (μ') are the partitions obtained from (λ) and (μ) by two successive removals of a single node. Again we can apply the orthogonality theorem for S_{n-2} to each term in the product $\sum_{(\lambda')} \sum_{(\mu')}$ and thus evaluate a'.

Similarly the contribution of the factor 2β to $a_{(\mu)}$ is

$$a'' = \sum_{\substack{\alpha, \beta, \gamma, \ldots \\ \alpha + 2\beta + \cdots = n}} \frac{1}{\alpha! 2^\beta \beta! \cdots} (2\beta) \chi^{(\lambda)}_{(1^\alpha 2^\beta \ldots)} \chi^{(\mu)}_{(1^\alpha 2^\beta \ldots)}$$

$$= \sum_{\substack{\alpha, \beta > 0, \gamma, \ldots \\ \alpha + 2\beta + \cdots = n}} \frac{1}{\alpha! 2^{\beta-1} (\beta - 1)!} \chi^{(\lambda)}_{(1^\alpha 2^\beta \ldots)} \chi^{(\mu)}_{(1^\alpha 2^\beta \ldots)}. \qquad (7\text{–}182)$$

Now we apply to (λ) and (μ) the branching law for the regular removal of a 2-cycle (cf. the discussion preceding Eq. 7–56), giving

$$a'' = \sum_{\substack{\alpha, \beta > 0, \gamma, \ldots \\ \alpha + 2\beta + \cdots = n}} \frac{1}{\alpha! 2^{\beta-1} (\beta - 1)! \cdots} \sum_{(\lambda')} (\pm) \, \chi^{(\lambda')}_{(1^\alpha 2^{\beta-1} \ldots)} \sum_{(\mu')} \chi^{(\mu')}_{(1^\alpha 2^{\beta-1} \ldots)}$$

$$= \sum_{\substack{\alpha, \beta', \gamma, \ldots \\ \alpha + 2\beta' + \cdots = n-2}} \frac{1}{\alpha! 2^{\beta'} \beta'! \cdots} \sum_{(\lambda')} (\pm) \, \chi^{(\lambda')}_{(1^\alpha 2^{\beta'} \ldots)} \sum_{(\mu')} (\pm) \, \chi^{(\mu')}_{(1^\alpha 2^{\beta'} \ldots)}.$$

$$(7\text{–}183)$$

The partitions (λ') and (μ') in (7–183) are obtained from (λ) and (μ) by regular removal of a 2-cycle. The character appears with a plus sign if a positive (horizontal) strip is removed, and with a minus sign if a negative (vertical) strip is removed. Again Eq. (7–183) can be evaluated by using the orthogonality theorem for S_{n-2}.

The general procedure should be clear from the examples given above.

———————

Problems. (1) Complete the calculations of this section and derive the Clebsch-Gordan series for

(a) $(n - 2, 1^2) \times (\lambda)$,   (b) $(n - 2, 2) \times (\lambda)$.

(2) Show that the formulas for (a) and (b) in Problem 1 can be obtained from each other by means of Eq. (7–167). [Hint: Construct the triple product $(\lambda) \times (n-1,1) \times (n-1,1)$.]

7–14 Clebsch-Gordan (CG) coefficients for the symmetric group. Symmetry properties. Recursion formulas.

In this section we wish to consider two problems:

(a) Derivation of symmetry properties of the CG-coefficients for the symmetric group.

(b) Deduction of recursion formulas for the CG-coefficients.

In treating the first problem our goal is to obtain relations like those found in Section 5–9 [Eq. (5–151)] for simply reducible groups. We cannot use the method of Section 5–9 since the symmetric group for $n > 4$ is not simply reducible. The Kronecker product of two irreducible representations will in general contain representations more than once, i.e., $(\lambda\mu\nu) > 1$. But the symmetric groups are members of a special class of groups: All representations can be expressed in real form (Section 7–7). For such groups the distinction between the CG-coefficient and (3-j)-coefficient in Eq. (5–140) reduces to the trivial factor $[j_3]$, since every representation coincides with its complex conjugate. We shall therefore proceed as in Section 5–7.

For *any* group having only *real* irreducible representations, the matrix representatives $D^{(\lambda)}_{ij}(R)$ are all real, and the matrices $D^{(\lambda)}(R)$ are real orthogonal matrices:

$$D^{(\lambda)}_{ij}(R)D^{(\lambda)}_{ij'}(R) = \delta_{jj'}. \tag{7–184}$$

We shall use the convention of summing over repeated Latin indices as we did in Section 5–7.

We shall write the CG-coefficient of Section 5–7 as

$$(\mu j, \nu l | \lambda \tau_\lambda s) = S^{\lambda \tau_\lambda}_{s}{}^{\mu}_{j}{}^{\nu}_{l}, \tag{7–185}$$

and denote the dimension of the λ-representation by n_λ. In this notation, Eq. (5–114) becomes

$$D^{(\mu)}_{ij}(R)D^{(\nu)}_{kl}(R)S^{\lambda\tau_\lambda}_{s}{}^{\mu}_{j}{}^{\nu}_{l} = D^{(\lambda\tau_\lambda)}_{s's}(R)S^{\lambda\tau_\lambda}_{s'}{}^{\mu}_{i}{}^{\nu}_{k}, \tag{7–186}$$

where, as in Section 5–7, we choose the matrices $D^{(\lambda\tau_\lambda)}(R)$ to be the same for all the equivalent representations contained in $D^{(\mu)} \times D^{(\nu)}$:

$$D^{(\lambda\tau_\lambda)}(R) = D^{(\lambda)}(R). \tag{7–187}$$

The equations (7–186) for the CG-coefficients are linear equations with real coefficients, and hence the CG-coefficients can be chosen to be real.

Thus the matrix of the CG-coefficients will be a real orthogonal matrix:

$$S^{\lambda\tau\lambda\ \mu\ \nu}_{\ \ i\ \ j\ k}S^{\epsilon\tau\epsilon\ \mu\ \nu}_{\ \ i'\ \ j\ k} = \delta_{\lambda\epsilon}\,\delta_{\tau_\lambda\tau_\epsilon}\,\delta_{ii'}, \tag{7–188}$$

$$\sum_{\lambda\tau_\lambda} S^{\lambda\tau\lambda\ \mu\ \nu}_{\ \ i\ \ j\ k}S^{\lambda\tau\lambda\ \mu,\ \nu}_{\ \ i\ \ j'\ k'} = \delta_{jj'}\,\delta_{kk'}. \tag{7–188a}$$

We can now shift factors in Eq. (7–186) and obtain Eqs. (5–115) and (5–116):

$$S^{\lambda'\tau'_\lambda\ \mu\ \nu}_{\ \ t\ \ i\ k}D^{(\mu)}_{ij}(R)D^{(\nu)}_{kl}(R)S^{\lambda\tau\lambda\ \mu\ \nu}_{\ \ s\ \ j\ l} = D^{(\lambda\tau\lambda)}_{s's}(R)\,\delta_{\lambda\lambda'}\,\delta_{\tau_\lambda\tau'_\lambda},\,\delta_{ts'} \tag{7–189}$$

$$D^{(\mu)}_{ij}(R)D^{(\nu)}_{kl}(R) = \sum_{\lambda\tau_\lambda} S^{\lambda\tau\lambda\ \mu\ \nu}_{\ \ s'\ \ i\ k}D^{(\lambda\tau\lambda)}_{s's}(R)S^{\lambda\tau\lambda\ \mu\ \nu}_{\ \ s\ \ j\ l}. \tag{7–190}$$

Since $D^{(\lambda)}(R)$ is an orthogonal matrix, shifting it to the left of Eq. (7–186) gives

$$D^{(\lambda)}_{ts}(R)D^{(\mu)}_{ij}(R)D^{(\nu)}_{kl}(R)S^{\lambda\tau\lambda\ \mu\ \nu}_{\ \ s\ \ j\ l} = S^{\lambda\tau\lambda\ \mu\ \nu}_{\ \ t\ \ i\ k}, \tag{7–191}$$

which is analogous to Eq. (5–146). We can also use the orthogonality theorem to shift $D^{(\lambda)}$ to the left in Eq. (7–186), and obtain

$$\sum_R D^{(\lambda)}_{ts}(R)D^{(\mu)}_{ij}(R)D^{(\nu)}_{kl}(R) = g \sum_{\tau_\lambda} S^{\lambda\tau\lambda\ \mu\ \nu}_{\ \ t\ \ i\ k}S^{\lambda\tau\lambda\ \mu\ \nu}_{\ \ s\ \ j\ l}, \tag{7–192}$$

which is analogous to Eq. (5–149). If we now set $s = t$, $j = i$, and $l = k$, and permute the factors on the left, we get in place of Eq. (5–150) the symmetry relations

$$\sum_{\tau_\lambda} [S^{\lambda\tau\lambda\ \mu\ \nu}_{\ \ t\ \ i\ k}]^2 = \sum_{\tau_\mu} [S^{\mu\tau_\mu\ \lambda\ \nu}_{\ \ i\ \ t\ k}]^2 = \sum_{\tau_\nu} [S^{\nu\tau_\nu\ \mu\ \lambda}_{\ \ k\ \ i\ t}]^2 = \text{etc.} \tag{7–193}$$

The presence of the summations in Eq. (7–193) prevents us from going ahead, as we did in Section 5–9, to adjust phases and impose symmetry conditions on the CG-coefficients. But we can still accomplish this because we now have $(\lambda\mu\nu) > 1$ and are not restricted to changes in phase of our basis functions. As we pointed out in Section 5–8 [Eq. (5–117)], we can keep the matrix of CG-coefficients real and orthogonal, without altering the representation matrices $D^{(\lambda)}(R)$, if we take any linear combinations of the basis functions $\Psi^{(\lambda\tau_\lambda)}_s$,

$$\Psi'^{(\lambda\tau_\lambda)}_s = \sum_{\tau'_\lambda} c_{\tau_\lambda\tau'_\lambda}\Psi^{(\lambda\tau'_\lambda)}_s, \tag{7–194}$$

where the coefficients $c_{\tau_\lambda\tau'_\lambda}$ form a real orthogonal matrix:

$$\sum_{\tau_\lambda} c_{\tau_\lambda\tau'_\lambda}c_{\tau_\lambda\tau''_\lambda} = \delta_{\tau'_\lambda\tau''_\lambda}. \tag{7–195}$$

We now show how this additional freedom in the choice of basis functions can be used to impose symmetry conditions on the CG-coefficients.

We shift the factor $D^{(\nu)}_{kl}(R)$ to the right side of Eq. (7–186), obtaining

$$D^{(\mu)}_{ij}(R)S^{\lambda\tau\lambda}_{s}{}^{\mu}_{j}{}^{\nu}_{l} = S^{\lambda\tau\lambda}_{s'}{}^{\mu}_{i}{}^{\nu}_{k}D^{(\lambda)}_{s's}(R)D^{(\nu)}_{kl}(R), \qquad (7\text{–}196)$$

and use an equation similar to (7–190) to expand the product

$$D^{(\lambda)}_{s's}(R)D^{(\nu)}_{kl}(R).$$

We get

$$D^{(\mu)}_{ij}(R)S^{\lambda\tau\lambda}_{s}{}^{\mu}_{j}{}^{\nu}_{l} = \sum_{\epsilon\tau\epsilon} S^{\lambda\tau\lambda}_{s'}{}^{\mu}_{i}{}^{\nu}_{k}S^{\epsilon\tau\epsilon}_{t'}{}^{\lambda}_{s'}{}^{\nu}_{k}D^{(\epsilon)}_{t't}(R)S^{\epsilon\tau\epsilon}_{t}{}^{\lambda}_{s}{}^{\nu}_{l}. \qquad (7\text{–}197)$$

The last factor on the right can be brought to the left side, giving

$$D^{(\mu)}_{ij}(R)S^{\lambda\tau\lambda}_{s}{}^{\mu}_{j}{}^{\nu}_{l}S^{\epsilon\tau\epsilon}_{t}{}^{\lambda}_{s}{}^{\nu}_{l} = S^{\lambda\tau\lambda}_{s'}{}^{\mu}_{i}{}^{\nu}_{k}S^{\epsilon\tau\epsilon}_{t'}{}^{\lambda}_{s'}{}^{\nu}_{k}D^{(\epsilon)}_{t't}(R). \qquad (7\text{–}198)$$

From Eq. (7–198) we see that the matrix M with

$$M_{jt} = S^{\lambda\tau\lambda}_{s}{}^{\mu}_{j}{}^{\nu}_{l}S^{\epsilon\tau\epsilon}_{t}{}^{\lambda}_{s}{}^{\nu}_{l} \qquad (7\text{–}199)$$

satisfies the equation

$$D^{(\mu)}_{ij}(R)M_{jt} = M_{it'}D^{(\epsilon)}_{t't}(R), \qquad (7\text{–}200)$$

so that

$$D^{(\mu)}M = MD^{(\epsilon)}. \qquad (7\text{–}200\text{a})$$

From Schur's lemmas, M is the null matrix if $D^{(\mu)}$ and $D^{(\epsilon)}$ are inequivalent ($\mu \neq \epsilon$), and it is a multiple of the unit matrix if $\mu = \epsilon$. We can therefore write the expression in Eq. (7–199) as

$$S^{\lambda\tau\lambda}_{s}{}^{\mu}_{j}{}^{\nu}_{l}S^{\epsilon\tau\epsilon}_{t}{}^{\lambda}_{s}{}^{\nu}_{l} = \sqrt{\frac{n_\lambda}{n_\mu}}\, m^{(\lambda\mu)\nu}_{\tau_\lambda\tau_\mu}\, \delta_{\epsilon\mu}\, \delta_{jt}. \qquad (7\text{–}201)$$

In Eq. (7–201) we have inserted the factor containing the dimensions of the representations to simplify later equations. The matrix $m^{(\lambda\mu)\nu}_{\tau_\lambda\tau_\mu}$ is a square matrix since $(\mu\nu\lambda) = (\lambda\nu\mu)$. The superscripts indicate that the matrix m will depend on λ, μ, ν and on the fact that we have interchanged the first and second columns in the CG-coefficients.

We shift the second factor from the left side of Eq. (7–201):

$$\frac{S^{\lambda\tau\lambda}_{s}{}^{\mu}_{j}{}^{\nu}_{l}}{\sqrt{n_\lambda}} = \sum_{\tau_\mu} m^{(\lambda\mu)\nu}_{\tau_\lambda\tau_\mu}\frac{S^{\mu\tau_\mu}_{j}{}^{\lambda}_{s}{}^{\nu}_{l}}{\sqrt{n_\mu}}. \qquad (7\text{–}202)$$

Equation (7–202), with the fixed matrix $m^{(\lambda\mu)\nu}$ defined by Eq. (7–201), is valid for all s, j, l. We multiply Eq. (7–202) by the similar equation for $S^{\lambda \tau'_\lambda}_{s}{}^{\mu}_{j}{}^{\nu}_{l}/\sqrt{n_\lambda}$ and find

$$\delta_{\tau_\lambda \tau'_\lambda} = \sum_{\tau_\mu} m^{(\lambda\mu)\nu}_{\tau_\lambda \tau_\mu} \, m^{(\lambda\mu)\nu}_{\tau'_\lambda \tau_\mu}. \tag{7–203}$$

Since the CG-coefficients are real, the matrix $m^{(\lambda\mu)\nu}$ is a real orthogonal matrix and can be used to transform the basis of a representation (cf. Eq. 7–194).

The entire derivation can be repeated, shifting $D^{(\mu)}_{ij}(R)$ to the right in Eq. (7–186). If we do this, we obtain the equations

$$S^{\lambda \tau_\lambda}_{s}{}^{\mu}_{l}{}^{\nu}_{j} S^{\epsilon \tau_\epsilon}_{t}{}^{\mu}_{l}{}^{\lambda}_{s} = \sqrt{\frac{n_\lambda}{n_\nu}} \, \overleftrightarrow{m}^{\lambda\mu\nu}_{\tau_\lambda \tau_\nu} \, \delta_{\epsilon\nu} \, \delta_{jt}, \tag{7–201a}$$

$$\frac{S^{\lambda \tau_\lambda}_{s}{}^{\mu}_{l}{}^{\nu}_{j}}{\sqrt{n_\lambda}} = \sum_{\tau_\nu} \overleftrightarrow{m}^{\lambda\mu\nu}_{\tau_\lambda \tau_\nu} \frac{S^{\nu \tau_\nu}_{j}{}^{\mu}_{l}{}^{\lambda}_{s}}{\sqrt{n_\nu}}, \tag{7–202a}$$

$$\delta_{\tau_\lambda \tau'_\lambda} = \sum_{\tau_\nu} \overleftrightarrow{m}^{\lambda\mu\nu}_{\tau_\lambda \tau_\nu} \overleftrightarrow{m}^{\lambda\mu\nu}_{\tau'_\lambda \tau_\nu}. \tag{7–203a}$$

We consider first the case where $\lambda \neq \mu \neq \nu$. Assume that we are given the CG-coefficients for some choice of basis. Then Eq. (7–201) provides us with a matrix $m^{(\lambda\mu)\nu}_{\tau_\lambda \tau_\mu}$ which we use to transform the basis functions $\Psi^{(\mu\tau_\mu)}_j$:

$$\sum_{\tau_\mu} m^{(\lambda\mu)\nu}_{\tau_\lambda \tau_\mu} \Psi^{(\mu\tau_\mu)}_j \to \Psi^{(\mu\tau_\lambda)}_j \qquad \text{for all} \quad j. \tag{7–204}$$

This transformation induces the transformation

$$\sum_{\tau_\mu} m^{(\lambda\mu)\nu}_{\tau_\lambda \tau_\mu} S^{\mu\tau_\mu}_{j}{}^{\lambda}_{s}{}^{\nu}_{l} \to S^{\mu\tau_\lambda}_{j}{}^{\lambda}_{s}{}^{\nu}_{l} \qquad \text{for all} \quad j, s, l, \tag{7–204a}$$

so that, with this new basis, Eq. (7–202) becomes

$$\frac{S^{\lambda \tau_\lambda}_{s}{}^{\mu}_{j}{}^{\nu}_{l}}{\sqrt{n_\lambda}} = \frac{S^{\mu\tau_\lambda}_{j}{}^{\lambda}_{s}{}^{\nu}_{l}}{\sqrt{n_\mu}}. \tag{7–205}$$

We can *independently* use the matrix $\overleftrightarrow{m}^{\lambda\mu\nu}$ defined by Eq. (7–201a) to transform the basis functions $\Psi^{(\nu\tau_\nu)}_j$. The new CG-coefficients will then satisfy the equations

$$\frac{S^{\lambda \tau_\lambda}_{s}{}^{\mu}_{j}{}^{\nu}_{l}}{\sqrt{n_\lambda}} = \frac{S^{\nu\tau_\lambda}_{l}{}^{\mu}_{j}{}^{\lambda}_{s}}{\sqrt{n_\nu}}. \tag{7–205a}$$

Since we have assumed $\mu \neq \nu \neq \lambda$, the order of the factors $\psi^{(\mu)}_j \psi^{(\nu)}_i$, etc., in the product functions is irrelevant. We can therefore make

$$S^{\lambda\tau\lambda}{}_{s}{}^{\mu}{}_{j}{}^{\nu}{}_{l} = S^{\lambda\tau\lambda}{}_{s}{}^{\nu}{}_{l}{}^{\mu}{}_{j},$$

$$S^{\mu\tau\lambda}{}_{j}{}^{\lambda}{}_{s}{}^{\nu}{}_{l} = S^{\mu\tau\lambda}{}_{j}{}^{\nu}{}_{l}{}^{\lambda}{}_{s}, \tag{7-205b}$$

$$S^{\nu\tau\lambda}{}_{l}{}^{\mu}{}_{j}{}^{\lambda}{}_{s} = S^{\nu\tau\lambda}{}_{l}{}^{\lambda}{}_{s}{}^{\mu}{}_{j}.$$

Equations (7-205), (7-205a) and (7-205b) replace Eq. (5-151), which was derived for simply reducible groups in Section 5-9.

We must now examine the special cases. First suppose $\lambda \neq \mu = \nu$. We can still carry out the first basis transformation of Eq. (7-204a) and obtain

$$\frac{S^{\lambda\tau\lambda}{}_{s}{}^{\mu}{}_{j}{}^{\mu}{}_{l}}{\sqrt{n_\lambda}} = \frac{S^{\mu\tau\lambda}{}_{j}{}^{\lambda}{}_{s}{}^{\mu}{}_{l}}{\sqrt{n_\mu}}, \tag{7-206}$$

but the second basis transformation is no longer independent and would undo the work of the first. In fact, this second transformation is unnecessary. Since $\mu = \nu$, the symmetry of the CG-coefficients under interchange of the second and third columns is determined by whether the product functions $\Psi_s^{(\lambda\tau\lambda)}$ are contained in $[D^{(\mu)} \times D^{(\mu)}]$ or $\{D^{(\mu)} \times D^{(\mu)}\}$. So our first step should be to assign the components $\Psi_s^{(\lambda\tau\lambda)}$ to the symmetrized or antisymmetrized square of $D^{(\mu)}$. Then

$$S^{\lambda\tau\lambda}{}_{s}{}^{\mu}{}_{j}{}^{\mu}{}_{l} = \delta_{\tau\lambda} S^{\lambda\tau\lambda}{}_{s}{}^{\mu}{}_{l}{}^{\mu}{}_{j}, \tag{7-206a}$$

where $\delta_{\tau\lambda} = +1$ if the $\Psi_s^{(\lambda\tau\lambda)}$ are contained in the symmetrized product, and $\delta_{\tau\lambda} = -1$ if they are contained in the antisymmetrized product. The matrix $m^{(\lambda\mu)\mu}$ should be defined *after* this adjustment. Since $\lambda \neq \mu$, the order of the factors in the product $\psi_s^{(\lambda)}\psi_l^{(\mu)}$ is irrelevant, so we can set

$$S^{\mu\tau\lambda}{}_{j}{}^{\lambda}{}_{s}{}^{\mu}{}_{l} = S^{\mu\tau\lambda}{}_{j}{}^{\mu}{}_{l}{}^{\lambda}{}_{s}. \tag{7-206b}$$

Combining Eqs. (7-206), (7-206a) and (7-206b), we obtain the complete set of symmetry relations:

$$\frac{S^{\lambda\tau\lambda}{}_{s}{}^{\mu}{}_{j}{}^{\mu}{}_{l}}{\sqrt{n_\lambda}} = \delta_{\tau\lambda} \frac{S^{\lambda\tau\lambda}{}_{s}{}^{\mu}{}_{l}{}^{\mu}{}_{j}}{\sqrt{n_\lambda}} = \frac{S^{\mu\tau\lambda}{}_{j}{}^{\lambda}{}_{s}{}^{\mu}{}_{l}}{\sqrt{n_\mu}} = \frac{S^{\mu\tau\lambda}{}_{j}{}^{\mu}{}_{l}{}^{\lambda}{}_{s}}{\sqrt{n_\mu}}$$

$$= \delta_{\tau\lambda} \frac{S^{\mu\tau\lambda}{}_{l}{}^{\lambda}{}_{s}{}^{\mu}{}_{j}}{\sqrt{n_\mu}} = \delta_{\tau\lambda} \frac{S^{\mu\tau\lambda}{}_{l}{}^{\mu}{}_{j}{}^{\lambda}{}_{s}}{\sqrt{n_\mu}}. \tag{7-206c}$$

Finally we consider the case where $\lambda = \mu = \nu$. We have

$$S^{\lambda\tau\lambda}{}_{s}{}^{\lambda}{}_{j}{}^{\lambda}{}_{l} = \delta_{\tau\lambda} S^{\lambda\tau\lambda}{}_{s}{}^{\lambda}{}_{l}{}^{\lambda}{}_{j}, \tag{7-207}$$

where $\delta_{\tau_\lambda} = +1$ if the product functions $\Psi_s^{(\lambda\tau\lambda)}$ are contained in the symmetrized square, and $\delta_{\tau_\lambda} = -1$ if they are contained in the antisymmetrized square. For this case, the linear equations (7–191) for the CG-coefficients become

$$D_{ts}^{(\lambda)}(R)D_{ij}^{(\lambda)}(R)D_{kl}^{(\lambda)}(R)S^{\lambda\tau\lambda}{}_s{}^{\lambda}_j{}^{\lambda}_l = S^{\lambda\tau\lambda}{}_t{}^{\lambda}_i{}^{\lambda}_k \qquad (7\text{–}191\text{a})$$

or

$$[A_{tik,sjl}(R) - \delta_{ts}\,\delta_{ij}\,\delta_{kl}]S^{\lambda\tau\lambda}{}_s{}^{\lambda}_j{}^{\lambda}_l = 0. \qquad (7\text{–}191\text{b})$$

The coefficients in Eq. (7–191b) are symmetric under the simultaneous interchange $t \leftrightarrow i$, $s \leftrightarrow j$. Thus if $S^{\lambda\tau\lambda}{}_s{}^{\lambda}_j{}^{\lambda}_l$ is a solution, so is $S^{\lambda\tau\lambda}{}_j{}^{\lambda}_s{}^{\lambda}_l$. The sum and difference of these two solutions are also solutions, and at most one of them is zero. So we can certainly make

$$S^{\lambda\tau\lambda}{}_s{}^{\lambda}_j{}^{\lambda}_l = \epsilon_{\tau_\lambda}S^{\lambda\tau\lambda}{}_j{}^{\lambda}_s{}^{\lambda}_l, \qquad (7\text{–}207\text{a})$$

where $\epsilon_{\tau_\lambda} = \pm 1$. Combining Eqs. (7–207) and (7–207a), we get the set of symmetry relations

$$S^{\lambda\tau\lambda}{}_s{}^{\lambda}_j{}^{\lambda}_l = \delta_{\tau_\lambda} S^{\lambda\tau\lambda}{}_s{}^{\lambda}_l{}^{\lambda}_j = \epsilon_{\tau_\lambda}S^{\lambda\tau\lambda}{}_j{}^{\lambda}_s{}^{\lambda}_l$$

$$= \epsilon_{\tau_\lambda}\,\delta_{\tau_\lambda}S^{\lambda\tau\lambda}{}_j{}^{\lambda}_l{}^{\lambda}_s = \delta_{\tau_\lambda}S^{\lambda\tau\lambda}{}_l{}^{\lambda}_j{}^{\lambda}_s = \epsilon_{\tau_\lambda}\,\delta_{\tau_\lambda}S^{\lambda\tau\lambda}{}_l{}^{\lambda}_s{}^{\lambda}_j. \qquad (7\text{–}207\text{b})$$

These symmetries of the Clebsch-Gordan coefficients exist for *any* group having *only integer* (*real*) representations.

For the symmetric group we shall use the Young-Yamanouchi orthogonal representation of Section 7–7. The basis functions for this representation will be denoted by $[\lambda i]$, where λ denotes the partition and i is the Yamanouchi symbol (Y-symbol) labeling the rows of the representation (λ). If several independent functions belonging to the same row of (λ) appear in an argument, we distinguish them by primes on the Y-symbol.

There are some special cases for which we can easily determine the CG-coefficients. First, we know that $(\lambda) \times (\mu)$ contains the identity representation once if and only if $(\lambda) = (\mu)$. The expansion of the completely symmetric function in terms of products $[\lambda i][\lambda j]'$ is almost obvious. The basis functions are the components in the unitary vector space of the representation, and the scalar product of a vector with itself gives an invariant, i.e., is completely symmetric. Thus,

$$[(n), 1] = \frac{1}{\sqrt{n_\lambda}} \sum_i [\lambda i][\lambda i]' \qquad (7\text{–}208)$$

or

$$S^{(n)}{}_1{}^{\lambda}_i{}^{\lambda}_j = \frac{1}{\sqrt{n_\lambda}}\,\delta_{ij}, \qquad (7\text{–}209)$$

where n_λ is the dimension of the representation (λ). The fact that (7-208) is completely symmetric is also verified directly by applying any permutation R:

$$R \sum_i [\lambda i][\lambda i]' = \sum_{i,j,k} [\lambda j][\lambda k]' D^\lambda_{ji}(R) D^\lambda_{ki}(R)$$

$$= \sum_{j,k} [\lambda j][\lambda k]' \sum_i D^\lambda_{ji}(R) D^\lambda_{ki}(R)$$

$$= \sum_{j,k} [\lambda j][\lambda k]' \, \delta_{jk} = \sum_j [\lambda j][\lambda j]'.$$

Secondly, we know that $(\lambda) \times (\tilde\lambda)$ contains the antisymmetric representation (1^n). We define the basis function $[\tilde\lambda \tilde i]$ to be that basis function of $(\tilde\lambda)$ whose tableau is obtained from the tableau of $[\lambda i]$ when we interchange row and column. For example,

$$[\lambda i] = \begin{matrix} 124 \\ 35 \\ 6 \end{matrix} \qquad [\tilde\lambda \tilde i] = \begin{matrix} 136 \\ 25 \\ 4 \end{matrix}$$

As we pointed out in Section 7-7, we can restrict ourselves to transpositions on successive letters. For such permutations, it is easily seen that

$$D^{\tilde\lambda}_{\tilde i \tilde j} = D^\lambda_{ij} \qquad \text{for} \quad i \neq j, \tag{7-210}$$

$$D^{\tilde\lambda}_{\tilde i \tilde i} = -D^\lambda_{ii}. \tag{7-210a}$$

In addition, we note that the nondiagonal elements D^λ_{ij} $(i \neq j)$ are different from zero only if the Y-symbols for i and j are obtained from each other by a transposition of neighboring numbers in the symbol. For example, (23) applied to the $[\lambda i]$ given above (Y-symbol [321211]) gives a matrix element $\sqrt{3}/2$ to the function with Y-symbol [321121]. We now define a diagonal matrix with diagonal elements Λ^λ_i as follows:

$\Lambda^\lambda_i = \pm 1$ depending on whether the Y-symbol i is obtained from the Y-symbol with the letters in natural order by an even or odd number of transpositions. The completely antisymmetric function is

$$[(1^n), 1] = \frac{1}{\sqrt{n_\lambda}} \sum_i \Lambda^\lambda_i [\lambda i][\tilde\lambda \tilde i]'. \tag{7-211}$$

In terms of Clebsch-Gordan coefficients,

$$S^{(1^n)}_{1} {}^{\lambda\ \tilde\lambda}_{i\ \tilde k} = \frac{1}{\sqrt{n_\lambda}} \Lambda^\lambda_i \, \delta_{ik}. \tag{7-211a}$$

There are certain symmetries which are a special feature of the symmetric group. These are derived by using an argument related to that which led to (7–211) and (7–211a). We know that

$$(\lambda) \times (1^n) = (\tilde{\lambda}), \qquad [(\lambda) \neq (\tilde{\lambda})],$$

so there is a one-to-one correspondence of basis functions of (λ) and $(\tilde{\lambda})$. From (7–211a) and (7–205a),

$$\frac{S_{k\;i\;1}^{\tilde{\lambda}\;\lambda\;(1^n)}}{\sqrt{n_\lambda}} = \frac{S_{1\;i\;\tilde{k}}^{(1^n)\;\lambda\;\tilde{\lambda}}}{\sqrt{n_{(1^n)}}} = \frac{1}{\sqrt{n_\lambda}}\Lambda_i^\lambda\,\delta_{ik},$$

whence

$$S_{k\;i\;1}^{\tilde{\lambda}\;\lambda\;(1^n)} = \Lambda_i^\lambda\,\delta_{ik}, \tag{7-212}$$

and

$$[\tilde{\lambda}\tilde{i}] = \Lambda_i^\lambda[\lambda i][(1^n),\,1]. \tag{7-213}$$

We express the function $[\gamma i]$ in terms of functions from $(\tilde{\alpha}) \times (\tilde{\beta})$ [remember (7–174)]:

$$[\gamma i] = \sum_{j,k} S_{i\;\tilde{j}\;\tilde{k}}^{\gamma\;\tilde{\alpha}\;\tilde{\beta}}[\tilde{\alpha}\tilde{j}][\tilde{\beta}\tilde{k}]'$$

$$= \sum_{j,k} S_{i\;\tilde{j}\;\tilde{k}}^{\gamma\;\tilde{\alpha}\;\tilde{\beta}}\Lambda_j^\alpha\Lambda_k^\beta[\alpha j][\beta k]'[(1^n),\,1][(1^n),\,1]';$$

$$[\gamma i] = \sum_{j,k} S_{i\;\tilde{j}\;\tilde{k}}^{\gamma\;\tilde{\alpha}\;\tilde{\beta}}\Lambda_j^\alpha\Lambda_k^\beta[\alpha j][\beta k]', \tag{7-214}$$

where we first use (7–213), and then note that $[(1^n),\,1][(1^n),\,1]$ is a completely symmetric function. The CG-coefficient can be shifted to the left side, giving

$$[\alpha j][\beta k]' = \Lambda_j^\alpha\Lambda_k^\beta\sum_{\gamma,i} S_{i\;\tilde{j}\;\tilde{k}}^{\gamma\;\tilde{\alpha}\;\tilde{\beta}}[\gamma i]. \tag{7-215}$$

We also have

$$[\alpha j][\beta k]' = \sum_{\gamma,i} S_{i\;j\;k}^{\gamma\;\alpha\;\beta}[\gamma i], \tag{7-215a}$$

so that, comparing coefficients of the orthonormal basis functions in the last two equations, we find

$$S_{i\;\tilde{j}\;\tilde{k}}^{\gamma\;\tilde{\alpha}\;\tilde{\beta}} = \Lambda_j^\alpha\Lambda_k^\beta S_{i\;j\;k}^{\gamma\;\alpha\;\beta}. \tag{7-216}$$

We obtain other forms of this symmetry relation by replacing representations by their conjugates or repeating the argument, starting from relations like

$$[\tilde{\gamma}\tilde{i}] = \sum_{j,k} S_{i\;j\;\tilde{k}}^{\tilde{\gamma}\;\alpha\;\tilde{\beta}}[\alpha j][\tilde{\beta}\tilde{k}]'.$$

We find the symmetry relations

$$S^{\tilde{\gamma}}_{i} {}^{\alpha}_{j} {}^{\tilde{\beta}}_{k} = \Lambda^{\gamma}_{i} \Lambda^{\beta}_{k} S^{\gamma}_{i} {}^{\alpha}_{j} {}^{\beta}_{k}, \tag{7-216a}$$

$$S^{\tilde{\gamma}}_{i} {}^{\tilde{\alpha}}_{j} {}^{\beta}_{k} = \Lambda^{\gamma}_{i} \Lambda^{\alpha}_{j} S^{\gamma}_{i} {}^{\alpha}_{j} {}^{\beta}_{k}. \tag{7-216b}$$

The symmetry relations [(7–205), (7–205a and b)] and [(7–216), (7–216a and b)] enable us to shorten the work of finding Clebsch-Gordan coefficients: If we order the partitions in descending order from (n) to (1^n), we need only the coefficients for products $(\lambda) \times (\mu)$ where (μ) is below (λ) and (μ) is no lower than the self-conjugate partitions of n.

Now we wish to see what process replaces the well-known recursion formulas which exist for the rotation group. In treating this problem, it is not necessary to indicate explicitly which set of product functions $\Psi_s^{(\lambda\tau\lambda)}$ we are using. So instead of following the notation of Eq. (7–189) we can write the transformation of $D^\alpha \times D^\beta$ to diagonal form as

$$\sum_{a,e,c,g} S^{\gamma}_{i} {}^{\alpha}_{a} {}^{\beta}_{e} D^{\alpha}_{ac}(R) D^{\beta}_{eg}(R) S^{\lambda}_{k} {}^{\alpha}_{c} {}^{\beta}_{g} = D^{\gamma}_{ik}(R) \, \delta_{\gamma\lambda}. \tag{7-217}$$

We need more complete designations of the rows of the various representations, and we shall therefore use double subscripts in the CG-coefficients, for example, $S^{\gamma}_{ij} {}^{\alpha}_{ab} {}^{\beta}_{ef}$. The first letter in the pair of subscripts denotes the first letter in the Y-symbol, and the second letter stands for the remainder of the Y-symbol. In this notation, Eq. (7–217) becomes

$$\sum_{\substack{ab,cd \\ ef,gh}} S^{\gamma}_{ij} {}^{\alpha}_{ab} {}^{\beta}_{ef} D^{\alpha}_{ab,cd} D^{\beta}_{ef,gh} S^{\lambda}_{kl} {}^{\alpha}_{cd} {}^{\beta}_{gh} = D^{\gamma}_{ij,kl} \, \delta_{\gamma\lambda}. \tag{7-217a}$$

We first restrict ourselves to permutations of the symmetric group on $n - 1$ letters, i.e., to permutations which leave the last letter unchanged. For such permutations, Eq. (7–77) in the present notation is

$$D^{\alpha}_{ab,cd} = D^{\alpha_a}_{bd} \, \delta_{ac}, \tag{7-218}$$

where α_a is the representation of the symmetric group on $n - 1$ letters which is obtained from α by removing the last letter from line a of the pattern. (This is the branching rule.) Substituting expressions like (7–218) for all the D's in (7–217a), we have

$$\sum_{\substack{ab,cd \\ ef,gh}} S^{\gamma}_{ij} {}^{\alpha}_{ab} {}^{\beta}_{ef} D^{\alpha_a}_{bd} D^{\beta_e}_{fh} \, \delta_{ac} \, \delta_{eg} S^{\lambda}_{kl} {}^{\alpha}_{cd} {}^{\beta}_{gh} = D^{\gamma_i}_{jl} \, \delta_{ik} \, \delta_{\gamma\lambda} \tag{7-219}$$

or

$$\sum_{\substack{ab,d \\ ef,h}} S^{\gamma}_{ij} {}^{\alpha}_{ab} {}^{\beta}_{ef} D^{\alpha_a}_{bd} D^{\beta_e}_{fh} S^{\lambda}_{kl} {}^{\alpha}_{ad} {}^{\beta}_{eh} = D^{\gamma_i}_{jl} \, \delta_{ik} \, \delta_{\gamma\lambda}. \tag{7-219a}$$

We can also write an equation like (7–217a) for the symmetric group on $n - 1$ letters:

$$\sum_{db,fh} S_p^{\gamma'\,\alpha'\,\beta'}{}_{b\,f} D^{\alpha'}_{bd} D^{\beta'}_{fh} S_q^{\lambda'\,\alpha'\,\beta'}{}_{d\,h} = D^{\gamma'}_{pq}\,\delta_{\gamma'\lambda'}, \qquad (7\text{–}220)$$

where the primes are a reminder that these are partitions of $n - 1$, and the subscripts are Y-symbols with $n - 1$ places corresponding to the second subscripts in (7–217a). Shifting in (7–220), we have

$$D^{\alpha'}_{bd} D^{\beta'}_{fh} = \sum_{\gamma',p,q} S_p^{\gamma'\,\alpha'\,\beta'}{}_{b\,f} D^{\gamma'}_{pq} S_q^{\gamma'\,\alpha'\,\beta'}{}_{d\,h}. \qquad (7\text{–}220\text{a})$$

We substitute (7–220a) in (7–219a), choosing $\alpha' = \alpha_a$, $\beta' = \beta_e$, and obtain

$$\sum_{\substack{ab,d,ef,h \\ \gamma',p,q}} S_{ij}^{\gamma\,\alpha\,\beta}{}_{ab\,ef} S_p^{\gamma'\,\alpha_a\,\beta_e}{}_{b\,f} D^{\gamma'}_{pq} S_q^{\gamma'\,\alpha_a\,\beta_e}{}_{d\,h} S_{kl}^{\lambda\,\alpha\,\beta}{}_{ad\,eh} = D_{jl}^{\gamma_i}\,\delta_{ik}\,\delta_{\gamma\lambda}. \qquad (7\text{–}221)$$

We first transfer the last factor to the right side:

$$\sum_{\substack{b,f \\ \gamma',p,q}} S_{ij}^{\gamma\,\alpha\,\beta}{}_{ab\,ef} S_p^{\gamma'\,\alpha_a\,\beta_e}{}_{b\,f} D^{\gamma'}_{pq} S_q^{\gamma'\,\alpha_a\,\beta_e}{}_{d\,h} = \sum_l D_{jl}^{\gamma_i} S_{il}^{\gamma\,\alpha\,\beta}{}_{ad\,eh};$$

then we can once more shift the last factor to the right side:

$$\sum_{b,f,p} S_{ij}^{\gamma\,\alpha\,\beta}{}_{ab\,ef} S_p^{\gamma'\,\alpha_a\,\beta_e}{}_{b\,f} D^{\gamma'}_{pq} = \sum_{l,d,h} D_{jl}^{\gamma_i} S_{il}^{\gamma\,\alpha\,\beta}{}_{ad\,eh} S_q^{\gamma'\,\alpha_a\,\beta_e}{}_{d\,h}. \qquad (7\text{–}222)$$

We can now define a matrix

$$M_{jp}\begin{bmatrix} \gamma & & \alpha & \beta \\ & \gamma' & & \\ i & & a & e \end{bmatrix} = \sum_{b,f} S_{ij}^{\gamma\,\alpha\,\beta}{}_{ab\,ef} S_p^{\gamma'\,\alpha_a\,\beta_e}{}_{b\,f}. \qquad (7\text{–}223)$$

Then Eq. (7–222) states that

$$\sum_p M_{jp} D^{\gamma'}_{pq} = \sum_l D_{jl}^{\gamma_i} M_{lq}$$

or

$$MD^{\gamma'} = D^{\gamma_i} M. \qquad (7\text{–}224)$$

Applying Schur's lemmas, we have $M = 0$ if $\gamma' \neq \gamma_i$, and $M = $ multiple of the unit matrix if $\gamma' = \gamma_i$; hence,

$$\sum_{b,f} S_{ij}^{\gamma\,\alpha\,\beta}{}_{ab\,ef} S_p^{\gamma'\,\alpha_a\,\beta_e}{}_{b\,f} = \delta_{\gamma'\gamma_i}\,\delta_{jp}\,K\begin{bmatrix} \gamma & \alpha & \beta \\ i & a & e \end{bmatrix}, \qquad (7\text{–}225)$$

or, shifting the second factor to the right side, we have

$$S^{\gamma}_{ij\ ab\ ef}{}^{\alpha\ \beta} = K\begin{bmatrix} \gamma & \alpha & \beta \\ i & a & e \end{bmatrix} S^{\gamma i\ \alpha_a \beta_e}_{j\ b\ f}. \tag{7-226}$$

Equation (7–226) states that the matrix $K\begin{bmatrix} \gamma & \alpha & \beta \\ i & a & e \end{bmatrix}$ enables us to generate the CG-coefficients for n from those for $n-1$; thus (7–226) replaces the corresponding recursion formula for the rotation group. We now investigate the properties of K. We write, in analogy to (7–226),

$$S^{\gamma'}_{i'j'\ ab\ ef}{}^{\alpha\ \beta} = K\begin{bmatrix} \gamma' & \alpha & \beta \\ i' & a & e \end{bmatrix} S^{\gamma_{i'}\ \alpha_a \beta_e}_{j'\ b\ f},$$

multiply the two equations and sum over ab, ef:

$$\sum_{ab,ef} S^{\gamma'}_{i'j'\ ab\ ef}{}^{\alpha\ \beta} S^{\gamma}_{ij\ ab\ ef}{}^{\alpha\ \beta} = \sum_{a,e} K\begin{bmatrix} \gamma & \alpha & \beta \\ i & a & e \end{bmatrix} K\begin{bmatrix} \gamma' & \alpha & \beta \\ i' & a & e \end{bmatrix}$$

$$\times \sum_{b,f} S^{\gamma i\ \alpha_a \beta_e}_{j\ b\ f} S^{\gamma_{i'}\ \alpha_a \beta_e}_{j'\ b\ f}. \tag{7-227}$$

Applying the unitarity condition to both sides, we obtain

$$\delta_{\gamma\gamma'}\ \delta_{ii'}\ \delta_{jj'} = \sum_{a,e} K\begin{bmatrix} \gamma & \alpha & \beta \\ i & a & e \end{bmatrix} K\begin{bmatrix} \gamma' & \alpha & \beta \\ i' & a & e \end{bmatrix} \delta(\gamma_i \gamma'_{i'})\ \delta_{jj'},$$

or setting $j = j'$ and $\gamma_i = \gamma'_{i'}$,

$$\sum_{a,e} K\begin{bmatrix} \gamma & \alpha & \beta \\ i & a & e \end{bmatrix} K\begin{bmatrix} \gamma' & \alpha & \beta \\ i' & a & e \end{bmatrix} = \delta_{\gamma\gamma'}\ \delta_{ii'}. \tag{7-228}$$

It is important to note that (7–228), which is the first part of the condition that K be a unitary matrix, refers only to those γ's which branch from the same representation γ_i on $n-1$ symbols. To prove that K is unitary we must derive the second condition corresponding to (7–188a). We now multiply (7–226) by the similar expression

$$S^{\gamma}_{ij\ a'b'\ e'f'}{}^{\alpha\ \beta} = K\begin{bmatrix} \gamma & \alpha & \beta \\ i & a' & e' \end{bmatrix} S^{\gamma i\ \alpha_{a'}\ \beta_{e'}}_{j\ b'\ f'}$$

and sum over γ, i, j:

$$\sum_{\gamma,i,j} K\begin{bmatrix} \gamma & \alpha & \beta \\ i & a & e \end{bmatrix} K\begin{bmatrix} \gamma & \alpha & \beta \\ i & a' & e' \end{bmatrix} S^{\gamma i\ \alpha_a \beta_e}_{j\ b\ f} S^{\gamma i\ \alpha_{a'}\ \beta_{e'}}_{j\ b'\ f'}$$

$$= \sum_{\gamma,i,j} S^{\gamma}_{ij\ ab\ ef}{}^{\alpha\ \beta} S^{\gamma}_{ij\ a'b'\ e'f'}{}^{\alpha\ \beta} = \delta_{aa'}\ \delta_{bb'}\ \delta_{ee'}\ \delta_{ff'}, \tag{7-229}$$

where we use the unitary condition (7–188a) in the last step. Now we multiply both sides by $S_k^{\gamma'}{}_b^{\alpha_a}{}_f^{\beta_e} S_k^{\gamma'}{}_{b'}^{\alpha_{a'}}{}_{f'}^{\beta_{e'}}$ and sum over b, f, b', f':

$$\sum_{\gamma,i,j} K\begin{bmatrix} \gamma & \alpha & \beta \\ i & a & e \end{bmatrix} K\begin{bmatrix} \gamma & \alpha & \beta \\ i & a' & e' \end{bmatrix} \sum_{b,f} S_j^{\gamma_i}{}_b^{\alpha_a}{}_f^{\beta_e} S_k^{\gamma'}{}_b^{\alpha_a}{}_f^{\beta_e} \sum_{b',f'} S_k^{\gamma'}{}_{b'}^{\alpha_{a'}}{}_{f'}^{\beta_{e'}} S_j^{\gamma_i}{}_{b'}^{\alpha_{a'}}{}_{f'}^{\beta_{e'}}$$

$$= \sum_{\gamma,i,j} K\begin{bmatrix} \gamma & \alpha & \beta \\ i & a & e \end{bmatrix} K\begin{bmatrix} \gamma & \alpha & \beta \\ i & a' & e' \end{bmatrix} \delta(\gamma', \gamma_i)\, \delta_{jk}$$

$$= \sum_{\substack{\gamma,i \\ \gamma_i = \gamma'}} K\begin{bmatrix} \gamma & \alpha & \beta \\ i & a & e \end{bmatrix} K\begin{bmatrix} \gamma & \alpha & \beta \\ i & a' & e' \end{bmatrix}$$

$$= \delta_{aa'}\, \delta_{ee'} \sum_{b,f} S_k^{\gamma'}{}_b^{\alpha_a}{}_f^{\beta_e} S_k^{\gamma'}{}_b^{\alpha_{a'}}{}_f^{\beta_{e'}}$$

$$= \delta_{aa'}\, \delta_{ee'} \sum_{b,f} (S_k^{\gamma'}{}_b^{\alpha_a}{}_f^{\beta_e})^2 = \delta_{aa'}\, \delta_{ee'};$$

that is,

$$\sum_{\substack{\gamma,i \\ \gamma_i = \gamma'}} K\begin{bmatrix} \gamma & \alpha & \beta \\ i & a & e \end{bmatrix} K\begin{bmatrix} \gamma & \alpha & \beta \\ i & a' & e' \end{bmatrix} = \delta_{aa'}\, \delta_{ee'}. \tag{7–230}$$

This is precisely the second condition for the unitarity of K. We thus have the unitary conditions (7–228) and (7–230) which are necessary to determine the matrix K. So far we have restricted ourselves to permutations which left the last symbol unchanged. We next apply Eq. (7–190) for the transposition $(n, n-1)$. (It is now necessary to introduce a triple subscript notation: The first two subscripts are the first two letters in the Y-symbol and the third subscript represents the remainder of the Y-symbol.) Since $(n, n-1)$ affects only the first two places in the Y-symbol, only matrix elements like $D_{abr,cdr}^\alpha$ appear. Furthermore, we know from (7–111) that the matrix elements for $(n, n-1)$ are

$$D_{abr,cdr}^\alpha = f_{ab}^\alpha\, \delta_{ac}\, \delta_{bd} + g_{ab}^\alpha\, \delta_{ad}\, \delta_{bc}, \tag{7–231}$$

where f_{ab}^α is the reciprocal of the axial distance from a to b, [instead of f we used the symbol σ in (7–111)] and $g_{ab}^\alpha = \sqrt{1 - (f_{ab}^\alpha)^2}$. We substitute (7–231) in the equation

$$\sum_{ab,ef} S_{ijt}^\gamma{}_{abr}^\alpha{}_{efs}^\beta D_{abr,cdr}^\alpha D_{efs,ghs}^\beta = \sum_{k,l} D_{ijt,klv}^\gamma S_{klt}^\gamma{}_{cdr}^\alpha{}_{ghs}^\beta, \tag{7–232}$$

and obtain

$$\sum_{ab,ef} S^{\gamma}_{ijt}{}^{\alpha}_{abr}{}^{\beta}_{efs} [f^{\alpha}_{ab}f^{\beta}_{ef} \, \delta_{ac} \, \delta_{bd} \, \delta_{eg} \, \delta_{fh} + f^{\alpha}_{ab}g^{\beta}_{ef} \, \delta_{ac} \, \delta_{bd} \, \delta_{eh} \, \delta_{gf}$$

$$+ g^{\alpha}_{ab}f^{\beta}_{ef} \, \delta_{ad} \, \delta_{bc} \, \delta_{eg} \, \delta_{fh} + g^{\alpha}_{ab}g^{\beta}_{ef} \, \delta_{ad} \, \delta_{bc} \, \delta_{eh} \, \delta_{gf}]$$

$$= \sum_{k,l} [f^{\gamma}_{ij} \, \delta_{ik} \, \delta_{jl} + g^{\gamma}_{ij} \, \delta_{il} \, \delta_{jk}]S^{\gamma}_{klt}{}^{\alpha}_{cdr}{}^{\beta}_{ghs} \qquad (7\text{--}232a)$$

or

$$S^{\gamma}_{ijt}{}^{\alpha}_{cdr}{}^{\beta}_{ghs}f^{\alpha}_{cd}f^{\beta}_{gh} + S^{\gamma}_{ijt}{}^{\alpha}_{cdr}{}^{\beta}_{hgs}f^{\alpha}_{cd}g^{\beta}_{hg}$$

$$+ S^{\gamma}_{ijt}{}^{\alpha}_{dcr}{}^{\beta}_{ghs}g^{\alpha}_{dc}f^{\beta}_{gh} + S^{\gamma}_{ijt}{}^{\alpha}_{dcr}{}^{\beta}_{hgs}g^{\alpha}_{dc}g^{\beta}_{hg}$$

$$= f^{\gamma}_{ij}S^{\gamma}_{ijt}{}^{\alpha}_{cdr}{}^{\beta}_{ghs} + g^{\gamma}_{ij}S^{\gamma}_{jit}{}^{\alpha}_{cdr}{}^{\beta}_{ghs}. \qquad (7\text{--}232b)$$

We now introduce the K's by substituting from (7–226):

$$K\begin{bmatrix} \gamma & \alpha & \beta \\ i & c & g \end{bmatrix} S^{\gamma i}_{jt}{}^{\alpha c}_{dr}{}^{\beta g}_{hs} [f^{\alpha}_{cd}f^{\beta}_{gh} - f^{\gamma}_{ij}] + K\begin{bmatrix} \gamma & \alpha & \beta \\ i & c & h \end{bmatrix} S^{\gamma i}_{jt}{}^{\alpha c}_{dr}{}^{\beta h}_{gs}f^{\alpha}_{cd}g^{\beta}_{hg}$$

$$+ K\begin{bmatrix} \gamma & \alpha & \beta \\ i & d & g \end{bmatrix} S^{\gamma i}_{jt}{}^{\alpha d}_{cr}{}^{\beta g}_{hs}g^{\alpha}_{dc}f^{\beta}_{gh}$$

$$+ K\begin{bmatrix} \gamma & \alpha & \beta \\ i & d & h \end{bmatrix} S^{\gamma i}_{jt}{}^{\alpha d}_{cr}{}^{\beta h}_{gs}g^{\alpha}_{dc}g^{\beta}_{hg}$$

$$= K\begin{bmatrix} \gamma & \alpha & \beta \\ j & c & g \end{bmatrix} S^{\gamma j}_{it}{}^{\alpha c}_{dr}{}^{\beta g}_{hs}g^{\gamma}_{ij}. \qquad (7\text{--}233)$$

Once the CG-coefficients for $n - 1$ particles are known, Eqs. (7–233) and the unitary conditions on K enable one to find K and to determine the CG-coefficients for n particles by means of (7–226).

The actual computation of the recursion matrix K is tedious, but it is much simpler than the direct calculation of the CG-coefficients. In the process of finding K, as soon as we reach a case where $(\gamma\alpha\beta) > 1$, Eq. (7–233) will have $(\gamma\alpha\beta)$ solutions. There is no general criterion for the selection of the solutions.

For low values of n, the CG-coefficients can be computed directly or, even more easily, from the matrix K. We shall not give any of the general formulas obtained for the K's, but quote one of the simpler results. The inner product $(n - 1, 1) \times (n - 1, 1)$ contains $(n - 1, 1)$ once. If we denote the basis function having the letter i in the second row of the pattern by f_i (or F_i), then

$$F_n = \frac{1}{\sqrt{(n - 1)(n - 2)}} \left[(n - 2)f_nf'_n - \sum_{i=2}^{n-1} f_if'_i \right]. \qquad (7\text{--}234)$$

Tables of the CG-coefficients can be given explicitly for low n without necessitating too much work. For example, the complete table for $(2, 1) \times (2, 1) = (3) + (2, 1) + (1^3)$ is

	$[211] \cdot [211]'$	$[211] \cdot [121]'$	$[121] \cdot [211]'$	$[121] \cdot [121]'$	
$[321]$	0	$1/\sqrt{2}$	$-1/\sqrt{2}$	0	
$[211]$	$1/\sqrt{2}$	0	0	$-1/\sqrt{2}$	
$[121]$	0	$-1/\sqrt{2}$	$-1/\sqrt{2}$	0	
$[111]$	$1/\sqrt{2}$	0	0	$1/\sqrt{2}$	$(7\text{--}235)$

As an example of a more complicated case, we give the table of CG-coefficients for

$$(3, 1^2) \times (3, 1^2) = (5) + (4, 1) + 2(3, 2) + (3, 1^2)$$
$$+ 2(2^2, 1) + (2, 1^3) + (1^5)$$

(Table 7–5). The two sets of functions for the product $(3, 2)$ were selected by taking a particular solution of the recursion equations, and can be replaced by any independent pair of linear combinations.

TABLE 7–5

	a = 123, 4, 5						b = 124, 3, 5						c = 125, 3, 4					
	aa'	ab'	ac'	ad'	ae'	af'	ba'	bb'	bc'	bd'	be'	bf'	ca'	cb'	cc'	cd'	ce'	cf'
(5):																		
[11111]	$\frac{1}{\sqrt6}$						$\frac{1}{\sqrt6}$						$\frac{1}{\sqrt6}$					
(4, 1):																		
[21111]	$\frac{1}{\sqrt6}$						$\frac{1}{\sqrt6}$						$-\frac{1}{\sqrt6}$					
[12111]	$\sqrt{\tfrac{5}{18}}$						$-\sqrt{\tfrac{5}{72}}$	$-\sqrt{\tfrac{1}{24}}$					$-\sqrt{\tfrac{1}{24}}$	$\sqrt{\tfrac{5}{72}}$				
[11211]		$-\sqrt{\tfrac{5}{72}}$	$\sqrt{\tfrac{1}{24}}$				$-\sqrt{\tfrac{5}{72}}$	$\sqrt{\tfrac{5}{36}}$					$\sqrt{\tfrac{1}{24}}$	$\sqrt{\tfrac{5}{36}}$				
[11121]			$-\sqrt{\tfrac{5}{72}}$	$\sqrt{\tfrac{1}{24}}$				$-\sqrt{\tfrac{5}{36}}$	$\sqrt{\tfrac{1}{24}}$					$-\sqrt{\tfrac{5}{36}}$	$\sqrt{\tfrac{5}{72}}$			
(3, 2):																		
[22111]	$\sqrt{\tfrac{1}{3}}$							$-\sqrt{\tfrac{1}{12}}$						$-\sqrt{\tfrac{1}{12}}$				
[22111]'	$-\sqrt{\tfrac{1}{18}}$						$\sqrt{\tfrac{1}{72}}$	$-\sqrt{\tfrac{5}{24}}$					$-\sqrt{\tfrac{5}{24}}$	$-\sqrt{\tfrac{1}{72}}$				
[21211]		$-\sqrt{\tfrac{1}{12}}$					$-\sqrt{\tfrac{1}{12}}$	$\sqrt{\tfrac{1}{6}}$						$-\sqrt{\tfrac{1}{6}}$				
[21211]'		$\sqrt{\tfrac{1}{72}}$	$\sqrt{\tfrac{5}{24}}$				$\sqrt{\tfrac{1}{72}}$	$-\tfrac{1}{6}$					$\sqrt{\tfrac{5}{24}}$	$-\tfrac{1}{6}$				
[21121]			$-\sqrt{\tfrac{1}{12}}$						$-\sqrt{\tfrac{1}{6}}$				$\sqrt{\tfrac{1}{6}}$	$-\sqrt{\tfrac{1}{12}}$				
[21121]'			$\sqrt{\tfrac{1}{72}}$	$\sqrt{\tfrac{5}{24}}$					$\tfrac{1}{6}$	$\sqrt{\tfrac{5}{24}}$			$\tfrac{1}{6}$	$-\sqrt{\tfrac{1}{72}}$				
[12211]	$-\tfrac{1}{4}$	$\sqrt{\tfrac{5}{48}}$					$-\tfrac{1}{4}$	$-\sqrt{\tfrac{1}{32}}$	$-\sqrt{\tfrac{5}{96}}$				$\sqrt{\tfrac{5}{48}}$	$-\sqrt{\tfrac{5}{96}}$	$\sqrt{\tfrac{1}{32}}$			
[12211]'	$-\sqrt{\tfrac{1}{6}}$						$-\sqrt{\tfrac{1}{6}}$	$-\sqrt{\tfrac{1}{12}}$						$-\sqrt{\tfrac{1}{12}}$				
(3, 1²):																		
[32111]								$\tfrac{1}{2}$						$-\tfrac{1}{2}$				
[31211]		$-\tfrac{1}{2}$											$\tfrac{1}{2}$					
[31121]				$-\tfrac{1}{2}$						$-\tfrac{1}{2}$								
[13211]	$\tfrac{1}{2}$						$-\tfrac{1}{2}$											
[13121]		$\tfrac{1}{2}$														$-\tfrac{1}{2}$		
[11321]								$\tfrac{1}{2}$						$\tfrac{1}{2}$				

TABLE 7–5

$d = 134,\,2,\,5$						$e = 135,\,2,\,4$						$f = 145,\,2,\,3$					
da'	db'	dc'	dd'	de'	df'	ea'	eb'	ec'	ed'	ee'	ef'	fa'	fb'	fc'	fd'	fe'	ff'
			$\frac{1}{\sqrt6}$						$\frac{1}{\sqrt6}$								$\frac{1}{\sqrt6}$
			$\frac{1}{\sqrt6}$						$-\frac{1}{\sqrt6}$								$-\frac{1}{\sqrt6}$
			$-\sqrt{\frac{5}{72}}$	$-\sqrt{\frac{1}{24}}$				$-\sqrt{\frac{1}{24}}$	$\sqrt{\frac{5}{72}}$								$-\sqrt{\frac{5}{18}}$
			$-\sqrt{\frac{5}{36}}$		$-\sqrt{\frac{1}{24}}$				$-\sqrt{\frac{5}{36}}$	$-\sqrt{\frac{5}{72}}$					$-\sqrt{\frac{1}{24}}$	$-\sqrt{\frac{5}{72}}$	
$-\sqrt{\frac{5}{72}}$	$-\sqrt{\frac{5}{36}}$					$\sqrt{\frac{1}{24}}$	$-\sqrt{\frac{5}{36}}$							$\sqrt{\frac{1}{24}}$	$\sqrt{\frac{5}{72}}$		
			$-\sqrt{\frac{1}{12}}$						$-\sqrt{\frac{1}{12}}$								$\sqrt{\frac{1}{3}}$
			$\sqrt{\frac{1}{72}}$	$-\sqrt{\frac{5}{24}}$				$-\sqrt{\frac{5}{24}}$	$-\sqrt{\frac{1}{72}}$								$\sqrt{\frac{1}{18}}$
			$-\sqrt{\frac{1}{6}}$						$\sqrt{\frac{1}{6}}$	$\sqrt{\frac{1}{12}}$					$\sqrt{\frac{1}{12}}$		
			$\frac{1}{6}$		$-\sqrt{\frac{5}{24}}$				$\frac{1}{6}$	$\sqrt{\frac{1}{72}}$				$-\sqrt{\frac{5}{24}}$		$\sqrt{\frac{1}{72}}$	
$-\sqrt{\frac{1}{12}}$	$-\sqrt{\frac{1}{6}}$					$\sqrt{\frac{1}{6}}$								$-\sqrt{\frac{1}{12}}$			
$\sqrt{\frac{1}{72}}$	$\frac{1}{6}$					$\sqrt{\frac{5}{24}}$	$\frac{1}{6}$							$\sqrt{\frac{5}{24}}$	$-\sqrt{\frac{1}{72}}$		
			$\sqrt{\frac{1}{32}}$	$\sqrt{\frac{5}{96}}$	$\sqrt{\frac{5}{48}}$			$\sqrt{\frac{5}{96}}$	$-\sqrt{\frac{1}{32}}$	$\frac{1}{4}$					$\sqrt{\frac{5}{48}}$	$\frac{1}{4}$	
			$\sqrt{\frac{1}{12}}$						$\sqrt{\frac{1}{12}}$	$-\sqrt{\frac{1}{6}}$						$-\sqrt{\frac{1}{6}}$	
				$\frac{1}{2}$					$-\frac{1}{2}$							$-\frac{1}{2}$	
					$\frac{1}{2}$	$\frac{1}{2}$								$\frac{1}{2}$			
											$\frac{1}{2}$					$-\frac{1}{2}$	
														$\frac{1}{2}$			
$-\frac{1}{2}$																	
	$-\frac{1}{2}$						$-\frac{1}{2}$										

TABLE 7-6

CHARACTER TABLES OF SYMMETRIC GROUPS
(through $n = 7$)

$n = 3$

Class Partition	1 (1^3)	3 $(1, 2)$	2 (3)
(3)	1	1	1
(2, 1)	2	0	-1
(1^3)	1	-1	1

$n = 4$

Class Partition	1 (1^4)	6 $(1^2, 2)$	8 $(1, 3)$	3 (2^2)	6 (4)
(4)	1	1	1	1	1
(3, 1)	3	1	0	-1	-1
(2^2)	2	0	-1	2	0
$(2, 1^2)$	3	-1	0	-1	1
(1^4)	1	-1	1	1	-1

$n = 5$

Class Partition	1 (1^5)	10 $(1^3, 2)$	20 $(1^2, 3)$	15 $(1, 2^2)$	30 $(1, 4)$	20 $(2, 3)$	24 (5)
(5)	1	1	1	1	1	1	1
(4, 1)	4	2	1	0	0	-1	-1
(3, 2)	5	1	-1	1	-1	1	0
$(3, 1^2)$	6	0	0	-2	0	0	1
$(2^2, 1)$	5	-1	-1	1	1	-1	0
$(2, 1^3)$	4	-2	1	0	0	1	-1
(1^5)	1	-1	1	1	-1	-1	1

(*continued*)

TABLE 7-6 (continued)

$n = 6$

Class / Partition	1 (1^6)	15 $(1^4, 2)$	40 $(1^3, 3)$	45 $(1^2, 2^2)$	90 $(1^2, 4)$	120 $(1, 2, 3)$	144 $(1, 5)$	15 (2^3)	90 $(2, 4)$	40 (3^2)	120 (6)
(6)	1	1	1	1	1	1	1	1	1	1	1
$(5, 1)$	5	3	2	1	1	0	0	-1	-1	-1	-1
$(4, 2)$	9	3	0	1	-1	0	-1	3	-1	0	0
$(4, 1^2)$	10	2	1	-2	0	-1	0	-2	0	1	-1
(3^2)	5	1	-1	1	-1	1	0	-3	1	2	0
$(3, 2, 1)$	16	0	-2	0	0	0	1	0	0	-2	0
(2^3)	5	-1	-1	1	1	-1	0	3	-1	2	0
$(3, 1^3)$	10	-2	1	-2	0	1	0	2	0	1	-1
$(2^2, 1^2)$	9	-3	0	1	1	0	1	-3	-1	0	0
$(2, 1^4)$	5	-3	2	1	-1	0	0	1	1	-1	-1
(1^6)	1	-1	1	1	-1	-1	1	-1	1	1	-1

(continued)

TABLE 7–6 (continued)

$n = 7$

Class Partition	1 (1^7)	21 $(1^5, 2)$	70 $(1^4, 3)$	105 $(1^3, 2^2)$	210 $(1^3, 4)$	420 $(1^2, 2, 3)$	504 $(1^2, 5)$	105 $(1, 2^3)$	630 $(1, 2, 4)$	280 $(1, 3^2)$	840 $(1, 6)$	210 $(2^2, 3)$	504 $(2, 5)$	420 $(3, 4)$	720 (7)
(7)	1	1	1	1	1	1	1	1	1	1	1	1	1	1	1
$(6, 1)$	6	4	3	2	2	1	1	0	0	0	0	-1	-1	-1	-1
$(5, 2)$	14	6	2	2	0	0	-1	2	0	-1	-1	2	1	0	0
$(5, 1^2)$	15	5	3	-1	1	-1	0	-3	-1	0	0	-1	0	1	1
$(4, 3)$	14	4	-1	2	-2	1	-1	0	0	2	0	-1	-1	1	0
$(4, 2, 1)$	35	5	-1	-1	-1	-1	0	1	1	-1	1	-1	0	-1	0
$(3^2, 1)$	21	1	-3	1	-1	1	1	-3	-1	0	0	1	1	-1	0
$(4, 1^3)$	20	0	2	-4	0	0	0	0	0	2	0	2	0	0	-1
$(3, 2^2)$	21	-1	-3	1	1	-1	1	3	-1	0	0	1	-1	1	0
$(3, 2, 1^2)$	35	-5	-1	-1	1	1	0	-1	1	-1	-1	-1	0	1	0
$(2^3, 1)$	14	-4	-1	2	2	-1	-1	0	0	2	0	-1	1	-1	0
$(3, 1^4)$	15	-5	3	-1	-1	1	0	3	-1	0	0	-1	0	-1	1
$(2^2, 1^3)$	14	-6	2	2	0	0	-1	-2	0	-1	1	2	-1	0	0
$(2, 1^5)$	6	-4	3	2	-2	-1	1	0	0	0	0	-1	1	1	-1
(1^7)	1	-1	1	1	-1	-1	1	-1	1	1	-1	1	-1	-1	1

CHAPTER 8

CONTINUOUS GROUPS

8–1 Summary of results for finite groups. Up to now we have dealt exclusively with finite groups, i.e., groups consisting of a finite number of elements. The point symmetry groups which we have discussed were finite subgroups of the rotation-reflection group in three dimensions, and the representations which we have found were finite subgroups of the matrix groups in n dimensions. The symmetry groups of many physical systems consist of an *infinite* rather than a finite number of elements. For example, a crystalline lattice is unchanged by a triple infinity of translations; the Hamiltonian of an electron in a central field is invariant under all rotations. Thus physical problems require that we examine the theory of the representations of groups with an infinite number of elements.

But there is still another reason for considering such infinite groups. In determining representations of the point groups, we found a homomorphic mapping of the elements of the point group H on a group of matrices H' in n dimensions. But the point group H is a subgroup of the rotation-reflection group G. If we find a representation of G in terms of n-dimensional matrices G', then we automatically obtain a representation of the subgroup H in terms of a subgroup H' of G'.

Our problem is to carry over to infinite groups the theorems which we have derived concerning representations of finite groups. To clarify the problems, we first summarize our results for finite groups in a form which is easily extended to infinite groups.

A finite group G of order g consists of g elements $R_1, \ldots, R_g$. We may consider the g elements of the group as a set of "points"—the *group manifold*. Or we can label a set of points with the integers 1 to g, and associate a group element with each point in this space; thus to the point labeled a we associate the group element R_a. (In other words, the elements are labeled by a parameter which takes on g values.) In the product $R_b R_a$, if we keep a fixed and let R_b run through the group G, the product R_c runs through the group. The content of the group multiplication table is the fixing, for all values of a and b, of the parameter value c to which the product $R_b R_a = R_c$ belongs. We may say that the group table defines a function

$$c = \phi(a, b). \tag{8–1}$$

The parameter of the product element is a function of the parameters of the factors.

We can define functions on the group manifold. Thus to each element R_a (or to each value a of the discrete parameter) we can assign a number $f(R_a)$ or $f(a)$. Our functions are then defined on g points. (Earlier we called such a function a vector in our g-dimensional space.) For example, in our discussion of representations, the quantities $D_{ij}^{(\mu)}(R_a) = D_{ij}^{(\mu)}(a)$, for fixed μ, i, j, were functions defined on the group manifold. Similarly, the characters $\chi^{(\mu)}(R_a) = \chi^{(\mu)}(a)$, for given μ, yielded functions defined on the group manifold. The characters had the special property that the function had the same value on all elements belonging to the same class. Thus $\chi^{(\mu)}(a)$ is a *class function*, i.e., a function such that

$$\chi^{(\mu)}(\phi(a, b)) = \chi^{(\mu)}(\phi(b, a)) \tag{8-2}$$

for all a and b.

In Section 3–11 we proved that every representation of a finite group is equivalent to a unitary representation. We were then able to show that any representation is fully reducible and expressible in terms of irreducible unitary representations. An irreducible unitary representation $D^{(\mu)}$ of degree n_μ defined a set of n_μ^2 unitary-orthogonal functions $D_{ij}^{(\mu)}(a)(i, j = 1, \ldots, n_\mu)$ on the group manifold. The functions defined by the various inequivalent irreducible representations were mutually orthogonal:

$$\sum_a D_{ij}^{*(\mu)}(a)\, D_{kl}^{(\nu)}(a) = \delta_{ik}\, \delta_{jl}\, \delta_{\mu\nu} \frac{g}{n_\mu} = \frac{\delta_{ik}\, \delta_{jl}\, \delta_{\mu\nu}}{n_\mu} \sum_a 1. \tag{3-143}$$

Similarly, for the characters of the inequivalent irreducible representations,

$$\sum_a \chi^{*(\mu)}(a)\chi^{(\nu)}(a) = \delta_{\mu\nu} \cdot g = \delta_{\mu\nu} \sum_a 1. \tag{3-147}$$

Also, we showed that the functions $D_{ij}^{(\mu)}(a)$ formed a complete set; i.e., any function $f(a)$ defined on the group manifold can be expanded in terms of the $D_{ij}^{(\mu)}(a)$. Similarly, any class function $g(a)$ can be expanded in terms of the complete set of class functions $\chi^{(\mu)}(a)$.

In the derivation of the above theorems, two steps depended on the fact that we were dealing with a finite group. First, the sum over the group meant the addition of a finite number of quantities. When we deal with infinite or continuous groups, we must understand how to replace the finite sum by an infinite sum or an integral. Secondly, in the derivation of the orthogonality theorems (3–139) and in Section 3–11 [Eq. (3–101)] we made use of the result that if $f(R)$ is a function defined on the group manifold, then

$$\sum_R f(R) = \sum_R f(SR), \tag{8-3}$$

where S is any element of the group. This statement, which was evident from the group table of the finite group, must somehow be translated to apply to infinite or continuous groups if we are to establish the orthogonality theorems.

8–2 Infinite discrete groups. First we shall consider infinite discrete groups. The elements of the group, R_a, are labeled by a subscript a which runs through, say the integers $1, 2, \ldots, \infty$. The group manifold is the countable (denumerable) set of "points" R_a (or we may label the integer points $a = 1, 2, \ldots, \infty$ and assign the element R_a to the point a). We could also label our elements with all integers, positive, negative, and zero, or with any denumerable set. (By a denumerable set we mean a set which can be put into one-to-one correspondence with the set of positive integers.) Thus the set

$$
\begin{array}{ccccccc}
0 & 1 & -1 & 2 & -2 & \cdots \\
\downarrow & \downarrow & \downarrow & \downarrow & \downarrow & & \text{etc.} \\
1 & 2 & 3 & 4 & 5 & \cdots
\end{array}
$$

achieves the correspondence between the set of all integers and the set of positive integers. The law of combination $R_b R_a = R_c$ provides us with a function which determines the parameter of the product of two elements in terms of the parameters of the factors, that is,

$$c = \phi(a, b). \tag{8–1}$$

For example, the set of coordinate transformations

$$R_n: x' = x + n, \tag{8–4}$$

where n is an integer, form a group. The inverse of R_n is $(R_n)^{-1} = R_{-n}$, and the identity element has the parameter zero. The group is abelian:

$$c = \phi(a, b) = \phi(b, a) = a + b. \tag{8–5}$$

The group consisting of all integers, with ordinary addition as the law of combination, is isomorphic to this transformation group.

The discrete infinite groups exhibit one peculiarity—we never need more than one discrete parameter to label the elements. For example, consider the group of transformations:

$$R_{mn}: \begin{array}{l} x' = x + m, \\ y' = y + n, \end{array} \qquad m, n \ \text{integers.} \tag{8–6}$$

Written in this fashion, the identity is the element R_{00}; the inverse of

R_{mn} is $(R_{mn})^{-1} = R_{-m,-n}$, and the product $R_{mn}R_{m'n'}$ is $R_{m+m',n+n'}$. But the apparent dependence on two parameters is easily eliminated. The points of the group manifold are the lattice points in two dimensions:

$$
\begin{array}{ccccccccc}
-2,1 & & -1,1 & \leftarrow & 0,1 & \leftarrow & 1,1 & & 2,1 \\
& & & \downarrow & & & \uparrow & & \uparrow \\
-2,0 & & -1,0 & & 0,0 & \rightarrow & 1,0 & & 2,0 \\
& & & \downarrow & & & & & \uparrow \\
-2,-1 & & -1,-1 & \rightarrow & 0,-1 & \rightarrow & 1,-1 & \rightarrow & 2,-1
\end{array}
$$

But we can relabel the elements with positive integers in the following order:

$$
\begin{array}{cccccccccc}
0,0 & 1,0 & 1,1 & 0,1 & -1,1 & -1,0 & -1,-1 & 0,-1 & 1,-1 & 2,-1 \\
\downarrow & \downarrow & \downarrow & \downarrow & \downarrow & \downarrow & \downarrow & \downarrow & \downarrow & \downarrow \\
1 & 2 & 3 & 4 & 5 & 6 & 7 & 8 & 9 & 10
\end{array}
$$
etc.,

and re-express the law of combination in terms of our new single parameter.

Another example is the group of transformations

$$R_r: \quad x' = rx, \tag{8-7}$$

where r is a positive rational number. The rational numbers can be put into one-to-one correspondence with the positive integers. For example, write the rational number r in its lowest terms as $r = m/n$, where m and n are positive integers. Then enumerate all those r's for which $m + n = 2$, that is, $m = 1$, $n = 1$, $r = 1$; next, $m + n = 3$ gives $\frac{1}{2}$, $\frac{2}{1}$ (we take the smallest m first); then, $m + n = 4$ gives $\frac{1}{3}$, $\frac{3}{1}$ (we omit any pair m,n which have a common factor, since r would have appeared earlier in its lowest terms), etc. In similar fashion, we can always express the numbering of a discrete infinite group in terms of the positive integers.

Other examples of discrete infinite groups are

$$
\begin{cases}
x' = x + n, \\
y' = y + 2n,
\end{cases}
\quad n \text{ integral;}
$$

$$
\begin{cases}
x' = rx, \\
y' = ry,
\end{cases}
\quad r \text{ rational,} \quad \neq 0.
$$

We must point out another peculiarity of these denumerable groups. Consider the group of transformations

$$x' = x + n, \quad n \text{ integral.}$$

If we adjoin the transformation $x' = -x$ and form all products, we obtain the group of transformations

$$x' = \pm x + n.$$

This new group can again be labeled with the positive integers, by following each transformation $x' = x + n$ by the transformation $x' = -x + n$ in the enumeration. We may say that a discrete set is so full of holes that there is always room to add one, two, or even a denumerable set of elements, and still leave the set denumerable. We shall see that continuous groups are very different in this respect.

8-3 Continuous groups. Lie groups. A group is said to be *continuous* if some generalized definition of "nearness" or continuity is imposed on the elements of the group manifold. We demand that a "small change" in one of the factors of a product produce a small change in the product. In its most general form, which we shall not discuss, this leads to the definition of a *topological group*, i.e., one for which the group manifold forms a *topological space*. We shall restrict ourselves to the simpler case where the elements of the group manifold can be labeled by a finite set of continuously varying parameters or a set of functions. For example, the set of transformations

$$x' = ax + b \tag{8-8}$$

forms a group. The two parameters a and b vary continuously from $-\infty$ to $+\infty$, and we say that the group is a two-parameter continuous group. In general, an *r-parameter continuous group* has its elements labeled by r continuously varying *real* parameters $a_1, \ldots, a_r$, so that the elements of the group are $R(a_1, \ldots, a_r) = R(a)$. Groups whose elements can be labeled by a *finite* number of continuously varying parameters are said to be *finite continuous groups*. The range of variation of the parameters is not specified; they may vary between $-\infty$ and $+\infty$ or be confined to some finite domain. If the domain of variation of the parameters is finite, the group manifold is said to be *closed*.

For an *r*-parameter group, continuity is expressed in terms of distances in the parameter space. Two group elements $R(a)$ and $R(a')$ are "near" to each other if the distance $[\sum_1^r (a_i - a_i')^2]^{1/2}$ is small. If the elements of the group are labeled by a set of functions in some function space, the "nearness" of group elements is expressed in terms of the distance in the function space (cf. Sections 3–1 and 3–12).

The requirements that the elements $R(a)$ form a continuous group are the same as for finite groups. First there must be a set of parameter values a^0 such that

$$R(a^0)R(a) = R(a)R(a^0) = R(a) \tag{8-9}$$

for all a. $R(a^0)$ is the identity element of the group. For convenience in general arguments, we shall take $a^0 = 0$. Next, for any value of a, we can find a value $\bar{a}$ such that

$$R(\bar{a})R(a) = R(a)R(\bar{a}) = R(0). \tag{8-10}$$

Then $R(\bar{a})$ is the element inverse to $R(a)$:

$$R(\bar{a}) = [R(a)]^{-1}. \tag{8-11}$$

The product of two elements of the set must also belong to the set; given the parameter values a and b, we can find a set of parameter values c such that

$$R(c) = R(b)R(a). \tag{8-12}$$

The parameters c are *real* functions of the *real* parameters a and b:

$$c_k = \phi_k(a_1, \ldots, a_r; b_1, \ldots, b_r), \qquad k = 1, \ldots, r, \tag{8-13}$$

or, symbolically

$$c = \phi(a; b). \tag{8-13a}$$

So far the requirements are the same as for finite or denumerable groups, but now we require in addition that the parameters of a product be analytic functions of the parameters of the factors; i.e., the function in Eq. (8–13) shall possess derivatives of all orders with respect to both arguments. Similarly, we require that the $\bar{a}$ in Eq. (8–10) be analytic functions of the a. We then get an *r-parameter Lie group*.

When we say that we have an r-parameter group, we imply that the r parameters are *essential*; this means that we cannot find a set of continuous parameters $\alpha_1, \ldots, \alpha_m$, with $m < r$, which suffice to label the elements of the group. If this were possible, it would mean that the assignment of a set of values to the $\alpha_1, \ldots, \alpha_m$ defines the group element, so that the r parameters $a_1, \ldots, a_r$ which specify the same group element must be connected to the α's by relations of the form

$$\alpha_1 = \omega_1(a_1, \ldots, a_r), \ldots, \alpha_m = \omega_m(a_1, \ldots, a_r), \tag{8-14}$$

where the ω's are some functions of the a's. For any given set of values of the α's, any solution of the equations (8–14) gives a set of a's which label the same element. Thus, if the parameters are not essential, we have an infinite number of continuously varying values of the parameters $a_1, \ldots, a_r$ which correspond to one and the same group element. As a trivial example, consider the transformations $x' = x + a + b$. The two parameters a and b are not essential, for we can find a single parameter c

which when varied continuously gives all the transformations of the group. Thus, for a given value of c, any values of a and b which satisfy $a + b = c$ give the same transformation. Hence the group is actually a one-parameter group.

We shall be interested mainly in groups of transformations, and now rewrite all our results in this form. An *r-parameter Lie group* of *transformations* is a group of transformations

$$x'_i = f_i(x_1, \ldots, x_n; a_1, \ldots, a_r), \qquad i = 1, \ldots, n, \qquad (8\text{–}15)$$

or, symbolically,

$$x' = f(x; a), \qquad (8\text{–}15a)$$

for which the functions f_i are analytic functions of the parameters a. The r real parameters a_j are assumed to be essential. If they are not, then, as we have shown above, we can find, in the neighborhood of any set of values $a_1, \ldots, a_r$ of the parameters, other parameter sets which give the same transformation. In other words, if the parameters are not essential, there exist parameter values $a_1 + \epsilon_1, \ldots, a_r + \epsilon_r$, where the ϵ's are arbitrarily small quantities which are functions of the $a_1, \ldots, a_r$, such that

$$f_i(x; a) = f_i(x; a + \epsilon) \qquad (8\text{–}16)$$

for all values of x. Expanding in terms of the small functions ϵ_k, we have

$$0 = \sum_{k=1}^{r} \epsilon_k(a) \frac{\partial f_i(x; a)}{\partial a_k} + \text{higher terms in the } \epsilon_k \qquad (i = 1, \ldots, n).$$
$$(8\text{–}17)$$

If we let the ϵ_k approach zero, their ratios in the limit approach a set of functions of a, so that we may write (since the higher terms in Eq. (8–17) go to zero)

$$\sum_{r} \chi_k(a) \frac{\partial f_i(x; a)}{\partial a_k} = 0 \qquad \text{for all } x \text{ and } a \qquad (i = 1, \ldots, n), \qquad (8\text{–}18)$$

where the $\chi_k(a)$ are a set of r functions of the a's. The necessary and sufficient condition for the r parameters $a_1, \ldots, a_r$ to be essential is that it shall be impossible to find r functions $\chi_k(a)$ which satisfy Eq. (8–18) for all values of i, x, and a.

The transformations must satisfy all the group requirements. Thus, given a transformation labeled by the parameter set a [Eq. (8–15a)], we can find a parameter set $\bar{a}$ such that

$$x'' = f(x'; \bar{a}) = f\big(f(x; a); \bar{a}\big) = x. \qquad (8\text{–}19)$$

This means that Eqs. (8–15) must be solvable for the x_i in terms of the x'_i, the condition being that the Jacobian is different from zero:

$$
\begin{vmatrix}
\dfrac{\partial f_1}{\partial x_1} & \cdots & \dfrac{\partial f_1}{\partial x_n} \\
\vdots & & \vdots \\
\dfrac{\partial f_n}{\partial x_1} & \cdots & \dfrac{\partial f_n}{\partial x_n}
\end{vmatrix} \neq 0. \tag{8-20}
$$

If we perform in succession two transformations of the set

$$
\begin{aligned}
x_i' &= f_i(x_1, \ldots, x_n; \, a_1, \ldots, a_r), \\
x_i'' &= f_i(x_1', \ldots, x_n'; \, b_1, \ldots, b_r),
\end{aligned} \tag{8-21}
$$

we require that the resulting transformation also be a member of the set. In other words, there must exist a set of parameter values $c_1, \ldots, c_r$ such that

$$
x_i'' = f_i(x_1, \ldots, x_n; \, c_1, \ldots, c_r). \tag{8-22}
$$

The parameters c must be functions of the parameters a and b:

$$
c_k = \phi_k(a_1, \ldots, a_r; \, b_1, \ldots, b_r). \tag{8-23}
$$

We assume that the functions ϕ_k are analytic, and that the $\bar{a}$ in Eq. (8–19) are analytic functions of the a. There must also exist a set of parameter values a^0 which corresponds to the identity transformation

$$
x' = f(x; a^0) = x. \tag{8-24}
$$

In general arguments we shall take the a^0 equal to zero.

The group requirements impose severe restrictions on the functions f_i. Writing out in full the statements of Eqs. (8–21) through (8–23), we obtain

$$
\begin{aligned}
x_i'' = f_i(x'; b) &= f_i(f_1(x; a), \ldots, f_n(x; a); b), \\
&= f_i(x; c), \\
&= f_i(x; \phi(a; b));
\end{aligned} \tag{8-25}
$$

hence, symbolically,

$$
f(f(x; a); b) = f(x; \phi(a; b)) \tag{8-26}
$$

is an identity in x, a, b.

We can consider (8–15) from another viewpoint which may clarify the difference between finite and infinite continuous groups. Starting with (8–15), we can differentiate the x'''s with respect to the x's and obtain a set of equations from which the finite set of parameters a can be eliminated. We will then be left with a finite set of partial differential equations for the x'''s which no longer contain any arbitrary elements. Furthermore, the general solution of this set of partial differential equations will depend on just r arbitrary constants, i.e., we get back (8–15). For example, consider the continuous group

$$x'_i = \sum_{j=1}^n a_{ij}x_j + a_i, \qquad i = 1, \ldots, n.$$

Since the transformations are linear, all second derivatives of x'_i vanish, so our defining set of partial differential equations is

$$\frac{\partial^2 x'_i}{\partial x_j \, \partial x_k} = 0, \qquad i, j, k = 1, \ldots, n,$$

and contains no arbitrary elements. If the solutions of a finite set of partial differential equations containing no arbitrary elements depend on a finite number of parameters and form a group, then we say that the group is a finite, continuous group. A case where these conditions are not satisfied, so that we get an *infinite continuous group*, is the following: Our differential equations are

$$\frac{\partial x'_i}{\partial x_k} = 0, \qquad i \neq k = 1, \ldots, n.$$

The solutions are $x'_i = F_i(x_i)$, $i = 1, \ldots, n$, and clearly form a group; but the functions F_i are arbitrary so that we cannot label the transformations by a finite number of parameters.

8–4 Examples of Lie groups. Before continuing this discussion, let us look at some examples of continuous groups:

EXAMPLES.

(1) $x' = ax, a \neq 0$:

$$\text{Identity element: } a = 1.$$
$$\text{Inverse element: } \bar{a} = \frac{1}{a}.$$
$$\text{Product element: } c = ba.$$

This is a one-parameter abelian group; c is an analytic function of a and b.

(2) $x' = a_1 x + a_2, a_1 \neq 0$:

$$\text{Identity element: } a_1 = 1, \qquad a_2 = 0.$$
$$\text{Inverse element: } \bar{a}_1 = \frac{1}{a_1}, \qquad a_2 = \frac{-a_2}{a_1}.$$
$$\text{Product element: } c_1 = b_1 a_1, \qquad c_2 = b_2 + b_1 a_2.$$

This is a two-parameter, nonabelian group.

(3) *Linear group in two dimensions, GL(2):*

$$x' = a_1 x + a_2 y, \qquad \begin{vmatrix} a_1 & a_2 \\ a_3 & a_4 \end{vmatrix} \neq 0.$$
$$y' = a_3 x + a_4 y,$$

The four parameters are essential. [Prove this, using the criterion of Eq. (8–18).] If we consider x, y as components of a vector r, the transformations can be written in matrix notation:

$$r' = Ar, \qquad \begin{bmatrix} x' \\ y' \end{bmatrix} = \begin{bmatrix} a_1 & a_2 \\ a_3 & a_4 \end{bmatrix} \begin{bmatrix} x \\ y \end{bmatrix}.$$

The linear group in two dimensions is isomorphic to the group of 2×2 matrices, with matrix multiplication as the law of combination.

$$\text{Identity element: } A = \begin{bmatrix} 1 & 0 \\ 0 & 1 \end{bmatrix} = 1.$$

Inverse element: $A = A^{-1}$

Product element: $C = BA$.

The linear group in two dimensions is a four-parameter, nonabelian group.

(4) *Linear group in n dimensions, GL(n):*

$$x' = \sum_j a_{ij} x_j; \qquad i = 1, \ldots, n, \qquad |a_{ij}| \neq 0,$$

or, in matrix notation,

$$r' = Ar, \quad \det A \neq 0.$$

The results of Example 3 apply. The linear group in n dimensions is nonabelian $(n > 1)$. The number of essential parameters is n^2. The parameters can vary over an infinite range, so $GL(n)$ is not closed.

(5) *Special linear group (unimodular group) in two dimensions SL(2):* This group is obtained from Example 3 by requiring that the determinant of the transformation be unity: $a_1 a_4 - a_2 a_3 = 1$. This restriction provides one functional relation between the four parameters. Thus we have a three-parameter group. The group properties are maintained since a transformation with determinant unity has an inverse with determinant unity, and the product of two unimodular transformations is again unimodular.

(6) *Unimodular group in n dimensions, SL(n):* We restrict Example 4 to transformations with determinant equal to unity. The number of essential parameters is $n^2 - 1$.

(7) *Projective group in one dimension:*

$$x' = \frac{a_1 x + a_2}{a_3 x + a_4}.$$

Problem: Discuss the group of Example 7, and find the parameters of the inverse and product. How many parameters are essential?

EXAMPLES (*cont.*).

(8) *Orthogonal group in two dimensions, $O(2)$:* We consider only those transformations of Example 3 which leave $x^2 + y^2$ invariant:

$$x'^2 + y'^2 = (a_1 x + a_2 y)^2 + (a_3 x + a_4 y)^2 = x^2 + y^2,$$

$$a_1^2 + a_3^2 = 1, \qquad a_2^2 + a_4^2 = 1, \qquad a_1 a_2 + a_3 a_4 = 0.$$

The four parameters are subjected to three functional relations, so that we have a one-parameter group. This is the group of rotations about the z-axis, and can be written as

$$\begin{aligned} x' &= x \cos \phi - y \sin \phi, \\ y' &= x \sin \phi + y \cos \phi, \end{aligned} \qquad 0 \leq \phi \leq 2\pi,$$

where ϕ is the angle of rotation about the z-axis. The group is abelian; the angle of the resultant of two transformations is the sum of the angles of the individual transformations.

(9) *Orthogonal group in three dimensions, $O(3)$:* We restrict the linear group in three dimensions to those transformations which leave $x^2 + y^2 + z^2$ invariant. This invariance condition imposes six conditions on the nine parameters. Thus we have a three-parameter group.

(10) *Orthogonal group in n dimensions, $O(n)$:* We restrict the transformations of the general linear group to those which leave $\sum_{i=1}^{n} x_i^2$ invariant. Thus we impose $n + (n/2)(n - 1)$ conditions on the n^2 parameters, which leaves us with $n(n - 1)/2$ essential parameters.

So far we have considered only real transformations of real variables. If in Example 4 we consider the x_i as complex variables and the a_{ij} as complex coefficients, the number of essential (real) parameters is $2n^2$ (since the real and imaginary parts of a_{ij} are independent parameters).

(11) *Unitary group in two dimensions, $U(2)$:*

$$\begin{aligned} x_1' &= a_{11} x_1 + a_{12} x_2, \\ x_2' &= a_{21} x_1 + a_{22} x_2, \end{aligned} \qquad x, a, \text{ complex,} \qquad \det A \neq 0.$$

We require that $|x_1|^2 + |x_2|^2$ be invariant under the transformations of the group. Then

$$|x_1'|^2 + |x_2'|^2 = |a_{11}x_1 + a_{12}x_2|^2 + |a_{21}x_1 + a_{22}x_2|^2 = |x_1|^2 + |x_2|^2;$$

$$|a_{11}|^2 + |a_{21}|^2 = 1, \qquad |a_{12}|^2 + |a_{22}|^2 = 1, \qquad a_{11}a_{12}^* + a_{21}a_{22}^* = 0.$$

Thus we have four functional relations (the last relation is actually a pair) between the eight parameters, so the group depends on four essential real parameters. This result is simply the writing out of the requirement that the matrix be unitary, $AA^\dagger = 1$.

(12) *Unitary group in n dimensions, U(n):*

$$r' = Ar, \qquad AA^\dagger = 1.$$

The unitary condition imposes $n + 2n(n - 1)/2$ conditions on the $2n^2$ real parameters, leaving us with n^2 essential real parameters. Since the unitary conditions require that $\sum_j |a_{ij}|^2 = 1$, we see that $|a_{ij}|^2 \leq 1$ for all i and j. Thus the parameters of $U(n)$ are restricted to vary over a finite range, and the unitary group is closed. Consequently, all the subgroups of the unitary group [such as the real orthogonal group $O(n)$ and the unitary unimodular group] are also closed.

(13) *Group of rigid motions in three-dimensional space:*

$$x_i' = \sum_{j=1}^{3} a_{ij}x_j + a_i, \qquad i = 1, 2, 3,$$

where we require that $\sum_{i=1}^{3} (x_i^{(1)} - x_i^{(2)})^2$ be invariant. The group depends on six essential parameters. It is the combination of the rotation and translation groups in three dimensions.

(14) If we combine the group in Example 13 with the transformations $x_i' = ax_i$ and form all products, we obtain the seven-parameter similitude group.

(15) If we combine the group of Example 13 with all transformations of reciprocal radii (see Fig. 8–1), we obtain the ten-parameter group of

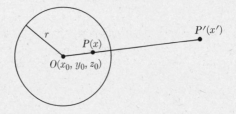

FIGURE 8–1

conformal transformations: $P(x) \rightarrow P'(x')$, where $OP(OP') = r^2$; the center x_0, y_0, z_0 of the sphere can be any point in space, and the radius r of the sphere takes on all values greater than zero.

8-5 Isomorphism. Subgroups. Mixed continuous groups. Using these examples, we can consider some of the concepts discussed earlier for finite groups. The transformations of an r-parameter continuous group are functions of n variables x_i and r parameters a_k. The structure of the group is independent of the number of variables x_i, and depends only on the number of parameters a_k and the functions ϕ of Eq. (8–23). Two groups can be isomorphic even though the numbers of coordinate variables are different, provided that the number of parameters and the law of combination of parameters is the same in both groups. A trivial example is the isomorphism of the two groups:

$$x' = ax \qquad \text{and} \qquad \begin{cases} x' = ax, \\ y' = ay. \end{cases}$$

Or, if we take the unimodular group in two dimensions (three parameters),

$$x' = ax + by,$$
$$y' = cx + dy,$$

we can define a new variable $\xi = x/y$ for which

$$\xi' = \frac{a\xi + b}{c\xi + d},$$

and obtain the one-dimensional projective group.

Next we look at the concepts of conjugate elements, classes, subgroups, and invariant subgroups. The group of Example 2,

$$R_a: \quad x' = a_1 x + a_2,$$

is a two-parameter nonabelian group. Which elements of the group are conjugate to a given element R_a? To answer this, we keep R_a fixed and form $R_b R_a R_b^{-1}$ as R_b varies over the group:

$$R_b: \quad x' = b_1 x + b_2, \qquad R_b^{-1}: \quad x' = \frac{1}{b_1}(x - b_2);$$

$$x' = \frac{1}{b_1}(x - b_2), \qquad x'' = a_1 x' + a_2 = \frac{a_1}{b_1}(x - b_2) + a_2,$$

$$x''' = b_1 x'' + b_2 = b_1 \left[\frac{a_1}{b_1}(x - b_2) + a_2 \right] + b_2$$

$$= a_1 x + b_2 + a_2 b_1 - a_1 b_2.$$

So $R_b R_a R_b^{-1}$ is the transformation

$$x' = a_1 x + b_2 + a_2 b_1 - a_1 b_2.$$

If we keep a_1, a_2 fixed and vary b_1, b_2, the coefficient of x remains equal to a_1, while the constant term can take on all possible values. So the transformations $x' = a_1 x + a_2$, with fixed a_1, form a class. For each value of a_1 we obtain a class of conjugate elements. There is a continuous infinity of classes in the group.

The group of translations $x' = x + a_2$ is a one-parameter abelian subgroup of the above group. Since the coefficient of x is the same in all translations, the subgroup consists of a single class of the original group, and is therefore an invariant subgroup. On the other hand, the one-parameter abelian subgroup $x' = a_1 x$ is not invariant.

The unimodular group in n dimensions is an invariant subgroup of the linear group, since the determinant of the matrix product ASA^{-1} is the same as the determinant of S.

The unimodular group in two dimensions contains the two-parameter subgroup

$$x' = ax + by, \qquad y' = \frac{1}{a} y.$$

This subgroup is not abelian and not invariant. The unimodular group also contains the one-parameter subgroup

$$x' = ax, \qquad y' = \frac{1}{a} y,$$

which is abelian but not invariant, and the one-parameter rotation group, which is also abelian but not invariant.

Problem: Prove the above statements about subgroups of the unimodular group.

We have already seen that subgroups can be obtained by requiring the invariance of some function of the coordinates. For example, by requiring the invariance of $x^2 + y^2$ we obtained from the linear group in two dimensions the subgroup of orthogonal transformations. The same result held for n dimensions. If, in a plane, we take the coordinates of two points, x, y, and ξ, η to form the expression $x\eta - y\xi$ and require this expression to be invariant under linear transformations, we obtain

$$
\begin{aligned}
x'\eta' - y'\xi' &= (ax + by)(c\xi + d\eta) - (cx + dy)(a\xi + b\eta) \\
&= (ad - bc)(x\eta - y\xi) = (x\eta - y\xi) \quad \text{if} \quad ad - bc = 1.
\end{aligned}
$$

$$(8\text{-}27)$$

Thus $x\eta - y\xi$ is invariant under transformations with $ad - bc = 1$, i.e., under the unimodular group.

Problem. Consider the linear group in four dimensions. If for any two points x, ξ, we require the invariance of the bilinear form $x_1\xi_2 - x_2\xi_1 + x_3\xi_4 - x_4\xi_3$, we obtain a subgroup of the linear group. How many parameters characterize this group? Derive the general result in $2n$ dimensions. [This group is the symplectic group $Sp(2n)$.]

There frequently occur groups which, in addition to continuous parameters, require a discrete label to characterize all their elements. Such groups are said to be *mixed* continuous groups. For example, the group $G: x' = \pm x + a$ (discussed earlier for the case where a was integral) is a mixed continuous group in one parameter if a varies continuously from $-\infty$ to $+\infty$. The group manifold consists of two disconnected pieces. The transformations $x' = x + a$ form a subgroup H, all of whose transformations can be reached by continuous variation of the parameter a from 0 (the identity transformation) to its final value. The transformations $x' = -x + a$ cannot be reached continuously from the identity. They are the residue class obtained by taking the product of H with the inversion $I: x' = -x$. So

$$G = H + HI,$$

and H is an invariant subgroup of G; the order of the factor group is 2.

Similarly, the orthogonal group consists of two parts, namely of the transformations with determinant Δ equal to unity and those with $\Delta = -1$. The first set forms a subgroup $O^+(n)$ of the orthogonal group (proper rotations) and can be reached by a continuous path from the identity element. This subgroup is invariant under the full rotation group. We note that, for continuous groups, there is no way of crowding the various pieces together in order to label all elements, using only one continuous parameter. There are no holes in the continuum into which we can insert additional elements.

8–6 One-parameter groups. Infinitesimal transformations. In Eq. (8–23) we have expressed the parameters c of a product of transformations in terms of the parameters a and b of the factors. This equation must be solvable for a in terms of b and c or, for b, in terms of a and c. This requires that neither of the Jacobians $|\partial\phi_k/\partial a_l|$, $|\partial\phi_k/\partial b_l|$ shall vanish. Also, Eq. (8–26) expresses the restrictions placed on the transformation functions f_i if they are to form a group. As written, Eq. (8–26) is an identity in x,

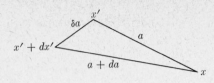

FIGURE 8–2

a, b. But it can also be expressed as an identity in x or x', and any two of the three parameters a, b, c.

The transformation $x' = f(x; a)$ takes all points of the space from their initial positions x to final positions x'. It would be more natural to consider the gradual shift of the points of the space as we vary the parameters continuously from their initial values $a = 0$. This leads us to the concept of infinitesimal transformations. We illustrate the method first for a one-parameter group in one variable x (see Fig. 8–2). Suppose that the transformation with parameter a takes x to x'. The neighboring parameter value $a + da$ will take the points x to points $x' + dx'$ (since f is an analytic function of a). But we can also find a parameter value δa very close to zero (i.e., to the identity) which takes x' to $x' + dx'$. Thus we have two alternative paths from x to $x' + dx'$:

$$x' + dx' = f(x; a + da) \tag{8–28}$$

or

$$x' = f(x; a), \qquad x' + dx' = f(x'; \delta a). \tag{8–29}$$

Expanding the last equation, we have

$$dx' = \left(\frac{\partial f(x'; a)}{\partial a}\right)_{a=0} \cdot \delta a = u(x')\,\delta a. \tag{8–30}$$

Equation (8–23) states that

$$a + da = \phi(a; \delta a), \tag{8–31}$$

so that

$$da = \left(\frac{\partial \phi(a; b)}{\partial b}\right)_{b=0} \cdot \delta a \tag{8–32}$$

or

$$\delta a = \psi(a)\,da. \tag{8–33}$$

Substituting (8–33) in (8–30), we have

$$dx' = u(x')\psi(a)\,da, \tag{8–34}$$

$$\frac{dx'}{u(x')} = \psi(a)\,da. \tag{8–34a}$$

We integrate (8–34a) from $a = 0$ to a. The initial value of x' is x. Calling the integral of $1/u(x')$ the function $U(x')$, we have

$$U(x') - U(x) = \int_0^a \psi(a) \, da. \tag{8–35}$$

If we introduce new variables $y = U(x)$ and let $\int_0^a \psi(a) \, da = t$, we have

$$y' - y = t. \tag{8–36}$$

But as we have seen for finite groups, a transformation of coordinates or introduction of new variables merely transforms all the elements of the group by the same transformation. Thus we have shown that a one-parameter continuous group is equivalent to a group of translations, and must be abelian. We note that the result does not apply to a mixed group, since our last step assumed that the element could be reached by a continuous path from the identity. (We took the simple case of one variable, but the proof for a one-parameter group in any number of variables is easily given.)

_____

Problem. Carry out the proof for transformations in two variables.

_____

We chose to give the proof for a one-parameter group of transformations. The results concerning the structure of the group are independent of the particular realization. We have thus shown that for a one-parameter Lie group one can always introduce a *canonical parameter* t such that

$$R(t_1)R(t_2) = R(t_1 + t_2),$$
$$[R(t)]^{-1} = R(-t). \tag{8–37}$$

We now carry out the analogous expansions in the general case:

$$x'_i = f_i(x_1, \ldots, x_n; a_1, \ldots, a_r), \qquad i = 1, \ldots, n, \tag{8–38}$$

$$x'_i + dx'_i = f_i(x'_1, \ldots, x'_n; \delta a_1, \ldots, \delta a_r); \tag{8–38a}$$

$$dx'_i = \sum_{k=1}^r \left[\frac{\partial f_i(x'_1, \ldots, x'_n; a_1, \ldots, a_r)}{\partial a_k} \right]_{a=0} \delta a_k = \sum_{k=1}^r u_{ik}(x') \, \delta a_k; \tag{8–39}$$

$$a_l + da_l = \phi_l(a_1, \ldots, a_r; \delta a_1, \ldots, \delta a_r), \tag{8–40}$$

$$da_l = \sum_{m=1}^r \left[\frac{\partial \phi_l(a_1, \ldots, a_r; b_1, \ldots, b_r)}{\partial b_m} \right]_{b=0} \delta a_m = \sum_{m=1}^r \Theta_{lm}(a) \, \delta a_m. \tag{8–41}$$

At $a = 0$, $\Theta_{lm}(0) = \delta_{lm}$.

Solving (8–41) for the δa's in terms of the da's, we have

$$\delta a_k = \sum_{l=1}^{r} \psi_{kl}(a) \, da_l, \qquad (8\text{–}42)$$

where the matrices Ψ and Θ satisfy

$$\Psi\Theta = 1, \qquad \psi_{kl}(0) = \delta_{kl}; \qquad (8\text{–}42a)$$

substituting in (8–39), we get

$$dx'_i = \sum_{k,l=1}^{r} u_{ik}(x')\psi_{kl}(a) \, da_l \qquad (8\text{–}43)$$

or

$$\frac{\partial x'_i}{\partial a_l} = \sum_{k=1}^{r} u_{ik}(x')\psi_{kl}(a). \qquad (8\text{–}44)$$

In Eq. (8–44) we may consider the x''s as functions of the parameters a. The coordinates x are the initial values of the x''s for $a = 0$.

If only one of the parameters is changed from zero, we have a one-parameter subgroup and obtain a particular infinitesimal transformation. Any infinitesimal transformation is a linear combination of the r independent infinitesimal transformations. If we examine the change of a function $F(x)$ under the infinitesimal transformation (8–38), we find

$$dF = \sum_{i=1}^{n} \frac{\partial F}{\partial x_i} dx_i = \sum_{i=1}^{n} \frac{\partial F}{\partial x_i} \sum_{l=1}^{r} u_{il}(x) \, \delta a_l$$

$$= \sum_{l=1}^{r} \delta a_l \left(\sum_{i=1}^{n} u_{il}(x) \frac{\partial}{\partial x_i} \right) F = \sum_{l=1}^{r} \delta a_l X_l F. \qquad (8\text{–}45)$$

The operators

$$X_\rho = \sum_{i=1}^{n} u_{i\rho}(x) \frac{\partial}{\partial x_i} \qquad (8\text{–}46)$$

are called the *infinitesimal operators* of the group. The operator $1 + \sum_\rho X_\rho \, \delta a_\rho$ is close to the identity operator. When we choose the function F to be one of the variables x_i, we find

$$x'_i = \left[1 + \sum_\rho X_\rho \, \delta a_\rho \right] x_i = x_i + \sum_\rho u_{i\rho}(x) \, \delta a_\rho,$$

so that we get back Eq. (8–39).

We note that if we neglect higher-order terms in the infinitesimals δa, the infinitesimal transformations commute with one another. In fact, the result of two successive infinitesimal transformations is just the sum of the two (provided we neglect higher terms).

Consider as an example the group $x' = ax + b$. The identity element has parameters $a = 1$, $b = 0$. The infinitesimal transformations are

$$x' = (1 + \delta a)x + \delta b = x + x \cdot \delta a + \delta b; \qquad dx = x \cdot \delta a + \delta b.$$

Thus the infinitesimal operators of the group are $X_1 = x(\partial/\partial x)$ and $X_2 = \partial/\partial x$. We note that the *commutator* $[X_1, X_2] = X_1 X_2 - X_2 X_1 = -(\partial/\partial x) = -X_2$ gives no new operator, but is again one of the infinitesimal transformations of the group.

As another example, take the group

$$\begin{cases} x' = ax, \\ y' = by. \end{cases}$$

The identity element has $a = b = 1$. The infinitesimal transformations are

$$x' = (1 + \delta a)x = x + x \cdot \delta a,$$
$$y' = (1 + \delta b)y = y + y \cdot \delta b,$$

so the infinitesimal operators of the group are

$$X_1 = x\frac{\partial}{\partial x}, \qquad X_2 = y\frac{\partial}{\partial y}.$$

Again we note that the commutator $[X_1, X_2]$ gives nothing new, i.e.,

$$[X_1, X_2] = 0.$$

For the one-parameter group

$$\begin{cases} x' = ax, \\ y' = \dfrac{1}{a}y, \end{cases}$$

the identity element has $a = 1$; the infinitesimal transformation is

$$x' = (1 + \delta a)x = x + x \cdot \delta a,$$
$$y' = (1 + \delta a)^{-1}y = (1 - \delta a)y = y - y \cdot \delta a,$$

and the infinitesimal operator is $X = x(\partial/\partial x) - y(\partial/\partial y)$.

Next consider the linear group in two dimensions,

$$x' = ax + by,$$
$$y' = cx + dy.$$

The identity element has $a = d = 1$ and $b = c = 0$. The infinitesimal transformations are

$$x' = (1 + \delta a)x + \delta b\, y = x + x \cdot \delta a + y \cdot \delta b,$$
$$y' = \delta c\, x + (1 + \delta d)y = y + x \cdot \delta c + y \cdot \delta d.$$

The four infinitesimal operators of the group are

$$X_1 = x\frac{\partial}{\partial x}, \qquad X_2 = y\frac{\partial}{\partial x}, \qquad X_3 = x\frac{\partial}{\partial y}, \qquad X_4 = y\frac{\partial}{\partial y}.$$

If we take the commutators of the infinitesimal operators, we find

$$[X_1, X_2] = -X_2, \qquad [X_1, X_3] = X_3, \qquad [X_1, X_4] = 0,$$
$$[X_2, X_3] = X_4 - X_1, \qquad [X_2, X_4] = -X_2, \qquad [X_3, X_4] = X_3.$$

We see that all commutators can be expressed as linear combinations of the infinitesimal operators themselves.

For the rotation group in two dimensions,

$$x' = x \cos \phi - y \sin \phi,$$
$$y' = x \sin \phi + y \cos \phi,$$

we obtain the infinitesimal transformations by expanding in ϕ around $\phi = 0$:

$$x' = x - y \cdot \delta\phi,$$
$$y' = x \cdot \delta\phi + y.$$

The infinitesimal operator is $X = x(\partial/\partial y) - y(\partial/\partial x)$, the angular-momentum operator.

The orthogonal transformations are characterized by the statement that the transpose $\tilde{A}$ of the matrix A of the transformation is its inverse:

$$A\tilde{A} = 1.$$

The proper rotations have determinant unity, so that the infinitesimal rotations have a matrix of the form

$$A = 1 + B,$$

where 1 is the unit matrix and the matrix B has all its elements in the neighborhood of zero. The orthogonality condition requires

$$1 = A\tilde{A} = (1 + B)(1 + \tilde{B}) \approx 1 + B + \tilde{B},$$

or $B + \widetilde{B} = 0$. Thus B must be a skew-symmetric matrix which has three independent components:

$$B = \begin{bmatrix} 0 & \zeta & -\eta \\ -\zeta & 0 & \xi \\ \eta & -\xi & 0 \end{bmatrix}, \qquad \text{where} \quad \xi, \eta, \zeta \quad \text{are constants.}$$

The infinitesimal rotations are

$$dx = y\zeta - z\eta,$$
$$dy = -x\zeta + z\xi,$$
$$dz = x\eta - y\xi,$$

and the infinitesimal operators are

$$X_1 = z\frac{\partial}{\partial y} - y\frac{\partial}{\partial z}, \qquad X_2 = x\frac{\partial}{\partial z} - z\frac{\partial}{\partial x}, \qquad X_3 = y\frac{\partial}{\partial x} - x\frac{\partial}{\partial y},$$

the angular-momentum operators for the three coordinate directions. The commutators of the infinitesimal operators give the familiar equations:

$$[X_1, X_2] = X_3, \qquad [X_2, X_3] = X_1, \qquad [X_3, X_1] = X_2.$$

8-7 Structure constants. We now show that the property found in these particular examples is general: the commutators of the infinitesimal operators are linearly expressible in terms of the infinitesimal operators. We started in (8-38) from the requirement that the transformations f form a group with r essential parameters. This means that the u_{ik} in (8-39) are linearly independent. Then, using the law of combination of the parameters [Eq. (8-41)], we arrived at (8-44), which we repeat here, using x instead of x':

$$\frac{\partial x_i}{\partial a_\lambda} = u_{ik}(x)\psi_{\kappa\lambda}(a), \qquad \begin{matrix} i = 1, \ldots, n, \\ \kappa, \lambda = 1, \ldots, r, \end{matrix} \qquad (8\text{-}47)$$

where we introduce the convention of summing over repeated indices. Equation (8-47) describes the motion of the point x from its initial position $x(0)$ (when $a = 0$). If the equations (8-47) are to yield the transformation equations (8-37) with arbitrary initial conditions, we must have

$$\frac{\partial^2 x_i}{\partial a_\lambda \, \partial a_\mu} = \frac{\partial^2 x_i}{\partial a_\mu \, \partial a_\lambda} \qquad (8\text{-}48)$$

or

$$\frac{\partial}{\partial a_\lambda}[u_{i\kappa}\psi_{\kappa\mu}] - \frac{\partial}{\partial a_\mu}[u_{i\kappa}\psi_{\kappa\lambda}] = 0,$$

$$u_{i\kappa}\left[\frac{\partial \psi_{\kappa\mu}}{\partial a_\lambda} - \frac{\partial \psi_{\kappa\lambda}}{\partial a_\mu}\right] + \psi_{\kappa\mu}\frac{\partial u_{i\kappa}}{\partial a_\lambda} - \psi_{\kappa\lambda}\frac{\partial u_{i\kappa}}{\partial a_\mu} = 0. \qquad (8\text{-}48a)$$

The $u_{i\kappa}$ are functions of x and, through the x's, depend on a. So

$$\frac{\partial u_{i\kappa}}{\partial a_\lambda} = \frac{\partial u_{i\kappa}}{\partial x_j}\frac{\partial x_j}{\partial a_\lambda} = \frac{\partial u_{i\kappa}}{\partial x_j}u_{j\nu}\psi_{\nu\lambda}, \tag{8–49}$$

where we use (8–47) in the last step. Substituting in (8–48a), we have

$$u_{i\kappa}\left[\frac{\partial\psi_{\kappa\mu}}{\partial a_\lambda} - \frac{\partial\psi_{\kappa\lambda}}{\partial a_\mu}\right] + \left[u_{j\nu}\frac{\partial u_{i\kappa}}{\partial x_j} - u_{j\kappa}\frac{\partial u_{i\nu}}{\partial x_j}\right]\psi_{\kappa\mu}\psi_{\nu\lambda} = 0. \tag{8–50}$$

From (8–42a), $\psi_{\kappa\mu}\Theta_{\mu\lambda} = \delta_{\kappa\lambda}$, so that

$$u_{j\sigma}\frac{\partial u_{i\tau}}{\partial x_j} - u_{j\tau}\frac{\partial u_{i\sigma}}{\partial x_j} = \left[\frac{\partial\psi_{\kappa\mu}}{\partial a_\lambda} - \frac{\partial\psi_{\kappa\lambda}}{\partial a_\mu}\right]\Theta_{\mu\tau}\Theta_{\lambda\sigma}u_{i\kappa}$$

$$= c^\kappa_{\tau\sigma}(a)u_{i\kappa}(x), \tag{8–51}$$

where

$$c^\kappa_{\tau\sigma}(a) = \left[\frac{\partial\psi_{\kappa\mu}}{\partial a_\lambda} - \frac{\partial\psi_{\kappa\lambda}}{\partial a_\mu}\right]\Theta_{\mu\tau}\Theta_{\lambda\sigma}. \tag{8–52}$$

If we differentiate (8–51) with respect to a_ρ, and apply the operator $(\partial x_\kappa/\partial a_\rho)(\partial/\partial x_\kappa)$, which acts only on the x's, we find

$$\frac{\partial c^\kappa_{\tau\sigma}}{\partial a_\rho}u_{i\kappa} = 0. \tag{8–53}$$

Since the $u_{i\kappa}(x)$ are linearly independent, we conclude that the $c^\kappa_{\tau\sigma}$ are independent of a, and are constants. Then Eqs. (8–51) and (8–52) become

$$u_{j\sigma}\frac{\partial u_{i\tau}}{\partial x_j} - u_{j\tau}\frac{\partial u_{i\sigma}}{\partial x_j} = c^\kappa_{\tau\sigma}u_{i\kappa}, \tag{8–51a}$$

$$\frac{\partial\psi_{\kappa\mu}}{\partial a_\lambda} - \frac{\partial\psi_{\kappa\lambda}}{\partial a_\mu} = c^\kappa_{\tau\sigma}\psi_{\tau\mu}\psi_{\sigma\lambda}. \tag{8–52a}$$

These conditions are a consequence of our requirement that (8–47) be integrable. The infinitesimal operators which are defined to be

$$X_\rho = \sum_i u_{i\rho}\frac{\partial}{\partial x_i} \tag{8–46}$$

have the property that their commutators

$$[X_\rho, X_\sigma] = X_\rho X_\sigma - X_\sigma X_\rho \tag{8–54}$$

satisfy

$$[X_\rho, X_\sigma] = u_{i\rho}\frac{\partial}{\partial x_i}u_{j\sigma}\frac{\partial}{\partial x_j} - u_{j\sigma}\frac{\partial}{\partial x_j}u_{i\rho}\frac{\partial}{\partial x_i}$$

$$= \left[u_{i\rho}\frac{\partial u_{j\sigma}}{\partial x_i} - u_{i\sigma}\frac{\partial u_{j\rho}}{\partial x_i}\right]\frac{\partial}{\partial x_j};$$

and using (8–51a), we have

$$[X_\rho, X_\sigma] = c^\kappa_{\rho\sigma} u_{j\kappa} \frac{\partial}{\partial x_j} = c^\kappa_{\rho\sigma} X_\kappa. \tag{8–55}$$

Equation (8–55) states that all commutators are linearly expressible in terms of the infinitesimal operators, where the coefficients $c^\kappa_{\rho\sigma}$ are the *structure constants* of the Lie group. Clearly,

$$c^\kappa_{\rho\sigma} = -c^\kappa_{\sigma\rho}. \tag{8–56}$$

If we substitute (8–55) into the Jacobi identity

$$[[X_\rho, X_\sigma], X_\tau] + [[X_\sigma, X_\tau], X_\rho] + [[X_\tau, X_\rho], X_\sigma] = 0, \tag{8–57}$$

we find

$$c^\mu_{\rho\sigma} c^\nu_{\mu\tau} + c^\mu_{\sigma\tau} c^\nu_{\mu\rho} + c^\mu_{\tau\rho} c^\nu_{\mu\sigma} = 0. \tag{8–58}$$

To recapitulate: Starting from the group of transformations (8–38), we arrived at (8–47), then at (8–51a), (8–52a), and finally at (8–56), (8–58). Lie proved the remarkable result that this procedure can be reversed, i.e., if we have constants satisfying (8–56) and (8–58), we can find u's and ψ's satisfying (8–51a) and (8–52a), and we can then find functions which are integrals of (8–47) and form a group.

8–8 Lie algebras. By using Eq. (8–46) we can write Eq. (8–47) in the form

$$\frac{\partial x_i}{\partial a_\lambda} = \psi_{\kappa\lambda}(a) X_\kappa x_i, \qquad \psi_{\kappa\lambda}(0) = \delta_{\kappa\lambda}. \tag{8–47a}$$

Any transformation of the group can be reached by letting the parameters a_λ vary along a line

$$a_\lambda = s_\lambda \tau, \qquad \lambda = 1, \ldots, r, \tag{8–59}$$

where s_λ is a real vector. For $\tau = 0$ we get the identity transformation, while different values of τ give different transformation operators $S(\tau)$:

$$x_i(\tau) = S(\tau) x_i(0), \qquad S(0) = 1. \tag{8–60}$$

Substituting in Eq. (8–47a), we have

$$\frac{dx_i}{d\tau} = \frac{\partial x_i}{\partial a_\lambda} \frac{da_\lambda}{d\tau} = s_\lambda \psi_{\kappa\lambda}(s_\rho \tau) X_\kappa x_i \tag{8–61}$$

or

$$\frac{dS(\tau)}{d\tau} x_i(0) = s_\lambda \psi_{\kappa\lambda}(s_\rho \tau) X_\kappa S(\tau) x_i(0); \tag{8–61a}$$

so the transformation operator $S(\tau)$ satisfies the differential equation

$$\frac{dS(\tau)}{d\tau} = s_\lambda \psi_{\kappa\lambda}(s_\rho\tau)X_\kappa S(\tau). \tag{8-61b}$$

At $\tau = 0$,

$$\left.\frac{dS}{d\tau}\right|_{\tau=0} = s_\kappa X_\kappa; \tag{8-62}$$

thus the operator $S(\tau)$ has the Taylor expansion

$$S(\tau) = 1 + \tau s_\kappa X_\kappa + \cdots, \tag{8-63}$$

that is,

$$x_i(\tau) = (1 + \tau s_\kappa X_\kappa + \cdots)x_i(0). \tag{8-63a}$$

From Eq. (8–61b) we see that the infinitesimal operator $s_\kappa X_\kappa$ determines the transformation operators $S(\tau)$, while Eq. (8–63) shows that $S(\tau)$ determines the infinitesimal operator $s_\kappa X_\kappa$.

If we now consider a second vector t_λ giving the infinitesimal operator $t_\lambda X_\lambda$, the corresponding transformation operators $T(\tau)$ will satisfy equations similar to (8–61) through (8–63). The product $S(\tau)T(\tau)$ will have the Taylor expansion

$$(1 + \tau s_\kappa X_\kappa + \cdots)(1 + \tau t_\kappa X_\kappa + \cdots) = 1 + \tau(s_\kappa + t_\kappa)X_\kappa + \cdots \tag{8-64}$$

Thus the product $S(\tau)T(\tau)$ corresponds to the infinitesimal operator $(s_\kappa + t_\kappa)X_\kappa$, i.e., to the sum of the infinitesimal operators corresponding to $S(\tau)$ and $T(\tau)$. The commutator of $S(\tau)$ and $T(\tau)$ is the transformation operator $S^{-1}(\tau)T^{-1}(\tau)S(\tau)T(\tau)$, which has the Taylor expansion

$$(1 - \tau s_\kappa X_\kappa + \cdots)(1 - \tau t_\lambda X_\lambda + \cdots)(1 + \tau s_\mu X_\mu + \cdots)$$
$$\times (1 + \tau t_\nu X_\nu + \cdots) = 1 + \tau^2[s_\kappa X_\kappa, t_\lambda X_\lambda] + \cdots, \tag{8-65}$$

so that the corresponding infinitesimal operator is the commutator of the infinitesimal operators corresponding to $S(\tau)$ and $T(\tau)$. [We see from Eq. (8–65) that the variable τ^2 should be used in the differential equation for the commutator.] If $S(\tau)$ and $T(\tau)$ commute, their commutator is the identity operator, and the corresponding infinitesimal operators have $[s_\kappa X_\kappa, t_\lambda X_\lambda] = 0$.

These results enable us to describe various properties of the Lie group in terms of the infinitesimal operators. If the group G is abelian so that all of its elements commute with one another, then

$$[X_\kappa, X_\lambda] = 0, \qquad \kappa, \lambda = 1, \ldots, r,$$

or, in terms of the structure constants,

$$c_{\kappa\lambda}^{\rho} = 0, \qquad \kappa, \lambda, \rho = 1, \ldots, r. \tag{8-66}$$

If H is a p-parameter subgroup of G, we can select p infinitesimal operators corresponding to elements in the subgroup H, while the remaining $r - p$ infinitesimal operators correspond to elements in $G - H$. Since H is a group, the commutators of the infinitesimal operators $X_1, \ldots, X_p$ must be expressible in terms of $X_1, \ldots, X_p$ alone, so that

$$c_{\kappa\lambda}^{\rho} = 0, \qquad \kappa, \lambda = 1, \ldots, p; \qquad \rho = p + 1, \ldots, r. \tag{8-67}$$

If the subgroup H is invariant and S is any element in H, then $T^{-1}ST$ is also in H for any element T of the whole group G. But then $S^{-1}T^{-1}ST$ is also in H; so from Eq. (8-65) we see that the commutator $[s_\kappa X_\kappa, t_\lambda X_\lambda]$ must be expressible as a linear combination of the infinitesimal operators of H alone. Thus, for an invariant subgroup,

$$c_{\kappa\lambda}^{\rho} = 0, \qquad \kappa = 1, \ldots, p; \qquad \rho = p + 1, \ldots, r. \tag{8-68}$$

If the group G is the direct product of H and $G - H$, then

$$c_{\kappa\lambda}^{\rho} = 0, \quad \text{for} \quad \kappa = 1, \ldots, p; \qquad \rho = p + 1, \ldots, r,$$
$$\text{and for} \quad \kappa = p + 1, \ldots, r, \qquad \rho = 1, \ldots, p. \tag{8-68a}$$

The group G will be simple (have no proper invariant subgroups) if Eq. (8-68) cannot be satisfied for any choice of basis X_ρ. Similarly, the group G will be semisimple (have no abelian invariant subgroups) if the equation

$$c_{\kappa\lambda}^{\rho} = 0, \qquad \kappa, \lambda, \rho = 1, \ldots, p, \tag{8-68b}$$

and Eq. (8-68) cannot be satisfied for any choice of basis X_ρ.

The structure of an abstract Lie group is contained entirely in Eq. (8-23), which gives the law of combination of the real parameters a_i and b_i of the factors, according to which the real parameters c_i of the product are obtained. In the realization of the abstract Lie group as a group of transformations, the variables x_i could be real or complex, so that the infinitesimal operators may be complex, but any relations which describe the group structure must involve only real coefficients. Thus the structure constants $c_{\kappa\lambda}^{\rho}$ must be real numbers.

We have found that for an r-parameter transformation group there are r linearly independent infinitesimal operators X_ρ. Linear combinations of these quantities can be formed to give an r-dimensional vector space. If we are considering problems concerned with the structure of the Lie group, we should take only linear combinations with real coefficients.

Thus we should consider the real vector space of quantities $\sum_\rho a_\rho X_\rho$, where the a_ρ are real numbers. Equation (8–55) defines a "product" in this space, since the structure constants are real. If we now abstract from our particular realization of the Lie group, we conclude that the r-parameter Lie group has associated with it a real r-dimensional vector space of quantities $\sum_\rho a_\rho X_\rho$ which is closed under a multiplication defined by Eqs. (8–55) through (8–57); this is the *Lie algebra* of the Lie group.

A real Lie algebra consists of quantities $A, B, \ldots$ from which linear combinations $aA + bB$ with real coefficients can be formed. The product of the elements A and B is $[A, B]$ and is contained in the real vector space. The product satisfies Eqs. (8–56) through (8–57):

$$[A, B] = - [B, A], \tag{8–56a}$$

$$[A, [B, C]] + [B, [C, A]] + [C, [A, B]] = 0. \tag{8–57a}$$

All the elements A are expressible in terms of a set of r basis vectors X_ρ as

$$A = \sum_\rho a_\rho X_\rho. \tag{8–69}$$

If we take linear combinations of the quantities A, B with complex coefficients and define $[A + iB, C]$ to be equal to $[A, C] + i[B, C]$, we obtain the *complex extension* of the real Lie algebra. It may be convenient to work with the complex extension, but statements concerning the structure of the Lie group should be made only on the basis of the real Lie algebra.

We can also pose the inverse problem: Given a real Lie algebra with preassigned structure constants $c_{\rho\sigma}^\kappa$ [which satisfy Eq. (8–59)], construct the Lie group which has this algebra as its Lie algebra. Stated in terms of transformations, the problem would be to find the finite transformations by integration, starting from preassigned commutation relations of the infinitesimal operators. We state the result without proof: To every Lie algebra there corresponds a Lie group; the structure constants determine the Lie group locally (i.e., in the neighborhood of the identity element).

The terminology used for groups is taken over for Lie algebras. Thus a Lie algebra G is said to be abelian if Eq. (8–66) is satisfied. The algebra H is a subalgebra of G if Eq. (8–67) holds, and is an invariant subalgebra if Eq. (8–68) is valid. The Lie algebra G is the direct sum of the algebras H and $G - H$ if Eq. (8–68a) holds. Similarly, the Lie algebra G is simple if Eq. (8–68) cannot be satisfied and semisimple if both Eqs. (8–68) and (8–68b) cannot be satisfied.

8–9 Structure of Lie algebras. To deduce all possible structures of Lie algebras is a formidable mathematical problem. We shall only indicate some of the features of the procedure.

The problem is to find for each value of r all possible real solutions of Eq. (8-58),

$$c_{\rho\sigma}^{\mu}c_{\mu\tau}^{\nu} + c_{\sigma\tau}^{\mu}c_{\mu\rho}^{\nu} + c_{\tau\rho}^{\mu}c_{\mu\sigma}^{\nu} = 0, \qquad (8\text{-}58)$$

subject to Eq. (8-56):

$$c_{\rho\sigma}^{\kappa} = -c_{\sigma\rho}^{\kappa}. \qquad (8\text{-}56)$$

The problem is difficult because the set of equations (8-58) is quadratic in the unknowns. In addition, many of the solutions will be equivalent to one another, for if we replace the basis X_ρ by another basis

$$X_\rho' = a_{\rho\nu}X_\nu, \qquad (8\text{-}70)$$

with a nonsingular matrix $a_{\rho\nu}$, we get a new set of structure constants $c_{\rho\sigma}'^{\kappa}$ which again satisfy Eqs. (8-56) and (8-58):

$$c_{\rho\sigma}'^{\mu}X_\mu' = [X_\rho', X_\sigma'] = [a_{\rho\nu}X_\nu, a_{\sigma\lambda}X_\lambda]$$
$$= a_{\rho\nu}a_{\sigma\lambda}[X_\nu, X_\lambda] = a_{\rho\nu}a_{\sigma\lambda}c_{\nu\lambda}^{\kappa}X_\kappa = c_{\rho\sigma}'^{\mu}a_{\mu\kappa}X_\kappa,$$

so that

$$c_{\rho\sigma}'^{\mu}a_{\mu\kappa} = a_{\rho\nu}a_{\sigma\lambda}c_{\nu\lambda}^{\kappa}. \qquad (8\text{-}71)$$

Multiplying by the inverse of the matrix a, we have

$$c_{\rho\sigma}'^{\mu} = a_{\rho\nu}a_{\sigma\lambda}c_{\nu\lambda}^{\kappa}a_{\kappa\mu}^{-1}. \qquad (8\text{-}71a)$$

Problem. Prove that the quantities X_ρ' defined by Eq. (8-70) satisfy Eqs. (8-56) and (8-58).

For $r = 1$, all elements of the Lie algebra are multiples of one basis vector X, so all commutators vanish. The corresponding Lie group is a one-parameter abelian group.

For $r = 2$, we have two basis elements X_1, X_2 and $[X_1, X_2] = aX_1 + bX_2$. If $a = b = 0$, then $[X_1, X_2] = 0$, and the algebra is abelian and the direct sum of the algebras generated by X_1 and by X_2. If, say $a \neq 0$, we can replace the basis elements by $X_1' = aX_1 + bX_2$, $X_2' = (1/a)X_2$ for which $[X_1', X_2'] = X_1'$. We see that the subalgebra generated by X_1' is invariant and abelian, and hence the Lie algebra is not semi-simple. Thus for $r = 2$, we obtain no semisimple Lie algebras. An example of a Lie group of transformations corresponding to the Lie algebra with $[X_1, X_2] = 0$ is $x' = x + a, y' = y + b$. It is a direct product of one-parameter groups and, according to Section 8-6, it can therefore always be expressed as a translation group in two dimensions. Clearly for an r-dimensional Lie algebra for which $[X_i, X_j] = 0$ for all i, j, the Lie group is (locally) isomorphic to the r-dimensional translation group. A transformation group for which the Lie algebra has $[X_1, X_2] = X_1$ is $x' = ax + b$, with $X_1 = \partial/\partial x$ and $X_2 = x(\partial/\partial x)$. The translations $x' = x + b$ which are generated by X_1 form an abelian invariant subgroup.

Problem. Integrate Eq. (8–47a) for the above case and get back the finite transformations of the group.

This elementary procedure is already quite complicated for $r = 3$ because Eq. (8–58) gives three quadratic relations which must be satisfied by the structure constants. But one can still use the following simple method: We set $\sigma = \mu$ in Eq. (8–71a) and sum over μ:

$$c'^{\mu}_{\rho\mu} = a_{\rho\nu}a^{-1}_{\kappa\mu}a_{\mu\lambda}c^{\kappa}_{\nu\lambda} = a_{\rho\nu}\,\delta_{\kappa\lambda}c^{\kappa}_{\nu\lambda} = a_{\rho\nu}c^{\lambda}_{\nu\lambda}; \qquad (8\text{--}72)$$

the quantities $c^{\lambda}_{\nu\lambda}$ transform like a vector when we change basis. Since the matrix $a_{\rho\nu}$ is an arbitrary nonsingular matrix, we can always transform the vector $c^{\lambda}_{\nu\lambda}$ so that one component is equal to unity and the others are zero. The only case in which this is not possible is that in which $c^{\lambda}_{\nu\lambda}$ is the null vector. Thus we can always choose our basis so that either: (a) $c^{\lambda}_{1\lambda} = 1$, $c^{\lambda}_{i\lambda} = 0$; $\nu \neq 1$; or (b) $c^{\lambda}_{\nu\lambda} = 0$ for all ν.

If we substitute in Eqs. (8–71a), we find that for case (a) the quadratic equations reduce to linear equations, giving $c^1_{23} = c^1_{31} = c^1_{21} = 0$. Combining this with the equations $c^{\lambda}_{\nu\lambda} = 0$ for $\nu = 2, 3$, we find $c^3_{23} = c^2_{32} = 0$. Thus X_3 commutes with X_1 and X_2, and gives an abelian invariant subgroup; hence the Lie group is not semisimple. We obtain two structures:

$$[X_1, X_2] = [X_2, X_3] = [X_3, X_1] = 0, \qquad (8\text{--}73)$$

$$[X_1, X_2] = X_1, \qquad [X_2, X_3] = [X_3, X_1] = 0. \qquad (8\text{--}74)$$

For case (b), Eqs. (8–71a) reduce to identities. By making further real linear transformations one finds, aside from the structures given previously, only two independent types:

$$[X_1, X_2] = X_3, \qquad [X_2, X_3] = X_1, \qquad [X_3, X_1] = X_2, \qquad (8\text{--}75)$$

$$[X_1, X_2] = X_3, \qquad [X_2, X_3] = -X_1, \qquad [X_3, X_1] = -X_2. \qquad (8\text{--}76)$$

Both algebras are simple.

Problems. (1) Complete the derivation outlined here and show that there are only four independent structures for a real Lie algebra with $r = 3$.

(2) Show that the real Lie algebra

$$[X_1, X_2] = X_3, \qquad [X_2, X_3] = -X_1, \qquad [X_3, X_1] = X_2 \qquad (8\text{-}76a)$$

has the same structure as Eq. (8–76).

(3) Prove that the algebras of Eqs. (8–75) and (8–76) are simple.

A Lie group having the structure of Eq. (8–75) is the real orthogonal group in three dimensions, $O(3)$. This group leaves the quadratic form $x^2 + y^2 + z^2$ invariant. The infinitesimal operators are

$$X_1 = z\frac{\partial}{\partial y} - y\frac{\partial}{\partial z}, \qquad X_2 = x\frac{\partial}{\partial z} - z\frac{\partial}{\partial x}, \qquad X_3 = y\frac{\partial}{\partial x} - x\frac{\partial}{\partial y}. \tag{8-77}$$

The group of real linear transformations which leave the quadratic form $x^2 + y^2 - z^2$ invariant has the structure (8–76). The infinitesimal operators are

$$X_1 = z\frac{\partial}{\partial y} + y\frac{\partial}{\partial z}, \qquad X_2 = x\frac{\partial}{\partial z} + z\frac{\partial}{\partial x}, \qquad X_3 = y\frac{\partial}{\partial x} - x\frac{\partial}{\partial y}. \tag{8-78}$$

This group is the two-dimensional "Lorentz group."

It is important to emphasize that we are considering only real Lie algebras. If we go to the complex extension, the algebras described by Eqs. (8–75) and (8–76) have the same structure. For example, if we replace X_2 by $-iX_2$ and X_3 by $-iX_3$, Eqs. (8–76a) are converted to (8–75). Correspondingly, if we replace z by iz in (8–78), we obtain the infinitesimal operators of Eq. (8–77). The substitution replaces the quadratic form $x^2 + y^2 - z^2$ by $x^2 + y^2 + z^2$. This introduction of an imaginary variable for the "time coordinate" is customary in treating Lorentz groups, but it can lead to erroneous conclusions about group structure. The parameters of the real orthogonal group $O(3)$ are restricted to a finite domain, so that the group manifold is closed, while the parameters of the "Lorentz group" of Eq. (8–78) vary over an infinite range.

As another example of the distinction between a real Lie algebra and its complex extension, we consider first the group $O(4)$ of real orthogonal transformations in four dimensions. This is the group of real linear transformations which leave the quadratic form $x^2 + y^2 + z^2 + t^2$ invariant. The six infinitesimal operators can be chosen as

$$A_1 = z\frac{\partial}{\partial y} - y\frac{\partial}{\partial z}, \qquad A_2 = x\frac{\partial}{\partial z} - z\frac{\partial}{\partial x}, \qquad A_3 = y\frac{\partial}{\partial x} - x\frac{\partial}{\partial y};$$

$$B_1 = x\frac{\partial}{\partial t} - t\frac{\partial}{\partial x}, \qquad B_2 = y\frac{\partial}{\partial t} - t\frac{\partial}{\partial y}, \qquad B_3 = z\frac{\partial}{\partial t} - t\frac{\partial}{\partial z}. \tag{8-79}$$

Evaluating the commutators, we obtain the relations

$$[A_1, A_2] = A_3, \qquad [A_2, A_3] = A_1, \qquad [A_3, A_1] = A_2, \qquad (8\text{-}80)$$

$$[B_1, B_2] = A_3, \qquad [B_2, B_3] = A_1, \qquad [B_3, B_1] = A_2, \qquad (8\text{-}80a)$$

$$[A_1, B_1] = [A_2, B_2] = [A_3, B_3] = 0, \qquad (8\text{-}80b)$$

$$[A_1, B_2] = B_3, \qquad [A_1, B_3] = -B_2, \qquad (8\text{-}80c)$$

$$[A_2, B_1] = -B_3, \qquad [A_2, B_3] = B_1, \qquad (8\text{-}80d)$$

$$[A_3, B_1] = B_2, \qquad [A_3, B_2] = -B_1 \qquad (8\text{-}80e)$$

which describe the structure of the corresponding Lie algebra. If we make a linear transformation to the basis consisting of the quantities

$$J_i = \frac{A_i + B_i}{2}, \qquad K_i = \frac{A_i - B_i}{2} \qquad (8\text{-}81)$$

we find that

$$[J_1, J_2] = J_3, \qquad [J_2, J_3] = J_1, \qquad [J_3, J_1] = J_2, \qquad (8\text{-}82)$$

$$[K_1, K_2] = K_3, \qquad [K_2, K_3] = K_1, \qquad [K_3, K_1] = K_2, \qquad (8\text{-}82a)$$

$$[J_i, K_j] = 0. \qquad (8\text{-}82b)$$

Consequently $O(4)$ is locally isomorphic to the direct product of two groups each of which is isomorphic to $O(3)$.

Next we consider the Lorentz group of the real transformations which leave the quadratic form $x^2 + y^2 + z^2 - t^2$ invariant. The infinitesimal operators can be chosen as

$$A_1 = z \frac{\partial}{\partial y} - y \frac{\partial}{\partial z}, \qquad A_2 = x \frac{\partial}{\partial z} - z \frac{\partial}{\partial x}, \qquad A_3 = y \frac{\partial}{\partial x} - x \frac{\partial}{\partial y};$$

$$B_1 = x \frac{\partial}{\partial t} + t \frac{\partial}{\partial x}, \qquad B_2 = y \frac{\partial}{\partial t} + t \frac{\partial}{\partial y}, \qquad B_3 = z \frac{\partial}{\partial t} + t \frac{\partial}{\partial z}. \qquad (8\text{-}83)$$

The structure of the corresponding Lie algebra is given by the equations

$$[A_1, A_2] = A_3, \qquad [A_2, A_3] = A_1, \qquad [A_3, A_1] = A_2, \qquad (8\text{-}84)$$

$$[B_1, B_2] = -A_3, \qquad [B_2, B_3] = -A_1, \qquad [B_3, B_1] = -A_2, \qquad (8\text{-}84a)$$

$$[A_1, B_1] = [A_2, B_2] = [A_3, B_3] = 0, \qquad (8\text{-}84b)$$

$$[A_1, B_2] = B_3, \qquad [A_1, B_3] = -B_2, \qquad (8\text{-}84c)$$

$$[A_2, B_1] = -B_3, \qquad [A_2, B_3] = B_1, \qquad (8\text{-}84d)$$

$$[A_3, B_1] = B_2, \qquad [A_3, B_2] = -B_1. \qquad (8\text{-}84e)$$

If we now try the substitution (8–81), we find that the Lie algebra does not split into a direct sum. In fact, when we discuss the Lorentz group in a later chapter, we shall give a direct proof that it is simple.

Problem. Consider the Lorentz group with two "space variables" and two "time variables," i.e., the group of real transformations which leave $x^2 + y^2 - z^2 - t^2$ invariant. Find the analogs of Eqs. (8–79) and (8–80) through (8–80e). Show that the Lie algebra is a direct sum of two Lie algebras, each having the structure (8–76).

If we were permitted to make complex substitutions, the substitution $B_i \to iB_i$ would make Eqs. (8–84) through (8–84e) identical with Eqs. (8–80) through (8–80e). For the Lorentz group, the quantities

$$J_i = \frac{A_i + iB_i}{2}, \qquad K_i = \frac{A_i - iB_i}{2} \tag{8–85}$$

would satisfy Eqs. (8–82) through (8–82b). It is clear that all three real Lie algebras just discussed have the same complex extension, yet their structures are completely different.

We also note that $O(4)$ is closed (cf. Example 12 in Section 8–4), while the parameters of the Lorentz groups vary over an infinite range.

8–10 Structure of compact semisimple Lie groups and their algebras.
In this section we shall cite without proofs the theorems which enable one to find the structure of semisimple Lie groups. We have used the term "closed" to describe Lie groups whose parameters vary over a finite range. The group manifold itself (i.e., the set of elements of the Lie group) is then said to be *compact*. In general, a set M is compact if every infinite subset of M contains a sequence which converges to an element of M. For example, any region of finite extension in a euclidean space is compact (Bolzano-Weierstrass theorem). On the other hand, a region of euclidean space extending to infinity is not compact since an infinite sequence of points $p_1, p_2, \ldots$ for which one or more of the coordinates of p_i go to infinity does not converge.

One can prove that a continuous function defined on a compact set is bounded. (This can also be used as an alternative definition of compactness.) It is then possible to give a proper definition of integration on the set. If the parameters of a Lie group vary over an infinite range, it is easy to construct continuous functions of the parameters which are not bounded.

The Lie algebra of a compact Lie group is also said to be compact. We found in the previous section that for $r = 3$, the compact group $O(3)$ and the noncompact Lorentz group of Eq. (8–78) had the same complex extension. Again we found that the complex extensions of the real Lie algebras of the groups which left the forms $x^2 + y^2 + z^2 + t^2$, $x^2 + y^2 + z^2 - t^2$, and $x^2 + y^2 - z^2 - t^2$ invariant were the same, but only the first group was compact. These are special cases of a general result: Every complex semisimple Lie algebra has precisely one compact real form.

We can associate a linear transformation with each element A of a Lie algebra: For any element S the commutator $[A, S]$ is again an element of the algebra, so we define the operator p_A which when applied to S gives a vector

$$p_A S = [A, S]. \tag{8-86}$$

If we choose a particular basis in which $A = a_\mu X_\mu$, $S = s_\mu X_\mu$, we can find the matrix of p_A in this basis:

$$(p_A S)_\alpha X_\alpha = c_{\nu\beta}^\alpha a_\nu s_\beta X_\alpha = (p_A)_{\alpha\beta} s_\beta X_\alpha,$$

$$(p_A)_{\alpha\beta} = c_{\nu\beta}^\alpha a_\nu. \tag{8-87}$$

If we now apply the operator p_B corresponding to the element B, we get $p_B p_A S = [B, [A, S]]$ and find

$$(p_B p_A)_{\gamma\beta} = c_{\mu\alpha}^\gamma c_{\nu\beta}^\alpha b_\mu a_\nu. \tag{8-88}$$

The trace of the transformation $p_B p_A$ is

$$\text{tr}\,(p_B p_A) = c_{\mu\alpha}^\beta c_{\nu\beta}^\alpha b_\mu a_\nu = g_{\mu\nu} b_\mu a_\nu, \tag{8-89}$$

where

$$g_{\mu\nu} = g_{\nu\mu} = c_{\mu\alpha}^\beta c_{\nu\beta}^\alpha. \tag{8-90}$$

Equation (8–89) enables us to associate with any two elements of the algebra a symmetric bilinear form which we call the *scalar product* (A, B) of A and B:

$$(A, B) = g_{\mu\nu} b_\mu a_\nu. \tag{8-91}$$

If we make the basis transformation (8–70), we find by using (8–71a) and (8–90) that

$$g'_{\mu\nu} = a_{\mu\alpha} a_{\nu\beta} g_{\alpha\beta} = a_{\mu\alpha} g_{\alpha\beta} \tilde{a}_{\beta\nu}, \tag{8-92}$$

so that $g_{\mu\nu}$ transforms like a symmetric tensor of rank two, while the scalar product (8–91) is invariant. Thus $g_{\mu\nu}$ acts as the metric matrix in our vector space. By using (8–58) we can show that the quantities

$$c_{\lambda\mu\nu} = g_{\lambda\rho} c_{\mu\nu}^\rho \tag{8-93}$$

are antisymmetric under any interchange of indices.

Problem. Prove that $c_{\lambda\mu\nu}$ is antisymmetric under any interchange of indices.

Since $g_{\mu\nu}$ is a real symmetric matrix, it can always be brought to diagonal form by real transformations i.e., by a change of basis [Eqs. (8–70) and (8–92)]. In this special basis, $g_{\mu\nu}$ will have the form

$$g_{\mu\nu} = \epsilon_\mu\, \delta_{\mu\nu}, \qquad (8\text{–}93\text{a})$$

with $\epsilon_\mu = +1$ for $\mu = 1, \ldots, k$, $\epsilon_\mu = -1$ for $\mu = k + 1, \ldots, l$, $\epsilon_\mu = 0$ for $\mu = l + 1, \ldots, r$. If $g_{\mu\nu}$ is nonsingular ($\det g \neq 0$), no zeros will appear in the canonical form (8–93), and we will have k diagonal elements equal to $+1$ and $r - k$ equal to -1. A theorem of Cartan states the following:

The necessary and sufficient condition for a Lie algebra to be semisimple is that $\det g \neq 0$.

This condition can be restated in other forms. If $\det g = 0$, the equations $g_{\mu\nu}a_\nu = 0$ have a nontrivial solution. For this vector A we then have $g_{\mu\nu}a_\nu b_\mu = 0$ for any b_μ so that $(A, B) = 0$ for all B. Hence we can restate Cartan's criterion as:

The necessary and sufficient condition for a Lie algebra to be semisimple is that there exist no element A of the algebra which is orthogonal to the whole algebra [in terms of the scalar product (8–91)].

If the matrix $g_{\mu\nu}$ is negative definite, the diagonal form (8–93) will have all its diagonal elements equal to -1. One can prove the following:

The necessary and sufficient condition for a semisimple Lie algebra to be compact is that the matrix $g_{\mu\nu}$ be negative definite [that is, $(A, A) < 0$ for every element A of the algebra].

Finally we note that for a compact semisimple Lie algebra we can choose the basis so that $g_{\mu\nu} = -\delta_{\mu\nu}$. In this basis, according to Eq. (8–93), the structure constants $c^\rho_{\mu\nu}$ are antisymmetric under any interchange of indices.

The actual analysis of the structure of compact semisimple Lie algebras is beyond the scope of this book, but the summary given here should enable the reader to follow the discussion in the books of Pontriagin and Racah.

8–11 Linear representations of Lie groups. The linear representations of Lie groups are defined in the same way as those for finite groups (cf. Chapter 3). We associate with each element R of the group a linear operator $D(R)$. The operators $D(R)$ act on the vectors ψ in a euclidean space of finite dimension or a Hilbert space in which a positive definite scalar product (ψ_1, ψ_2) is defined. They must satisfy the usual requirements

$$D(R_1)D(R_2) = D(R_1R_2), \qquad (8\text{–}94)$$

$$D(E) = 1. \qquad (8\text{–}94\text{a})$$

We now require that the operators $D(R)$ be *bounded operators* [i.e.,

$(D(R)\psi, \psi)$ shall be finite for all ψ] and that $(D(R)\psi, \phi)$ be a continuous function of the parameters of the group element R.

We can attack the problem of finding the representations of a Lie group directly or we can instead consider the related problem of finding the representations of its Lie algebra. Here we associate with each element A of the algebra a linear operator $D(A)$ in a Hilbert space and require that

$$D(A + B) = D(A) + D(B), \tag{8-95}$$

$$D(\alpha A) = \alpha D(A), \tag{8-95a}$$

$$D([A, B]) = D(A)D(B) - D(B)D(A) = [D(A), D(B)]. \tag{8-95b}$$

But we are really interested in finding the representations of the Lie group and use the representation of the algebra as an intermediate step. We must therefore examine the representations of the algebra and show that they can be extended to give a representation of the Lie group which satisfies the requirements stated above.

For a one-parameter subgroup of the Lie group, we can introduce a canonical parameter t so that

$$R(t_1 + t_2) = R(t_1)R(t_2), \qquad R(0) = E. \tag{8-37}$$

We shall denote the operator representative of the element $R(t)$ by the same symbol and can regard (8-37) as an operator equation. If we differentiate (8-37) with respect to t_1 and then set $t_1 = 0$, $t_2 = t$, we find

$$\frac{dR}{dt} = \frac{dR}{dt}\bigg|_{t=0} \cdot R(t) = RR(t). \tag{8-96}$$

The operator

$$R = \frac{dR}{dt}\bigg|_{t=0} = \lim_{t \to 0} \frac{R(t) - 1}{t} \tag{8-97}$$

in the representation space is associated with the infinitesimal operator of the Lie group which generated $R(t)$ [cf. Eq. (8-62)]. We assume that

$$\lim_{t \to 0} \frac{R(t) - 1}{t}\psi$$

exists for a set of vectors ψ which is everywhere dense in the Hilbert space, so that (by continuity) the operator R is well defined. Then Eq. (8-96) is meaningful and can be used to express $R(t)$ in terms of R:

$$R(t) = \exp(Rt). \tag{8-98}$$

If we proceed by constructing representations of the Lie algebra, we must require that the representatives of the elements of the Lie group [which can be found by using (8-98)] be bounded operators.

A representation of the Lie group will be unitary if the operators $R(t)$ are unitary:

$$(R(t)\psi, R(t)\phi) = (\psi, \phi). \tag{8–99}$$

Differentiating with respect to t and setting $t = 0$, we find

$$(R\psi, \phi) + (\psi, R\phi) = 0$$

for all ψ, ϕ, so

$$R + R^\dagger = 0 \tag{8–100}$$

Thus unitary operators for the group are associated with anti-Hermitian operators for the Lie algebra. If we let $R = iH$, then

$$H = H^\dagger, \qquad R(t) = \exp(iHt). \tag{8–100a}$$

For example, for $O(3)$, the orthogonal group in three dimensions, the operators of a representation of the Lie algebra must satisfy Eq. (8–75):

$$[X_1, X_2] = X_3, \qquad [X_2, X_3] = X_1, \qquad (X_3, X_1) = X_2. \tag{8–75}$$

If we set $iX_\mu = J_\mu$, then

$$[J_1, J_2] = iJ_3, \qquad [J_2, J_3] = iJ_1, \qquad [J_3, J_1] = iJ_2. \tag{8–101}$$

If the operators J_μ are Hermitian, the representation of the Lie group will be unitary. As we shall see later, all representations of $O(3)$ are equivalent to unitary representations, so for this case it is convenient to change to the Hermitian operators J_μ.

8–12 Invariant integration. In order to define characters and derive the orthogonality theorems, we must establish for continuous groups some analog of Eq. (8–3). The essential point in the derivation of Eq. (8–3) was that we attached equal weights to all elements R of the finite group. Thus, if we sum a function over some subset $\mathfrak{M}$, the quantity $\sum_{\mathfrak{M}} f(R)$ is obviously equal to $\sum_{S\mathfrak{M}} f(S^{-1}R)$, where $S\mathfrak{M}$ is the set of elements obtained from $\mathfrak{M}$ by left translation with the element S. From the equation

$$\sum_{\mathfrak{M}} f(R) = \sum_{S\mathfrak{M}} f(S^{-1}R) \tag{8–102}$$

we find, by choosing $\mathfrak{M}$ to be the whole group G, that

$$\sum_{G} f(R) = \sum_{G} f(S^{-1}R). \tag{8–3}$$

For a Lie group we must find something to replace the statement that the weight attached to an element A is equal to the weight attached to the element BA which is obtained from A by left translation. We wish to associate with a set of elements in the neighborhood of A a *volume (measure)* $d\tau_A$ such that the measure $d\tau_{BA}$ of the set of elements which is obtained from these elements by left translation with B is equal to $d\tau_A$:

$$d\tau_{BA} = d\tau_A. \tag{8–103}$$

Once this *left-invariant measure* is established, the analogs of Eqs. (8–102) and (8–3),

$$\int_{\mathfrak{M}} d\tau_A f(A) = \int_{B\mathfrak{M}} d\tau_{BA} f(B^{-1}A) = \int_{B\mathfrak{M}} d\tau_A f(B^{-1}A), \tag{8–104}$$

$$\int_G d\tau_A f(A) = \int_G d\tau_A f(B^{-1}A) \tag{8–104a}$$

follow immediately. The set of elements $\mathfrak{M}$ in the neighborhood of A have parameter values close to those for A. If we denote the parameters of A by $a: a_1, \ldots, a_r$, the set of elements will occupy a volume in the parameter space which we denote by da. When we make a left translation of this set with the element B having parameters b, the resulting set $B\mathfrak{M}$ will have parameters in the neighborhood of the values $c_k = \phi_k(a, b)$ of Eq. (8–23). Thus the parameters of the set $B\mathfrak{M}$ will occupy a volume dc in the parameter space which can be calculated from Eq. (8–23). To make the measure of the group elements $\mathfrak{M}$ and $B\mathfrak{M}$ the same, we choose a density function $\rho(a)$ such that

$$d\tau_{\mathfrak{M}} = \rho(a)\, da = \rho(c)\, dc = d\tau_{B\mathfrak{M}}. \tag{8–105}$$

The density function ρ is easily found. In the neighborhood of the identity, we can arbitrarily fix the value of $\rho(0)$. The set in the neighborhood of the identity is carried by the left translation with B into a region of the parameter space in the neighborhood of b. From (8–23)

$$b_k = \phi_k(0; b), \tag{8–106}$$

$$db_k = \sum_{l=1}^{r} \left[\frac{\partial \phi_k(a; b)}{\partial a_l}\right]_{a=0} da_l, \tag{8–107}$$

so the volume elements db and da in the parameter space are related by the equation

$$db = \begin{vmatrix} \left.\dfrac{\partial \phi_1(a;b)}{\partial a_1}\right|_{a=0} & \cdots & \left.\dfrac{\partial \phi_r(a;b)}{\partial a_1}\right|_{a=0} \\ \vdots & & \vdots \\ \left.\dfrac{\partial \phi_1(a;b)}{\partial a_r}\right|_{a=0} & \cdots & \left.\dfrac{\partial \phi_r(a;b)}{\partial a_r}\right|_{a=0} \end{vmatrix} da = J(b)\, da. \tag{8–108}$$

Setting

$$\rho(b) = \frac{\rho(0)}{J(b)},$$ (8–109)

we have

$$\rho(b)\,db = \rho(0)\,da.$$ (8–110)

Thus the density function $\rho(b)$ is determined for all b by making a left translation with the element B. Our definition is consistent, for if we go from a volume element in the neighborhood of any parameter values a to some other parameter values c via the transformation with parameters b: $c = \phi(a; b)$, we can perform the operation in two steps, first applying the inverse of the transformation with parameters a (which brings us back to the origin) and then with parameters c which takes us to c. Since our definition holds for each step, it gives a consistent definition for the whole process.

We give some simple examples of the calculation of the density function $\rho(a)$. In the group $x' = x + a$,

$$c = \phi(a; b) = a + b; \qquad \left.\frac{\partial\phi(a; b)}{\partial a}\right|_{a=0} = 1,$$

so the density function is a constant. In integrating a function $f(a)$ over the group, we must form $\int da\,f(a)$.

Next we take the group $x' = ax$:

$$c = \phi(a; b) = ab; \qquad \left.\frac{\partial\phi(a; b)}{\partial a}\right|_{a=1} = b.$$

The density function is $\rho(b) = 1/b$. The same result applies to the group

$$\begin{cases} x' = ax, \\ y' = \dfrac{1}{a}\,y. \end{cases}$$

In fact, we see from our argument that isomorphic groups always have the same density function.

Another example is the group $x' = a_1 x + a_2$ for which

$$c_1 = \phi_1(a; b) = b_1 a_1; \qquad c_2 = \phi_2(a; b) = b_2 + b_1 a_2;$$

$$\left.\frac{\partial\phi_1}{\partial a_1}\right|_{\substack{a_1=1\\a_2=0}} = b_1; \qquad \frac{\partial\phi_1}{\partial a_2} = \frac{\partial\phi_2}{\partial a_1} = 0; \qquad \left.\frac{\partial\phi_2}{\partial a_2}\right|_{\substack{a_1=1\\a_2=0}} = b_1; \qquad J(b) = b_1^2.$$

The density function is $\rho(b) = 1/b_1^2$; in performing integrations, we form $\int (da_1\, da_2/a_1^2)$.

For the rotation group in two dimensions, $c = a + b$ (the parameter is the angle of rotation), so the density is uniform. To integrate over the group we form $\int d\phi$.

Problem. Calculate the density function for the linear group in two dimensions.

There is a peculiarity about our definition of the density function in the case of the group $x' = a_1 x + a_2$. We defined the density so that if $c = \phi(a; b)$, then $\rho(c)\, dc = \rho(a)\, da$. In terms of group operations, our definition was based on left translations with a group element b. We found the density function $1/a_1^2$. Suppose that instead we had defined our density function by using right translations, i.e., by making $\rho(c)\, dc = \rho(a)\, da$ where now $c = \phi(b; a)$. This is easily done and gives $\rho(a) = 1/a_1$. In other words, if we define a density function so that the measure is invariant under left translation, it will *not* be invariant under right translation. Thus, in general, we find a *left-invariant measure* and a *right-invariant measure* for the group, and these two measures are not the same.

For compact groups the two measures are the same, and we can therefore establish a single invariant measure on the group. To prove this we consider a set $\mathfrak{M}$ in the neighborhood of the identity. We perform a left translation with an element B which takes each element A of $\mathfrak{M}$ into BA and moves the set $\mathfrak{M}$ into a set $B\mathfrak{M}$ in the neighborhood of B. Next we make a right translation with B^{-1}, bringing the element BA to $A' = BAB^{-1}$ and the set $B\mathfrak{M}$ to $B\mathfrak{M}B^{-1}$. The set $B\mathfrak{M}B^{-1}$ is again in the neighborhood of the identity, and the parameters of A' are obtained from those of A by a linear transformation $D(B)$. Thus a linear transformation $D(B)$ is associated with each element B of the group, and we see from $A' = BAB^{-1}$ that the matrices $D(B)$ form a representation of the group.

The two definitions of measure will coincide if the volume of parameter space occupied by $B\mathfrak{M}B^{-1}$ and $\mathfrak{M}$ are equal. This will be the case if the absolute value of the determinant of $D(B)$ is equal to unity. We now prove that this is the case if the group is compact: If the element B is of finite order so that $B^m = E$, then

$$[D(B)]^m = D(B^m) = 1,$$

and $|\det D(B)| = 1$. If B is not of finite order, we consider the sequence of elements B^n $(n = 1, 2, \ldots)$. If the group is compact, this sequence of elements has a limit β which is in the group manifold. The function det $D(B)$ is a continuous function on the compact group and is therefore bounded, so det $D(\beta)$ is finite. Also det $D(\beta) \neq 0$ since the matrices of a representation are nonsingular. If $|\det D(B)| > 1$, the sequence

$$|\det D(B^n)| = |\det D(B)|^n, \tag{8-111}$$

which should approach det $D(\beta)$, will actually tend toward infinity. If $|\det D(B)| < 1$, the sequence (8–111) will tend to zero and not to $|\det D(\beta)|$. Thus for a compact group, $|\det D(B)| = 1$.

8–13 Irreducible representations of Lie groups and Lie algebras. The Casimir operator. If we restrict ourselves to compact Lie groups, the invariant integration established in the preceding section enables us to carry over all the theorems which were derived in Chapter 3 for finite groups. The integral of a continuous function over the compact group is well defined, and we can repeat the proof of Section 3–11 to prove that every representation of a compact group is equivalent to a unitary representation. The proof of the orthogonality relations and the theorems concerning characters also go through as before. One can finally prove that every representation of a compact group is fully reducible to a sum of irreducible representations, all of which have finite dimensions, and that the regular representation contains all irreducible representations.

For groups which are not compact, difficulties arise. For example, the translation group $x' = x + a$ has the representation

$$a \rightarrow \begin{bmatrix} 1 & e^a \\ 0 & 1 \end{bmatrix} \tag{8-112}$$

which is clearly reducible, but does not decompose into a sum of representations. This same difficulty will arise if the Lie group contains an invariant abelian subgroup. For semisimple Lie groups one can prove that every representation of finite degree is fully reducible.

In the process of finding the irreducible representations of a semisimple Lie algebra, a theorem of Casimir is extremely useful. For such algebras, the metric matrix $g_{\mu\nu}$ of Eq. (8–90) is nonsingular and has an inverse $g^{\mu\nu}$,

$$g^{\mu\rho}g_{\rho\nu} = \delta_{\mu\nu}, \tag{8-113}$$

which is clearly also a symmetric matrix. If we denote the operators corresponding to the basis elements of the algebra by X_μ, the Casimir operator is defined as

$$C = g^{\rho\sigma}X_\rho X_\sigma. \tag{8-114}$$

If we take the commutator of C with any operator of the representation, we find

$$[C, X_\tau] = g^{\rho\sigma}[X_\rho X_\sigma, X_\tau]$$

$$= g^{\rho\sigma} X_\rho [X_\sigma, X_\tau] + g^{\rho\sigma}[X_\rho, X_\tau]X_\sigma$$

$$= g^{\rho\sigma} c_{\sigma\tau}^\lambda X_\rho X_\lambda + g^{\rho\sigma} c_{\rho\tau}^\lambda X_\lambda X_\sigma$$

$$= g^{\rho\sigma} c_{\sigma\tau}^\lambda X_\rho X_\lambda + g^{\sigma\rho} c_{\sigma\tau}^\lambda X_\lambda X_\rho$$

$$= g^{\rho\sigma} c_{\sigma\tau}^\lambda [X_\rho X_\lambda + X_\lambda X_\rho].$$

From Eq. (8–93) we find

$$c_{\sigma\tau}^\lambda = g^{\lambda\nu} c_{\nu\sigma\tau}, \tag{8–115}$$

so that

$$[C, X_\tau] = g^{\rho\sigma} g^{\lambda\nu} c_{\nu\sigma\tau} [X_\rho X_\lambda + X_\lambda X_\rho].$$

Since $c_{\nu\sigma\tau} = -c_{\sigma\nu\tau}$, the factor in front of the brackets is antisymmetric under interchange of ρ and λ. The bracket is symmetric in ρ and λ, so the product must be zero,

$$[C, X_\tau] = 0, \tag{8–116}$$

and the operator C therefore commutes with all operators of the representation.

If we now consider an irreducible representation, the operator C commutes with all operators of the representation and, by Schur's lemma, is a multiple of the unit operator. Thus the operator C has a fixed numerical value in a given irreducible representation, and its value can be used to characterize the irreducible representation.

For compact groups $g^{\rho\sigma} = -\delta^{\rho\sigma}$ in a suitable basis, so that the Casimir operator becomes

$$C = \sum_\rho X_\rho^2. \tag{8–117}$$

For the rotation group $O(3)$ with the operators J_μ of Eq. (8–101), the Casimir operator is

$$C = J_1^2 + J_2^2 + J_3^2, \tag{8–118}$$

i.e., the square of the "total angular momentum." It commutes with J_1, J_2, J_3, and its value characterizes the irreducible representation.

In general, more operators are required to characterize an irreducible representation completely. For example, for the group $O(4)$ of Eqs. (8–82), the two operators

$$F = \sum J_\mu^2, \qquad G = \sum K_\mu^2 \tag{8–119}$$

commute with all the operators. In terms of the operators A_μ, B_μ,

$$F = \sum A_\mu^2 + \sum B_\mu^2, \qquad G = \sum A_\mu B_\mu. \qquad (8\text{–}120)$$

Problems. (1) Construct the Casimir operator for the group defined by Eq. (8–76).

(2) Find the analogs of F and G in Eq. (8–120) for the Lorentz group defined by Eq. (8–83) and for the group which leaves $x^2 + y^2 - z^2 - t^2$ invariant.

The number of operators required to give a complete set is equal to the *rank* of the algebra, which is defined as follows: For any element A, we look for all independent solutions of the equation

$$[A, X] = 0. \qquad (8\text{–}121)$$

This equation always has at least the one solution $X = A$. We now vary the element A to reduce the number of independent solutions of (8–121) to a minimum. This minimum number l is called the rank of the algebra, and l operators of the Casimir type are required to characterize an irreducible representation.

8–14 Multiple-valued representations. Universal covering group. In defining representations of continuous groups, we required the matrix elements of the representation to be continuous functions on the group manifold. Among the continuous functions defined on the group G, there may be functions which are multiple-valued. Thus the possibility of multiple-valued representations arises. A representation of G will be called an m-valued representation if m different operators $D_1(R), \ldots, D_m(R)$ are associated with each element of the group, and all these operators must be retained if the representation is to be continuous.

We could of course have associated several operators with each element of a finite or discrete group, but the absence of any continuity requirement would permit us to split them into sets with a single operator for each group element.

For a continuous group we consider a continuous function $f(R)$ defined on the group. (In particular, $f(R)$ may be a matrix element of a representation.) We now move from the point R along some curve on the group manifold: We associate with each value of a real variable τ a point $g(\tau)$ on the group manifold where $g(\tau)$ is a continuous function of τ. For $\tau = 0$, $g(0) = R$, so our curve starts from R. We now consider closed curves, i.e., curves for which $g(1) = R$. We look at the values of $f[g(\tau)]$ along the closed curve. It may happen that, as τ goes from 0 to 1, the continuously varying function f does not return to its original value. We now take all possible closed curves $g(\tau)$. If, on our return to R, we find m different values of the function f, we say that the function f is *m-valued*.

It is clear that we can always select the function f so that it is single-valued, since we can choose $f(R) = 1$ for all R. What we are interested in is the maximum possible multivaluedness of continuous functions on the group. This number is a property of the group manifold or, for a Lie group, of the parameter space. If the closed curve $g(\tau)$ can be varied continuously so that it contracts to the point R, the continuous function f must return to its original value. If this is the case for all closed curves on the group, the group manifold is simply connected, and every continuous function on the group must be single-valued.

The closed curve $g(\tau)$ can be contracted continuously to the point R if there is a sequence of curves $g(\tau, \lambda)$ where g is a continuous function of the variables τ and λ such that $g(\tau, 0) = g(\tau)$ and $g(\tau, 1) = R$. Similarly, two curves $g_1(\tau)$ and $g_2(\tau)$ can be continuously deformed into each other if $g_1(\tau) = g(\tau, 0)$ and $g_2(\tau) = g(\tau, 1)$. If there are m closed curves which cannot be deformed into one another, the manifold is *m-connected*, and m-valued continuous functions can exist.

We shall have to examine the connectivity of individual continuous groups when we study them in detail. Here we give a few simple examples to illustrate the preceding definitions. The rotation group in two dimensions is parametrized by the angle of rotation ϕ, so the group manifold consists of the points on a circle, i.e., a two-dimensional sphere. The function $f(\phi) = e^{il\phi}$, with l a real number, is a continuous function on the group. If l is an integer, the function f is single-valued; if l is a rational number $= s/t$ in its lowest terms, f is t-valued; if l is irrational, the function f is infinitely multivalued. The space consisting of the points on a circle is infinitely connected: Curves like $\phi = g(\tau) = \tau(1 - \tau)$ are closed curves which can be contracted to a point. The curve $\phi = g(\tau) = 2\pi\tau$ is a single closed loop and cannot be deformed continuously to a point. The sequence of closed curves $\phi = g_n(\tau) = 2\pi n\tau$ are closed curves of n loops and cannot be deformed into one another.

As another example, we show that the n-dimensional euclidean space is simply connected. Any closed curve starting from the origin can be written as $\mathbf{r} = \mathbf{r}(\tau)$, where $\mathbf{r}$ is the vector from the origin to a point on the curve and $\mathbf{r}(0) = \mathbf{r}(1) = 0$. The family of curves $\mathbf{r}(\tau, \lambda) = (1 - \lambda)\mathbf{r}(\tau)$ deforms the closed curve continuously from $\mathbf{r}(\tau)$ for $\lambda = 0$ to the point $\mathbf{r} = 0$ for $\lambda = 1$. We also note that the removal of individual points from the space does not change the connectivity (provided the dimension of the space is greater than one).

If we can set up a one-to-one continuous correspondence between the points of two spaces, they have the same connectivity: If the points R'

are given in terms of the points R by the function $R' = h(R)$, then a curve $R = g(\tau)$ determines a corresponding curve $R' = h[g(\tau)]$.

Problem. Show that for $n > 2$, the n-dimensional sphere $\sum_{i=1}^{n} x_i^2 = 1$ is simply connected. [Hint: Use the stereographic projection to set up a one-to-one correspondence with the $(n - 1)$-dimensional euclidean space.]

If the group manifold is m-connected, we may expect that some of the irreducible representations will be m-valued. The orthogonality theorems and the properties of characters were derived on the tacit assumption that we were dealing with single-valued functions, and they will not be valid if some of the representations are multiple-valued. On the other hand, these multiple-valued representations cannot simply be ignored since they are important for many physical problems. This difficulty is overcome by considering the *universal covering group.* It can be shown that for any multiply connected group G there exists a simply connected group $\widetilde{G}$ (the universal covering group of G) such that $\widetilde{G}$ can be mapped homomorphically on G: The group $\widetilde{G}$ contains a *discrete* invariant subgroup N such that G is isomorphic to $\widetilde{G}/N$.

For example, if G is the two-dimensional rotation group, $\widetilde{G}$ is the group of the real numbers x, with addition as the law of combination. All the numbers x are shifted by multiples of 2π to bring them into the interval $0 - 2\pi$. The homomorphism is $x \to \phi = x - 2\pi[x/2\pi]$, $-\infty < x < \infty$, $0 \leq \phi \leq 2\pi$, where $[x/2\pi]$ is the largest integer contained in $x/2\pi$. The functions $e^{il\phi}$ will be single-valued functions on $\widetilde{G}$.

Every irreducible representation of the group G (whether single-valued or multiple-valued) is a single-valued representation of $\widetilde{G}$. To find all the irreducible representations of the group G, we study the group $\widetilde{G}$. For the simply connected group $\widetilde{G}$ all representations are single-valued, so that the orthogonality and completeness theorems are valid.

CHAPTER 9

AXIAL AND SPHERICAL SYMMETRY

Two of the most important symmetries which occur in physical problems are axial and spherical symmetry. To treat such problems we must find the representations of the real orthogonal group in two and three dimensions.

9–1 The rotation group in two dimensions. We consider the group of rotations about a fixed axis (z-axis) i.e., the two-dimensional pure rotation group: $\mathcal{C}_\infty$. The transformations are labeled by a single continuous parameter ϕ, the angle of rotation measured from, say the x-axis:

$$x' = x \cos \phi - y \sin \phi,$$
$$y' = x \sin \phi + y \cos \phi. \tag{9-1}$$

As we showed in Chapter 8, this group is abelian, the volume element is $d\phi$, and the infinitesimal operator is $J_z = x(\partial/\partial y) - y(\partial/\partial x) = \partial/\partial \phi$. Since the group is abelian, all irreducible representations are one-dimensional, so that the matrices are the same as the characters. Thus for any two angles ϕ_1 and ϕ_2, the characters must satisfy

$$\chi(\phi_1 + \phi_2) = \chi(\phi_1)\chi(\phi_2). \tag{9-2}$$

If we require the representation to be continuous, the solution must be of the form
$$\chi^{(m)}(\phi) = e^{im\phi}. \tag{9-3}$$

If, in addition, we want the representation to be single-valued, we must have $\chi(2\pi) = \chi(0)$, so that $m = 0, 1, 2$, etc. [If we admit 2-, 3-, ... -valued representations, then we could also obtain $\chi(\phi) = e^{im\phi/2}, e^{im\phi/3}, \ldots$ etc. All these representations are admissible for the two-dimensional rotation group. However if we start from some three-dimensional physical problem and go to the subgroup of rotations in two dimensions, only the one- and two-valued representations will occur.] These results also follow from the customary quantum-mechanical method using the infinitesimal operator $\partial/\partial \phi$. Since the representation is one-dimensional, we must have $\partial \chi/\partial \phi = \lambda \chi$, so (9-3) results.

The orthogonality theorem takes the form

$$\int_0^{2\pi} d\phi \chi^{*(m)}(\phi)\chi^{(m')}(\phi) = 2\pi \, \delta_{mm'}, \tag{9-4}$$

and we see from the theory of Fourier expansions that the characters of the single-valued representations form a complete set; there are no other inequivalent irreducible representations which are single-valued. The basis function for the representation $D^{(m)}$ is $\psi^{(m)} = e^{-im\phi}$. Every single-valued (periodic) function can be expanded in terms of these basis functions. We note that the characters of $D^{(m)}$ and $D^{(-m)}$ are complex conjugates. We should therefore (cf. p. 118) combine these into a two-dimensional representation.

If we consider a linear molecule in the approximation where the nuclei are kept fixed on the axis of the molecule, the electrons move in the axially symmetric field of the nuclei. The electronic states of the molecule are classified according to the irreducible representations of the axial symmetry group. States with the quantum number $\Lambda = |m| = 0$ are called Σ-states; those with $\Lambda = 1$ are called Π-states; $\Lambda = 2$ gives Δ-states, etc. The field produced by the nuclei is not only axially symmetric but is also invariant under reflection (σ_v) in any plane passing through the axis of the molecule. Thus the symmetry group of physical interest is the group $\mathfrak{C}_{\infty v}$. This group can be obtained from $\mathfrak{C}_\infty$ by adjoining the reflection in the xz-plane, σ_v. (Note that our procedure is identical with that used in Chapter 4.) The reflection changes the sign of ϕ, so that

$$\sigma_v e^{i\Lambda\phi} = e^{-i\Lambda\phi}, \qquad \sigma_v e^{-i\Lambda\phi} = e^{i\Lambda\phi}. \tag{9-5}$$

Thus, for $\Lambda \neq 0$, the representations $D^{(\Lambda)}$ and $D^{(-\Lambda)}$ are combined into a two-dimensional irreducible representation. The group is no longer abelian. The rotations with ϕ and $-\phi$ form a class. For $\Lambda \neq 0$, the matrices of the Λ-representation are (using basis functions $e^{\pm i\Lambda\phi}$)

$$C(\phi): \begin{bmatrix} e^{i\Lambda\phi} & 0 \\ 0 & e^{-i\Lambda\phi} \end{bmatrix}; \qquad \sigma_v: \begin{bmatrix} 0 & 1 \\ 1 & 0 \end{bmatrix}. \tag{9-6}$$

The matrices for other reflections can be obtained by transforming by a suitable rotation. Since all vertical reflection planes are equivalent (they can be brought into coincidence by a rotation), the reflections are all in the same class and have the same character. Also, from (9-6) we see that $\chi^{(\Lambda)}(\phi) = 2 \cos \Lambda\phi$. For $\Lambda = 0$, the reflection does not change the form of the basis function, but since $(\sigma_v)^2$ is the identity, σ_v can only multiply the basis function by ± 1. Thus the one-dimensional Σ-representation of $\mathfrak{C}_\infty$ gives rise to two one-dimensional representations of $\mathfrak{C}_{\infty v}$, Σ^+ and Σ^-, which are invariant under rotation and even and odd, respectively, under reflection. The character table is given in Table 9-1.

If the linear molecule is symmetric about its mass center (for example, if we have a homonuclear diatomic molecule), then the potential acting on the electrons is also invariant under reflection in the plane which

TABLE 9–1

$\mathcal{C}_{\infty v}$	E	$C(\phi)$	σ_v
A_1, Σ^+: $z; x^2 + y^2; z^2$	1	1	1
A_2, Σ^-: R_z	1	1	-1
E_1, Π: $(x, y); (xz, yz); (R_x, R_y)$	2	$2\cos\phi$	0
E_2, Δ: $(x^2 - y^2, xy)$	2	$2\cos 2\phi$	0
$\vdots$		$\vdots$	

passes through the mass center and is perpendicular to the axis of the molecule (σ_h). We then obtain the symmetry group $D_{\infty h}$. Since the original group contains the rotation through 180° about the vertical axis, the addition of a reflection in the horizontal plane leads to the inversion. We may write $D_{\infty h} = \mathcal{C}_{\infty v} \times \mathcal{C}_i$ or $D_{\infty h} = \mathcal{C}_{\infty v} + I \cdot \mathcal{C}_{\infty v}$. The character table for $D_{\infty h}$ (Table 9–2) is then easily obtained from Table 9–1. The notation g or u is used for states which are even or odd under inversion. The elements $IC(\phi)$ are rotation reflections $S(\phi)$; the elements $I\sigma_v$ are rotations through 180° about horizontal axes.

In Tables 9–1 and 9–2 we have assigned the components of the electric dipole and quadrupole moments to the various irreducible representations. In order to assign the components of an axial vector we consider the prototype, the cross product. The z-component R_z is $xy' - x'y \propto \sin(\phi' - \phi)$. From the second form, we see that R_z is invariant under rotations about the z-axis, and changes sign upon reflection in a vertical plane. From the first form we see that it is invariant under inversion. The y- and x-com-

TABLE 9–2

$D_{\infty h}$	E	$C(\phi)$	σ_v	I	$IC(\phi)$	$I\sigma_v$
Σ_g^+: $x^2 + y^2; z^2$	1	1	1	1	1	1
Σ_u^+: z	1	1	1	-1	-1	-1
Σ_g^-: R_z	1	1	-1	1	1	-1
Σ_u^-: zR_z	1	1	-1	-1	-1	1
Π_g: $(R_x, R_y); (xz, yz)$	2	$2\cos\phi$	0	2	$2\cos\phi$	0
Π_u: (x, y)	2	$2\cos\phi$	0	-2	$-2\cos\phi$	0
Δ_g: $(x^2 - y^2, xy)$	2	$2\cos 2\phi$	0	2	$2\cos 2\phi$	0
Δ_u:	2	$2\cos 2\phi$	0	-2	$-2\cos 2\phi$	0
$\vdots$	$\vdots$		$\vdots$			$\vdots$

ponents R_y, R_x are $xz' - zx'$ and $yz' - zy'$. They involve the basis functions $e^{\pm i\phi}$ and are therefore coupled to each other by rotation about the z-axis or reflection in the vertical axis. Since they contain products of two coordinates, they are invariant under inversion. It is on this basis that the assignments are made to the table.

The selection rules for the various types of transitions can now be calculated by the procedure used in Chapter 4.

Problem. Find the selection rules for electric and magnetic dipole and electric quadrupole transitions for the symmetry groups $C_{\infty v}$ and $D_{\infty h}$.

9–2 The rotation group in three dimensions. As we saw in the previous chapter, the group $O(3)$ of real orthogonal transformations in three dimensions is a three-parameter group. It consists of all transformations with real coefficients which leave $x^2 + y^2 + z^2$ invariant. The matrix A of an orthogonal transformation must satisfy

$$A\widetilde{A} = 1, \tag{9–7}$$

where $\widetilde{A}$ is the transpose of the matrix $A (\widetilde{A}_{ij} = A_{ji})$. Note that a real orthogonal matrix is unitary. Since the determinant of the transposed matrix, $\det \widetilde{A}$, is the same as the determinant of A, we obtain from (9–7):

$$(\det A)^2 = 1 \quad \text{or} \quad \det A = \pm 1. \tag{9–8}$$

We shall first restrict ourselves to the *proper* orthogonal transformations $O^+(3)$ for which $\det A = +1$. (These correspond to pure rotations.) We consider the problem of finding eigenvalues and eigenvectors for the matrix A. We try to find a vector u and a constant λ such that the effect of the transformation A on u is merely multiplication by λ, leaving its direction unchanged:

$$Au = \lambda u, \quad \sum_{j=1}^{3} A_{ij}u_j = \lambda u_i \quad (i = 1, 2, 3). \tag{9–9}$$

The condition that these equations have a nonvanishing solution is the eigenvalue equation, $\det(A - \lambda 1) = 0$, which when written out is a cubic equation in λ with real coefficients (since A is a real matrix). The roots (eigenvalues) are therefore either real or occur in pairs of complex conjugates. Furthermore, since A is unitary,

$$(u, u) = (Au, Au) = (\lambda u, \lambda u) = |\lambda|^2(u, u),$$

so that $|\lambda| = 1$. Thus for any real orthogonal transformation with determinant $+1$, the eigenvalues are

$$\lambda^{(1)} = 1, \qquad \lambda^{(2)} = e^{i\phi}, \qquad \lambda^{(3)} = e^{-i\phi} \qquad (0 \leq \phi \leq \pi). \quad (9\text{--}10)$$

We shall label the corresponding eigenvectors as $u^{(1)}$, $u^{(2)}$, $u^{(3)}$. The first of these satisfies the equation $Au^{(1)} = u^{(1)}$, so that it is completely unaffected by the transformation, and therefore will correspond to the direction of the axis of rotation. The eigenvectors are mutually unitary-orthogonal:

$$(u^{(i)}, u^{(j)}) = \delta_{ij}. \qquad (9\text{--}11)$$

Since $Au^{(1)} = u^{(1)}$,

$$\tilde{A}u^{(1)} = \tilde{A}Au^{(1)} = u^{(1)},$$

whence

$$(A - \tilde{A})u^{(1)} = 0,$$

or

$$(A_{12} - A_{21})u_2^{(1)} + (A_{13} - A_{31})u_3^{(1)} = 0,$$
$$(A_{21} - A_{12})u_1^{(1)} \qquad\qquad + (A_{23} - A_{32})u_3^{(1)} = 0,$$
$$(A_{31} - A_{13})u_1^{(1)} + (A_{32} - A_{23})u_2^{(1)} \qquad\qquad = 0. \quad (9\text{--}12)$$

Solving these equations, we find

$$u_1^{(1)} : u_2^{(1)} : u_3^{(1)} = (A_{23} - A_{32}) : (A_{31} - A_{13}) : (A_{12} - A_{21}). \quad (9\text{--}13)$$

Problem. Prove that a proper orthogonal transformation (real or complex) in an odd-dimensional space always possesses an axis (i.e., a line whose points are left unchanged).

Equation (9–13) determines the direction of the axis of rotation in terms of the coefficients of the transformation. The eigenvector $u^{(1)}$ can always be chosen to be a real vector. The angle ϕ can also be determined from the matrix A, since the sum of the eigenvalues is the trace of A:

$$A_{11} + A_{22} + A_{33} = 1 + e^{i\phi} + e^{-i\phi} = 1 + 2\cos\phi. \quad (9\text{--}14)$$

Since

$$Au^{(2)} = e^{i\phi}u^{(2)}, \qquad A^{*}u^{(2)^{*}} = Au^{(2)^{*}} = e^{-i\phi}u^{(2)^{*}},$$

it follows that

$$u^{(3)} = u^{(2)^{*}}. \qquad (9\text{--}15)$$

If we choose the (complex) vectors $u^{(1)}$, $u^{(2)}$, $u^{(3)}$ as coordinate vectors, the matrix A becomes the diagonal matrix

$$\Lambda = \begin{bmatrix} 1 & 0 & 0 \\ 0 & e^{i\phi} & 0 \\ 0 & 0 & e^{-i\phi} \end{bmatrix}. \tag{9-16}$$

The matrix which transforms A to the diagonal form Λ can be found as follows: Start from the eigenvalue equation

$$\sum_j A_{ij} u_j^{(k)} = \lambda^{(k)} u_i^{(k)} = \sum_j u_i^{(j)} (\lambda^{(k)} \delta_{jk}). \tag{9-17}$$

Define the matrices U and Λ:

$$U_{jk} = u_j^{(k)}, \qquad \Lambda_{jk} = \lambda^{(k)} \delta_{jk}. \tag{9-18}$$

Thus Λ is just the diagonal matrix of (9–16), while U is a matrix whose columns are the components of the eigenvectors. From (9–11) we see that U is a unitary matrix. Introducing the definitions (9–18) in (9–17), we have $AU = U\Lambda$, and multiplying by the inverse of U (it has an inverse since it is unitary) yields $U^{-1}AU = \Lambda$. So the matrix U_{jk} brings A to the diagonal form Λ. Our new coordinate vectors $u^{(1)}$, $u^{(2)}$, $u^{(3)}$ are complex. Since we are interested in transformations of real coordinates, it is convenient to go over to real coordinate vectors. (Of course, this means that now our transformation matrix will no longer be diagonal.) This can be done by means of the unitary matrix V,

$$V = \begin{bmatrix} 1 & 0 & 0 \\ 0 & \dfrac{1}{\sqrt{2}} & \dfrac{i}{\sqrt{2}} \\ 0 & \dfrac{1}{\sqrt{2}} & \dfrac{-i}{\sqrt{2}} \end{bmatrix}, \tag{9-19}$$

and $V^{-1}\Lambda V$ is

$$\begin{bmatrix} 1 & 0 & 0 \\ 0 & \dfrac{1}{\sqrt{2}} & \dfrac{1}{\sqrt{2}} \\ 0 & \dfrac{-i}{\sqrt{2}} & \dfrac{i}{\sqrt{2}} \end{bmatrix} \begin{bmatrix} 1 & 0 & 0 \\ 0 & e^{i\phi} & 0 \\ 0 & 0 & e^{-i\phi} \end{bmatrix} \begin{bmatrix} 1 & 0 & 0 \\ 0 & \dfrac{1}{\sqrt{2}} & \dfrac{i}{\sqrt{2}} \\ 0 & \dfrac{1}{\sqrt{2}} & \dfrac{-i}{\sqrt{2}} \end{bmatrix}$$

$$= \begin{bmatrix} 1 & 0 & 0 \\ 0 & \cos\phi & -\sin\phi \\ 0 & \sin\phi & \cos\phi \end{bmatrix} = R, \tag{9-20}$$

which we recognize as a rotation about the first axis through the angle ϕ; thus the parameter ϕ which appears in the eigenvalues is the angle of rotation. The matrix A was transformed to the final form by the transformation UV. From (9–19),

$$UV = \begin{bmatrix} u_1^{(1)} & \dfrac{1}{\sqrt{2}}\,(u_1^{(2)} + u_1^{(3)}) & \dfrac{i}{\sqrt{2}}\,(u_1^{(2)} - u_1^{(3)}) \\[2mm] u_2^{(1)} & \dfrac{1}{\sqrt{2}}\,(u_2^{(2)} + u_2^{(3)}) & \dfrac{i}{\sqrt{2}}\,(u_2^{(2)} - u_2^{(3)}) \\[2mm] u_3^{(1)} & \dfrac{1}{\sqrt{2}}\,(u_3^{(2)} + u_3^{(3)}) & \dfrac{i}{\sqrt{2}}\,(u_3^{(2)} - u_3^{(3)}) \end{bmatrix}. \qquad (9\text{–}21)$$

Thus UV is the product of two unitary transformations and is therefore unitary. Furthermore, from (9–21) and (9–15) we see that UV is real, so that UV is a real orthogonal matrix. Hence we have shown that any real matrix with angle ϕ can be transformed into any other matrix with the same ϕ by a real orthogonal transformation (i.e., by a rotation). Rotations through the same angle about any axis are equivalent and belong to the same class.

The direction of the axis of rotation and the angle ϕ provide us with three parameters for characterizing the rotation. We can take as parameters the components $u_1^{(1)}\phi$, $u_2^{(1)}\phi$, $u_3^{(1)}\phi$ of the vector whose length is ϕ and whose direction is that of the axis of rotation. The parameters of the rotation group are thus the points inside a sphere of radius π about the origin. To each point inside this sphere there corresponds the rotation, through an angle equal to the distance from the origin, about the direction of the radius vector drawn from the origin. We fix the sense of rotation according to the right-hand rule. These parameters give a unique assignment except for $\phi = \pi$. Two diametrically opposite points give the same rotation. Hence we must picture the sphere as being sewed onto itself, so that the points at opposite ends of a diameter coincide.

The density function for group integration can be calculated in the manner discussed in the previous chapter. Our parameters are the three cartesian coordinates of points inside a sphere of radius π or polar coordinates inside this sphere. The range of variation of the parameters is finite, so that the definitions of density by left and right translations are necessarily the same. The parameters of the identity are 0, 0, 0. If we consider the parameters ξ, η, ζ in the neighborhood of the identity, the matrix of the rotation corresponding to these infinitesimal values of the parameters is

$$S = \begin{bmatrix} 1 & -\zeta & \eta \\ \zeta & 1 & -\xi \\ -\eta & \xi & 1 \end{bmatrix}. \qquad (9\text{–}22)$$

We now wish to find the parameters of the rotation which results from following this infinitesimal rotation with a rotation through some angle ϕ. Then we will determine the Jacobian of the new parameters in terms of the ξ, η, ζ and obtain the density function. Since all directions of rotation are equivalent, it is clear that the density function can depend only on ϕ. We may therefore choose our rotation in the simple form of R in (9–20). Then RS is

$$\begin{bmatrix} 1 & -\zeta & \eta \\ \zeta \cos\phi + \eta \sin\phi & \cos\phi - \xi \sin\phi & -\xi \cos\phi - \sin\phi \\ \zeta \sin\phi - \eta \cos\phi & \sin\phi + \xi \cos\phi & -\xi \sin\phi + \cos\phi \end{bmatrix}. \quad (9\text{–}23)$$

Since ξ, η, ζ are small, we retain only first powers throughout. We use (9–14) to find the angle ϕ' of the product transformation,

$$1 + 2 \cos\phi' = 1 + 2 \cos\phi - 2\xi \sin\phi; \qquad \phi' = \phi + \xi, \quad (9\text{–}24)$$

and (9–13) to find the axis of rotation,

$$u_1^{(1)'} = -2\xi \cos\phi - 2 \sin\phi; \qquad u_2^{(1)'} = \zeta \sin\phi - \eta(1 + \cos\phi);$$

$$u_3^{(1)'} = -\eta \sin\phi - \zeta(1 + \cos\phi). \quad (9\text{–}25)$$

The vector $u^{(1)'}$ must be normalized to unity. Its length in (9–25) (to first order) is $2\xi \cos\phi + 2 \sin\phi$, so our normalized $u^{(1)'}$ has components

$$1, \qquad -\frac{\zeta}{2} + \frac{\eta}{2}\frac{(1 + \cos\phi)}{\sin\phi}, \qquad \frac{\eta}{2} + \frac{\zeta}{2}\frac{(1 + \cos\phi)}{\sin\phi}.$$

Combining this with (9–24), we find for the parameters of the product (to first order)

$$\phi + \xi, \qquad \phi\left[-\frac{\zeta}{2} + \frac{\eta}{2}\frac{(1 + \cos\phi)}{\sin\phi}\right], \qquad \phi\left[\frac{\eta}{2} + \frac{\zeta}{2}\frac{(1 + \cos\phi)}{\sin\phi}\right].$$

The Jacobian of the transformation is

$$J = \begin{vmatrix} 1 & 0 & 0 \\ 0 & \dfrac{\phi}{2}\dfrac{(1 + \cos\phi)}{\sin\phi} & \dfrac{-\phi}{2} \\ 0 & \dfrac{\phi}{2} & \dfrac{\phi}{2}\dfrac{(1 + \cos\phi)}{\sin\phi} \end{vmatrix} = \frac{\phi^2}{2(1 - \cos\phi)},$$

so that the density function $\rho(\phi)$ is

$$\rho(\phi) = \frac{2}{\phi^2}(1 - \cos\phi). \quad (9\text{–}26)$$

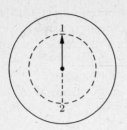

FIGURE 9–1

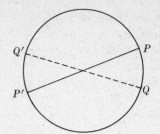

FIGURE 9–3

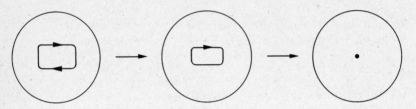

FIGURE 9–2

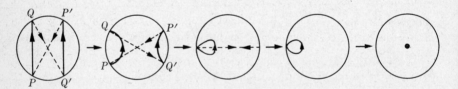

FIGURE 9–4

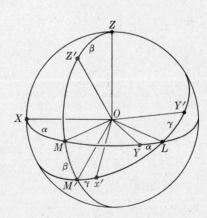

FIGURE 9–5

To integrate a function over the group we must form the integral

$$\int \phi^2 \, d\phi \, d\Omega \left[\frac{2}{\phi^2} (1 - \cos \phi) \right] f(\phi, \Omega) = 2 \int d\phi \, d\Omega f(\phi, \Omega)(1 - \cos \phi),$$

$$(9\text{-}27)$$

where Ω is the direction of the axis of rotation. All rotations through the same angle ϕ belong to the same class. Thus, if we are dealing with a class function and, in particular, the character of a representation, our integral is just

$$8\pi \int_0^\pi d\phi \chi(\phi)(1 - \cos \phi). \tag{9-27a}$$

The total "volume" of the group is $8\pi \int_0^\pi d\phi(1 - \cos \phi) = 8\pi^2$. The orthogonality theorem for the characters of irreducible representations of the rotation group can be written as

$$\frac{1}{\pi} \int_0^\pi d\phi(1 - \cos \phi)\chi^{(\mu)^*}(\phi)\chi^{(\nu)}(\phi) = \delta_{\mu\nu}. \tag{9-28}$$

We stated in Chapter 8 that the group $O^+(3)$ is simple. One can visualize this result by the following pictorial method: The parameter space is the interior of a sphere of radius π. Equal and opposite vectors from the origin correspond to inverse rotations. If $O^+(3)$ contains a proper invariant subgroup, this subgroup must contain some rotation other than the identity, e.g., the rotation indicated as 1 in Fig. 9–1. But if an invariant subgroup contains the rotation 1, it must also contain all transforms of 1, i.e., all the vectors on the sphere represented by the broken line. Thus the invariant subgroup contains the vector 2 (which is inverse to 1) and also a neighborhood of 2. If we take the product of 1 with the vectors on the dashed sphere in the neighborhood of 2, we obtain elements of the subgroup which are arbitrarily close to the identity (the origin of the sphere). Their transforms will fill out a small sphere around the origin. By taking successive products we fill out the whole sphere, so the invariant subgroup coincides with $O^+(3)$.

The connectivity of $O^+(3)$ can also be found in this pictorial fashion. To determine the connectivity of the parameter space we must find the number of types of closed paths which cannot be continuously deformed into one another. Closed curves of the first type can be contracted to a point, as shown in Fig. 9–2. (We draw all the curves as plane curves to avoid confusion.) The second type of closed curve is exemplified by a line along the diameter of the sphere (see Fig. 9–3). The points P, P' are the same point, so this is a closed curve. If we try to deform the curve continuously, any movement of one "end" to the point Q shifts the "other end" to the coincident point Q' which is diametrically opposite to Q. All other closed curves coincide with these types (are *homotopic* to them).

For example, a closed curve which has two places where jumps to antipodal points occur can always be contracted to a point, as shown in Fig. 9–4. Similarly one sees that all closed curves with an odd number of antipodal jumps are homotopic, and all closed curves with an even number are homotopic. Thus the group manifold is doubly connected, and we can have (at most) double-valued representations.

In the preceding discussion, we chose our parameters to give a one-to-one correspondence to the rotations. However, for most considerations this is not necessary, and other parameters are more convenient. The transformation (rotation) which takes the points in three-dimensional space from their initial to their final positions can be described conveniently in terms of the Euler angles α, β, γ. In Fig. 9–5 the coordinate axes OX, OY, OZ are fixed permanently. The transformation is described as the result of three successive, simple operations:

(1) Perform a rotation through angle γ about the Z-axis (γ is positive if the points on the positive X-axis move toward the positive Y-axis, and can vary between $\pm \pi$). As a result of this rotation, the points originally on the line OY are brought to the *line of nodes OL*.

(2) Rotate through an angle β about the line OL, so that points on OZ move toward OM as shown in the figure; β varies between 0 and π. As a result of this operation, points originally on OZ are brought to the line OZ'.

(3) Rotate through α about the line OZ'. The sign convention is the same as in step 1, and again $-\pi \leq \alpha \leq \pi$.

The assignment of parameters to rotations is not unique; thus for $\beta = 0$ we have a rotation about the Z-axis through the angle $\alpha + \gamma$.

We denote the transformation by the symbol $R(\alpha, \beta, \gamma)$. Now we show that it can also be described in terms of rotations about OY and OZ. The first step was a rotation about OZ which we call R_γ. The second rotation through angle β about the line OL will be denoted by R'_β. This rotation is the transform of a rotation R_β through angle β about OY; in fact $R'_\beta = R_\gamma R_\beta R_\gamma^{-1}$ so that $R'_\beta R_\gamma = R_\gamma R_\beta$. Similarly, if R'_α represents the rotation through angle α about OZ', and R_α the rotation through angle α about OZ, we find $R'_\alpha R'_\beta R_\gamma = R_\gamma R_\beta R_\alpha$. Thus an alternative description of $R(\alpha, \beta, \gamma)$ is:

(1) Rotate about OZ through α. (2) Rotate about OY through β. (3) Rotate about OZ through γ. As usual, we associate with the transformation R, $x' = Rx$, an operator O_R such that $O_R f(x) = f(R^{-1}x)$, whence

$$O_{R(\alpha,\beta,\gamma)} f(x) = f[R(-\gamma, -\beta, -\alpha)x]. \qquad (9\text{--}29)$$

Problem. Find the matrix of the transformation $R(\alpha, \beta, \gamma)$.

9–3 Continuous single-valued representations of the three-dimensional rotation group.   The continuous single-valued representations of the rotation group are actually well known to us in a different guise. Consider the Laplace equation

$$\nabla^2 \psi = 0. \tag{9-30}$$

The Laplacian is invariant under rotations, so that if ψ is a solution, then $O_R \psi$ is also a solution. We consider the solutions ψ which are homogeneous polynomials of degree l in x, y, z. Since the rotation R is a linear transformation of x, y, z, $O_R \psi$ will again be a homogeneous polynomial of degree l. Thus homogeneous polynomials of degree l which satisfy Laplace's equation are transformed among themselves by rotations and therefore give a basis for a representation of the rotation group. To find the degree of the representation based on polynomials of degree l, we must determine the number of independent polynomials. The most general polynomial of degree l in x, y, z can be written as

$$P = \sum_{a,b} c_{ab}(x + iy)^a (x - iy)^b z^{l-a-b}. \tag{9-31}$$

If this polynomial is to satisfy the Laplace equation $\nabla^2 P = 0$, then

$$0 = \sum_{a,b} c_{ab}[4ab(x + iy)^{a-1}(x - iy)^{b-1} z^{l-a-b}$$
$$+ (l - a - b)(l - a - b - 1)(x + iy)^a (x - iy)^b z^{l-a-b-2}],$$

and the coefficients must satisfy the recursion formula

$$4(a + 1)(b + 1)c_{a+1,b+1} + (l - a - b)(l - a - b - 1)c_{ab} = 0. \tag{9-32}$$

From (9–32) we see that coefficients c_{ab} with a fixed difference $a - b$ between subscripts are related to one another and can all be expressed in terms of a single coefficient of this type. Since $a - b$ can take on all values from $-l$ to $+l$, there are $2l + 1$ independent homogeneous lth-degree polynomial solutions of the Laplace equation. When expressed in polar coordinates r, θ, ϕ, the lth-degree polynomial is r^l times a function of θ and ϕ. These angle functions are the familiar spherical harmonics which we shall choose in the form $Y_m^l(\theta, \phi) = (1/\sqrt{2\pi})P_m^l(\theta)e^{im\phi}$ $(m = -l, \ldots, +l)$, and, for $m \geq 0$,

$$P_{-m}^l(\theta) = P_m^l(\theta) = \frac{N_m^l}{2^l l!} \sin^m \theta \frac{d^{l+m} \sin^{2l} \theta}{d(\cos \theta)^{l+m}}, \tag{9-33}$$

where $N_m^l = \{[(l - m)!/(l + m)!][(2l + 1)/2]\}^{1/2}$ is a normalization factor chosen so that

$$\int_0^\pi (P_m^l)^2 \sin \theta \, d\theta = 1. \tag{9-33a}$$

The following are normalized spherical harmonics $P_m^l(l = 0, \ldots, 4)$:

$$P_0^0 = 1;$$

$$P_0^1 = \sqrt{\tfrac{3}{2}} \cos \theta, \qquad P_1^1 = \sqrt{\tfrac{3}{4}} \sin \theta;$$

$$P_0^2 = \sqrt{\tfrac{5}{2}} \, (\tfrac{3}{2} \cos^2 \theta - \tfrac{1}{2}), \qquad P_1^2 = \sqrt{\tfrac{15}{4}} \sin \theta \cos \theta,$$

$$P_2^2 = \frac{\sqrt{15}}{4} \sin^2 \theta;$$

$$P_0^3 = \sqrt{\tfrac{7}{2}} \, (\tfrac{5}{2} \cos^3 \theta - \tfrac{3}{2} \cos \theta), \qquad P_1^3 = \sqrt{\tfrac{21}{2}} \, (\tfrac{5}{4} \cos^2 \theta - \tfrac{1}{4}) \sin \theta,$$

$$P_2^3 = \frac{\sqrt{105}}{4} \sin^2 \theta \cos \theta, \qquad\qquad P_3^3 = \frac{\sqrt{70}}{8} \sin^3 \theta;$$

$$P_0^4 = \sqrt{\tfrac{9}{2}} \, (\tfrac{35}{8} \cos^4 \theta - \tfrac{15}{4} \cos^2 \theta + \tfrac{3}{8}),$$

$$P_1^4 = 3 \, \frac{\sqrt{10}}{8} \, (7 \cos^3 \theta - 3 \cos \theta) \sin \theta,$$

$$P_2^4 = 3 \, \frac{\sqrt{5}}{8} \, (7 \cos^2 \theta - 1) \sin^2 \theta, \qquad P_3^4 = 3 \, \frac{\sqrt{70}}{8} \cos \theta \sin^3 \theta,$$

$$P_4^4 = 3 \, \frac{\sqrt{35}}{16} \sin^4 \theta.$$

From (9-33) we see that

$$P_0^l = N_0^l, \qquad P_m^l = 0 \qquad \text{for} \quad m \neq 0. \tag{9-34}$$

For a given l, the spherical harmonics give the basis for a $(2l + 1)$-dimensional representation $D^{(l)}$ of the rotation group. The rows of the representation are labeled by the index m, from $-l$ to $+l$. We shall use the Euler angles as parameters to characterize rotations and write the elements of the rotation group as $R(\alpha, \beta, \gamma)$. Thus $R(\alpha, 0, 0)$ is a rotation through angle α about the Z-axis. These special rotations leave θ unchanged and replace ϕ by $\phi + \alpha$. Hence, using Eq. (9-29), we have

$$O_{R(\alpha,0,0)} Y_m^l = O_{R(\alpha,0,0)} \frac{P_m^l}{\sqrt{2\pi}} \, (\theta) e^{im\phi} = \frac{P_m^l(\theta)}{\sqrt{2\pi}} \, e^{im(\phi-\alpha)} = e^{-im\alpha} Y_m^l. \tag{9-35}$$

In general,

$$O_{R(\alpha,\beta,\gamma)} Y_m^l(\theta, \phi) = \sum_{m'} Y_{m'}^l(\theta, \phi) \, D_{m'm}^{(l)}(\alpha, \beta, \gamma), \tag{9-36}$$

where $D^{(l)}_{m'm}(\alpha, \beta, \gamma)$ is the matrix of $R(\alpha, \beta, \gamma)$ in the representation $D^{(l)}$ based on the spherical harmonics of order l. From (9–35) and (9–36) we see that the matrices representing rotations about the Z-axis are diagonal:

$$D^{(l)}_{m'm}(\alpha, 0, 0) = e^{-im\alpha} \delta_{m'm},$$

$$D^{(l)}(\alpha, 0, 0) = \begin{bmatrix} e^{-il\alpha} & & & \\ & e^{-i(l-1)\alpha} & & \\ & & \ddots & \\ & & & e^{il\alpha} \end{bmatrix}. \quad (9\text{–}37)$$

We wish to prove that the representations $D^{(l)}$ defined above are irreducible. To do this we need only show that any matrix which commutes with all matrices of the representation is necessarily a multiple of the unit matrix. This can be done by considering the diagonal matrices (9–37) and the matrix of $R(0, \beta, 0)$ which is a rotation through angle β about the Y-axis. Consider the points in the ZX-plane; for such points the azimuth ϕ is zero. The rotation $R(0, \beta, 0)$ takes the point $\theta, 0$ to the point $\theta + \beta, 0$. Thus, again using (9–29), we obtain

$$O_{R(0,\beta,0)}P^l_m(\theta) = P^l_m(\theta - \beta) = \sum_{m'} P^l_{m'}(\theta)\, D^{(l)}_{m'm}(0, \beta, 0). \quad (9\text{–}38)$$

If we set $\theta = 0$, we get

$$P^l_m(-\beta) = \sum_{m'} D^{(l)}_{m'm}(0, \beta, 0)P^l_{m'}(0). \quad (9\text{–}39)$$

Using Eq. (9–34), we find

$$P^l_m(-\beta) = D^{(l)}_{0m}(0, \beta, 0)N^l_0. \quad (9\text{–}40)$$

Since, in general, $P^l_m(-\beta)$ is not zero, we conclude that the elements in the $(m = 0)$-row of the matrix $D^{(l)}(0, \beta, 0)$ are different from zero.

If the matrix A commutes with $D^{(l)}(\alpha, 0, 0)$ of (9–37), then

$$[A\, D^{(l)}(\alpha, 0, 0)]_{mm'} = [D^{(l)}(\alpha, 0, 0)A]_{mm'}; \quad A_{mm'}e^{-im'\alpha} = e^{-im\alpha}A_{mm'};$$

hence A must be a diagonal matrix, $A_{mm'} = a_m \delta_{mm'}$. If A commutes with $D^{(l)}(0, \beta, 0)$, then

$$[A\, D^{(l)}(0, \beta, 0)]_{0k} = [D^{(l)}(0, \beta, 0)A]_{0k} \quad \text{for all} \quad k,$$

or

$$a_0\, D^{(l)}_{0k}(0, \beta, 0) = D^{(l)}_{0k}(0, \beta, 0)a_k.$$

Since $D^{(l)}_{0k}(0, \beta, 0) \neq 0$, $a_0 = a_k$ for all k, all diagonal elements of A are the same, and A is a multiple of the unit matrix. Thus each of the representations $D^{(l)}(l = 0, 1, 2, \ldots)$ is irreducible. Furthermore, their dimensions are different, so they are not equivalent.

Since the character of a rotation depends only on the angle of rotation and not on the direction of the rotation axis, we can find the characters from the simple form of (9–37). We see from (9–37) that

$$\chi^{(l)}(\phi) = \sum_{m=-l}^{l} e^{im\phi} = 1 + 2 \cos \phi + \cdots + 2 \cos l\phi = \frac{\sin (l + \frac{1}{2})\phi}{\sin \phi/2}.$$

(9–41)

We verify that the orthogonality theorem (9–28) is satisfied:

$$\frac{1}{\pi} \int_0^\pi d\phi (1 - \cos \phi) \frac{\sin (l + \frac{1}{2})\phi \sin (l' + \frac{1}{2})\phi}{\sin^2 \phi/2}$$

$$= \frac{2}{\pi} \int_0^\pi d\phi \sin (l + \frac{1}{2})\phi \sin (l' + \frac{1}{2})\phi = \delta_{ll'}. \quad (9\text{–}42)$$

We now show that the representations $D^{(l)}$ form a complete set; i.e., there can be no other independent irreducible representations which are continuous and single-valued, for the character of such a representation, $\chi(\phi)$, would have to be orthogonal to all $\chi^{(l)}(\phi)$,

$$\int_0^\pi d\phi (1 - \cos \phi) \chi^{(l)}(\phi) \chi(\phi) = 0 \qquad \text{for all} \quad l;$$

or, if differences are taken for successive l's,

$$\int_0^\pi d\phi (1 - \cos \phi) [\chi^{(l+1)}(\phi) - \chi^{(l)}(\phi)] \chi(\phi) = 0 \qquad \text{for all} \quad l.$$

But $\chi^{(0)}(\phi) = 1$, and $\chi^{(l+1)}(\phi) - \chi^{(l)}(\phi) = 2 \cos l\phi$, so we would have $\int_0^\pi d\phi [(1 - \cos \phi) \chi(\phi)] \cos l\phi = 0$ for all l. Thus all Fourier coefficients of $(1 - \cos \phi) \chi(\phi)$ would have to vanish. Since the functions $\cos l\phi$ form a complete set in the interval 0 to π, we conclude that $\chi(\phi)$ would be zero. Summarizing, we have shown that the representations $D^{(0)}$, $D^{(1)}, \ldots$, etc., constitute a complete set of (single-valued) irreducible representations. Any single-valued representation can be expressed as a sum of the $D^{(l)}$. For later discussion we may also note that

$$D^{(l)}(\alpha, \beta, \gamma) = D^{(l)}(0, 0, \gamma) \, D^{(l)}(0, \beta, 0) \, D^{(l)}(\alpha, 0, 0);$$

$$D_{m'm}^{(l)}(\alpha, \beta, \gamma) = e^{-im'\gamma} D_{m'm}^{(l)}(0, \beta, 0) e^{-im\alpha}, \qquad (9\text{–}43)$$

and we need only find $D_{m'm}^{(l)}(0, \beta, 0)$ to obtain the complete matrix.

We also note the relation of the spherical harmonics Y_m^l to the coefficients of the representation:

$$O_{R(\phi,\theta,\gamma)} Y_m^l(\theta,\phi) = \sum_{m'} Y_{m'}^l(\theta,\phi)\, D_{m'm}^{(l)}(\phi,\theta,\gamma) = Y_m^l(0,-\gamma),$$

where the first equation is (9–36), and the second is obtained by means of (9–29). From (9–33) and (9–34),

$$Y_m^l(0,-\gamma) = \sqrt{\frac{2l+1}{4\pi}}\, e^{-im\gamma}\, \delta_{0m} = \sqrt{\frac{2l+1}{4\pi}}\, \delta_{0m}.$$

So we have

$$\sum_{m'} Y_{m'}^l(\theta,\phi)\, D_{m'm}^{(l)}(\phi,\theta,\gamma) = \sqrt{\frac{2l+1}{4\pi}}\, \delta_{0m}.$$

Multiplying by $D_{km}^{(l)*}(\phi,\theta,\gamma)$ and summing over m, we find

$$\sqrt{\frac{4\pi}{2l+1}}\, Y_k^l(\theta,\phi) = D_{k0}^{(l)*}(\phi,\theta,\gamma). \qquad (9\text{–}44)$$

9–4 Splitting of atomic levels in crystalline fields (single-valued representations).

Before going on to discuss the double-valued representations of the rotation group, we wish to make some application of our results to the splitting of atomic levels in crystalline fields. (Later we shall work out the same problem for two-valued representations.) First we note that the Hamiltonian of an electron in the central field of an atom is invariant not only under all rotations, but also under inversion in the origin. In other words, the symmetry group of the electron in an atom is the group $O(3)$ obtained by adjoining the inversion I to the group of pure rotations. Since $I^2 = E$, the matrix representative of I can be only plus or minus the identity matrix. Just as in Chapter 4, each representation of the pure rotation group gives rise to two representations of the full rotation-reflection group; in place of $D^{(l)}$ we now have the two representations $D^{(l+)}$ and $D^{(l-)}$:

$$D^{(l+)}(IR(\alpha,\beta,\gamma)) = D^{(l+)}(R(\alpha,\beta,\gamma)) = D^{(l-)}(R(\alpha,\beta,\gamma))$$
$$= -D^{(l-)}(IR(\alpha,\beta,\gamma)). \quad (9\text{–}45)$$

The representations $D^{(l+)}$ in which the matrix of I is $+1$ are called positive representations, while the $D^{(l-)}$ in which the matrix of I is -1 are called negative representations. Similarly, levels belonging to $D^{(l+)}(D^{(l-)})$ are said to be positive (negative) levels. So the single-valued representations of the full rotation group are: $0^+, 0^-; 1^+, 1^-; 2^+, 2^-;$ etc.

TABLE 9–3

$l =$	0	1	2	3	4	5
$\chi^{(l)}(C_2)$	1	−1				
$\chi^{(l)}(C_3)$	1	0	−1			
$\chi^{(l)}(C_4)$	1	1	−1	−1		
$\chi^{(l)}(C_6)$	1	2	1	−1	−2	−1

Each level of the free atom will belong to one of the irreducible representations of the full rotation group (if there is no accidental degeneracy). If the atom is put into a crystal, the electrons will be perturbed by the crystalline field, i.e., by the electric field which is produced at the position of the atom by all the other atoms in the crystal. This electric field will have the symmetry of one of the crystal point groups. Thus we are faced with a perturbation problem like those treated earlier in Chapter 6.

In the present case, the levels of the unperturbed system are classified according to the representations of the full rotation group; e.g., a level belonging to the l^+-representation will be $(2l + 1)$-fold degenerate. When the atom is placed in a crystal, the degenerate level will split into terms belonging to the various irreducible representations of the crystal point group. Just as in Chapter 6, we first find the characters of the elements of the crystal point group in the l^+-representation. Then by using Eq. (3–150) we find the number of times each irreducible representation of the crystal point group is contained in the l^+-representation, and thus determine how the level splits in the crystalline field.

Since the crystal point groups can contain rotations only through angles equal to π, $2\pi/3$, $\pi/2$, and $\pi/3$, we first compute $\chi^{(l)}(\phi)$ for these elements, using Eq. (9–41). Since

$$\chi^{(l)}\left(\frac{2\pi}{n}\right) = \frac{\sin\,(l + \frac{1}{2})(2\pi/n)}{\sin\,(\pi/n)}\,,$$

we see that $l = n$ repeats the result for $l = 0$, so that we need to tabulate the characters only for $0 \le l < n$. The results are given in Table 9–3. We also note that

$$\chi^{(l)}\left(\frac{2\pi}{n}\right) = -\chi^{(n-l-1)}\left(\frac{2\pi}{n}\right) = -\chi^{(mn-l-1)}\left(\frac{2\pi}{n}\right), \qquad (9\text{–}46)$$

where m is any integer.

First we consider a crystal group consisting of pure rotations only; as an example we choose the octahedral group O. From Table 4–19 we

TABLE 9–4

Characters of classes of O in the $(2l+1)$-dimensional representation $D^{(l)}$ of the rotation group						Resolution of $D^{(l)}$ into irreducible representations of O	Number of terms
l	E	C_3	C_4^2	C_2	C_4		
0	1	1	1	1	1	A_1	1
1	3	0	−1	−1	1	F_1	1
2	5	−1	1	1	−1	$E + F_2$	2
3	7	1	−1	−1	−1	$A_2 + F_1 + F_2$	3
4	9	0	1	1	1	$A_1 + E + F_1 + F_2$	4
5	11	−1	−1	−1	1	$E + 2F_1 + F_2$	4
6	13	1	1	1	−1	$A_1 + A_2 + E + F_1 + 2F_2$	6
12	25	1	1	1	1	$2A_1 + A_2 + 2E + 3F_1 + 3F_2$	11

see that O contains the classes

$$E; \quad C_3(8); \quad C_4^2(3); \quad C_2(6); \quad C_4(6).$$

The elements C_4^2 and C_2 are not equivalent within the group O; however, in the full rotation group they are equivalent since they are rotations through the same angle π. We use Table 9–3 to find the characters of these elements in the representation $D^{(l)}$ and then use Eq. (3–150) to find the representations of O into which $D^{(l)}$ is split (Table 9–4). We note that the table exhibits a number of general properties. Since the group O contains rotations C_2, C_3, C_4, whose least common multiple is 12, we see from Table 9–3 that for $l = 12$ the characters of all rotations will be +1. In addition, the identity will have the character $\chi^{(12)}(E) = 25$. Thus for $l = 12$, the characters are the sum of two sets: (1) all characters equal to +1, i.e., the identity representation, and (2) $\chi(E) = 24 =$ order of the group O, all other characters zero; this is the regular representation which we denote by reg. As shown in Section 3–17, each irreducible representation is contained in the regular representation a number of times equal to its dimension. So for the group O, reg $= A_1 + A_2 + 2E + 3F_1 + 3F_2$, and $D^{(12)} = A_1 + $ reg.

Similarly, for $l = 12m$, we obtain the regular representation m times, plus the identity representation. For $k < 12$,

$$D^{(12m+k)} = m \, (\text{reg}) + D^{(k)}. \tag{9–47}$$

Also from Eq. (9–46), we see that since 12 is a multiple of 2, 3, and 4,

$$\chi^{(l)}\left(\frac{2\pi}{n}\right) + \chi^{(11-l)}\left(\frac{2\pi}{n}\right) = 0, \qquad \chi^{(l)}(E) + \chi^{(11-l)}(E) = 24,$$

so that

$$D^{(l)} + D^{(11-l)} = \text{reg.} \tag{9–48}$$

For this reason we needed to tabulate the results in Table 9–4 through $l = 5$ only; for example, for $l = 10$, $D^{(10)} + D^{(1)} = \text{reg}$, so $D^{(10)} = A_1 + A_2 + 2E + 2F_1 + 3F_2$. Another interesting result is that no representation can appear more times in $D^{(l)}(l < 12)$ than it does in the regular representation. This result follows from Eq. (9–48) and the fact that the sums of representations have only positive coefficients.

The levels belonging to the various $D^{(l)}$ are usually denoted as S, P, D, etc., for $l = 0$, 1, 2, etc. We see from Table 9–4 that, under the symmetry group O, P-levels are not split. (This is physically evident, since x, y, and z are equivalent under cubic symmetry.) For $l \geq 2$, all terms are split by the crystalline field.

So far we have not considered inversions. If we take the full rotation-reflection group, our representations are 0^+, 0^-, etc. Since O does not contain the inversion I, $D^{(l\pm)}$ will be split in the same way.

Suppose now that we consider the group $O_h = O \times \mathcal{C}_i$ (cubic holohedral symmetry). In O_h, each representation of O is split into two representations according to how we assign $\chi(I) = \pm 1$. All the results of Table 9–4 can still be used for this case. We need only assign all positive (negative) states of the rotation-reflection group to positive (negative) representations of the crystal point group. For example, in O_h,

$$D^{(4+)} = A_1^+ + E^+ + F_1^+ + F_2^+; \qquad D^{(2-)} = E^- + F_2^-.$$

We can now outline the general procedure. If the crystal group contains the inversion I, we need only consider the invariant subgroup which does not contain I, and then assign positive (negative) states of the full rotation group to positive (negative) representations of the crystal group. Thus, if we are considering the splitting of levels by a crystalline field of symmetry $T_h = T \times \mathcal{C}_i$, we need only study the group T. Since in the full rotation group all axes are two-sided, the elements C_3 and C_3^2 of T have the same character in $D^{(l)}$. The results are given in Table 9–5. Note that in T the representation E is actually the sum of two one-dimensional representations and therefore appears only once in the regular representation. Again we have an equation analogous to Eq. (9–48). For the group T,

$$D^{(l)} + D^{(5-l)} = \text{reg.} \tag{9–49}$$

TABLE 9–5

Characters of classes of T in the $(2l+1)$-dimensional representation $D^{(l)}$ of the rotation group				Resolution of $D^{(l)}$ into irreducible representations of T	Number of terms
l	E	C_2	C_3	C_3^2		
0	1	1	1	1	A	1
1	3	-1	0	0	F	1
2	5	1	-1	-1	$E + F$	2
3	7	-1	1	1	$A + 2F$	3
4	9	1	0	0	$A + E + 2F$	4
5	11	-1	-1	-1	$E + 3F$	4
6	13	1	1	1	$2A + E + 3F = A + \text{reg}$	6

Comparing the table for T with that for O, we see that the results for T could have been obtained from those for O by identifying A_1 and A_2, and F_1 and F_2.

Finally, as indicated in Table 4–19, the group T_d is isomorphic to the group O, so that our results for O can be applied to it. We have now found the splitting of atomic levels in all crystals of the regular system (cf. Section 2–9).

Next we consider the hexagonal system (system VI in Section 2–9). The holohedral group is $D_{6h} = D_6 \times \mathcal{C}_i$, so we need only tabulate for D_6. (The isomorphic groups $\mathcal{C}_{6v}$ and D_{3h} will have the same representa-

TABLE 9–6

Characters of classes of the hexagonal group D_6 in the $(2l+1)$-dimensional representation $D^{(l)}$ of the rotation group							Resolution of $D^{(l)}$ into irreducible representations of the hexagonal group	Number of terms
l	E	C_6^3	C_6^2	C_6	C_2	$C_{2'}$		
0	1	1	1	1	1	1	A_1	1
1	3	-1	0	2	-1	-1	$A_2 + E_1$	2
2	5	1	-1	1	1	1	$A_1 + E_1 + E_2$	3
3	7	-1	1	-1	-1	-1	$A_2 + B_1 + B_2 + E_1 + E_2$	5
4	9	1	0	-2	1	1	$A_1 + B_1 + B_2 + E_1 + 2E_2$	6
5	11	-1	-1	-1	-1	-1	$A_2 + B_1 + B_2 + 2E_1 + 2E_2$	7
6	13	1	1	1	1	1	$2A_1 + A_2 + B_1 + B_2 + 2E_1 + 2E_2 = A_1 + \text{reg}$	9

TABLE 9-7

Characters of classes of the tetragonal group D_4 in the $(2l+1)$-dimensional representation $D^{(l)}$ of the rotation group					Resolution of $D^{(l)}$ into irreducible representations of the tetragonal group	Number of terms
l	E	C_4^2	C_4	C_2	$C_{2'}$		
0	1	1	1	1	1	A_1	1
1	3	−1	1	−1	−1	$A_2 + E$	2
2	5	1	−1	1	1	$A_1 + B_1 + B_2 + E$	4
3	7	−1	−1	−1	−1	$A_2 + B_1 + B_2 + 2E$	5
4	9	1	1	1	1	$2A_1 + A_2 + B_1 + B_2 + 2E = A_1 + \text{reg}$	7

tions.) The elements C_6^3, C_2, $C_{2'}$ are all rotations through the angle π and are all in the same class in the full rotation group; C_6^2 is a rotation through $2\pi/3$. The results are given in Table 9-6. The groups $\mathcal{C}_{6h} = \mathcal{C}_6 \times \mathcal{C}_i$ and $\mathcal{C}_6$ need no special treatment; since they are abelian, each $(2l+1)$-dimensional representation of the rotation-reflection group splits into $2l+1$ simple one-dimensional representations.

Next, we consider the tetragonal system (system V of Section 2-9). The holohedral group is $D_{4h} = D_4 \times \mathcal{C}_i$. We treat D_4 (our results will also apply to the isomorphic groups $\mathcal{C}_{4v}$ and D_{2d}) and obtain Table 9-7. The groups $\mathcal{C}_{4h} = \mathcal{C}_4 \times \mathcal{C}_i$ and $\mathcal{C}_4$ are abelian, so each level splits into simple levels.

Finally, in the rhombic system (system III of Section 2-9), we consider the abelian group V; each term splits completely into $2l+1$ simple terms.

The reader should now be able to treat special groups by the procedure which we have outlined.

Problem. Construct a table to show how levels of the full rotation-reflection group split in a field with symmetry D_{3d}.

9-5 Construction of crystal eigenfunctions.

In Chapter 6 we also considered the problem of finding proper zero-order wave functions for the perturbed problem. The same problem arises here, and is completely solvable by symmetry considerations so long as no representation appears more than once in the resolution of $D^{(l)}$. If the same irreducible representation of the crystal group appears m times in the resolution of $D^{(l)}$, we must solve a secular equation of order m in order to find the proper zero-order wave functions. We illustrate the procedure for finding proper zero-order functions for an electron in crystalline fields of various symmetries. (Some of this work is a repetition of the assignment of dipole and quadrupole moments to representations of the crystal group.)

We choose the free-atom wave functions in the real form

$$\sqrt{2}\, P_m^l(\theta)\ {\cos \atop \sin}\, m\phi.$$

Then all reflections merely multiply the function by ± 1. A rotation through π about the X-axis changes ϕ into $-\phi$, and θ into $\pi - \theta$; we see from the definition of $P_m^l(\theta)$ that this operation multiplies P_m^l by $(-1)^{l-m}$:

$$P_m^l(\pi - \theta) = (-1)^{l-m} P_m^l(\theta). \tag{9–50}$$

It leaves $\cos m\phi$ unchanged and multiplies $\sin m\phi$ by $(-1)^m$. In any case, the rotation through π about the X-axis does not couple different wave functions.

We consider first the tetragonal groups as exemplified by D_4. All elements of the group can be generated from the 4-fold rotation C_4 about the Z-axis and the 2-fold rotation C_2 about the X-axis. As we saw above, C_2 does not couple different wave functions. The rotation C_4 does not affect θ, but it replaces ${\cos \atop \sin}\, m\phi$ by ${\cos \atop \sin}\, m(\phi + \pi/2)$. Thus, for m even, it can at most produce a change of sign, while for m odd, it transforms $\cos m\phi$ and $\sin m\phi$ into each other.

The function P_0^l is invariant under C_4, while C_2 multiplies it by $(-1)^l$; so P_0^l belongs to the representation A_1 (cf. Table 4–16) for even l, and to A_2 for odd l. The same result applies to $\sqrt{2}\, P_{4\mu}^l \cos 4\mu\phi$ ($\mu = 1, 2, \ldots$) since $\cos 4\mu(\phi + \pi/2) = \cos 4\mu\phi$. The function $\sqrt{2}\, P_{4\mu}^l \sin 4\mu\phi$ is also unaffected by C_4; C_2 multiplies it by $(-1)^{l+1}$, so it belongs to A_2 for even l and to A_1 for odd l. Proceeding in this way, we obtain Table 9–8 listing

TABLE 9–8

EIGENFUNCTIONS FOR IRREDUCIBLE REPRESENTATIONS
OF THE TETRAGONAL GROUP

Representation		Eigenfunctions
odd l	even l	
A_2	A_1	$P_0^l;\ \sqrt{2}\, P_{4\mu}^l \cos 4\mu\phi$
A_1	A_2	$\sqrt{2}\, P_{4\mu}^l \sin 4\mu\phi$
B_1	B_2	$\sqrt{2}\, P_{4\mu+2}^l \cos (4\mu + 2)\phi$
B_2	B_1	$\sqrt{2}\, P_{4\mu+2}^l \sin (4\mu + 2)\phi$
E	E	$\sqrt{2}\, P_{2\mu+1}^l\ {\cos \atop \sin}\, (2\mu + 1)\phi$

the assignments of wave functions to the various representations of the tetragonal group. Using the table, we give the eigenfunctions explicitly for the first few values of l:

$$l = 0. \qquad A_1: P_0^0 \qquad\qquad\qquad l = 1. \qquad A_2: P_0^1$$

$$E : \sqrt{2}\, P_1^1 \frac{\cos}{\sin}\, \phi$$

$$l = 2. \qquad A_1: P_0^2 \qquad\qquad\qquad l = 3. \qquad A_2: P_0^3$$

$$B_2: \sqrt{2}\, P_2^2 \cos 2\phi \qquad\qquad B_1: \sqrt{2}\, P_2^3 \cos 2\phi$$

$$B_1: \sqrt{2}\, P_2^2 \sin 2\phi \qquad\qquad B_2: \sqrt{2}\, P_2^3 \sin 2\phi$$

$$E : \sqrt{2}\, P_1^2 \frac{\cos}{\sin}\, \phi \qquad\qquad E : \begin{array}{l} \sqrt{2}\, P_1^3 \frac{\cos}{\sin}\, \phi \\[4pt] \sqrt{2}\, P_3^3 \frac{\cos}{\sin}\, 3\phi \end{array}$$

$$l = 4. \qquad A_1: P_0^4, \quad \sqrt{2}\, P_4^4 \cos 4\phi$$

$$A_2: \sqrt{2}\, P_4^4 \sin 4\phi$$

$$B_2: \sqrt{2}\, P_2^4 \cos 2\phi$$

$$B_1: \sqrt{2}\, P_2^4 \sin 2\phi$$

$$E : \sqrt{2}\, P_1^4 \frac{\cos}{\sin}\, \phi, \quad \sqrt{2}\, P_3^4 \frac{\cos}{\sin}\, 3\phi \qquad (9\text{--}51)$$

For $l < 3$, each irreducible representation appears only once in the resolution, so that we have obtained the proper zero-order wave functions. For $l = 3$, $D^{(3)}$ contains the representation E twice. The proper zero-order functions for A_2, B_1, and B_2 are those tabulated, but for E we must now solve the secular equation. Note that only $\cos \phi$ and $\cos 3\phi$ ($\sin \phi$ and $\sin 3\phi$) are coupled to each other since they belong to the same row of the representation. We shall leave the solution of the secular equation as a problem.

Problem. Solve the secular equation and find the zero-order functions for the E-levels formed for $l = 3$.

Our main result is that the proper zero-order functions are no longer given by symmetry considerations, but depend in detail on the nature of the perturbing field. Similarly, for $l = 4$, we would have to solve a secular equation for the A_1-term and the E-term.

Next, we find the zero-order functions in a crystal having hexagonal symmetry, as exemplified by the group D_6. The argument is similar to

TABLE 9–9

EIGENFUNCTIONS FOR IRREDUCIBLE REPRESENTATIONS
OF THE HEXAGONAL GROUP

Representation		Eigenfunctions
odd l	even l	
A_2	A_1	$P_0^l;\ \sqrt{2}\,P_{6\mu}^l \cos 6\mu\phi$
A_1	A_2	$\sqrt{2}\,P_{6\mu}^l \sin 6\mu\phi$
B_1	B_2	$\sqrt{2}\,P_{6\mu+3}^l \cos (6\mu + 3)\phi$
B_2	B_1	$\sqrt{2}\,P_{6\mu+3}^l \sin (6\mu + 3)\phi$
E_2	E_2	$\sqrt{2}\,P_{6\mu\pm2}^l {\cos \atop \sin} (6\mu \pm 2)\phi$
E_1	E_1	$\sqrt{2}\,P_{6\mu\pm1}^l {\cos \atop \sin} (6\mu \pm 1)\phi$

that used for the tetragonal group. Now $\cos m\phi$ and $\sin m\phi$ are coupled unless m is a multiple of 3. The table of assignments (cf. Table 4–17) is given in Table 9–9. Again, we give the functions for the first few values of l:

$l = 0.\quad A_1: P_0^0$

$l = 1.\quad A_2: P_0^1$

$\qquad\qquad\qquad\qquad\qquad\quad E_1: \sqrt{2}\,P_1^1 {\cos \atop \sin} \phi$

$l = 2.\quad A_1: P_0^2$

$\qquad E_2: \sqrt{2}\,P_2^2 {\cos \atop \sin} 2\phi$

$\qquad E_1: \sqrt{2}\,P_1^2 {\cos \atop \sin} \phi$

$l = 3.\quad A_2: P_0^3$

$\qquad B_1: \sqrt{2}\,P_3^3 \cos 3\phi$

$\qquad B_2: \sqrt{2}\,P_3^3 \sin 3\phi$

$\qquad E_2: \sqrt{2}\,P_2^3 {\cos \atop \sin} 2\phi$

$\qquad E_1: \sqrt{2}\,P_1^3 {\cos \atop \sin} \phi$

$l = 4.\quad A_1: P_0^4$

$\qquad B_2: \sqrt{2}\,P_3^4 \cos 3\phi$

$\qquad B_1: \sqrt{2}\,P_3^4 \sin 3\phi$

$\qquad E_2: \sqrt{2}\,P_2^4 {\cos \atop \sin} 2\phi,\quad \sqrt{2}\,P_4^4 {\cos \atop \sin} 4\phi$

$\qquad E_1: \sqrt{2}\,P_1^4 {\cos \atop \sin} \phi$ \hfill (9–52)

For $l = 4$, we would have to solve the secular equation for the E_2-levels.

The angular distribution of the charge density for a degenerate state is found by taking the sum of the squares of the absolute values of the wave functions belonging to the various values of m which are degenerate with one another, so that the charge density is the sum of all the spherical harmonics belonging to the given l, and is therefore spherically symmetric. In the crystalline field the levels belonging to a given l are separated from one another, so that the charge distribution is no longer spherically symmetric. The asymmetry of the charge distribution will be related to the crystal structure. For the tetragonal group the crystal symmetry first makes itself felt in the P-state. For the A_2-level, the charge is concentrated along the Z-(principal) axis, since the density varies as $\cos^2\theta$; for the E-level the density varies as $\sin^2\theta$ and the charge is concentrated in the XY-plane. In the free atom these two states would be degenerate with each other, but the nonequivalence of Z with X and Y in the crystal lifts the degeneracy. Starting from a D-state, we get the levels B_2 and B_1, with charge densities varying as $\sin^4\theta\cos^2 2\phi$ and $\sin^4\theta\sin^2 2\phi$, respectively. In both cases the charge is concentrated in the equatorial plane, but for B_2 the density is high along the crystal axes X and Y, while for B_1 it is high along the lines $\phi = 45°$ and $135°$. Similar considerations apply to the charge density under hexagonal symmetry. The hexagonal pattern is first shown clearly in the levels resulting from an F-state of the free atom. The wave functions of the B_1- and B_2-levels are both concentrated in the hexagonal base planes; but for B_1 the density varies as $\cos^2 3\phi$ and is high for $\phi = m\pi/3$, while the density for B_2 varies as $\sin^2 3\phi$ and is shifted through $30°$ relative to the pattern for B_1.

Finally we consider the case of cubic symmetry (group O). The essential feature of this symmetry is the equivalence of x, y, and z. The procedure for finding the zero-order functions belonging to various irreducible representations is the following: For a given l, express the functions $r^l P_m^l(\theta) \frac{\cos}{\sin} m\phi$ as polynomials in x, y, z. Perform all possible permutations of x, y, and z. Polynomials which are transformed into one another by these permutations belong to the same irreducible representation. For $l = 0$, we have $P_0^0 = 1$, belonging to the identity representation A_1. For $l = 1$, (P-state), $rP_0^1 = z$ is transformed into x and y by permutations of the coordinates, so the P-level is not split by the crystalline field. To decide whether this set belongs to F_1 or F_2, we consider the effect of the operation C_4, a rotation through $\pi/2$ about the Z-axis, which changes x into y, y into $-x$, z into z; the character is 1, so the functions belong to F_1. For $l = 2$, we start with $r^2 P_0^2(\theta) = \sqrt{\frac{5}{2}}\,(\frac{3}{2}z^2 - \frac{1}{2})$. Permutations generate $\frac{1}{2}(3x^2 - 1)$ and $\frac{1}{2}(3y^2 - 1)$, which can both be expressed as linear combinations of $r^2 P_2^2 \cos 2\phi$ and $r^2 P_0^2$:

$$3x^2 - 1 = 3\sin^2\theta\cos^2\phi - 1 = \tfrac{3}{2}\sin^2\theta + \tfrac{3}{2}\sin^2\theta\cos 2\phi - 1$$
$$= \tfrac{3}{2}\sin^2\theta\cos 2\phi - (\tfrac{3}{2}\cos^2\theta - \tfrac{1}{2}).$$

TABLE 9–10

Term	Representation	Crystal eigenfunction
D_γ	E	$(2\gamma)_1 = P_0^2$ $(2\gamma)_2 = \sqrt{2}\, P_2^2 \cos 2\phi$
D_ϵ	F_2	$(2\epsilon)_1 = \sqrt{2}\, P_2^2 \sin 2\phi$ $(2\epsilon)_2 = \sqrt{2}\, P_1^2 \cos \phi$ $(2\epsilon)_3 = \sqrt{2}\, P_1^2 \sin \phi$
F_β	A_2	$(3\beta) = \sqrt{2}\, P_2^3 \sin 2\phi$
F_δ	F_1	$(3\delta)_1 = P_0^3$ $(3\delta)_2 = \sqrt{2}(\sqrt{\frac{5}{8}}\, P_3^3 \cos 3\phi - \sqrt{\frac{3}{8}}\, P_1^3 \cos \phi)$ $(3\delta)_3 = \sqrt{2}(\sqrt{\frac{5}{8}}\, P_3^3 \sin 3\phi + \sqrt{\frac{3}{8}}\, P_1^3 \sin \phi)$
F_ϵ	F_2	$(3\epsilon)_1 = \sqrt{2}\, P_2^3 \cos 2\phi$ $(3\epsilon)_2 = \sqrt{2}(\sqrt{\frac{3}{8}}\, P_3^3 \cos 3\phi + \sqrt{\frac{5}{8}}\, P_1^3 \cos \phi)$ $(3\epsilon)_3 = \sqrt{2}(\sqrt{\frac{3}{8}}\, P_3^3 \sin 3\phi - \sqrt{\frac{5}{8}}\, P_1^3 \sin \phi)$
G_α	A_1	$(4\alpha) = \sqrt{\frac{7}{12}}\, P_0^4 + \sqrt{\frac{10}{12}}\, P_4^4 \cos 4\phi$
G_γ	E	$(4\gamma)_1 = \sqrt{2}\, P_2^4 \cos 2\phi$ $(4\gamma)_2 = \sqrt{\frac{5}{12}}\, P_0^4 - \sqrt{\frac{14}{12}}\, P_4^4 \cos 4\phi$
G_δ	F	$(4\delta)_1 = \sqrt{2}\, P_4^4 \sin 4\phi$ $(4\delta)_2 = \sqrt{\frac{7}{4}}\, P_1^4 \cos \phi - \sqrt{\frac{1}{4}}\, P_3^4 \cos 3\phi$ $(4\delta)_3 = \sqrt{\frac{7}{4}}\, P_1^4 \sin \phi + \sqrt{\frac{1}{4}}\, P_3^4 \sin 3\phi$
G_ϵ	F	$(4\epsilon)_1 = \sqrt{2}\, P_2^4 \sin 2\phi$ $(4\epsilon)_2 = \sqrt{\frac{1}{4}}\, P_1^4 \cos \phi + \sqrt{\frac{7}{4}}\, P_3^4 \cos 3\phi$ $(4\epsilon)_3 = \sqrt{\frac{1}{4}}\, P_1^4 \sin \phi - \sqrt{\frac{7}{4}}\, P_3^4 \sin 3\phi$

Similarly, if we start from $r^2 P_1^2 \cos \phi \propto zx$, permutations generate xy and yz which are proportional to

$$\sin^2 \theta \sin 2\phi (= P_2^2 \sin 2\phi) \quad \text{and} \quad \sin \theta \cos \theta \sin \phi \; (= P_1^2 \sin \phi),$$

respectively. Applying C_4, we find $C_4(xy) = -xy, C_4(xz) = yz, C_4(yz) = -xz$, so the character of C_4 is -1, and the functions belong to F_2.

We have tabulated the zero-order functions through $l = 4$ (Table 9–10). The notation for the crystal terms and eigenfunctions is that of Bethe.

The charge density again shows the influence of the crystal structure, even in zero order. Thus a D-electron of an atom in a cubic crystal will be in either of the crystal states E or F_2, depending on which has the lower energy. If it is in the E-state, its charge density will have the angular distribution

$$(P_0^2)^2 + 2(P_2^2)^2 \cos^2 2\phi = \tfrac{5}{4}(\tfrac{3}{2} \cos^2 \theta - \tfrac{1}{2})^2 + \tfrac{15}{8} \sin^4 \theta \cos^2 2\phi.$$

This charge density will have its maximum value along the 4-fold crystal axes. It will be zero along the 3-fold axes ($\cos \theta = 1/\sqrt{3}, \phi = \pi/4$). On the other hand, if the electron is in the F_2-state, its charge density must be the complement of that for the E-state (since their sum must be spherically symmetric), so the electron in the F_2-state will have zero charge density along the coordinate directions and a maximum along the 3-fold axes.

Problems. (1) Find the zero-order crystal wave functions for $l = 3$ under cubic symmetry.

(2) Describe the charge density for the cubic crystal states resulting from an F-state of the free atom.

9–6 Two-valued representations of the rotation group. The unitary unimodular group in two dimensions. The representations $D^{(l)}$ of the rotation group are, as we have shown, a complete set if we require our matrices to be single-valued functions of the group parameters. But we know that there are systems of functions whose transformation properties under rotation are different from those of the spherical harmonics which formed the bases for the representations $D^{(l)}$. Our representations were obtained by considering functions of the space coordinates x, y, z. We often deal with physical systems which have additional internal degrees of freedom (spin). Under a space rotation the coordinates of the physical system change, but in addition the transformation affects the internal coordinates. If we rotate through an angle $\delta\phi$ about the Z-axis, any function ψ of the coordinates and the internal variables undergoes an infinitesimal change $\delta\psi$ (proportional to $\delta\phi$). Thus

$$\delta\psi = iJ_z\psi \, \delta\phi, \tag{9–53}$$

where J_z is the Hermitian infinitesimal operator for rotation about the Z-axis. Similarly, we introduce infinitesimal operators J_x, J_y corresponding to rotations about the other coordinate directions. The commutation relations for the J_ν are given by Eq. (8–101). As is shown in texts on quantum mechanics, these commutation relations have the consequence that the eigenvalues of J_ν are a set of numbers running from $-j$ to $+j$ in unit steps, and that j must be integral or half-integral. The representations which we have discussed so far correspond to integral j (integral angular momentum). We now wish to consider the other case where the number of basis functions of the irreducible representation is even. We start from the simplest case of $j = \frac{1}{2}$ since, as we saw when we discussed coupled systems, we can derive all other cases by taking product representations.

We consider a representation in terms of a pair of complex variables u, v. Under space rotations these functions are transformed into linear combinations of u and v:

$$u' = au + bv, \qquad v' = cu + dv. \tag{9–54}$$

The coefficients of the transformation will depend on the particular rotation being considered. From the physical point of view, we want the probability density $|u|^2 + |v|^2$ to be invariant, so we are interested only in unitary transformations. Secondly, we note that if we take two pairs of functions u_1, v_1; u_2, v_2 which are both transformed according to (9–54), the function $u_1 v_2 - u_2 v_1$ is merely multiplied by $(ad - bc)$ as a result of the transformation. Thus $u_1 v_2 - u_2 v_1$ forms the basis for a one-dimensional representation of the rotation group which must coincide with the representation $D^{(0)}$ which we derived earlier. Since the basis function of $D^{(0)}$ is invariant under space rotations, we see that we must require $ad - bc = 1$. We are dealing with the group of *unitary unimodular* transformations in two dimensions, the group $\mathfrak{U}_2$. [Another symbol for this group is $SU(2)$.] For such transformations,

$$aa^* + bb^* = 1, \qquad cc^* + dd^* = 1,$$
$$a^*c + b^*d = 0, \qquad ad - bc = 1. \tag{9–55}$$

Solving the last equation for d and substituting in the others, we find

$$d = -\frac{a^*c}{b^*}, \qquad \frac{-aa^*c}{b^*} - bc = 1; \qquad c = -b^*, \qquad d = a^*.$$

We are led to consider the group $\mathfrak{U}_2$ of transformations,

$$\begin{aligned} u' &= au + bv \\ v' &= -b^*u + a^*v \end{aligned} \qquad (aa^* + bb^* = 1), \tag{9–56}$$

which has three independent real parameters. Now we must find the ir-reducible representations of the unitary group and then show how the representations of $\mathfrak{U}_2$ enable us to obtain representations of the rotation group.

The matrices (9–56) already provide us with one representation of $\mathfrak{U}_2$; i.e., the representation in terms of the matrices of $\mathfrak{U}_2$ themselves. We can construct other representations by taking symmetric product representations of this representation with itself. The symmetric products u^2, uv, v^2 ($= x_1, x_2, x_3$), being homogeneous polynomials in u and v, are transformed among themselves by the transformations (9–56). From (9–56), by taking products, we find

$$
\begin{aligned}
x_1' &= a^2 x_1 + 2ab x_2 + b^2 x_3, \\
x_2' &= -ab^* x_1 + (aa^* - bb^*) x_2 + a^* b x_3, \\
x_3' &= b^{*2} x_1 - 2a^* b^* x_2 + a^{*2} x_3.
\end{aligned}
\tag{9–57}
$$

We now rewrite these equations in terms of $x_1 \pm x_3$, x_2:

$$
\begin{aligned}
x_1' - x_3' &= (a^2 - b^{*2}) x_1 + 2(ab + a^* b^*) x_2 + (b^2 - a^{*2}) x_3 \\
&= \tfrac{1}{2}(a^2 - b^{*2} + b^2 - a^{*2})(x_1 + x_3) + 2(ab + a^* b^*) x_2 \\
&\quad + \tfrac{1}{2}(a^2 - b^{*2} - b^2 + a^{*2})(x_1 - x_3),
\end{aligned}
$$

$$
\begin{aligned}
x_1' + x_3' &= \tfrac{1}{2}(a^2 + a^{*2} + b^2 + b^{*2})(x_1 + x_3) + 2(ab - a^* b^*) x_2 \\
&\quad + \tfrac{1}{2}(a^2 - a^{*2} + b^{*2} - b^2)(x_1 - x_3),
\end{aligned}
$$

$$
\begin{aligned}
x_2' &= \tfrac{1}{2}(a^* b - ab^*)(x_1 + x_3) + (aa^* - bb^*) x_2 \\
&\quad - \tfrac{1}{2}(a^* b + ab^*)(x_1 - x_3).
\end{aligned}
\tag{9–58}
$$

Now let $x = (1/2)(x_1 - x_3)$, $y = (1/2i)(x_1 + x_3)$, $z = x_2$; then

$$
\begin{aligned}
x' &= \frac{1}{2}(a^2 - b^{*2} - b^2 + a^{*2}) x \\
&\qquad + \frac{i}{2}(a^2 - b^{*2} + b^2 - a^{*2}) y + (ab + a^* b^*) z,
\end{aligned}
$$

$$
\begin{aligned}
y' &= -\frac{i}{2}(a^2 + b^{*2} - b^2 - a^{*2}) x \\
&\qquad + \frac{1}{2}(a^2 + b^{*2} + b^2 + a^{*2}) y - i(ab - a^* b^*) z,
\end{aligned}
$$

$$
z' = -(a^* b + ab^*) x + i(a^* b - ab^*) y + (aa^* - bb^*) z.
\tag{9–59}
$$

We have succeeded in associating with each matrix of $\mathfrak{U}_2$ a transformation of the variables x, y, z. Moreover, we see that the coefficients in (9–59) are all real. If we square, add, and use Eq. (9–55), we find that $x'^2 + y'^2 + z'^2 = x^2 + y^2 + z^2$; hence (9–59) is a real orthogonal transformation of x, y, z, (with determinant unity) and therefore a pure rotation. Thus, if we are given a unitary unimodular transformation (9–56), we can use (9–59) to find a three-dimensional rotation associated with it. We now show that all rotations are associated with unitary transformations in this way. The rotations are characterized by the Euler angles α, β, γ. If we choose a unitary transformation with $a = e^{i\alpha/2}$, $b = 0$, Eq. (9–59) reduces to

$$x' = x \cos \alpha - y \sin \alpha, \qquad y' = x \sin \alpha + y \cos \alpha, \qquad z' = z, \quad (9\text{–}60)$$

i.e., to the rotation $R(\alpha, 0, 0)$ through angle α about the Z-axis:

$$\begin{bmatrix} e^{i\alpha/2} & 0 \\ 0 & e^{-i\alpha/2} \end{bmatrix} \rightarrow \begin{bmatrix} \cos \alpha & -\sin \alpha & 0 \\ \sin \alpha & \cos \alpha & 0 \\ 0 & 0 & 1 \end{bmatrix}. \quad (9\text{–}61)$$

If we choose $a = \cos \beta/2$, $b = \sin \beta/2$, Eq. (9–59) reduces to

$$z' = z \cos \beta - x \sin \beta, \qquad x' = z \sin \beta + x \cos \beta, \qquad y' = y, \quad (9\text{–}62)$$

i.e., to the rotation $R(0, \beta, 0)$ through angle β about the Y-axis:

$$\begin{bmatrix} \cos \dfrac{\beta}{2} & \sin \dfrac{\beta}{2} \\ -\sin \dfrac{\beta}{2} & \cos \dfrac{\beta}{2} \end{bmatrix} \rightarrow \begin{bmatrix} \cos \beta & 0 & \sin \beta \\ 0 & 1 & 0 \\ -\sin \beta & 0 & \cos \beta \end{bmatrix}. \quad (9\text{–}63)$$

Thus we can associate the general rotation

$$R(\alpha, \beta, \gamma) = R(\gamma, 0, 0) R(0, \beta, 0) R(\alpha, 0, 0)$$

with the unitary transformation

$$\begin{bmatrix} e^{i\gamma/2} & 0 \\ 0 & e^{-i\gamma/2} \end{bmatrix} \begin{bmatrix} \cos \dfrac{\beta}{2} & \sin \dfrac{\beta}{2} \\ -\sin \dfrac{\beta}{2} & \cos \dfrac{\beta}{2} \end{bmatrix} \begin{bmatrix} e^{i\alpha/2} & 0 \\ 0 & e^{-i\alpha/2} \end{bmatrix}, \quad (9\text{–}64)$$

or

$$\begin{bmatrix} \cos \dfrac{\beta}{2} e^{(i/2)(\alpha+\gamma)} & \sin \dfrac{\beta}{2} e^{(i/2)(\gamma-\alpha)} \\ -\sin \dfrac{\beta}{2} e^{(i/2)(\alpha-\gamma)} & \cos \dfrac{\beta}{2} e^{-(i/2)(\alpha+\gamma)} \end{bmatrix} \rightarrow R(\alpha, \beta, \gamma). \quad (9\text{–}65)$$

We have achieved a homomorphic mapping of the unitary group $\mathfrak{U}_2$ onto the rotation group. We must now determine how many elements of $\mathfrak{U}_2$ are mapped into the identity of the rotation group. From (9–65) we see that the two unitary matrices

$$1 = \begin{bmatrix} 1 & 0 \\ 0 & 1 \end{bmatrix}$$

and

$$-1 = \begin{bmatrix} -1 & 0 \\ 0 & -1 \end{bmatrix}$$

are mapped into the identity element of the rotation group. These two elements therefore form an invariant subgroup of $\mathfrak{U}_2$, and their products with any element of $\mathfrak{U}_2$ are mapped into the same element of the rotation group. So each element of the rotation group is associated with a pair of elements of the unitary group which differ merely in a change of sign of all their coefficients. The group manifold of $\mathfrak{U}_2$ is in one-to-one correspondence with the points

$$|a|^2 + |b|^2 = 1 \tag{9–56}$$

of the surface of a sphere in a four-dimensional space. Since the sphere is simply connected (cf. the problem in Section 8–14), the group manifold of $\mathfrak{U}_2$ is simply connected. $\mathfrak{U}_2$ has only single-valued representations (cf. Section 8–14), and is the universal covering group of $0^+(3)$.

If we obtain a representation of the group $\mathfrak{U}_2$, the matrices of this representation will be associated with the corresponding element of the rotation group. Since the square of the element $\begin{bmatrix} -1 & 0 \\ 0 & -1 \end{bmatrix}$ of the unitary group is $\begin{bmatrix} 1 & 0 \\ 0 & 1 \end{bmatrix}$, its representative in any representation must be plus or minus the identity matrix, i.e., $D(-1) = \pm D(1)$ for any representation of the unitary group. If the representation has $D(-1) = D(1)$, then $D(-u) = D(u)$ for any element u of the unitary group, so that a single matrix is associated with each element of the rotation group. Such single-valued representations of the rotation group must coincide with the $D^{(l)}$ which we found earlier. If on the other hand, in a representation of the unitary group, $D(-1) = -D(1)$, then $D(-u) = -D(u)$ for each element u of the unitary group, and each rotation has two matrices associated with it. These two-valued representations of the rotation group are not proper representations, and the orthogonality theorems which we derived earlier cannot be applied to them. As we shall see later, in treating a double-valued representation of the rotation group we actually go over to the unitary group (for which the representation is single-valued) and treat it instead.

Our first task is to find all irreducible representations of the unitary group. The technique for constructing the representations consists again in forming symmetric products, as we did for the special case at the start of this section. Starting from u, v and transforming according to (9–56), we consider the set of products

$$u^{2j},\ u^{2j-1}v,\ u^{2j-2}v^2,\ \ldots,\ uv^{2j-1},\ v^{2j}; \tag{9–66}$$

or

$$f_m = \frac{u^{j+m}v^{j-m}}{\sqrt{(j+m)!(j-m)!}} \qquad (m = -j, -j+1, \ldots, j-1, j), \tag{9–67}$$

where j is integral or half-integral. (The factors in (9–67) have been chosen to make the representations unitary.) For a fixed j, the homogeneous polynomials f_m are transformed among themselves by the linear transformations (9–56). They are therefore the basis for a $(2j+1)$-dimensional representation of the unitary group. For example, for $j = \frac{1}{2}$, we have $f_{1/2} = u$, $f_{-1/2} = v$, and our two-dimensional representation of the unitary group is the set of matrices (9–56). For $j = 1$, we have $f_1 = u^2$, $f_0 = uv$, $f_{-1} = v^2$, and we obtain the three-dimensional representation (9–57). We shall label the representations of the unitary group as $D^{(j)}(a, b)$, where a and b are the coefficients in (9–56) which characterize the group elements $(aa^* + bb^* = 1)$. To find the matrices of the representation, we apply the transformation (9–56), $R(a, b)$, to f_m:

$$R(a, b)f_m = \frac{1}{\sqrt{(j+m)!(j-m)!}}(au + bv)^{j+m}(-b^*u + a^*v)^{j-m}. \tag{9–68}$$

Expanding by the binomial theorem, we have

$$R(a, b)f_m = \sum_{\mu,\nu=0} \frac{1}{\sqrt{(j+m)!(j-m)!}} \frac{(j+m)!}{\mu!(j+m-\mu)!}$$

$$\times (au)^{j+m-\mu}(bv)^\mu \frac{(j-m)!}{\nu!(j-m-\nu)!}(-b^*u)^{j-m-\nu}(a^*v)^\nu$$

$$= \sum_{\mu,\nu=0} \frac{\sqrt{(j+m)!(j-m)!}}{(j+m-\mu)!\mu!(j-m-\nu)!\nu!}$$

$$\times a^{j+m-\mu}a^{*\nu}b^\mu(-b^*)^{j-m-\nu}u^{2j-\mu-\nu}v^{\mu+\nu}. \tag{9–69}$$

We need not indicate the upper bounds on μ and ν, since the factorials·in the denominator annihilate the result when we go outside the proper range. We now express the right-hand side in terms of the functions f_m.

Let $\nu = j - \mu - m'$; then $2j - \mu - \nu = j + m'$, $\mu + \nu = j - m'$, and

$$R(a, b)f_m = \sum_{m'} f_{m'} \sum_{\mu=0} \frac{[(j + m)!(j - m)!(j + m')!(j - m')!]^{1/2}}{(j + m - \mu)!\mu!(j - m' - \mu)!(m' - m + \mu)!}$$

$$\times\ a^{j+m-\mu}a^{*j-m'-\mu}b^{\mu}(-b^*)^{m'-m+\mu} = \sum_{m'} f_{m'} D^{(j)}_{m'm}(a, b),$$

$$(9\text{-}70)$$

or

$$D^{(j)}_{m'm}(a, b) = \sum_{\mu} \frac{[(j + m)!(j - m)!(j + m')!(j - m')!]^{1/2}}{(j + m - \mu)!\mu!(j - m' - \mu)!(m' - m + \mu)!}$$

$$\times\ a^{j+m-\mu}a^{*j-m'-\mu}b^{\mu}(-b^*)^{m'-m+\mu}. \qquad (9\text{-}71)$$

We note that the basis functions f_m are independent and that

$$\sum_m |f_m|^2 = \sum_m \frac{|u^{j+m}v^{j-m}|^2}{(j + m)!(j - m)!} = \frac{1}{(2j)!}\{|u|^2 + |v|^2\}^{2j}, \quad (9\text{-}72)$$

so that $\sum_m |f_m|^2$ is invariant under the unitary transformations, and our representations are unitary.

The explicit form (9-71) for the matrices $D^{(j)}_{m'm}(a, b)$ is quite complicated. For the special case of $m' = j$, the factor $(j - m' - \mu)!$ in the denominator makes all terms vanish except the one for $\mu = 0$:

$$D^{(j)}_{jm}(a, b) = \sqrt{\frac{(2j)!}{(j + m)!(j - m)!}}\ a^{j+m}(-b^*)^{j-m}. \qquad (9\text{-}73)$$

From (9-73) we see that the matrix coefficients $D^{(j)}_{jm}(a, b)$ are, in general, different from zero.

For the special case $a = e^{i\alpha/2}$, $b = 0$, only the term with $\mu = 0$ remains in (9-71), and we find

$$D^{(j)}_{m'm}(e^{i\alpha/2}, 0) = \delta_{m'm}\ e^{im\alpha}. \qquad (9\text{-}74)$$

Using the special matrices (9-73) and (9-74), we can show that the representations $D^{(j)}(a, b)$ are irreducible. The method is the same as the one used for the representations $D^{(l)}$ of the rotation group. If a matrix A commutes with the diagonal matrix $D^{(j)}(e^{i\alpha/2}, 0)$, it must have only diagonal elements: $A_{m'm} = a_m\ \delta_{m'm}$. If A commutes with the general unitary matrix $D^{(j)}(a, b)$, then, taking the (jm)-component of the products AD and DA, we have $a_j\ D^{(j)}_{jm} = D^{(j)}_{jm}a_m$ for all m. From Eq. (9-73), $D^{(j)}_{jm} \neq 0$, so $a_m = a_j$ for all m. Thus we have shown that a matrix which commutes with all matrices of the representation must be a multiple of the unit matrix. Therefore the representations $D^{(j)}(a, b)$ are irreducible.

For different j, the dimensions of the representations $D^{(j)}$ differ, and hence they are not equivalent to one another.

In order to find the characters in the various representations $D^{(j)}(a, b)$, we note that every unitary unimodular matrix can be brought to diagonal form by a unitary unimodular transformation, and that its eigenvalues are a pair of complex conjugate numbers. Thus every unitary unimodular matrix is equivalent to one of the form

$$\begin{bmatrix} e^{i\alpha/2} & 0 \\ 0 & e^{-i\alpha/2} \end{bmatrix},$$

and has the same character as

$$\begin{bmatrix} e^{i\alpha/2} & 0 \\ 0 & e^{-i\alpha/2} \end{bmatrix}$$

in any representation. From Eq. (9–74), we have

$$\chi^{(j)}(e^{i\alpha/2}, 0) = \sum_{m=-j}^{j} e^{im\alpha} = \frac{\sin (j + \frac{1}{2})\alpha}{\sin \alpha/2}, \qquad (9\text{–}75)$$

where α ranges between 0 and 2π.

The representations $D^{(j)}(a, b)$ form a complete set of irreducible representations of $\mathfrak{U}_2$. The character of any other representation would (when multiplied by the density function of the group) have to be orthogonal to all the characters (9–75). But by taking differences we find

$$\chi^{(0)}(e^{i\alpha/2}, 0) = 1, \qquad \chi^{(1/2)} = 2 \cos \frac{\alpha}{2}, \qquad \chi^{(1)} - \chi^{(0)} = 2 \cos \alpha,$$

$$\chi^{(3/2)} - \chi^{(1/2)} = 2 \cos \frac{3\alpha}{2}, \text{ etc.}$$

These functions form a complete set in the range 0 to 2π, so there can be no independent representation.

The representations $D^{(j)}(a, b)$ of the unitary group provide us with representations of the rotation group. Using Eqs. (9–65) and (9–70), we have

$$D_{m'm}^{(j)}(\alpha, \beta, \gamma) = D_{m'm}^{(j)} \left(\cos \frac{\beta}{2} e^{(i/2)(\alpha+\gamma)}, \sin \frac{\beta}{2} e^{(i/2)(\gamma-\alpha)} \right)$$

$$= \sum_{\mu} (-1)^{m'-m-\mu}$$

$$\times \frac{[(j + m)!(j - m)!(j + m')!(j - m')!]^{1/2}}{(j + m - \mu)!\mu!(j - m' - \mu)!(m' - m + \mu)!}$$

$$\times e^{im'\alpha}e^{im\gamma} \left(\cos \frac{\beta}{2} \right)^{2j+m-m'-2\mu} \left(\sin \frac{\beta}{2} \right)^{m'-m+2\mu}.$$

$$(9\text{–}76)$$

If desired, we can eliminate the factor $(-1)^{m'-m}$ in (9–71) and (9–76) by going over to equivalent representations by transforming with the matrix $\delta_{m'm}(-1)^m$.

The characters of the elements of the rotation group can be obtained from the special form (9–75) with $\alpha = \phi$ and $\beta = \gamma = 0$:

$$\chi^{(j)}(\phi) = \sum_{m=-j}^{j} e^{im\phi} = \frac{\sin\,(j + \frac{1}{2})\phi}{\sin\,\phi/2}. \qquad (9\text{–}77)$$

For integral j, the characters coincide with those of the representations $D^{(l)}$ obtained earlier, so that the $D^{(j)}$ for integer j are the same as the $D^{(l)}$. For half-integral j, each rotation has two matrices $\pm D^{(j)}(\alpha, \beta, \gamma)$ associated with it in the representation. It is important to keep in mind that the representations $D^{(j)}$ are *always* single-valued representations of the unitary group. The double-valuedness of the representations of the rotation group for half-integral j arises because two unitary matrices whose representatives differ in sign are associated with each rotation. This double-valuedness is inherent. If R and S are two rotations, we can write only

$$D^{(j)}(R)D^{(j)}(S) = \pm D^{(j)}(RS), \qquad j = \tfrac{1}{2}, \tfrac{3}{2} \text{ etc.}, \qquad (9\text{–}78)$$

and cannot choose the sign uniquely. As an example consider the group consisting of the identity E and the rotation C_2 through angle π about the Z-axis. The identity E is associated with the two unitary matrices $\begin{bmatrix} 1 & 0 \\ 0 & 1 \end{bmatrix}$ and $\begin{bmatrix} -1 & 0 \\ 0 & -1 \end{bmatrix}$, while C_2 is associated with $\begin{bmatrix} i & 0 \\ 0 & -i \end{bmatrix}$ and $\begin{bmatrix} -i & 0 \\ 0 & i \end{bmatrix}$ according to Eq. (9–61). If we choose as our representatives

$$D(E) = \begin{bmatrix} 1 & 0 \\ 0 & 1 \end{bmatrix}, \qquad D(C_2) = \begin{bmatrix} i & 0 \\ 0 & -i \end{bmatrix},$$

then

$$D(E)D(C_2) = \begin{bmatrix} i & 0 \\ 0 & -i \end{bmatrix} = D(EC_2) = D(C_2),$$

but

$$D(C_2)D(C_2) = \begin{bmatrix} -1 & 0 \\ 0 & -1 \end{bmatrix} = -D(C_2^2) = -D(E). \qquad (9\text{–}78a)$$

The $\pm$ sign in (9–78) is essential and cannot be eliminated by arbitrary assignment.

If we now consider the rotation-reflection group, we adjoin the inversion I to the rotation group. The corresponding process in the unitary group can be executed only by considering it as an abstract group to which we adjoin an element i such that i^2 is the identity element of the unitary group and i commutes with all the elements of the unitary group. In any representation of the unitary group, the matrix representing i is necessarily either plus or minus the unit matrix, so that the same is true for I. Note that the association $i \to I$ is the *one* case where a member of the rotation-reflection group is associated with a *single* element of the (augmented) unitary group. For integral j, we obtain the same results as before: The rotation R has the representative $D^{(j)}(R)$; if $i(I)$ has the representative (1), then $D^{(j)}(RI)$ has the representative $D^{(j)}(R)$; if $i(I)$ has the representative (-1) then $D^{(j)}(RI)$ has the representative $-D^{(j)}(R)$, in agreement with our earlier discussion of positive and negative representations. However, for half-integral j, the representative of R is the pair $\pm D^{(j)}(R)$, so that for either choice of the representative of I we obtain $D^{(j)}(RI) = \pm D^{(j)}(R)$.

9–7 Splitting of atomic levels in crystalline fields.   Double-valued representations of the crystal point groups.

Now we must consider the same perturbation problem that we examined for integral representations. How does a level of the atom split when the symmetry is lowered to that of one of the crystal point groups? We cannot proceed as we did before because the orthogonality theorems apply only to single-valued representations. We must first find the two-valued representations of the crystal groups, just as we have found the two-valued representations of the full rotation group.

The method usually presented for finding the two-valued representations of the crystal groups is quite artificial. We shall try to make it plausible. When we wish to find the two-valued representations of the group $\mathcal{C}_2$ [Eq. (9–78)], we consider the group $\mathcal{C}_2'$ of four unitary matrices, i.e.,

$$
\begin{array}{cc}
E & C_2 \\
\swarrow \quad \searrow & \swarrow \quad \searrow \\
E \qquad\qquad R & C_2 \qquad\qquad C_2R
\end{array}
$$

$$
\begin{bmatrix} 1 & 0 \\ 0 & 1 \end{bmatrix}, \quad
\begin{bmatrix} -1 & 0 \\ 0 & -1 \end{bmatrix}, \quad
\begin{bmatrix} i & 0 \\ 0 & -i \end{bmatrix}, \quad
\begin{bmatrix} -i & 0 \\ 0 & i \end{bmatrix} \qquad (9\text{–}79)
$$

and find its representations. Any representation in which the element R has the same character as E will be a single-valued representation of the group $\mathcal{C}_2$. If the character of R is the negative of the character of E, then we are dealing with a double-valued representation of the group $\mathcal{C}_2$. To obtain all representations of $\mathcal{C}_2$, we need only find the representations of the group $\mathcal{C}_2'$ of unitary matrices given in (9–79). This method for finding the representations of the point group is correct, but has the disadvantage that we lose the intuitively valuable picture of the elements as geometrical operations. Hence we try formally to consider the group of four operations (9–79) as a set of geometrical transformations. From (9–79) we note that $C_2^2 = R$; successive rotations through π about a given axis do not give the identity. We formally add to the ordinary rotation group an element R which corresponds to rotation through 2π about an axis, and take all possible products of R with the elements of the rotation group. We then obtain what is called the *double group* corresponding to the original rotation group.

As another example take the group D_2. This group is abelian and contains the four elements C_x, C_y, C_z, E in four classes. In order to find all representations of D_2 (integral and half-integral) we must go over to the subgroup of $\mathfrak{U}_2$ which is associated with D_2. This subgroup D_2' can be found from (9–61) and (9–63):

$$
\begin{array}{cccc}
E & C_z & C_y & C_x \\
\swarrow \searrow & \swarrow \searrow & \swarrow \searrow & \swarrow \searrow \\
E \quad\quad R & C_z \quad\quad C_z R & C_y \quad\quad C_y R & C_x \quad\quad C_x R
\end{array}
$$

$$
\begin{bmatrix} 1 & 0 \\ 0 & 1 \end{bmatrix}\begin{bmatrix} -1 & 0 \\ 0 & -1 \end{bmatrix} \quad \begin{bmatrix} i & 0 \\ 0 & -i \end{bmatrix}\begin{bmatrix} -i & 0 \\ 0 & i \end{bmatrix} \; \Big| \; \begin{bmatrix} 0 & 1 \\ -1 & 0 \end{bmatrix}\begin{bmatrix} 0 & -1 \\ 1 & 0 \end{bmatrix} \quad \begin{bmatrix} 0 & i \\ i & 0 \end{bmatrix}\begin{bmatrix} 0 & -i \\ -i & 0 \end{bmatrix}.
$$

$$(9\text{–}80)$$

We note that the group D_2' is not abelian:

$$
C_x C_y = \begin{bmatrix} -i & 0 \\ 0 & i \end{bmatrix} = C_z R, \qquad C_y C_x = \begin{bmatrix} i & 0 \\ 0 & -i \end{bmatrix} = C_z.
$$

Also $C_x^2 = C_y^2 = C_z^2 = R$, so successive rotations through 2π about any axis give the element R, which we now interpret as a rotation through angle 2π. From (9–80) we also verify that $R^2 = E$. Thus we may look upon the double group D_2' as a group of rotations in which we return to the identity only after rotation through 4π. We shall now derive all representations of the double groups corresponding to the various point groups.

To find all the representations of the groups $\mathfrak{C}_n$, we go over to the double groups $\mathfrak{C}_n'$ containing $2n$ elements $C_n, C_n^2, \ldots, C_n^n = R, C_n R, C_n^2 R, \ldots,$ $C_n^n R = C_n^{2n} = E$. The double groups $\mathfrak{C}_n'$ are cyclic abelian groups generated by an element C_n which now has period $2n$. Since the group is abelian, $\mathfrak{C}_n'$ has $2n$ one-dimensional representations. The basis functions for the various representations of $\mathfrak{C}_n'$ are

$$
1,\ e^{i\phi/2},\ e^{i\phi},\ e^{3i\phi/2},\ \ldots,\ e^{i(n-1/2)\phi}. \tag{9–81}
$$

Applying C_n to any one of these basis functions, we find

$$
C_n e^{im\phi} = e^{-2\pi im/n} e^{im\phi}. \tag{9–82}
$$

For integral m, $C_n^n e^{im\phi} = e^{im\phi}$ so that we obtain the single-valued representations tabulated in Chapter 4. For half-integral m, we have $C_n^n e^{im\phi} = -e^{im\phi}$ yielding double-valued representations of $\mathfrak{C}_n$.

For $\mathcal{C}_2'$, the basis functions 1 and $e^{i\phi}$ give the single-valued representations, while for the functions $e^{i\phi/2}$ and $e^{-i\phi/2}$ we find

$$C_2 e^{\pm i\phi/2} = e^{\pm(i/2)(\phi - \pi)} = \mp i e^{\pm i\phi/2}.$$

The (complex-conjugate) double-valued representations are

$\mathcal{C}_2'$	E	R	C_2	$C_2 R$
$E'\left\{\vphantom{\begin{matrix}1\\1\end{matrix}}\right.$	1	-1	i	$-i$
	1	-1	$-i$	i

$$(9\text{--}83)$$

We use a prime to denote a double-valued representation. We note a general property: in a single-valued representation any element S and the element SR have the same character, while in any two-valued representation their characters have opposite signs. Thus the characters of the two-valued representations are automatically orthogonal to those of the single-valued representations. The groups $\mathcal{C}_3'$, $\mathcal{C}_4'$, and $\mathcal{C}_6'$ can be treated in this same way.

Problem. Find the double-valued representations of the group $\mathcal{C}_3$.

In considering the single-valued representations of the point groups, we discussed the two-sidedness of axes. If a 2-fold axis (rotation U_2) existed perpendicular to the n-fold axis, then C_n^k and C_n^{n-k} were in the same class because $U_2 C_n^k U_2^{-1} = U_2 C_n^k U_2 = C_n^{n-k}$. Similarly, if a reflection plane passed through the n-fold axis, we obtained the same result. (This reduces to the previous case since $\sigma_v = I U_2$ and $\sigma_v C_n^k \sigma_v^{-1} = I U_2 C_n^k I U_2 = C_n^{n-k}$, because I commutes with all elements of the group.) In the double groups the inverse of U_2 is not U_2 but rather $U_2 R$, so that the element equivalent to C_n^k is $U_2 C_n^k U_2 R = C_n^{n-k} R$. In the groups D_n, the Z-axis was two-sided, so that C_n^k and C_n^{n-k} were in the same class. Now when we consider the double group D_n', the elements C_n^k and $C_n^{n-k} R$ are in one class, while C_n^{n-k} and $C_n^k R$ are in another. So, in general, the double group has twice as many elements in twice as many classes. The number of classes is not doubled in one special case: if n is even, the rotation through π, $C_n^{n/2}$ forms a class by itself in D_n, and also gives a single class $C_n^{n/2}$, $C_n^{n/2} R$ in the double group D_n'. The presence of a 2-fold axis with even n makes the number of classes in the double group less than twice the number of classes in the original group.

We take first the group D_2', which we discussed earlier. All axes are two-sided, so the double group D_2' has eight elements in 5 classes:

$$E; \quad R; \quad C_x, C_x R; \quad C_y, C_y R; \quad C_z, C_z R.$$

There are five irreducible representations of D_2', and $\sum_{i=1}^{5} n_i^2 = 8$. Of these representations, we have previously found the single-valued ones which assign the same matrix to E and R. There were four such representations, all one-dimensional. Thus we are left with one new double-valued representation whose dimension is 2: $2^2 + 1^2 + 1^2 + 1^2 + 1^2 = 8$. Its matrices are those in (9–80). The characters are

$$
\begin{array}{c|ccccc}
 & & & C_x & C_y & C_z \\
D_2' & E & R & C_x R & C_y R & C_z R \\
\hline
E' & 2 & -2 & 0 & 0 & 0
\end{array}
\tag{9-84}
$$

We note another general property of two-valued representations. Since $D(SR) = -D(S)$, it follows that $\chi(SR) = -\chi(S)$; so if S and SR are in the same class, $\chi(S)$ must be zero. This will occur for any rotation through π about a two-sided axis. The basis functions of the two-dimensional representation E' are $e^{\pm i\phi/2}$.

The group D_3 had 6 elements in 3 classes. The double group D_3' has 12 elements in 6 classes:

$$E; \quad R; \quad C_3, C_3^2 R; \quad C_3^2, C_3 R; \quad C_x(3); \quad C_x R(3).$$

Because the group does not contain a rotation through π about the two-sided Z-axis, we double both the number of elements and the number of classes, and obtain three new (double-valued) representations of order 1, 1, 2.

In a one-dimensional representation $\chi(C_3^2) = [\chi(C_3)]^2$, but since C_3^2 and $C_3 R$ are in the same class, $\chi(C_3^2) = -\chi(C_3)$, so that we must have $\chi(C_3) = -1$. Also, since $C_x^2 = R$ and $\chi(R) = -1$, $\chi(C_x) = \pm i$, giving us our two complex conjugate one-dimensional representations. Since we have only one two-dimensional representation, its characters must be real. Thus for C_x, the matrix must have each of the eigenvalues i, $-i$ once, yielding $\chi(C_x) = 0$. Since $C_3^3 = R$, the eigenvalues of C_3 are -1, $e^{\pm i\pi/3}$; in a real representation the pair of eigenvalues $e^{\pm i\pi/3}$ must appear, giving $\chi(C_3) = e^{\pi i/3} + e^{-\pi i/3} = 1$. The characters of the two-valued representations of D_3' are:

$$
\begin{array}{c|cccccc}
 & & & C_3 & C_3^2 & & \\
D_3' & E & R & C_3^2 R & C_3 R & C_x(3) & C_x R(3) \\
\hline
E_1' \left\{ \begin{array}{c} \\ \\ \end{array} \right. & \begin{array}{c} 1 \\ 1 \end{array} & \begin{array}{c} -1 \\ -1 \end{array} & \begin{array}{c} -1 \\ -1 \end{array} & \begin{array}{c} 1 \\ 1 \end{array} & \begin{array}{c} i \\ -i \end{array} & \begin{array}{c} -i \\ i \end{array} \\
E_2' & 2 & -2 & 1 & -1 & 0 & 0
\end{array}
\tag{9-85}
$$

An alternative method is to use Eq. (3–173). If we label the classes of D_3' in the order of (9–85), then

$$\mathcal{K}_3^2 = (C_3 + C_3^2 R)^2 = C_3^2 + C_3^4 R^2 + 2C_3^3 R$$
$$= C_3^2 + C_3 R + 2E = \mathcal{K}_4 + 2\mathcal{K}_1,$$

whence $c_{334} = 1$ and $c_{331} = 2$. Using Eq. (3–173), we find

$$4\chi_3^2 = \chi_1(2\chi_4 + 2\chi_1).$$

Since $\chi_4 = -\chi_3$ and $\chi_1 = n$, we have $4\chi_3^2 = n(-2\chi_3 + 2n)$. For $n = 1$, $4\chi_3^2 + 2\chi_3 = 2$, and $\chi_3 = -1$ or $\frac{1}{2}$. Since $\frac{1}{2}$ is excluded (in a one-dimensional representation χ must be a root of unity) we have $\chi_3 = -1$ in both one-dimensional representations. For $n = 2$, $4\chi_3^2 + 4\chi_3 = 8$, so $\chi_3 = 1$ or -2. The value -2 is excluded since the sum of the squares of the characters would exceed the order of the group. Thus we obtain our results by this alternative method, which we now apply to D_4' and D_6'.

The double group D_4' has 16 elements in 7 classes:

$$E \ \bigg|\ R \ \bigg|\ \begin{array}{c} C_4 \\ C_4^3 R \end{array} \ \bigg|\ \begin{array}{c} C_4^3 \\ C_4 R \end{array} \ \bigg|\ \begin{array}{c} C_4^2 \\ C_4^2 R \end{array} \ \bigg|\ \begin{array}{c} C_2(2) \\ C_2 R(2) \end{array} \ \bigg|\ \begin{array}{c} C_{2'}(2) \\ C_{2'} R(2) \end{array}$$

while D_4 had 8 elements in 5 classes. We have two two-dimensional double-valued representations. Since all axes are two-sided, we have $\chi_5 = \chi_6 = \chi_7 = 0$. Also,

$$\mathcal{K}_3^2 = (C_4 + C_4^3 R)^2 = C_4^2 + C_4^6 R^2 + 2C_4^4 R = \mathcal{K}_5 + 2\mathcal{K}_1,$$

so $c_{335} = 1$, $c_{331} = 2$. Since $\chi_5 = 0$, we have for $n = 2$, $\chi_3^2 = 2$, $\chi_3 = \pm\sqrt{2}$, giving us our two representations.

Similarly in D_6' we have 24 elements in 9 classes:

$$E \ \bigg|\ R \ \bigg|\ \begin{array}{c} C_6 \\ C_6^5 R \end{array} \ \bigg|\ \begin{array}{c} C_6^2 \\ C_6^4 R \end{array} \ \bigg|\ \begin{array}{c} C_6^5 \\ C_6 R \end{array} \ \bigg|\ \begin{array}{c} C_6^4 \\ C_6^2 R \end{array} \ \bigg|\ \begin{array}{c} C_6^3 \\ C_6^3 R \end{array} \ \bigg|\ \begin{array}{c} C_2(3) \\ C_2 R(3) \end{array} \ \bigg|\ \begin{array}{c} C_{2'}(3) \\ C_{2'} R(3) \end{array}$$

while D_6 had 12 elements in 6 classes. Since there are three two-valued representations and $n_1^2 + n_2^2 + n_3^2 = 12$, all have $n = 2$. Again $\chi_7 = \chi_8 = \chi_9 = 0$.

$\mathcal{K}_4^2 = \mathcal{K}_6 + 2\mathcal{K}_1$: so for $n = 2$, $\chi_4^2 = \chi_6 + 2$. But $\chi_6 = -\chi_4$, so $\chi_4^2 + \chi_4 = 2$ and $\chi_4 = 1$ or -2.

$\mathcal{K}_3\mathcal{K}_4 = \mathcal{K}_7 + \mathcal{K}_3$: Since $\chi_7 = 0$, we obtain for $n = 2$, $\chi_3\chi_4 = \chi_3$; hence $\chi_4 = 1$ unless $\chi_3 = 0$.

$\mathcal{K}_3^2 = \mathcal{K}_4 + 2\mathcal{K}_1$: so $\chi_3^2 = \chi_4 + 2$. For $\chi_4 = 1$, $\chi_3 = \pm\sqrt{3}$; for $\chi_3 = 0$, $\chi_4 = -2$.

We thus obtain the three representations which are given in our final tables (Table 9–11a through f).

The double group T' has 24 elements in 7 classes:

$$E \mid R \mid C_3(4) \mid C_3 R(4) \mid C_3^2(4) \mid C_3^2 R(4) \mid C_2(3), C_2 R(3),$$

TABLE 9–11

CHARACTER TABLES FOR TWO-VALUED REPRESENTATIONS
OF THE CRYSTAL POINT GROUPS

D_2'	E	R	C_x $C_x R$	C_y $C_y R$	C_z $C_z R$
E'	2	-2	0	0	0

(a)

D_3'	E	R	C_3 $C_3^2 R$	C_3^2 $C_3 R$	$C_x(3)$	$C_x R(3)$
$E_1' \big\{$	1	-1	-1	1	i	$-i$
	1	-1	-1	1	$-i$	i
E_2'	2	-2	1	-1	0	0

(b)

D_4'	E	R	C_4 $C_4^3 R$	C_4^3 $C_4 R$	$C_2(2)$ $C_2 R(2)$	$C_{2'}(2)$ $C_{2'} R(2)$	C_4^2 $C_4^2 R$
E_1'	2	-2	2	-2	0	0	0
E_2'	2	-2	-2	2	0	0	0

(c)

D_6'	E	R	C_6 $C_6^5 R$	C_6^2 $C_6^4 R$	C_6^5 $C_6 R$	C_6^4 $C_6^2 R$	C_6^3 $C_6^3 R$	$C_2(3)$ $C_2 R(3)$	$C_{2'}(3)$ $C_{2'} R(3)$
E_1'	2	-2	$\sqrt{3}$	1	$-\sqrt{3}$	-1	0	0	0
E_2'	2	-2	$-\sqrt{3}$	1	$\sqrt{3}$	-1	0	0	0
E_3'	2	-2	0	-2	0	2	0	0	0

(d)   (cont.)

TABLE 9–11 *(cont.)*

T'	E	R	$C_3(4)$	$C_3R(4)$	$C_3^2(4)$	$C_3^2R(4)$	$C_2(3),$ $C_2R(3)$
E'	2	-2	1	-1	-1	1	0
G' $\Big\{$	2	-2	ϵ	$-\epsilon$	$-\epsilon^2$	ϵ^2	0
	2	-2	ϵ^2	$-\epsilon^2$	$-\epsilon$	ϵ	0

(e)

O'	E	R	$C_3(4)$ $C_3^2R(4)$	$C_3^2(4)$ $C_3R(4)$	$C_4(3)$ $C_4^3R(3)$	$C_4^3(3)$ $C_4R(3)$	$C_4^2(3)$ $C_4^2R(3)$	$C_2(6)$ $C_2R(6)$
E_1'	2	-2	1	-1	$\sqrt{2}$	$-\sqrt{2}$	0	0
E_2'	2	-2	1	-1	$-\sqrt{2}$	$\sqrt{2}$	0	0
G'	4	-4	-1	1	0	0	0	0

(f)

while T had 12 elements in 4 classes, and we have therefore three new two-dimensional representations. Since complex representations occur in pairs, at least one of the representations must have real characters. If C_3 is to have a real character in a two-dimensional representation, we must choose the eigenvalues $e^{\pi i/3}$ and $e^{-\pi i/3}$, so $\chi(C_3) = 1$. (If we choose -1 twice, the sum of the squares of the characters will exceed the order of the group.) C_3^2 is brought to diagonal form together with C_3; hence $\chi(C_3^2) = e^{2\pi i/3} + e^{-2\pi i/3} = -1$, and the characters are 2; -2; 1; -1; -1; 1; 0. Now the simplest way to obtain the other two representations is to take the product of the representation just found with the one-dimensional representations which have the characters 1; 1; ϵ; ϵ; ϵ^2; ϵ^2; 1 and 1; 1; ϵ^2; ϵ^2; ϵ; ϵ; 1, respectively. We then obtain the sets of characters 2; -2; ϵ; $-\epsilon$; $-\epsilon^2$; ϵ^2; 0 and 2; -2; ϵ^2; $-\epsilon^2$; $-\epsilon$; ϵ; 0.

Problem. Find the characters of the two-valued representations of T' by using Eq. (3–173).

The double group O' has 48 elements in 8 classes:

E	R	$C_3(4)$ $C_3^2R(4)$	$C_3^2(4)$ $C_3R(4)$	$C_4(3)$ $C_4^3R(3)$	$C_4^3(3)$ $C_4R(3)$	$C_4^2(3)$ $C_4^2R(3)$	$C_2(6)$ $C_2R(6)$

while O had 24 elements in 5 classes, so there are three two-valued representations, and $n_1^2 + n_2^2 + n_3^2 = 24$. Two of the representations are two-dimensional, the other is four-dimensional. Since there is only one four-dimensional representation, its characters must be real. In any representation, $\chi_7 = \chi_8 = 0$; $\chi_6 = -\chi_5$ and $\chi_4 = -\chi_3$. In an irreducible representation, the sum of the squares of the characters must equal the order of the group, so $4^2 + (-4)^2 + 16\chi_3^2 + 12\chi_5^2 = 48$, or $4\chi_3^2 + 3\chi_5^2 = 4$. If C_4 is brought to diagonal form, the 4 diagonal elements must be chosen from the eigenvalues $e^{\pm i\pi/4}$ and $e^{\pm 3i\pi/4}$. The only way to get a real character without violating the last equation is to ensure that all four eigenvalues appear, in which case $\chi(C_4) = \chi_5 = 0$. Then we must have $\chi_3^2 = 1, \chi_3 = \pm 1$. But C_3 has the eigenvalues $-1, e^{\pm i\pi/3}$; the four elements of the matrix of C_3 must be chosen from these in such a way as to give a real trace. If we choose the pair $e^{\pm i\pi/3}$ twice, then χ_3 is too large to satisfy the condition given above. We must choose $-1, -1, e^{\pm i\pi/3}$, which yields $\chi_3 = -1$. So we get the set of characters $4; -4; -1; 1; 0; 0; 0; 0$. To obtain the two-dimensional representations, we make use of the theorem according to which their characters must be orthogonal to the character of the four-dimensional representation just obtained. Since they have $\chi(E) = 2, \chi(R) = -2$, we find $4 \cdot 2 + (-4)(-2) - 16\chi_3 = 0$, so $\chi_3 = 1$. Next we set the sum of the squares of the characters equal to the order of the group, i.e., $4 + 4 + 16 + 12\chi_5^2 = 48$, so that $\chi_5 = \pm\sqrt{2}$.

The two-valued representations of the point groups are collected in Table 9–11 for convenient reference.

We can now solve the problem of splitting of levels in a crystalline field for the case where the level belongs to a two-valued representation of the rotation group. First it is clear that only two-valued representations of the crystal group will appear in the resolution, since single- and double-valued representations always have orthogonal characters. We use Eq. (9–77) to find the characters of the elements of the double group in the half-integral representations $D^{(j)}$. In the tetragonal double group D'_4, we have from (9–77):

$Angle$

$\phi = 0: \quad \chi_1 = 2j + 1, \quad \chi_2 = -\chi_1;$

$\phi = \pi: \quad \chi_5 = \chi_6 = \chi_7 = 0;$

$$\phi = \frac{\pi}{2}: \quad \chi_3 = \frac{\sin (j + \frac{1}{2})\pi/2}{\sin \pi/4} = \begin{cases} \sqrt{2} & \text{for } j \equiv \frac{1}{2} \pmod 4, \\ 0 & \text{for } j \equiv \frac{3}{2}, \frac{7}{2} \pmod 4, \\ -\sqrt{2} & \text{for } j \equiv \frac{5}{2} \pmod 4; \end{cases}$$

$$\chi_4 = -\chi_3. \tag{9–86}$$

TABLE 9–12

Characters of classes of D_4' in the $(2j+1)$-dimensional representation of the rotation group			Resolution of $D^{(j)}$ into irreducible representations of D_4'	Number of terms
j	K_1	K_3		
$\frac{1}{2}$	2	$\sqrt{2}$	E_1'	1
$\frac{3}{2}$	4	0	$E_1' + E_2'$	2
$\frac{5}{2}$	6	$-\sqrt{2}$	$E_1' + 2E_2'$	3
$\frac{7}{2}$	8	0	$2E_1' + 2E_2'$	4
$4\lambda + j'$	$8\lambda + 2j' + 1$	same as for j'	$2\lambda(E_1' + E_2') +$ terms for j'	$j + \frac{1}{2}$

Since χ_1 and χ_3 are the only independent characters, we need record only these two in carrying out the resolution. The results obtained by means of (9–86) and Table 9–11(c) are given in Table 9–12.

For the hexagonal group D_6', the characters of the classes in the half-integral representations $D^{(j)}$ are:

Angle

$$\phi = 0: \quad \chi_1 = 2j + 1, \quad \chi_2 = -\chi_1;$$

$$\phi = \pi: \quad \chi_7 = \chi_8 = \chi_9 = 0;$$

$$\phi = \frac{\pi}{3}: \quad \chi_3 = \frac{\sin\left(j + \frac{1}{2}\right)\pi/3}{\sin \pi/6} = \begin{array}{ll} \sqrt{3} & \text{for } j \equiv \frac{1}{2}, \frac{3}{2} \pmod 6, \\ 0 & \text{for } j \equiv \frac{5}{2}, \frac{11}{2} \pmod 6, \\ -\sqrt{3} & \text{for } j \equiv \frac{7}{2}, \frac{9}{2} \pmod 6; \end{array}$$

$$\phi = \frac{2\pi}{3}: \quad \chi_4 = \frac{\sin\left(j + \frac{1}{2}\right)2\pi/3}{\sin \pi/3} = \begin{array}{ll} 1 & \text{for } j \equiv \frac{1}{2} \pmod 3, \\ -1 & \text{for } j \equiv \frac{3}{2} \pmod 3, \\ 0 & \text{for } j \equiv \frac{5}{2} \pmod 3; \end{array}$$

$$\chi_5 = -\chi_3, \quad \chi_6 = -\chi_4.$$

$$(9–87)$$

In performing the resolution, we need only record χ_1, χ_3, χ_4. Again using the character table for D_6' [Table 9–11(d)], we obtain the results listed in Table 9–13.

TABLE 9–13

Characters of classes of D_6' in the $(2j+1)$-dimensional representation of the rotation group				Resolution of $D^{(j)}$ into irreducible representations of D_6'	Number of terms
j	K_1	K_2	K_3		
$\frac{1}{2}$	2	$\sqrt{3}$	1	E_1'	1
$\frac{3}{2}$	4	$\sqrt{3}$	-1	$E_1' + E_3'$	2
$\frac{5}{2}$	6	0	0	$E_1' + E_2' + E_3'$	3
$\frac{7}{2}$	8	$-\sqrt{3}$	1	$E_1' + 2E_2' + E_3'$	4
$\frac{9}{2}$	10	$-\sqrt{3}$	-1	$E_1' + 2E_2' + 2E_3'$	5
$\frac{11}{2}$	12	0	0	$2E_1' + 2E_2' + 2E_3'$	6
$6\lambda + j'$	$2j+1$	same as for j'		$2\lambda(E_1' + E_2' + E_3') +$ terms for j'	$j + \frac{1}{2}$

For the reduction to the cubic double group O', the characters are:

Angle

$$\phi = 0: \quad \chi_1 = 2j + 1, \quad \chi_2 = -\chi_1;$$

$$\phi = \pi: \quad \chi_7 = \chi_8 = 0;$$

$$\phi = \frac{2\pi}{3}: \quad \chi_3 = \frac{\sin\,(j + \tfrac{1}{2})2\pi/3}{\sin\,\pi/3} = \begin{array}{ll} 1 & \text{for} \quad j \equiv \tfrac{1}{2} \pmod 3, \\ -1 & \text{for} \quad j \equiv \tfrac{3}{2} \pmod 3, \\ 0 & \text{for} \quad j \equiv \tfrac{5}{2} \pmod 3; \end{array}$$

$$\phi = \frac{\pi}{2}: \quad \chi_5 = \frac{\sin\,(j + \tfrac{1}{2})\pi/2}{\sin\,\pi/4} = \begin{array}{ll} \sqrt{2} & \text{for} \quad j \equiv \tfrac{1}{2} \pmod 4, \\ 0 & \text{for} \quad j \equiv \tfrac{3}{2}, \tfrac{7}{2} \pmod 4, \\ -\sqrt{2} & \text{for} \quad j \equiv \tfrac{5}{2} \pmod 4; \end{array}$$

$$\chi_4 = -\chi_3, \quad \chi_6 = -\chi_5.$$

(9–88)

The reduction table is given in Table 9–14.

Problem. Carry out the resolution of the two-valued representations of the rotation group for the crystal symmetries D_3 and T.

TABLE 9–14

Characters of classes of O' in the $(2j + 1)$-dimensional representation of the rotation group				Resolution of $D^{(j)}$ into irreducible representations of O'	Number of terms
j	K_1	K_2	K_3		
$\frac{1}{2}$	2	1	$\sqrt{2}$	E'_1	1
$\frac{3}{2}$	4	-1	0	G'	1
$\frac{5}{2}$	6	0	$-\sqrt{2}$	$E'_2 + G'$	2
$\frac{7}{2}$	8	1	0	$E'_1 + E'_2 + G'$	3
$\frac{9}{2}$	10	-1	$\sqrt{2}$	$E'_1 + 2G'$	3
$\frac{11}{2}$	12	0	0	$E'_1 + E'_2 + 2G'$	4
$6 + j'$				$E'_1 + E'_2 + 2G' +$ terms for j' with E'_1 and E'_2 interchanged	$4 +$ number of terms for j'
$12\lambda + j'$				$2\lambda(E'_1 + E'_2 + 2G') +$ terms for j'	$8\lambda +$ number of terms for j'

9–8 Coupled systems. Addition of angular momenta. Clebsch-Gordan coefficients.

When the physicist considers the problem of coupling two subsystems whose Hamiltonians are invariant under the rotation group, he proceeds by adding the "angular momenta" of the parts to give the total angular momentum of the system. We first wish to show the relation between this method, which uses the Lie algebra, and ours which uses the Lie group.

As in Section 6–4, we consider two subsystems, labeled 1 and 2, for which the Hamiltonians of the individual systems are invariant under the rotation group. With any rotation R_1 of the first system there is associated an operator O_{R_1} in the Hilbert space of system 1, and similarly for any rotation S_2 of the second system we have an operator O_{S_2} in the Hilbert space of the second system. Corresponding to the operators O_{R_1}, we will obtain infinitesimal operators $\mathbf{J}_1 : J_{1x}, J_{1y}, J_{1z}$ which act on vectors in the Hilbert space of system 1. Similarly for system 2, we have infinitesimal operators $\mathbf{J}_2 : J_{2x}, J_{2y}, J_{2z}$. Each set of operators satisfies the commutation rules

$$[J_{nx}, J_{ny}] = iJ_{nz}, \qquad [J_{ny}, J_{nz}] = iJ_{nx}, \qquad [J_{nz}, J_{nx}] = iJ_{ny}. \quad (9\text{–}89)$$

Moreover, since the operators act on functions of different variables,

$$[\mathbf{J}_1, \mathbf{J}_2] = 0. \qquad (9\text{–}90)$$

If we consider the two uncoupled systems together, their Hamiltonian will be invariant under any combined rotation R_1S_2, where R and S can be different rotations. The Hilbert space of the representation will be the product of the spaces for the subsystems (i.e., it will consist of products of functions for the subsystems). The operators $O_{R_1}O_{S_2}$ acting in this space will give an irreducible representation of this direct product. By setting S equal to the identity, we can obtain the infinitesimal operators $\mathbf{J}_1$, and by setting R equal to the identity, we obtain $\mathbf{J}_2$. So for the direct product we have the six independent infinitesimal operators $\mathbf{J}_1, \mathbf{J}_2$. When the systems are coupled to one another by adding terms to the Hamiltonian which depend on the distance between 1 and 2, the total Hamiltonian will no longer be invariant under separate rotations of 1 and 2. The symmetry group will be reduced from the direct product R_1S_2 to the group in which 1 and 2 are rotated through the same angle, R_1R_2. Instead of the operators $O_{R_1}O_{S_2}$ we must now consider the subgroup of operators $O_{R_1}O_{R_2}$. For this subgroup of the direct product there are only three infinitesimal operators,

$$\mathbf{J} = \mathbf{J}_1 + \mathbf{J}_2. \tag{9-91}$$

The operators $\mathbf{J}_1$ and $\mathbf{J}_2$ satisfy Eqs. (9–89) and (9–90), and the operators $\mathbf{J}$ necessarily satisfy Eq. (9–89). If the Hilbert spaces of 1 and 2 correspond to irreducible representations, our problem is to determine which irreducible representations of $\mathbf{J}$ are contained in the product space.

Similarly, if we couple r systems, we go over from the direct product $O_{R_1}O_{S_2}\cdots O_{T_r}$ to $O_{R_1}O_{R_2}\cdots O_{R_r}$, and from the $3r$ infinitesimal operators $\mathbf{J}_1, \mathbf{J}_2, \ldots, \mathbf{J}_r$ to the three operators

$$\mathbf{J} = \sum_{n=1}^{r} \mathbf{J}_n. \tag{9-92}$$

As shown in Section 8–13, the numerical values of the Casimir operators $\mathbf{J}^2, \mathbf{J}_1^2, \ldots, \mathbf{J}_r^2$ characterize the irreducible representations. For given eigenvalues of $\mathbf{J}_1^2, \ldots, \mathbf{J}_r^2$, one then calculates the eigenvalues of $\mathbf{J}^2$ in the manner used in texts on quantum mechanics.

The Clebsch-Gordan series for the rotation group can be easily found by using the characters

$$\chi^{(j)}(\phi) = \sum_{m=-j}^{j} e^{im\phi}. \tag{9-77}$$

The character of the direct product $D^{(j_1)} \times D^{(j_2)}$ of two irreducible representations is

$$\chi^{(j_1 \times j_2)}(\phi_1, \phi_2) = \sum_{m_1=-j_1}^{j_1} e^{im_1\phi_1} \sum_{m_2=-j_2}^{j_2} e^{im_2\phi_2}.$$

For the coupled systems we must consider only those group elements for which $\phi_1 = \phi_2 = \phi$, so that

$$\chi^{(j_1 \times j_2)}(\phi) = \sum_{m_1=-j_1}^{j_1} e^{im_1\phi} \sum_{m_2=-j_2}^{j_2} e^{im_2\phi}$$

$$= \sum_{m_1=-j_1}^{j_1} \sum_{m_2=-j_2}^{j_2} e^{i(m_1+m_2)\phi}$$

$$= \sum_{J=|j_1-j_2|}^{j_1+j_2} \sum_{M=-J}^{J} e^{iM\phi} = \sum_{J=|j_1-j_2|}^{j_1+j_2} \chi^{(J)}(\phi). \quad (9\text{--}93)$$

The Clebsch-Gordan series is therefore

$$D^{(j_1)} \times D^{(j_2)} = \sum_{J=|j_1-j_2|}^{j_1+j_2} D^{(J)}. \quad (9\text{--}94)$$

Each irreducible representation is contained at most once in the product of two irreducible representations. The rotation group is simply reducible (cf. Section 5–8).

One of the principal tools for physical applications is the use of the Clebsch-Gordan coefficients for the rotation group. These are the coefficients in the expansion of basis functions Ψ_M^J of $D^{(J)}$ in products $\psi_{m_1}^{j_1}\psi_{m_2}^{j_2}$ of basis functions of the irreducible representations $D^{(j_1)}$ and $D^{(j_2)}$. We discussed the general problem of symmetry of CG-coefficients in Sections 5–7 through 5–9. The general formula for the CG-coefficients for the rotation group (the vector-addition coefficients) has been derived in many ways. The method given here is probably the simplest of all.

Instead of working with the rotation group, we consider its covering group, the unimodular, unitary group $\mathfrak{U}_2$. From our discussion in Section 9–6, the treatment of $\mathfrak{U}_2$ gives all the necessary information about the rotation group. If the pair of variables u_1, u_2 transform according to Eq. (9–56), that is,

$$\begin{aligned} u_1' &= au_1 + bu_2 \\ u_2' &= -b^*u_1 + a^*u_2 \end{aligned} \qquad (aa^* + bb^* = 1), \qquad (9\text{--}56)$$

or, in matrix form,

$$u' = mu, \qquad u = \begin{bmatrix} u_1 \\ u_2 \end{bmatrix}, \qquad m = \begin{bmatrix} a & b \\ -b^* & a^* \end{bmatrix}, \qquad (9\text{--}95)$$

then the set

$$f_m = \frac{u_1^{j+m}u_2^{j-m}}{\sqrt{(j+m)!(j-m)!}} \qquad (m = -j, -j+1, \ldots, j-1, j) \qquad (9\text{--}67)$$

forms the basis for the irreducible representation $D^{(j)}$. We now consider a pair of variables x_1, x_2 which transform as follows:

$$x_1' = a^*x_1 + b^*x_2, \qquad x_2' = -bx_1 + ax_2; \qquad (9\text{-}96)$$

$$x' = m^*x. \qquad (9\text{-}96a)$$

Since $m^\dagger m = 1$, $m^* = \tilde{m}^{-1}$, so that

$$x' = \tilde{m}^{-1}x. \qquad (9\text{-}97)$$

The variables x_1, x_2 transform according to the adjoint representation (complex-conjugate representation; cf. Section 5–3). In some texts the *contravariant* variables x are said to transform *contragrediently* to the u's of Eq. (9–56), while any other variables which transform like the u's are said to transform *cogrediently* (or *covariantly*).

From Eqs. (9–56) and (9–97),

$$\tilde{x}'u' = \tilde{x}m^{-1}mu = \tilde{x}u, \qquad (9\text{-}98)$$

so that $\tilde{x}u = x_1u_1 + x_2u_2$ is invariant. We also notice that if we replace the variables (u_1, u_2) by $(u_2, -u_1)$, Eq. (9–56) changes into Eq. (9–96). We go from the variables (u_1, u_2) to the contravariant set $(u_2, -u_1)$ by the transformation

$$\begin{bmatrix} u_2 \\ -u_1 \end{bmatrix} = \begin{bmatrix} 0 & 1 \\ -1 & 0 \end{bmatrix} \begin{bmatrix} u_1 \\ u_2 \end{bmatrix} = g \begin{bmatrix} u_1 \\ u_2 \end{bmatrix}, \qquad (9\text{-}99)$$

$$g = \begin{bmatrix} 0 & 1 \\ -1 & 0 \end{bmatrix}. \qquad (9\text{-}100)$$

If we have a second set of covariant variables (v_1, v_2), the product

$$(v_1, v_2) \begin{bmatrix} u_2 \\ -u_1 \end{bmatrix} = (v_1, v_2) \begin{bmatrix} 0 & 1 \\ -1 & 0 \end{bmatrix} \begin{bmatrix} u_1 \\ u_2 \end{bmatrix} = \tilde{v}gu \qquad (9\text{-}101)$$

is invariant under the transformation m. We see from Eqs. (9–99) and (9–100) that g serves as a metric matrix.

We now consider functions like the f_m of Eq. (9–67), for $j = j_1$ and $j = j_2$,

$$\psi_{m_1}^{j_1} = \frac{u_1^{j_1+m_1} u_2^{j_1-m_1}}{\sqrt{(j_1 + m_1)!(j_1 - m_1)!}}, \qquad \psi_{m_2}^{j_2} = \frac{v_1^{j_2+m_2} v_2^{j_2-m_2}}{\sqrt{(j_2 + m_2)!(j_2 - m_2)!}}. \qquad (9\text{-}102)$$

For any J between $|j_1 - j_2|$ and $j_1 + j_2$ we construct the polynomial

$$A_J = (u_1v_2 - u_2v_1)^{j_1+j_2-J}(u_1x_1 + u_2x_2)^{j_1-j_2+J}(v_1x_1 + v_2x_2)^{j_2-j_1+J}, \qquad (9\text{-}103)$$

where x_1, x_2 are a pair of contravariant variables. The polynomial A_J is of degree $2j_1$ in u_1, u_2 (like the $\psi_{m_1}^{j_1}$); its degree in the variables v_1, v_2 is $2j_2$ (like the $\psi_{m_2}^{j_2}$). Its degree in the contravariant variables x_1, x_2 is $2J$. From Eqs. (9–98) and (9–101) we see that the quantities in parentheses in (9–103) are invariant under transformations of $\mathfrak{U}_2$ so that A_J is an invariant. If we expand the invariant A_J in powers of x_1, x_2, we obtain

$$A_J = \sum_{M=-J}^{J} W_M^J X_M^J, \tag{9–104}$$

where

$$X_M^J = \frac{x_1^{J+M} x_2^{J-M}}{\sqrt{(J+M)!(J-M)!}}. \tag{9–105}$$

The coefficients W_M^J are polynomials in u_1, u_2, v_1, v_2. Now we compare A_J with the invariant quantity

$$B_J = (u_1 x_1 + u_2 x_2)^{2J}$$

$$= (2J)! \sum_{M=-J}^{J} \frac{1}{(J+M)!(J-M)!} u_1^{J+M} x_1^{J+M} u_2^{J-M} x_2^{J-M}$$

$$= (2J)! \sum_{M=-J}^{J} \Psi_M^J X_M^J. \tag{9–106}$$

Since the W_M^J of Eq. (9–104) and the Ψ_M^J of Eq. (9–106) both transform contravariantly to the X_M^J, they both transform in the same way, so the W_M^J are a basis for $D^{(J)}$. To calculate the W_M^J we expand A_J, using the binomial theorem,

$$(u_1 v_2 - u_2 v_1)^{j_1+j_2-J}$$
$$= \sum_{\lambda}^{j_1+j_2-J} (-1)^{\lambda} \binom{j_1+j_2-J}{\lambda} (u_1 v_2)^{j_1+j_2-J-\lambda} (u_2 v_1)^{\lambda},$$

$$(u_1 x_1 + u_2 x_2)^{j_1-j_2+J} = \sum_{\mu}^{j_1-j_2+J} \binom{j_1-j_2+J}{\mu} (u_1 x_1)^{j_1-j_2+J-\mu} (u_2 x_2)^{\mu},$$

$$(v_1 x_1 + v_2 x_2)^{j_2-j_1+J} = \sum_{\nu}^{j_2-j_1+J} \binom{j_2-j_1+J}{\nu} (v_1 x_1)^{j_2-j_1+J-\nu} (v_2 x_2)^{\nu}.$$

$$A_J = \sum_{\lambda,\mu,\nu} (-1)^{\lambda} \binom{j_1+j_2-J}{\lambda} \binom{j_1-j_2+J}{\mu} \binom{j_2-j_1+J}{\nu}$$
$$\times u_1^{2j_1-\lambda-\mu} u_2^{\lambda+\mu} v_1^{j_2-j_1+J-\nu+\lambda} v_2^{j_1+j_2-J-\lambda+\nu} x_1^{2J-\mu-\nu} x_2^{\mu+\nu}, \tag{9–107}$$

and introduce the new summation variables $m_1 = j_1 - \lambda - \mu$, $m_2 = J - j_1 + \lambda - \nu$. Then

$$A_J = \sum_{m_1, m_2, \lambda} (-1)^\lambda \begin{pmatrix} j_1 + j_2 - J \\ \lambda \end{pmatrix} \begin{pmatrix} j_1 - j_2 + J \\ j_1 - \lambda - m_1 \end{pmatrix} \begin{pmatrix} j_2 - j_1 + J \\ j_2 - \lambda + m_2 \end{pmatrix}$$
$$\times u_1^{j_1 + m_1} u_2^{j_1 - m_1} v_1^{j_2 + m_2} v_2^{j_2 - m_2} x_1^{J + m_1 + m_2} x_2^{J - m_1 - m_2}. \qquad (9\text{-}108)$$

Now we use Eqs. (9-102) and (9-105) and find

$$A_J = \sum_{m_1, m_2, \lambda} (-1)^\lambda$$
$$\times \left[\frac{(j_1 + j_2 - J)!(j_1 - j_2 + J)!(j_2 - j_1 + J)!}{\lambda!(j_1 + j_2 - J - \lambda)!(j_1 - \lambda - m_1)!} \right.$$
$$\times \left. \frac{\{(j_1 + m_1)!(j_1 - m_1)!(j_2 + m_2)!(j_2 - m_2)!(J + M)!(J - M)!\}^{1/2}}{(J - j_2 + \lambda + m_1)!(j_2 - \lambda + m_2)!(J - j_1 + \lambda - m_2)!} \right]$$
$$\times \psi_{m_1}^{j_1} \psi_{m_2}^{j_2} X_{m_1 + m_2}^J. \qquad (9\text{-}109)$$

Setting $m_1 + m_2 = M$, we find the coefficient of X_M^J in Eq. (9-109):

$$W_M^J = (j_1 + j_2 - J)!(j_1 - j_2 + J)!(j_2 - j_1 + J)!$$
$$\times \sum_{\substack{m_1, m_2 \\ m_1 + m_2 = M}} c_{m_1 m_2}^J \psi_{m_1}^{j_1} \psi_{m_2}^{j_2}, \qquad (9\text{-}110)$$

where

$$c_{m_1 m_2}^J$$
$$= \sum_\lambda (-1)^\lambda \left[\frac{\{(j_1 + m_1)!(j_1 - m_1)!(j_2 + m_2)!(j_2 - m_2)!\}^{1/2}}{\lambda!(j_1 + j_2 - J - \lambda)!(j_1 - \lambda - m_1)!} \right.$$
$$\times \left. \frac{\{(J + M)!(J - M)!\}^{1/2}}{(J - j_2 + \lambda + m_1)!(j_2 - \lambda + m_2)!(J - j_1 + \lambda - m_2)!} \right].$$
$$(9\text{-}111)$$

The transformation from the products $\psi_{m_1}^{j_1} \psi_{m_2}^{j_2}$ to basis functions Ψ_M^J is made unitary by introducing a normalization factor ρ_J in Eq. (9-110), so that

$$\Psi_M^J = \rho_J \sum_{\substack{m_1, m_2 \\ m_1 + m_2 = M}} c_{m_1 m_2}^J \psi_{m_1}^{j_1} \psi_{m_2}^{j_2} \qquad (9\text{-}112)$$

is a unitary transformation with

$$|\rho_J|^2 \sum_{\substack{m_1, m_2 \\ m_1 + m_2 = M}} |c_{m_1 m_2}^J|^2 = 1. \qquad (9\text{-}113)$$

Since ρ_J is a number independent of M, we can evaluate it by choosing M conveniently in Eq. (9–113). We set $M = J$ in Eq. (9–113) and use Eq. (9–111). The factorial $(j_1 - \lambda - m_1)!$ destroys all terms with $\lambda > j_1 - m_1$ in $c^J_{m_1m_2}$, and the factorial $(J - j_1 + \lambda - m_2)! = (\lambda - \{j_1 - m_1\})!$ destroys all terms with $\lambda < (j_1 - m_1)$. So we are left with the one term for $\lambda = j_1 - m_1$, and $c^J_{m_1m_2}$ becomes

$$c^J_{m_1m_2} = (-1)^{j_1-m_1} \frac{\sqrt{(2J)!}}{(J - j_2 + j_1)!(J - j_1 + j_2)!}$$

$$\times \left[\frac{(j_1 + m_1)!(j_2 + m_2)!}{(j_1 - m_1)!(j_2 - m_2)!}\right]^{1/2}$$

$$= (-1)^{j_1-m_1}$$

$$\times \left[\frac{(2J)!}{(J - j_2 + j_1)!(J - j_1 + j_2)!}\right]^{1/2}$$

$$\times \left[\binom{j_1 + m_1}{j_2 - m_2}\binom{j_2 + m_2}{j_1 - m_1}\right]^{1/2} \quad (m_1 + m_2 = J). \quad (9\text{–}114)$$

For this special case,

$$\sum_{\substack{m_1, m_2 \\ m_1+m_2=J}} |c^J_{m_1m_2}|^2$$

$$= \frac{(2J)!}{(J - j_2 + j_1)!(J - j_1 + j_2)!} \sum_{\substack{m_1, m_2 \\ m_1+m_2=J}} \binom{j_1 + m_1}{j_2 - m_2}\binom{j_2 + m_2}{j_1 - m_1}.$$

$$(9\text{–}115)$$

This sum can be simplified by using the relation

$$\Gamma(z)\Gamma(1 - z) = \frac{\pi}{\sin \pi z}. \quad (9\text{–}116)$$

We interpret all factorials as Γ-functions. Then

$$\binom{x}{y} = \frac{x!}{y!(x - y)!} = \frac{\Gamma(x + 1)}{y!\Gamma(x - y + 1)}$$

$$= \frac{\pi}{\sin \pi(x + 1)} \frac{\sin \pi(x - y + 1)}{\pi} \frac{\Gamma(y - x)}{y!\Gamma(-x)}$$

$$= \frac{(-1)^y \Gamma(y - x)}{y!\Gamma(-x)} = (-1)^y \binom{y - x - 1}{y}. \quad (9\text{–}117)$$

Using this result, we find that Eq. (9–115) becomes

$$\sum_{\substack{m_1,m_2 \\ m_1+m_2=J}} |c_{m_1m_2}^J|^2 = \frac{(-1)^{j_1+j_2-J}(2J)!}{(J-j_2+j_1)!(J-j_1+j_2)!}$$

$$\times \sum_{\substack{m_1,m_2 \\ m_1+m_2=J}} \binom{j_2-j_1-J-1}{j_2-m_2}\binom{j_1-j_2-J-1}{j_1-m_1}.$$

$$(9\text{–}118)$$

Again from the binomial theorem, we obtain

$$(1+x)^r(1+x)^s = \sum_\alpha \binom{r}{\alpha}x^\alpha \sum_\beta \binom{s}{\beta}x^\beta = \sum_{\alpha,\beta}\binom{r}{\alpha}\binom{s}{\beta}x^{\alpha+\beta}$$

$$= (1+x)^{r+s} = \sum_\gamma \binom{r+s}{\gamma}x^\gamma,$$

so that

$$\binom{r+s}{\gamma} = \sum_{\substack{\alpha,\beta \\ \alpha+\beta=\gamma}} \binom{r}{\alpha}\binom{s}{\beta}.$$

$$(9\text{–}119)$$

Applying this relation to Eq. (9–118), we have

$$\sum_{\substack{m_1,m_2 \\ m_1+m_2=J}} |c_{m_1m_2}^J|^2 = \frac{(-1)^{j_1+j_2-J}(2J)!}{(J-j_2+j_1)!(J-j_1+j_2)!}\begin{bmatrix} -2J-2 \\ j_1+j_2-J \end{bmatrix}$$

$$= \frac{(2J)!}{(J-j_2+j_1)!(J-j_1+j_2)!}\begin{bmatrix} j_1+j_2+J+1 \\ j_1+j_2-J \end{bmatrix},$$

$$(9\text{–}120)$$

where we have again used Eq. (9–117) in the last step. From Eq. (9–113) we now find

$$\rho_J = \left[\frac{(2J+1)(J-j_2+j_1)!(J-j_1+j_2)!(j_1+j_2-J)!}{(j_1+j_2+J+1)!}\right]^{1/2}.$$

$$(9\text{–}121)$$

The Clebsch-Gordan coefficients are

$$(j_1m_1j_2m_2|JM) = \rho_J c_{m_1m_2}^J \qquad (M=m_1+m_2), \qquad (9\text{–}122)$$

with ρ_J given by Eq. (9–121) and $c_{m_1m_2}^J$ given by Eq. (9–111).

The preceding formulas are quite complicated, but they become considerably simplified in some special cases:

(a) $J = j_1 + j_2$: Only the term with $\lambda = 0$ remains in Eq. (9–111), and

$$c_{m_1 m_2}^J = \left[\frac{(J + M)!(J - M)!}{(j_1 + m_1)!(j_1 - m_1)!(j_2 + m_2)!(j_2 - m_2)!} \right]^{1/2},$$

$$\rho_J = \sqrt{\frac{(2j_1)!(2j_2)!}{(2J)!}},$$

$$(j_1 m_1 j_2 m_2 | J = j_1 + j_2, M) = \left[\frac{\binom{2j_1}{j_1 - m_1}\binom{2j_2}{j_2 - m_2}}{\binom{2J}{J - M}} \right]^{1/2}. \quad (9\text{–}123)$$

(b) $J = j_1 - j_2$: Only the term with $\lambda = j_2 - m_2$ remains in Eq. (9–111), and

$$c_{m_1 m_2}^J = (-1)^{j_2 + m_2} \left[\frac{(j_1 + m_1)!(j_1 - m_1)!}{(j_2 + m_2)!(j_2 - m_2)!(J + M)!(J - M)!} \right]^{1/2},$$

$$\rho_J = \sqrt{\frac{(2J + 1)!(2j_2)!}{(2j_1 + 1)!}}.$$

$$(9\text{–}124)$$

For $j_2 = \frac{1}{2}$, $J = j_1 \pm \frac{1}{2}$, so these two special cases give the complete set of coefficients.

CLEBSCH–GORDAN COEFFICIENTS FOR $j_2 = \frac{1}{2}$

	$m_2 = \frac{1}{2}$	$m_2 = -\frac{1}{2}$
$J = j_1 + \frac{1}{2}$	$\sqrt{\dfrac{j_1 + m_1 + 1}{2j_1 + 1}}$	$\sqrt{\dfrac{j_1 - m_1 + 1}{2j_1 + 1}}$
$j_1 - \frac{1}{2}$	$-\sqrt{\dfrac{j_1 - m_1}{2j_1 + 1}}$	$\sqrt{\dfrac{j_1 + m_1}{2j_1 + 1}}$

Complete discussions of the properties of the CG-coefficients and numerical tables are available.

We found that we arrive at the complex-conjugate representation by making the substitution $u_1 \to u_2$, $u_2 \to -u_1$. Under this substitution,

$$\psi_m^j \to (-1)^{j-m} \psi_{-m}^j, \quad (9\text{–}125)$$

and the function ψ^j_{-m} transforms according to $D^{(j)*}$. From Eq. (9–125) and Eq. (5–140) we find the relation between the $3j$- and CG-coefficients:

$$\begin{pmatrix} j_1 j_2 j_3 \\ m_1 m_2 m_3 \end{pmatrix} = (-1)^{j_3-m_3}(2j_3 + 1)^{-1/2}(j_1 m_1 j_2 m_2 | j_3, -m_3). \quad (9\text{–}126)$$

From this equation and the results of Section 5–9 one can find the symmetry properties of the CG-coefficients.

CHAPTER 10

LINEAR GROUPS IN n-DIMENSIONAL SPACE; IRREDUCIBLE TENSORS

In this chapter, we define tensors with respect to any group G of linear transformations in n-dimensional space. The tensors of rank r form a vector space of n^r dimensions and constitute the basis for a representation of the group G. By using permutation operators (Young symmetrizers), we can decompose this representation into irreducible representations of G.

For certain subgroups of the general linear group $GL(n)$, such as the orthogonal group $O(n)$ and the symplectic group $Sp(n)$, we can define a process of contraction of tensor indices which leads to a further reduction.

The methods used in this chapter are closely related to the treatment of the symmetric group in Chapter 7. In many cases the results of Chapter 7 will be restated in a new terminology.

10–1 Tensors with respect to $GL(n)$. In Section 5–1 we discussed the construction of product representations. Suppose that we are given a group G of linear transformations in an n-dimensional space R_n (where, in particular, the group G may be a faithful representation of some abstract group). A vector $\mathbf{x}$ in the space R_n has components $x_1, \ldots, x_n$. The transformations $\mathbf{a}$ of the group G transform $\mathbf{x}$ into $\mathbf{x}'$:

$$\mathbf{x}' = \mathbf{a}\mathbf{x}, \quad x_i' = a_{ij}x_j \quad (i = 1, \ldots, n). \tag{10-1}$$

We now consider the n^2 quantities $x_i y_j$ $(i, j = 1, \ldots, n)$ which can be formed by taking products of the components of two vectors $\mathbf{x}$ and $\mathbf{y}$ in R_n. When the transformation (10–1) is applied to vectors in R_n, the set of quantities $x_i y_j$ is subjected to the transformation

$$x_i' y_j' = a_{ik}a_{jl}x_k y_l. \tag{10-2}$$

We see that the n^2 quantities $x_i y_j$ transform according to $\mathbf{a} \times \mathbf{a}$, the Kronecker square of the transformation $\mathbf{a}$.

A set of n^2 quantities F_{ij}, whose law of transformation is

$$F_{ij}' = a_{ik}a_{jl}F_{kl}, \tag{10-3}$$

form a *tensor* $\mathfrak{F}_{..}$ *of rank two*.

The group G of transformations in the n-dimensional space R_n induces the group of transformations $\mathbf{a} \times \mathbf{a}$ in the n^2-dimensional space of tensors $\mathfrak{F}_{..}$. The tensor $\mathfrak{F}_{..}$ is described in a given coordinate basis by its n^2 components F_{ij}.

The n^2 quantities $x_i y_j$ formed from the components of any two vectors **x** and **y** of R_n are the components of a second-rank tensor. In particular we can choose for **x** the vector whose μth component is 1 while all other components are 0, and similarly, choose **y** with its νth component equal to 1 and all other components zero. The second-rank tensor $^{\mu\nu}(\mathbf{xy})$ constructed from these two vectors has one nonzero component:

$$^{\mu\nu}(\mathbf{xy})_{ij} = \delta_{i\mu}\,\delta_{j\nu}. \tag{10-4}$$

By varying μ, ν over the values $1, 2, \ldots, n$, we obtain n^2 independent second-rank tensors. All the second-rank tensors $\mathfrak{F}..$ are expressible as linear combinations of the basis tensors $^{\mu\nu}(\mathbf{xy})$.

Again we emphasize that the tensors $\mathfrak{F}..$ are defined *with respect to the group* G, since the law of transformation (10–3) is determined by (10–1). If we choose a different group of linear transformations in n dimensions, we obtain a *different* space of second-rank tensors.

In the same way, we can define rth-rank tensors with respect to the group G. When the **x**'s are subjected to the transformation (10–1), the n^r quantities $x_{i_1}^{(1)} x_{i_2}^{(2)} \ldots x_{i_r}^{(r)}$ $(i_\mu = 1, \ldots, n, \mu = 1, \ldots, r)$ formed from r vectors $\mathbf{x}^{(1)}, \mathbf{x}^{(2)}, \ldots, \mathbf{x}^{(r)}$ in R_n transform as follows:

$$x_{i_1}^{(1)\prime} x_{i_2}^{(2)\prime} \ldots x_{i_r}^{(r)\prime} = a_{i_1 j_1} a_{i_2 j_2} \ldots a_{i_r j_r} x_{j_1}^{(1)} x_{j_2}^{(2)} \ldots x_{j_r}^{(r)}. \tag{10-5}$$

The *rth-rank tensor* $\mathfrak{F}.....$ is a quantity which is described by n^r components $F_{i_1 i_2 \ldots i_r}$ in a given coordinate basis, and which transforms like the product of r vectors:

$$F'_{i_1 i_2 \ldots i_r} = a_{i_1 j_1} a_{i_2 j_2} \ldots a_{i_r j_r} F_{j_1 j_2 \ldots j_r}. \tag{10-6}$$

In other words, the transformation **a** in R_n induces the transformation $\mathbf{a} \times \mathbf{a} \times \cdots \times \mathbf{a}$ (with r factors) in the space of rth-rank tensors.

At first we shall choose for the group G the general linear group $GL(n)$ of all nonsingular linear transformations in n-dimensional space. Later in this chapter we shall consider some of the subgroups of $GL(n)$.

Problems. (1) Show that every rth-rank tensor can be expressed linearly in terms of products formed from r vectors in R_n.

(2) Show that the matrix elements for transformations of rth-rank tensors are homogeneous polynomials of degree r in the matrix elements of the transformations of the group G.

10–2 The construction of irreducible tensors with respect to $GL(n)$.
We start with second-rank tensors. Section 5–2 showed that the Kronecker square $\mathbf{a} \times \mathbf{a}$ is reducible. By permuting the indices i_1, i_2 of $F_{i_1 i_2}$ we ob-

tain the tensors $(F_{i_1 i_2} + F_{i_2 i_1})$ and $(F_{i_1 i_2} - F_{i_2 i_1})$ which form the bases for the symmetric and antisymmetric product representations, respectively. Thus, permuting the tensor indices and taking linear combinations decomposes the space of second-rank tensors into two invariant subspaces.

The process which was used in this simple case can be described as follows: To each permutation p of the symmetric group S_2 we associate an operator $\mathbf{p}$ which acts on second-rank tensors $F_{i_1 i_2}$. The operator $\mathbf{p}$ applied to F gives a tensor $\mathbf{p}F$, where

$$(\mathbf{p}F)_{i_1 i_2} = F_{i_{1'} i_{2'}}, \quad \text{for} \quad p = \begin{pmatrix} 1 \, 2 \\ 1'2' \end{pmatrix}. \tag{10–7}$$

The operator $\mathbf{p}$ acts on the subindices 1, 2. For example, consider a tensor $F_{i_1 i_2}$ for $n = 4$. The tensor component F_{34} has indices $i_1 = 3$, $i_2 = 4$. The permutation operator $\mathbf{p}$ for $p = (12)$ takes i_1 into i_2 and i_2 into i_1, that is, it takes 3 into 4 and 4 into 3:

$$(\mathbf{p}F)_{34} = F_{43}.$$

Similarly, for F_{23}, $i_1 = 2$, $i_2 = 3$, so that $(\mathbf{p}F)_{23} = F_{32}$. For the component F_{33}, $i_1 = i_2 = 3$, so $(\mathbf{p}F)_{33} = F_{33}$.

The symmetric and antisymmetric tensors $F_{i_1 i_2} \pm F_{i_2 i_1}$ can be written as

$$([\mathbf{e} \pm \mathbf{p}]F)_{i_1 i_2},$$

where $\mathbf{e}$ is the identity operator. We obtain the decomposition of the space of tensors $F_{i_1 i_2}$ by applying the operators $(\mathbf{e} \pm \mathbf{p})$. But, as we saw in Section 7–12, these operators are the Young symmetrizers which generate the irreducible representations of S_2.

The operator $\mathbf{p}$ *commutes* with the transformations (10–3) in the tensor space:

$$\begin{aligned}
(\mathbf{p}F')_{i_1 i_2} = F'_{i_2 i_1} &= a_{i_2 j_1} a_{i_1 j_2} F_{j_1 j_2} \\
&= a_{i_2 j_2} a_{i_1 j_1} F_{j_2 j_1} \\
&= a_{i_1 j_1} a_{i_2 j_2} (\mathbf{p}F)_{j_1 j_2}, \tag{10–8}
\end{aligned}$$

i.e., symbolically,

$$\mathbf{p}(\mathbf{a} \times \mathbf{a})F = (\mathbf{a} \times \mathbf{a})\mathbf{p}F. \tag{10–8a}$$

The reason for this is that the product $a_{i_1 j_1} a_{i_2 j_2}$ is *bisymmetric:* when the same permutation is applied to i_1, i_2 and to j_1, j_2, the product is unchanged.

The transformation law for rth-rank tensors is

$$F'_{i_1 i_2 \ldots i_r} = a_{i_1 j_1} a_{i_2 j_2} \ldots a_{i_r j_r} F_{j_1 j_2 \ldots j_r}, \tag{10–6}$$

which we abbreviate to

$$F'_{(i)} = a_{(i)(j)} F_{(j)}. \tag{10–6a}$$

To each permutation $p = \left(\begin{smallmatrix}1, & 2, & \cdots r \\ 1', & 2', & \cdots r'\end{smallmatrix}\right)$ of the symmetric group S_r we associate an operator $\mathbf{p}$ which acts on the subindices of the tensor $F_{i_1 i_2 \ldots i_r}$,

$$(\mathbf{p}F)_{i_1 i_2 \ldots i_r} = F_{i_1' i_2' \ldots i_r'} \equiv F_{p(i)}, \tag{10-9}$$

or, in abbreviated notation,

$$(\mathbf{p}F)_{(i)} = F_{p(i)}. \tag{10-9a}$$

Then

$$\begin{aligned}
(\mathbf{p}F')_{(i)} &= (F')_{p(i)} = a_{p(i)p(j)} F_{p(j)} \\
&= a_{p(i)p(j)} (\mathbf{p}F)_{(j)} \\
&= a_{(i)(j)} (\mathbf{p}F)_{(j)},
\end{aligned} \tag{10-10}$$

where we have used (10-9) in the second step and the fact that the tensor transformation is *bisymmetric* in the last step:

$$a_{p(i)p(j)} \equiv a_{i_1' j_1'} a_{i_2' j_2'} \ldots a_{i_r' j_r'} = a_{i_1 j_1} a_{i_2 j_2} \ldots a_{i_r j_r} \equiv a_{(i)(j)}. \tag{10-11}$$

Equation (10-10) states that the permutation operators $\mathbf{p}$ commute with all bisymmetric transformations in the tensor space. Therefore those tensors of rank r which have a particular symmetry will be transformed among themselves by the transformations (10-6). The whole space of rth-rank tensors is therefore reducible into subspaces consisting of tensors of different symmetry.

In order to obtain rth-rank tensors which have a definite symmetry type, we apply the Young symmetrizers (cf. Section 7-10) to the general tensor $F_{i_1 \ldots i_r}$. To each Young pattern $[\lambda_1 \ldots \lambda_k]$, with $\sum_{i=1}^{k} \lambda_i = r$, there corresponds a particular symmetry type of tensors of rank r. To indicate the symmetry type of a tensor we shall write its indices in the boxes of the Young diagram. For example, for $r = 2$, we write the symmetric tensors as

$$F_{\boxed{i_1}\boxed{i_2}}$$

and the antisymmetric tensors as

$$F_{\substack{\boxed{i_1} \\ \boxed{i_2}}}$$

For $r = 3$, there are three symmetry classes,

$$F_{\boxed{i_1}\boxed{i_2}\boxed{i_3}}, \quad F_{\substack{\boxed{i_1}\boxed{i_2} \\ \boxed{i_3}}}, \quad F_{\substack{\boxed{i_1} \\ \boxed{i_2} \\ \boxed{i_3}}},$$

corresponding to the partitions [3], [21], and [111]. The rth-rank tensors, with the symmetry described by the partition $[\lambda_1 \ldots \lambda_k]$, have the form

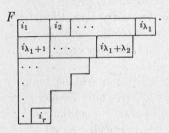

Tensors of this symmetry type are generated by applying to the general rth-rank tensor the Young operator $\mathbf{Y} = \mathbf{QP}$, where $\mathbf{P}$ is the operator for the horizontal permutations in the diagram and $\mathbf{Q}$ is the operator for the vertical permutations (cf. Section 7–10). Consequently the tensor F will be antisymmetric in all the indices which appear in the same column. Any tensor component for which an index appears twice in the same column is necessarily equal to zero.

To illustrate the procedure for constructing tensors of given symmetry, we start from a general tensor $G_{i_1 i_2 i_3 i_4}$ and construct the tensor $F_{\boxed{i_1}\boxed{i_2}\,\boxed{i_3}\boxed{i_4}}$.

The Young operator for this diagram is $\mathbf{Y} = \mathbf{QP}$, with

$$P = [e + (12)][e + (34)], \qquad Q = [e - (13)][e - (24)].$$

Then

$$(\mathbf{P}G)_{i_1 i_2 i_3 i_4} = G_{i_1 i_2 i_3 i_4} + G_{i_2 i_1 i_3 i_4} + G_{i_1 i_2 i_4 i_3} + G_{i_2 i_1 i_4 i_3},$$

and

$$F_{\boxed{i_1}\boxed{i_2}\,\boxed{i_3}\boxed{i_4}} = (\mathbf{QP}G)_{i_1 i_2 i_3 i_4} = G_{i_1 i_2 i_3 i_4} - G_{i_3 i_2 i_1 i_4} - G_{i_1 i_4 i_3 i_2} + G_{i_3 i_4 i_1 i_2}$$

$$+ G_{i_2 i_1 i_3 i_4} - G_{i_2 i_3 i_1 i_4} - G_{i_4 i_1 i_3 i_2} + G_{i_4 i_3 i_1 i_2}$$

$$+ G_{i_1 i_2 i_4 i_3} - G_{i_3 i_2 i_4 i_1} - G_{i_1 i_4 i_2 i_3} + G_{i_3 i_4 i_2 i_1}$$

$$+ G_{i_2 i_1 i_4 i_3} - G_{i_2 i_3 i_4 i_1} - G_{i_4 i_1 i_2 i_3} + G_{i_4 i_3 i_2 i_1}.$$

Problems. (1) Find the tensor $F_{\boxed{i_1}\boxed{i_2}\boxed{i_3}\,\boxed{i_4}}$ which is generated from the general tensor $G_{i_1 i_2 i_3 i_4}$.

(2) Use the results of Section 7–10 to resolve the general tensor $G_{i_1 i_2 i_3 i_4}$ into a sum of tensors of definite symmetry type.

For the general linear group $GL(n)$, the matrix elements a_{ij} are not subject to any restrictive conditions; so the only process of reduction of the tensor space is the symmetrization process which we have used. The rth-rank tensors of a given symmetry form the basis for an irreducible representation of $GL(n)$; in other words, they are *irreducible tensors with respect to $GL(n)$*. We shall see later that, for certain subgroups of $GL(n)$, a further reduction is possible.

Which of the symmetry types will be realized for given vaiues of n and r? If the Young pattern contains more than n rows, at least one index must be repeated in the first column, so that all the tensors of this symmetry type must be identically equal to zero. We can therefore restrict ourselves to patterns with a maximum of n rows, and write the partitions as $[\lambda_1 \ldots \lambda_n]$, with $\lambda_1 + \cdots + \lambda_n = r$, $\lambda_1 \geq \cdots \geq \lambda_n \geq 0$. If the pattern contains fewer than n rows, some of the λ's are equal to zero and can be omitted from the partition label.

Furthermore, we can show that every pattern with n rows or less is realized, i.e., there exist nonzero tensors of all such symmetry types. Consider a Young pattern T with $m \leq n$ rows. We start from a tensor G which (in a particular basis) has all its components equal to zero except for the one component

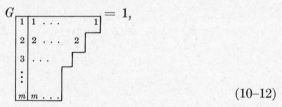

$$G = 1, \tag{10-12}$$

where we have arranged the indices according to the Young pattern T. When we apply the symmetrizer $\mathbf{Y} = \mathbf{QP}$ for the pattern T, the operator $\mathbf{P}$ leaves the tensor (10–12) unchanged (except for a multiplicative factor). The operator $\mathbf{Q}$ permutes the indices in each column separately. Hence, except for a numerical factor, the components of the tensor $F = \mathbf{Y}G$ are equal to $+1$ for indices obtained from those in (10–12) by an even permutation, to -1 if the arrangement is obtained by an odd permutation, and are zero otherwise. Thus all the irreducible subspaces for patterns with $m \leq n$ are realized.

The transformation properties of the general rth-rank tensor (10–6) were the same as those of the products of components of r vectors (10–5). For the symmetric and antisymmetric tensors of rank r, we can construct simple tensors from products of vector components. To obtain a completely symmetric rth-rank tensor, we choose all the vectors $\mathbf{x}^{(i)}$ in (10–5) to be the same vector $\mathbf{x}$. We then get a symmetric tensor with components

$x_{i_1} x_{i_2} \ldots x_{i_r}$. Collecting all factors with the same value of the index, we can rewrite the components as $x_1^{a_1} x_2^{a_2} \ldots x_n^{a_n}$, where

$$a_1 + a_2 + \cdots + a_n = r.$$

For example, for $r = 2$, we obtain the components x_i^2 $(i = 1, \ldots, n)$ and $x_i x_j$ $(i < j, \, i = 1, \ldots, n - 1)$.

An antisymmetric tensor of rank 2 can be constructed from two vectors $\mathbf{x}^{(1)}$, $\mathbf{x}^{(2)}$. We construct the matrix

$$\begin{bmatrix} x_1^{(1)} & x_1^{(2)} \\ x_2^{(1)} & x_2^{(2)} \\ \vdots & \vdots \\ x_{i_1}^{(1)} & x_{i_1}^{(2)} \\ \vdots & \vdots \\ x_{i_2}^{(1)} & x_{i_2}^{(2)} \\ \vdots & \vdots \\ x_n^{(1)} & x_n^{(2)} \end{bmatrix}. \tag{10–13}$$

The two-rowed minor

$$\begin{bmatrix} x_{i_1}^{(1)} & x_{i_1}^{(2)} \\ x_{i_2}^{(1)} & x_{i_2}^{(2)} \end{bmatrix}$$

is the (i_1, i_2)-component of an antisymmetric tensor $F_{\boxed{\begin{smallmatrix} i_1 \\ i_2 \end{smallmatrix}}}$. Similarly, the

components of a completely antisymmetric rth-rank tensor $F_{\boxed{\begin{smallmatrix} i_1 \\ \vdots \\ i_r \end{smallmatrix}}}$ can be

formed from the minors of degree r of the matrix

$$[\mathbf{x}^{(1)} \mathbf{x}^{(2)} \ldots \mathbf{x}^{(r)}] \equiv \begin{bmatrix} x_1^{(1)} & x_1^{(2)} & \cdots & x_1^{(r)} \\ \vdots & \vdots & & \vdots \\ x_n^{(1)} & x_n^{(2)} & \cdots & x_n^{(r)} \end{bmatrix}. \tag{10–14}$$

For $r = n$, we get a tensor with one independent component, the completely antisymmetric tensor of rank r. The transformation $\mathbf{a}$ in R_n multiplies this tensor by det $\mathbf{a}$. If we restrict ourselves to subgroups containing only unimodular matrices, the antisymmetric tensor $[\mathbf{x}^{(1)} \ldots \mathbf{x}^{(n)}]$ is invariant under all transformations and is a multiple of the *unit antisymmetric tensor of rank n*,

$$\epsilon_{i_1 i_2 \ldots i_n} = \begin{cases} +1 & \text{if } i_1 i_2 \cdots i_n \text{ is an even permuta-} \\ & \text{tion of } 1, \ldots, n, \\ -1 & \text{if } i_1 i_2 \cdots i_n \text{ is an odd permuta-} \\ & \text{tion of } 1, \ldots, n, \\ 0 & \text{if any index is repeated.} \end{cases} \qquad (10\text{–}15)$$

The tensor $\epsilon_{i_1 i_2 \ldots i_n}$ has the *same* numerical values for its components in any basis.

Problems. (1) Show that the tensors $[\mathbf{x}^{(1)} \ldots \mathbf{x}^{(r)}]$ form a basis for the antisymmetric tensors of rank r.

(2) Show that the tensors $x_{i_1} x_{i_2} \ldots x_{i_r}$ form a basis for the symmetric tensors of rank r.

10–3 The dimensionality of the irreducible representations of $GL(n)$. In this section, we shall determine the number of independent components of tensors of definite symmetry type. Before giving the general formula, we consider some simple examples.

For $r = 3$, the tensors $F_{\boxed{i_1}\boxed{i_2}\boxed{i_3}}$ are symmetric in all three indices, so that we get one independent component for each choice of i_1, i_2, i_3, irrespective of order. We select as the typical independent component the standard tableau for the set i_1, i_2, i_3 (cf. Section 7–3); i.e., we arrange the symbols so that $i_1 \leq i_2 \leq i_3$. For example, if $n = 1$, there is only one component, $F_{\boxed{1}\boxed{1}\boxed{1}}$. For $n = 2$, the i's can equal 1 or 2, and the four independent components are

$$F_{\boxed{1}\boxed{1}\boxed{1}}, \qquad F_{\boxed{1}\boxed{1}\boxed{2}}, \qquad F_{\boxed{1}\boxed{2}\boxed{2}}, \qquad F_{\boxed{2}\boxed{2}\boxed{2}}.$$

This procedure is easily extended to the completely symmetric rth-rank tensor for arbitrary n. We choose the independent components to correspond to standard tableaux, so that $i_1 \leq i_2 \leq \cdots \leq i_r$. Then

$$i_1 < i_2 + 1 < i_3 + 2 < \cdots < i_r + r - 1$$

are *different* integers from the set $1, 2, \ldots, (n + r - 1)$. The number of independent components is therefore equal to the number of ways of selecting r different integers from the set $1, 2, \ldots, (n + r - 1)$ and is equal to $\binom{n+r-1}{r}$. For $n = 2$, $r = 3$, we obtain $\binom{4}{3} = 4$ components, in agreement with the enumeration given above.

For $r = 3$, we next consider the symmetry type $F_{\boxed{i_1}\boxed{i_2}\atop\boxed{i_3}}$. There are two

standard tableaux for the partition [21] of the symmetric group S_3, $\boxed{1}\boxed{2}\atop\boxed{3}$

and $\boxed{1}\boxed{3}\atop\boxed{2}$. In the tensor $F_{\boxed{i_1}\boxed{i_2}\atop\boxed{i_3}}$, we can insert any of the numbers $1, 2, \ldots, n$

for each of the indices i_1, i_2, i_3. These are to be arranged in standard order:
If $i_1 = a < i_2 = b < i_3 = c$, the standard components are $F_{\boxed{a}\boxed{b}\atop\boxed{c}}$ and

$F_{\boxed{a}\boxed{c}\atop\boxed{b}}$. If some of the i's have the same value, we can still proceed in this

way. In the standard arrangements of any set of values for i_1, i_2, i_3, the
numbers must not decrease as we move to the right in any row, and must
increase as we move down a column. (Since the tensor is antisymmetric
in the arguments of a column, components with two equal indices in a
column will vanish.) If all indices are different, we get two standard
tableaux for each choice of three different numbers i_1, i_2, i_3 from the set
$1, 2, \ldots, n$, yielding $2 \cdot \binom{n}{3}$ independent components. If two of the indices
are equal (say 112 or 122), we get *one* standard tableau $\left(\frac{11}{2} \text{ or } \frac{12}{2}\right)$. We thus
obtain $n(n - 1)$ independent components of this type. Components
with all three indices equal must vanish. The dimensionality of the space
of tensors $F_{\boxed{i_1}\boxed{i_2}\atop\boxed{i_3}}$ is therefore equal to

$$2 \cdot \binom{n}{3} + n(n - 1) = \frac{n(n^2 - 1)}{3}.$$

Finally, for the antisymmetric tensors $F_{\boxed{i_1}\atop\boxed{i_2}\atop\boxed{i_3}}$, there is one standard tableau

for each choice of indices, and all indices must be different in a nonzero
component. Thus the dimensionality of the space of antisymmetric tensors
of rank 3 is $\binom{n}{3}$. By the same argument, the space of antisymmetric tensors
of rank r has dimension $\binom{n}{r}$.

Problem. Use this method to find the dimensionality of the space of tensors
$F_{\boxed{i_1}\boxed{i_2}\atop\boxed{i_3}\boxed{i_4}}$. List a set of independent components for the cases $n = 3$; $n = 4$.

A second method, which is very useful for low values of r, is based on the analysis of outer products (cf. Section 7–12). The components x_i of a vector in R_n form a tensor of rank $r = 1$ corresponding to the Young pattern $\square$. In terms of outer products, the products of the components x_i with the components y_j of a second vector form the outer product $\square \otimes \square$ which can be resolved into

$$\square \otimes \square = \square\square + \begin{array}{c}\square\\\square\end{array}, \qquad (10\text{–}16)$$

i.e., into the symmetric and antisymmetric tensors of rank two. The number of components on the left of (10–16) is n^2, which is equal to $n(n + 1)/2 + n(n - 1)/2$, the sum of the dimensions of the two irreducible representations on the right.

Next we consider the products of components of a symmetric tensor of rank 2 with the components of a vector:

$$\square\square \otimes \square = \square\square\square + \begin{array}{c}\square\square\\\square\end{array}. \qquad (10\text{–}17)$$

The number of independent components on the left is $n \cdot [n(n + 1)/2]$. On the right, the symmetric tensor of rank 3 has $\binom{n+2}{3}$ independent components. Subtracting, we find that the space of tensors $F_{\square\square \atop \square}$ with respect to $GL(n)$ has the dimensionality

$$^nN_{[21]} = \frac{n^2(n + 1)}{2} - \frac{n(n + 1)(2n + 1)}{6} = \frac{n(n^2 - 1)}{3}, \qquad (10\text{–}18)$$

which agrees with our direct computation. We use the symbol $^nN_{[\lambda_1\ldots\lambda_n]}$ to denote the dimensionality of the space of tensors with the symmetry $[\lambda_1 \ldots \lambda_n]$.

Similarly, from the outer product

$$[p] \otimes [1] = [p + 1] + [p, 1], \qquad (10\text{–}19)$$

we find

$$^nN_{[p,1]} = n \cdot {}^nN_{[p]} - {}^nN_{[p+1]}$$
$$= n \cdot \binom{n+p-1}{p} - \binom{n+p}{p+1}. \qquad (10\text{–}20)$$

The most powerful method for finding the dimensionality $^nN_{[\lambda]}$ is to use a branching theorem analogous to the one for the symmetric group (cf. Section 7–5). We shall outline the procedure and state the results.

The group $GL(n)$ contains many subgroups which are isomorphic to $GL(n - 1)$. For example, we obtain such a subgroup if we restrict ourselves to those transformations of $GL(n)$ which leave the component x_n unchanged. An irreducible representation of $GL(n)$ will decompose into a sum of irreducible representations of the subgroup $GL(n-1)$. Suppose that the irreducible representation of $GL(n)$ corresponds to the Young pattern $[\lambda_1 \ldots \lambda_n]$:

The independent components of the tensor are those which correspond to standard tableaux. If the index n appears in a standard tableau, it can occur only in the last cell of each column of the diagram, as indicated by the crosses in the figure. In other words, the index n can appear only in the overhang of each row beyond its successor. When we go to the subgroup $GL(n-1)$, the index n can be dropped from the tableau. We then have all the possible patterns on $(n-1)$ symbols which can be obtained from the original pattern. Thus the irreducible representation of $GL(n)$ corresponding to the pattern $[\lambda_1 \ldots \lambda_n]$ splits into the sum of irreducible representations of $GL(n-1)$ corresponding to the patterns $[\lambda_1' \ldots \lambda_{n-1}']$, where

$$\lambda_1 \geq \lambda_1' \geq \lambda_2 \geq \lambda_2' \geq \cdots \geq \lambda_{n-1} \geq \lambda_{n-1}' \geq \lambda_n. \qquad (10\text{–}21)$$

Equating the dimension of the representation of $GL(n)$ to the sum of the dimensions of the representations in its decomposition, we find the recursion formula

$$^{n}N_{[\lambda_1 \ldots \lambda_n]} = \sum_{\lambda_1' \ldots \lambda_{n-1}'} {}^{n-1}N_{[\lambda_1' \ldots \lambda_{n-1}']}, \qquad (10\text{–}22)$$

where the sum is extended over the range given by Eq. (10–21).

For the symmetric tensors, we have the recursion formula

$$^{n}N_{[r]} = {}^{n-1}N_{[r]} + {}^{n-1}N_{[r-1]} + \cdots + {}^{n-1}N_{[1]}, \qquad (10\text{–}23)$$

and for the antisymmetric tensors,

$$^{n}N_{[1^r]} = {}^{n-1}N_{[1^r]} + {}^{n-1}N_{[1^{r-1}]}. \qquad (10\text{–}24)$$

The branching formula can be used to obtain the general expression for $^{n}N_{[\lambda]}$. The result is

$$^{n}N_{[\lambda]} = \frac{D(l_1, \ldots, l_n)}{D(n-1, n-2, \ldots, 0)}, \qquad (10\text{–}25)$$

where $l_j = \lambda_j + n - j$ and D is the determinant defined in Eq. (7–23).

10–4 Irreducible representations of subgroups of $GL(n)$: $SL(n)$, $U(n)$, $SU(n)$.

In this section we show that the irreducible representations of $GL(n)$ remain irreducible when we go to certain subgroups of $GL(n)$.

The elements of the matrices of an irreducible representation of $GL(n)$ in terms of rth-rank tensors are homogeneous polynomials of degree r in the elements a_{ij} of the transformation $\mathbf{a}$ of Eq. (10–1). If the matrices representing a subgroup H are reducible, they can be brought to reduced form by an appropriate change of basis. This basis transformation will not reduce the matrices for all transformations in $GL(n)$, since we started from an irreducible representation of $GL(n)$. Under a change of basis, the matrix elements remain homogeneous polynomials of degree r in the a_{ij}. Thus the representation will be reducible for H if a certain set of homogeneous polynomials of degree r, $P_\nu(\mathbf{a})$, vanish for all $\mathbf{a}$ in H, but not for all $\mathbf{a}$ in $GL(n)$.

The group $GL(n)$ consists of all nonsingular linear transformations with complex coefficients. Suppose that H is the subgroup $GL'(n)$ of *real* linear transformations. Reducibility for $GL'(n)$ means that the set of polynomials $P_\nu(\mathbf{a})$ vanishes for all real values of its arguments a_{ij}. But from a standard theorem of algebra, if a set of polynomials vanishes for all *real* values of its arguments, it must vanish for *all* values. Therefore, if the representation is reducible for $GL'(n)$, it must be reducible for $GL(n)$. Conversely, an irreducible representation of $GL(n)$ remains irreducible when we restrict ourselves to real transformations.

Next we consider the unimodular group $SL(n)$. Any matrix $\mathbf{a}$ of $GL(n)$ can be written as $\mathbf{a} = \alpha\mathbf{b}$, where det $\mathbf{b} = 1$, by setting $\alpha = (\det \mathbf{a})^{1/n}$. Thus, to each matrix $\mathbf{a}$ of $GL(n)$ there corresponds a matrix $\mathbf{b}$ of $SL(n)$. Suppose that the polynomials P_ν vanish for transformations of the unimodular group: $P_\nu(\mathbf{b}) = 0$ for all $\mathbf{b}$ in $SL(n)$. For any transformation $\mathbf{a}$ of $GL(n)$, $P_\nu(\mathbf{a}) = \alpha^r P_\nu(\mathbf{b})$, where $\mathbf{b}$ is in $SL(n)$. Consequently $P_\nu(\mathbf{a}) = 0$, so that reducibility for $SL(n)$ implies reducibility for $GL(n)$. Conversely, an irreducible representation of $GL(n)$ remains irreducible for $SL(n)$.

Problem. Prove that an irreducible representation of $GL(n)$ remains irreducible when we go to the subgroup of real, unimodular transformations, $SL'(n)$.

The same results can be obtained by considering the representations of the Lie algebra of $GL(n)$. The infinitesimal matrices of $GL(n)$ are matrices having all elements zero, except for a 1 at the (ij)-position. The basis elements of the Lie algebra are the set X_{ij} $(i, j = 1, \ldots, n)$, with the commutation relations

$$[X_{ij}, X_{kl}] = \delta_{jk}X_{il} - \delta_{il}X_{kj}. \tag{10-26}$$

The elements X_{ij} can be represented by the differential operators

$$X_{ij} = x_i \frac{\partial}{\partial x_j}. \tag{10-27}$$

The Lie algebra consists of all elements $\sum_{ij} \alpha_{ij}X_{ij}$. For $GL'(n)$, the α_{ij}'s must be real, while for $GL(n)$ the α_{ij}'s can take on any complex values. Suppose that we have a representation $X_{ij} \rightarrow D^{ij}$ of the basis elements X_{ij} in terms of matrices D^{ij}. Then the general element of the algebra has the representative $\sum \alpha_{ij}D^{ij}$. If the representation is reducible for $GL'(n)$, we can find a basis in which the matrices $\sum \alpha_{ij}D^{ij}$ are in reduced form for all real values of the α_{ij}. In other words, if the representation is reducible for $GL'(n)$, a certain set of linear forms in the α_{ij} vanishes for all *real* values of α_{ij}. But if this is the case, the linear forms must vanish for *any* *complex* values of the α_{ij}, and the representation is reducible for $GL(n)$. Conversely, if the representation is irreducible for $GL(n)$, it must be irreducible for $GL'(n)$.

Problem. By considering the representation of the Lie algebra, prove that an irreducible representation of $GL(n)$ remains irreducible for $SL(n)$ and $SL'(n)$. (Hint: Show that the unimodular condition gives another linear relation between the α_{ij}'s.)

The unitary group $U(n)$ is defined by the condition $UU^\dagger = 1$. The elements in the neighborhood of the identity are $U = 1 + iS$, where S is an infinitesimal matrix. From $UU^\dagger = 1$, we find $S - S^\dagger = 0$. Thus the infinitesimal matrices for $U(n)$ are the Hermitian matrices. We choose as basis elements of the Lie algebra the n^2 matrices:

$k \neq j$; $X^{(kj)}$: 1 at (kj)- and (jk)-positions, all other elements = 0;

$\qquad\quad X'^{(kj)}$: i at (kj), $-i$ at (jk)-position, all other elements = 0.

$k = j$; $X^{(kk)}$: 1 at (kk)-position, all other elements = 0.

$$\tag{10-28}$$

The elements of the Lie algebra of $U(n)$ are all linear combinations of the basis elements (10–28) with *real* coefficients, whereas complex coefficients give the Lie algebra of $GL(n)$. By a repetition of the argument which we used for $GL'(n)$, we see that an irreducible representation of $GL(n)$ remains irreducible for $U(n)$.

Problem. Prove that irreducible representations of $GL(n)$ remain irreducible when we go to the unitary, unimodular subgroup $SU(n)$. [Hint: Use the result that irreducibility for $GL(n)$ implies irreducibility for $SL(n)$. Then make the transition from $SL(n)$ to $SU(n)$.]

As a result of the preceding argument, we know that the irreducible representations of $SL(n)$, $U(n)$, and $SU(n)$ are the representations which we found in Section 10–2. However, these representations may not be *independent* representations for these subgroups of $GL(n)$.

The Young diagram $[1^n]$ has one standard tableau, and the corresponding irreducible representation of $GL(n)$ is one-dimensional. The tensor $[\mathbf{x}^{(1)} \cdots \mathbf{x}^{(n)}]$ of Section 10–2 can serve as the basis vector for this representation. When we transform with the matrix $\mathbf{a}$, the tensor is multiplied by det $\mathbf{a}$.

Problem. Prove the last statement by writing out the transformation of the antisymmetric tensor $[\mathbf{x}^{(1)} \ldots \mathbf{x}^{(n)}]$.

Similarly, for the pattern $[2^n]$, there is only one standard tableau,

and one independent basis element. The representation is one-dimensional. When we transform with the matrix $\mathbf{a}$, the tensor is multiplied by $(\det \mathbf{a})^2$.

In general, for the pattern $[s^n]$, the representation is one-dimensional, and the tensor is multiplied by $(\det \mathbf{a})^s$ when we transform with the matrix $\mathbf{a}$.

Suppose we have a representation of $GL(n)$ with the pattern $[\lambda_1 \ldots \lambda_n]$. If we add a column of n boxes to the pattern, the only set of indices in standard order which can be inserted in the additional column is $1, 2, \ldots, n$. Thus the number of standard tableaux for the new representation,

$$[\lambda_1 + 1, \lambda_2 + 1, \ldots, \lambda_n + 1],$$

is the same as for $[\lambda_1 \lambda_2 \ldots \lambda_n]$. The only change in the representation matrices is that they are multiplied by the common factor det $\mathbf{a}$. Similarly, if we adjoin s columns of length n to the pattern $[\lambda_1 \ldots \lambda_n]$, the new pattern $[\lambda_1 + s, \lambda_2 + s, \ldots, \lambda_n + s]$ has the same number of standard tableaux as $[\lambda_1\lambda_2 \ldots \lambda_n]$. The matrices for this new representation differ from those for $[\lambda_1 \ldots \lambda_n]$ by the factor $(\det \mathbf{a})^s$.

If we are dealing with unimodular subgroups of $GL(n)$, such as $SL(n)$ or $SU(n)$, $\det \mathbf{a} = 1$, and the representations corresponding to the patterns $[\lambda_1 \ldots \lambda_n]$ and $[\lambda_1 + s, \ldots, \lambda_n + s]$ are equivalent. For the unimodular groups we need consider only those patterns which have *fewer* than n rows:

$$[\lambda_1\lambda_2 \ldots \lambda_n] \equiv [\lambda_1 - \lambda_n, \lambda_2 - \lambda_n, \ldots, \lambda_{n-1} - \lambda_n]. \quad (10\text{–}29)$$

There is a second equivalence which occurs for unimodular groups. The pattern $[1^{n-1}]$ has n standard tableaux. The number of basis functions for this representation is precisely the same as for $[1]$. These two representations are equivalent for unimodular transformations: $[1^{n-1}] \equiv [1]$.

Problem. Show that, for unimodular transformations, $[1^{n-1}] \equiv [1]$.

We shall state the general result without proof: For unimodular transformations,

$$[\lambda_1\lambda_2 \ldots \lambda_n] \equiv [\lambda_1 - \lambda_n, \lambda_1 - \lambda_{n-1}, \ldots, \lambda_1 - \lambda_2]. \quad (10\text{–}30)$$

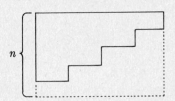

The pattern shown by the solid lines in the diagram is equivalent to the dotted pattern which completes the rectangle. In particular, $[1^{n-1}] \equiv [1]$ and $[1^{n-r}] \equiv [1^r]$.

10–5 The orthogonal group in n dimensions. Contraction. Traceless tensors. When we go from the linear group $GL(n)$ to the orthogonal subgroup $O(n)$, the representations in terms of tensors of a given symmetry will no longer be irreducible. The reason for this is that, in addition to the operation of symmetrization which we used for constructing irreducible representations of $GL(n)$, a new operation of *contraction* appears which commutes with the orthogonal transformations.

Suppose we consider the space of tensors of rank r, with components $F_{i_1 i_2 \ldots i_r}$. Under the linear group $GL(n)$, the only operations which commute with the Kronecker rth power of the transformation $\mathbf{a}$ are the permutations of the tensor indices. In Section 10–2 we used the permutation operators to reduce the space of rth-rank tensors into subspaces of tensors of given symmetry. For orthogonal transformations,

$$a_{ij}a_{ik} = a_{ji}a_{ki} = \delta_{jk}. \tag{10-31}$$

If, for example, we set the first two indices of $F_{i_1 i_2 \dots i_r}$ equal to each other and sum over all values of $i_1 = i_2$, we obtain the (12)-trace (contraction) of the tensor:

$$F^{(12)}_{i_3 \dots i_r} = F_{iii_3 \dots i_r} = \delta_{i_1 i_2} F_{i_1 i_2 i_3 \dots i_r}. \tag{10-32}$$

The contraction process gives a new tensor of rank $(r - 2)$, and the operation of contraction commutes with the transformation of the tensor:

$$F'_{i_1 i_2 \dots i_r} = a_{i_1 j_1} a_{i_2 j_2} \dots a_{i_r j_r} F_{j_1 j_2 \dots j_r}, \tag{10-6}$$

$$\begin{aligned} F'^{(12)}_{i_3 \dots i_r} = F'_{iii_3 \dots i_r} &= a_{ij_1} a_{ij_2} a_{i_3 j_3} \dots a_{i_r j_r} F_{j_1 j_2 j_3 \dots j_r} \\ &= \delta_{j_1 j_2} a_{i_3 j_3} \dots a_{i_r j_r} F_{j_1 j_2 j_3 \dots j_r} \\ &= a_{i_3 j_3} \dots a_{i_r j_r} F^{(12)}_{j_3 \dots j_r} = F^{(12)'}_{i_3 \dots i_r}. \end{aligned} \tag{10-33}$$

The contraction process can be applied to any pair of indices, so that there are $r(r - 1)/2$ traces $F^{(\alpha \beta)}_{\dots}$ ($\alpha < \beta; \alpha, \beta = 1, \dots, r$) of the rth-rank tensor.

We now select, from the space of rth-rank tensors, the subspace of tensors for which *all pair traces are zero*. From Eq. (10–33) we see that this subspace is invariant—the *traceless rth-rank tensors* are transformed among themselves under transformations induced by $O(n)$. In fact, we can show that every tensor $F_{i_1 \dots i_r}$ can be decomposed uniquely into a traceless tensor F^0 plus a tensor of the form

$$\Phi_{i_1 \dots i_r} = \delta_{i_1 i_2} G^{(12)}_{i_3 \dots i_r} + \cdots$$

$$+ \delta_{i_\alpha i_\beta} G^{(\alpha \beta)}_{i_1 \dots i_{\alpha-1} i_{\alpha+1} \dots i_{\beta-1} i_{\beta+1} \dots i_r} \quad \left(\frac{r(r-1)}{2} \text{ terms} \right). \tag{10-34}$$

To show this, we consider the subspace Σ of all tensors Φ. The condition that a tensor F be "orthogonal" to this subspace,

$$(F, \Phi) \equiv F_{i_1 \dots i_r} \Phi_{i_1 \dots i_r} = 0 \qquad \text{for all } \Phi, \tag{10-35}$$

means that all traces of F must vanish. [Choose the tensors Φ with only $G^{(12)}$ different from zero. Then (10–35) requires $F_{iii_3 \dots i_r} G^{(12)}_{i_3 \dots i_r} = 0$. Since the components of $G^{(12)}$ are arbitrary, we must have $F_{iii_3 \dots i_r} = 0$. Similarly, we choose Φ with only $G^{(\alpha \beta)}$ different from zero, and prove that all pair traces of F vanish.] Thus the set of traceless tensors F^0 forms a subspace perpendicular to Σ, and the whole space is the direct sum of these two subspaces:

$$F_{i_1\ldots i_r} = \Phi_{i_1\ldots i_r} + F^0_{i_1\ldots i_r}. \tag{10–36}$$

The space Σ defined by Eq. (10–34) is an invariant subspace. In fact, each of the terms is separately transformed into a similar term under orthogonal transformations:

$$a_{i_1j_1}a_{i_2j_2}\ldots a_{i_rj_r}\left(\delta_{j_1j_2}G^{(12)}_{j_3\ldots j_r}\right) = \delta_{i_1i_2}a_{i_3j_3}\ldots a_{i_rj_r}G^{(12)}_{j_3\ldots j_r}$$

$$= \delta_{i_1i_2}G^{(12)'}_{i_3\ldots i_r}. \tag{10–37}$$

Thus the decomposition (10–36) is invariant under orthogonal transformations.

We illustrate the decomposition (10–36) for $r = 2$. For second-rank tensors F_{ij},

$$F_{ij} = \frac{1}{n}F_{kk}\,\delta_{ij} + \left[F_{ij} - \frac{1}{n}F_{kk}\,\delta_{ij}\right]$$

$$= \Phi_{ij} + F^0_{ij}. \tag{10–38}$$

The tensor $F^0_{ij} = F_{ij} - (1/n)F_{kk}\,\delta_{ij}$ has zero trace. The tensor $G^{(12)} = (1/n)F_{kk}$ has rank $r - 2 = 0$.

For $r = 3$, we write $F_{i_1i_2i_3}$ as

$$F_{i_1i_2i_3} = F^0_{i_1i_2i_3} + H_{i_3}\,\delta_{i_1i_2} + K_{i_2}\,\delta_{i_1i_3} + L_{i_1}\,\delta_{i_2i_3}, \tag{10–39}$$

and require that the tensor F^0 be traceless:

$$\begin{array}{lll}
(12)\text{-trace:} & F_{iii_3} = nH_{i_3} + K_{i_3} + L_{i_3}, \\
(13)\text{-trace:} & F_{ii_2i} = H_{i_2} + nK_{i_2} + L_{i_2}, \\
(23)\text{-trace:} & F_{i_1ii} = H_{i_1} + K_{i_1} + nL_{i_1}.
\end{array}$$

Solving, we find

$$H_j = \frac{1}{n^2 + n - 2}[(n + 1)F^{(12)}_{..j} - F^{(13)}_{.j.} - F^{(23)}_{j..}],$$

$$K_j = \frac{1}{n^2 + n - 2}[-F^{(12)}_{..j} + (n + 1)F^{(13)}_{.j.} - F^{(23)}_{j..}], \tag{10–40}$$

$$L_j = \frac{1}{n^2 + n - 2}[-F^{(12)}_{..j} - F^{(13)}_{.j.} + (n + 1)F^{(23)}_{j..}].$$

Problem. Find the expression for the traceless part of a fourth-rank tensor $F_{i_1i_2i_3i_4}$ in terms of the components of the tensor.

10–6 The irreducible representations of $O(n)$. We now note that a permutation of the indices takes a traceless tensor into another traceless tensor. Thus we can start from the subspace of traceless tensors of rank r and apply the Young symmetrizers to obtain traceless tensors of a given symmetry type. In this way, we arrive at the irreducible representations of $O(n)$

An irreducible representation of $O(n)$ is thus associated with a Young diagram $[\lambda_1 \ldots \lambda_n]$, with $\lambda_1 + \lambda_2 + \cdots + \lambda_n = r$. However, not all Young diagrams are admissible. For a whole class of diagrams, the traceless tensors having the particular symmetry type are identically zero. The general theorem states the following:

THEOREM. *The traceless tensors corresponding to Young diagrams in which the sum of the lengths of the first two columns exceeds n must be identically zero.* In other words, the tensors

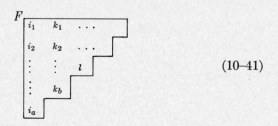

$$(10\text{–}41)$$

with $a + b > n$ must be of the form (10–34).

We shall not give the general proof. Instead we give the proofs for some particular simple cases. We start with $n = 2$. The symmetric tensor $F_{\boxed{i}\boxed{j}}$ has $a + b = 2$. Since this number does not exceed $n = 2$, we can construct traceless tensors with this symmetry. The independent components are F_{11} and F_{12}. The remaining components are $F_{22} = -F_{11}$ and $F_{21} = F_{12}$.

For the pattern $\boxed{\dfrac{i}{j}}$, $a + b = 2$, so that there are traceless tensors of this type. In fact, since the tensor is antisymmetric in i and j, its trace is necessarily zero.

For $r = 3$, the traceless tensor $F_{\boxed{i_1}\boxed{i_2}\boxed{i_3}}$ has independent components F_{111}, F_{112}. The other standard components are given by

$$F_{111} + F_{122} = 0, \qquad F_{112} + F_{222} = 0.$$

Since $n = 2$, the tensors for [111] are all zero.

The tensor $F_{\boxed{i}\boxed{j}\atop\boxed{k}}$ has only two independent nonzero components:

$F_{11\atop 2}$ and $F_{12\atop 2}$. But if we require that the tensor be traceless, we obtain the equations

$$0 = F_{11\atop 2} + F_{22\atop 2} = F_{11\atop 2}, \qquad 0 = F_{12\atop 2} + F_{11\atop 1} = F_{12\atop 2}.$$

Thus the traceless tensors $F_{\boxed{i}\boxed{j}\atop\boxed{k}}$ vanish identically, in agreement with the general theorem $(a + b = 3 > 2)$. We also see that the same argument applies to the traceless tensors $F_{\boxed{i}\boxed{j}\boxed{\cdots}\atop\boxed{k}}$ having any number of additional indices in the first row. For each fixed set of values of the additional indices, the argument given above shows that the tensor is zero if the (ij)- and (jk)-traces are zero.

From the general pattern (10–41), we see that the number of independent traces is very small. Since F is antisymmetric in $i_1, \ldots, i_a$, the trace on i_ν and l is the same (except for a possible change in sign) for all elements i_ν in the first column. Thus, in writing the trace conditions, we need to select any one index from each column. (For example, we can take traces only on pairs of indices in the first row of the diagram.)

For $n = 3$, we consider the traceless tensors [22], with $a + b = 4 > 3$. The zero-trace conditions are of two types: $\sum\limits_i F_{ii\atop kl} = 0$, with $k \neq l$, and $\sum\limits_i F_{ii\atop kk} = 0$. For the first type, when k and l are assigned, i can take on only a single value $\neq k, l$, so that we obtain equations like $F_{11\atop 23} = 0$, $F_{22\atop 13} = 0$, etc. For the second type, we find the equations

$$F_{22\atop 11} + F_{33\atop 11} = 0, \qquad F_{11\atop 22} + F_{33\atop 22} = 0, \qquad F_{11\atop 33} + F_{22\atop 33} = 0.$$

Since the tensor is antisymmetric in the arguments of a column, these equations can be rewritten as

$$F_{11\atop 22} + F_{11\atop 33} = 0, \qquad F_{11\atop 22} + F_{22\atop 33} = 0, \qquad F_{11\atop 33} + F_{22\atop 33} = 0,$$

for which the only solution is $F_{11\atop 22} = F_{11\atop 33} = F_{22\atop 33} = 0$. Thus we find that the traceless tensors of symmetry [22] are identically zero, in agreement with the general theorem.

Problem. For $n = 3$, apply this method to show that the traceless tensors of symmetry [31] vanish identically.

Do the same for the traceless tensors with symmetry [221] when $n = 3$ and $n = 4$. What happens for $n = 5$?

From the preceding general theorem, we conclude that only those diagrams are permissible for which the sum of the lengths of the first two columns is $\leq n$. The permissible diagrams can be paired into *associate diagrams* T and T' as follows: The length of the first column in T is less than or equal to $n/2$, $a \leq n/2$, the length of the first column in T' is $n - a$, and all other columns in T and T' have the same length. Since $a + b \leq n$ for a permissible diagram and $a \geq b$, we have $n - a \geq b$ and $(n - a) + b \leq n$, so that T' is a permissible diagram if T is. For example, for $n = 3$, the diagrams

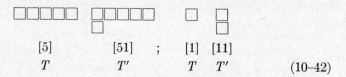

$$[5] \qquad\qquad [51] \quad ; \quad [1] \quad [11]$$
$$T \qquad\qquad T' \qquad\qquad T \quad T' \qquad\qquad (10\text{--}42)$$

are associated. (In general, [r] and [r, 1] are associated.)

For $n = 5$, the diagrams

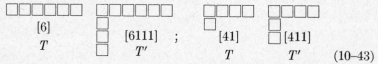

$$\begin{array}{cccc} [6] & & & \\ T & [6111] & ; & [41] \quad [411] \\ & T' & & T \qquad T' \end{array} \qquad (10\text{--}43)$$

are associated.

For $n = 4$, the diagrams

$$\begin{array}{cc} [5] & \\ T & [511] \\ & T' \end{array} \qquad\qquad (10\text{--}44)$$

are associated. But for this *even* value of n, we can have

$$T \qquad\qquad\equiv\qquad\qquad T' \qquad\qquad (10\text{--}45)$$

When n is even, $n = 2\nu$, and $a = \frac{1}{2} n = \nu$, the diagrams T and T' coincide, and T is said to be *self-associate*.

Thus, to describe the pattern T, we can use a symbol $(\mu_1, \mu_2, \ldots, \mu_\nu)$, with $\mu_1 \geq \mu_2 \geq \cdots \geq \mu_\nu$, where some of the μ's may be zero. The pattern T contains $\mu_1 + \mu_2 + \cdots + \mu_\nu = r$ indices, while T' contains a larger number unless T is self-associate. The permissible diagrams for a given value of n are exhausted by letting $r = 0, 1, 2 \ldots$ and choosing all non-negative μ's satisfying the equation

$$\mu_1 + \mu_2 + \cdots + \mu_\nu = r. \tag{10-46}$$

Here $n = 2\nu$ if n is even, and $n = 2\nu + 1$ if n is odd. For each solution of Eq. (10-46) we obtain a diagram T and then construct its associate T'. If n is odd, each diagram is obtained once. If n is even and the diagram T actually contains $\nu = n/2$ rows, T' will coincide with T.

The invariant subspaces defined by the permissible diagrams are not empty. By a procedure similar to that used in Section 10-2 [Eq. (10-12)], one can construct traceless tensors for any permissible diagram.

So far we have considered the full orthogonal group $O(n)$. If we restrict ourselves to the proper orthogonal group $O^+(n)$, for which det $\mathbf{a} = +1$, the representations corresponding to the associated diagrams T and T' become *equivalent*. For even n, the representation corresponding to a self-associate pattern T splits into two nonequivalent irreducible representations.

As a simple example, we note that for $n = 3$, the pattern $T = [1]$ describes vectors, while $T = [11]$ describes skew-symmetric tensors. Under proper orthogonal transformations both types of quantities transform in the same way, so that T and T' are equivalent. Under improper transformations (inversions), the tensors for $T = [1]$ change sign (polar vectors), while those for $T' = [11]$ do not (axial vectors).

For applications to physics, only the case of odd n will be of interest, so that the irreducible representations of $O^+(n)$ can be described completely by the symbol $(\mu_1, \mu_2, \ldots, \mu_\nu)$.

For $n = 3$, $\nu = 1$, so the irreducible representations of $O^+(3)$ are described by a single label (μ_1). The basis functions are the traceless symmetric tensors of rank μ_1. For $GL(3)$, the dimensionality of $[\lambda_1]$ was $\binom{\lambda_1 + 2}{\lambda_1} = (\lambda_1 + 1)(\lambda_1 + 2)/2$. The condition that the tensor be traceless imposes $\lambda_1(\lambda_1 - 1)/2$ conditions. Thus the number of independent functions for (μ_1) is

$$\frac{(\mu_1 + 1)(\mu_1 + 2)}{2} - \frac{\mu_1(\mu_1 - 1)}{2} = 2\mu_1 + 1.$$

The basis functions are the spherical harmonics of order μ_1. Thus quantities which transform like the spherical harmonics of order μ_1 are irreducible tensors [with respect to $O^+(3)$].

Problem. Prove that the spherical harmonics of order l form the basis for the representation (l) of $O^+(3)$.

TABLE 10–1

IRREDUCIBLE REPRESENTATIONS OF $O^+(5)$

$(\mu_1\mu_2)$	N	$(\mu_1\mu_2)$	N
(00)	1	(22)	35
(10)	5	(41)	154
(20)	14	(32)	105
(11)	10	(42)	220
(30)	30	(33)	84
(21)	35	(43)	231
(40)	55	(44)	165
(31)	81		

TABLE 10–2

IRREDUCIBLE REPRESENTATIONS OF $O^+(7)$

$(\mu_1\mu_2\mu_3)$	N	$(\mu_1\mu_2\mu_3)$	N
(000)	1	(321)	1617
(100)	7	(222)	294
(200)	27	(430)	3003
(110)	21	(421)	4550
(300)	77	(331)	2079
(210)	105	(322)	1386
(111)	35	(440)	3003
(400)	182	(431)	7722
(310)	330	(422)	4095
(220)	168	(332)	2310
(211)	189	(441)	8008
(410)	819	(432)	9009
(320)	693	(433)	1386
(311)	616	(333)	1386
(221)	378	(442)	10296
(420)	1911	(433)	6006
(411)	1560	(443)	9009
(330)	825	(444)	4719

For $n = 5$, $\nu = 2$, so that the irreducible representations of $O^+(5)$ (Table 10–1) are characterized by two integers μ_1, μ_2 with $\mu_1 \geq \mu_2$. The dimension of the representation $(\mu_1\mu_2)$ is given by the formula

$$N(\mu_1\mu_2) = \tfrac{1}{6}(\mu_1 - \mu_2 + 1)(\mu_1 + \mu_2 + 2)(2\mu_1 + 3)(2\mu_2 + 1). \quad (10\text{–}47)$$

For $n = 7$, $\nu = 3$, so the irreducible representations of $O^+(7)$ (Table 10–2) are characterized by three integers μ_1, μ_2, μ_3 with $\mu_1 \geq \mu_2 \geq \mu_3$.

10–7 Decomposition of irreducible representations of $U(n)$ with respect to $O^+(n)$.

In Section 10–5 we decomposed the space of tensors of rank r into the direct sum of two subspaces: the space of traceless tensors F^0 and the space Σ of tensors Φ having the form given by (10–34). Both subspaces were invariant under orthogonal transformations. In Section 10–6 we applied the Young symmetrizers to the traceless tensors $F^0_{i_1\ldots i_r}$ to obtain the irreducible representations of $O^+(n)$. The invariant subspace Σ of tensors Φ is *not* irreducible. By applying the trace operation we can resolve $\Phi_{i_1\ldots i_r}$ into a direct sum

$$\Phi = F^1 + \Phi', \quad (10\text{–}48)$$

where F^1 is the subspace of Σ whose tensors have all their "double traces" equal to zero,

$$F^1_{\ldots i \ldots i \ldots k \ldots k \ldots} = 0,$$

for all locations of the repeated indices i and k, and Φ' is a sum of terms of the form

$$\delta_{i_\alpha i_\beta} \, \delta_{i_\mu i_\nu} H^{(\alpha\beta,\mu\nu)}_{i_1 \ldots i_{\alpha-1} i_{\alpha+1} \ldots i_{\beta-1} i_{\beta+1} \ldots i_{\mu-1} i_{\mu+1} \ldots i_{\nu-1} i_{\nu+1} \ldots i_r}. \quad (10\text{–}49)$$

By an argument similar to that of Section 10–5, we can show that (10–48) is a decomposition of Σ into a direct sum of invariant subspaces F^1 and Φ'. Consequently the tensors F are decomposed into the direct sum

$$F = F^0 + F^1 + \Phi'. \quad (10\text{–}50)$$

The tensors F^1 are of the form

$$\delta_{i_\alpha i_\beta} G^{(\alpha\beta)}_{i_1 \ldots i_{\alpha-1} i_{\alpha+1} \ldots i_{\beta-1} i_{\beta+1} \ldots i_r}, \quad (10\text{–}51)$$

where the tensors $G^{(\alpha\beta)}$ have rank $v = r - 2$ and are traceless. Thus the irreducible representation of $U(n)$ by rth-rank tensors contains a representation of $O^+(n)$ in terms of traceless tensors of rank $r - 2$. Again by applying Young symmetrizers we can obtain irreducible representations of $O^+(n)$. This contraction process can be continued, so that the arbitrary tensor $F_{i_1 \cdots i_r}$ can be written as a sum of terms of the form

$$\delta_{i_{\alpha_1} i'_{\alpha_1}} \cdots \delta_{i_{\alpha_s} i'_{\alpha_s}} \Phi_{i\beta_1 \cdots i\beta_v} \qquad (2s + v = r), \qquad (10\text{-}52)$$

where the tensor $\Phi_{i_1 \cdots i_v}$ is traceless.

If we start from rth-rank tensors of given symmetry, the process of repeated contraction will give the decomposition into traceless tensors of rank $r, r - 2, r - 4, \ldots,$ etc.

Before considering the reduction process, we illustrate the use of repeated contractions. We consider the completely symmetric tensors of rank r which can be formed from the components A_i of a vector $\mathbf{A}$, and let $A_i A_i \equiv \mathbf{A}^2$. The possible tensors of rank 3 are

$$A_i A_j A_k, \qquad \mathbf{A}^2 A_i \, \delta_{jk}, \qquad (10\text{-}53)$$

where the factor $\mathbf{A}^2$ must be included in the second term since we must have homogeneous expressions of degree 3 in the components of $\mathbf{A}$. In addition, the second term of (10-53) is not symmetric, whereas the trace of $A_i A_j A_k$ must be symmetric in the indices. We symmetrize the second term, replacing it by

$$\mathbf{A}^2 (A_i \, \delta_{jk} + A_j \, \delta_{ki} + A_k \, \delta_{ij}). \qquad (10\text{-}54)$$

The (ij)-trace of (10-54) is $(n + 2)\mathbf{A}^2 A_k$, and the (ij)-trace of $A_i A_j A_k$ is $\mathbf{A}^2 A_k$. Thus the traceless symmetric tensor of rank 3 is

$$(n + 2) A_i A_j A_k - \mathbf{A}^2 (A_i \, \delta_{jk} + A_j \, \delta_{ki} + A_k \, \delta_{ij}). \qquad (10\text{-}55)$$

For $r = 4$, the possible symmetric tensors formed from $\mathbf{A}$ are:

$$\begin{aligned}
F &= A_i A_j A_k A_l, \\
\Phi &= \mathbf{A}^2 (A_i A_j \, \delta_{kl} + \cdots), \\
\Phi' &= (\mathbf{A}^2)^2 (\delta_{ij} \, \delta_{kl} + \delta_{ik} \, \delta_{jl} + \delta_{il} \, \delta_{jk}).
\end{aligned} \qquad (10\text{-}56)$$

The traceless tensor is

$$F^0 = (n + 2)(n + 4)F - (n + 2)\Phi + \Phi'. \qquad (10\text{-}57)$$

Problem. Find the traceless tensor of rank 5 formed from the components of a vector $\mathbf{A}$. Derive the general formula for arbitrary rank r.

The decomposition of a representation $[\lambda_1, \ldots, \lambda_n]$ of $U(n)$ into representations $(\mu_1, \ldots, \mu_v)$ of $O^+(n)$ means that the tensor F with symmetry

$[\lambda_1 \ldots \lambda_n]$ is written as a sum of terms (10–52) where the tensor Φ has symmetry $[\mu_1 \ldots \mu_\nu]$. Since the factors $\delta_{i_{\alpha_\nu} i'_{\alpha_\nu}}$ in (10–52) each have the symmetry [2], we may say that the tensor F with symmetry $[\lambda_1 \ldots \lambda_n]$ is obtained from the tensor Φ with symmetry $[\mu_1 \ldots \mu_\nu]$ by taking its outer product with s tensors, each having symmetry [2]:

$$[\lambda_1 \ldots \lambda_n] \quad \text{is contained in} \quad [\mu_1 \ldots \mu_\nu] \otimes \underbrace{[2] \otimes \cdots \otimes [2]}_{s \text{ factors}}. \quad (10\text{–}58)$$

For the special case of $s = 1$ [Eq. (10–34)], we get those patterns $(\mu_1, \ldots, \mu_\nu)$ for which

$$[\mu_1 \ldots \mu_\nu] \otimes [2] \quad\quad\quad (10\text{–}59)$$

contains $[\lambda_1 \ldots \lambda_n]$. We developed the expansion of outer products in Section 7–12. For the special case of (10–59), the pattern $(\mu_1, \ldots, \mu_\nu)$ is obtained from the pattern $[\lambda_1 \ldots \lambda_n]$ by a regular removal of two boxes. For example, for $[\lambda] = [4]$, a regular removal of two boxes leaves $[\mu'] = [2]$. A second removal of two boxes leaves $[\mu''] = [0]$. Thus the representation [4] of $U(n)$ is decomposed into

$$[4] = (400...) + (200...) + (000...) \quad\quad (10\text{–}60)$$

when we go to $O^+(n)$. [The dots in (10–60) denote zeros, since the symbols $(\mu_1, \ldots, \mu_\nu)$ are to contain ν terms.]

For the representation [21] of $U(n)$, a regular removal of two boxes leaves [1], so that the decomposition is

$$[21] = (21) + (10). \quad\quad\quad (10\text{–}61)$$

For the representation [22] of $U(n)$, a regular removal of two boxes leaves [2], and a second regular removal leaves [0], so that

$$[22] = (22) + (20) + (00). \quad\quad\quad (10\text{–}62)$$

Problem. Decompose the tensors with symmetry [21] into the parts indicated in Eq. (10–61).

Do the same for the tensors with symmetry [22].

The procedure becomes complicated for larger diagrams. Those diagrams in which the sum of the lengths of the first two rows exceeds n give no traceless tensor, and the equivalence of associated diagrams must be taken into account.

TABLE 10–3
DECOMPOSITION OF REPRESENTATIONS OF $U(5)$ INTO REPRESENTATIONS OF $O^+(5)$

r	$U(5)$ $[\lambda]$	$O^+(5)$ $(\mu_1\mu_2)$	$N[\lambda]$
0	[0]	(00)	1
1	[1]	(10)	5
2	[2]	(20) (00)	15
	[11]	(11)	10
3	[3]	(30) (10)	35
	[21]	(21) (10)	40
4	[4]	(40) (20) (00)	70
	[31]	(31) (20) (11)	105
	[22]	(22) (20) (00)	50
	[211]	(21) (11)	45
5	[41]	(41) (21) (30) (10)	224
	[32]	(32) (30) (21) (10)	175
	[311]	(31) (21) (11)	126
	[221]	(22) (21) (10)	75
	[2111]	(20) (11)	24
6	[42]	(42) (40) (31) (22) $(20)^2$ (00)	420
	[411]	(41) (31) (21) (11)	280
	[33]	(33) (31) (11)	175
	[321]	(32) (31) (22) (21) (20) (11)	280
	[3111]	(30) (21) (10)	70
7	[43]	(43) (41) (32) (30) (21) (10)	560
	[421]	(42) (41) (32) (31) (22) (30) $(21)^2$ (10)	700
	[331]	(33) (32) (31) (21) (11)	315
	[4111]	(40) (31) (20) (11)	160
	[322]	(32) (30) (22) (21) (10)	210
	[3211]	(31) (22) (21) (20) (11)	175
8	[44]	(44) (42) (40) (22) (20) (00)	490
	[431]	(43) (42) (33) (41) (32) $(31)^2$ (22) (21) (20) (11)	1050
	[422]	(42) (32) (40) (31) $(22)^2$ $(20)^2$ (00)	560
	[4211]	(41) (32) (31) (30) $(21)^2$ (11)	450
9	[441]	(44) (43) (42) (41) (32) (22) (30) (21) (10)	980
	[432]	(43) (42) (41) (33) $(32)^2$ (31) (30) (22) $(21)^2$ (10)	1120
	[4311]	(42) (40) (33) (32) $(31)^2$ (22) (21) (20) (11)	720
	[4221]	(41) (32) (31) (22) (30) $(21)^2$ (10)	480
10	[442]	(44) (43) $(42)^2$ (32) (40) (31) $(22)^2$ $(20)^2$ (00)	1176
	[4411]	(43) (33) (41) (32) (31) (21) (11)	700
	[4321]	(42) (41) (33) $(32)^2$ $(31)^2$ (30) $(22)^2$ $(21)^2$ (20) (11)	1024
	[4222]	(40) (31) (22) $(20)^2$ (00)	200

TABLE 10–4

DECOMPOSITION OF REPRESENTATIONS OF $U(7)$ INTO
REPRESENTATIONS OF $O^+(7)$

(for $r \leq 6$)

r	$U(7)$ $[\lambda]$	$O^+(7)$ $(\mu_1\mu_2\mu_3)$						$N[\lambda]$
0	[0]	(000)						1
1	[1]	(100)						7
2	[2]	(000)	(200)					28
	[11]	(110)						21
3	[3]	(100)	(300)					84
	[21]	(100)	(210)					112
	[111]	(111)						35
4	[4]	(000)	(200)	(400)				210
	[31]	(110)	(200)	(310)				378
	[22]	(000)	(200)	(220)				196
	[211]	(110)	(211)					210
5	[41]	(100)	(210)	(300)	(410)			1008
	[32]	(100)	(210)	(300)	(320)			882
	[311]	(111)	(210)	(311)				756
	[221]	(100)	(210)	(221)				490
	[2111]	(111)	(211)					224
6	[42]	(000)	$(200)^2$	(220)	(310)	(400)	(420)	2646
	[411]	(110)	(211)	(310)	(411)			2100
	[33]	(110)	(310)	(330)				1176
	[321]	(110)	(200)	(211)	(220)	(310)	(321)	2352
	[222]	(000)	(200)	(220)	(222)			490
	[3111]	(111)	(211)	(311)				840
	[2211]	(110)	(211)	(221)				588
	[21111]	(111)	(210)					140

Instead of attempting the general derivation, we present tables for the decomposition for $n = 5$ and $n = 7$ (Tables 10–3 and 10–4). Most of the results can be obtained by the simple procedure given above.

10–8 The symplectic group $Sp(n)$. Contraction. Traceless tensors. The orthogonal group in n dimensions, $O(n)$, is the group of linear transformations **a** which leave the scalar product

$$(\mathbf{xy}) = x_1 y_1 + \cdots + x_n y_n \tag{10–63}$$

invariant. More generally, $O(n)$ could be defined as the group of linear transformations which leave invariant a positive definite *symmetric* bilinear form. By a suitable change of basis such a form can be written in the *canonical* form (10–63).

The *symplectic group in n dimensions*, $Sp(n)$, is the set of all linear transformations **a** under which a nondegenerate *skew-symmetric* bilinear form is invariant. In other words, a nondegenerate bilinear form

$$\{xy\} = g_{ik}x_iy_k \qquad (g_{ik} = -g_{ki}), \tag{10–64}$$

the *skew product* of the vectors **x** and **y**, is unchanged by the transformations **a** of $Sp(n)$. The matrix $\mathbf{G} = (g_{ik})$ of Eq. (10–64) is skew-symmetric:

$$\widetilde{\mathbf{G}} = -\mathbf{G}. \tag{10–64a}$$

Taking the determinant of the matrices in (10–64a), we find det $G = (-1)^n$ det G. If n is odd, det $G = 0$, and the bilinear form (10–64) is degenerate. The symplectic group can therefore be defined only for *even-dimensional* spaces ($n = 2\nu$, ν integral). The condition that (10–64) be invariant under the transformation **a** is

$$\widetilde{\mathbf{a}}\mathbf{G}\mathbf{a} = \mathbf{G}. \tag{10–65}$$

The coordinate basis in the n-dimensional space can always be selected so that the skew product (10–64) takes on a simple *canonical* form. We start with an arbitrary nonzero vector $\mathbf{e}_1$ in the n-dimensional space. Since the skew product (10–64) is nondegenerate, we can find a vector **y** such that $\{\mathbf{e}_1\mathbf{y}\} \neq 0$. Multiplying **y** by a factor, we obtain a vector $\mathbf{e}_{1'}$ for which $\{\mathbf{e}_1\mathbf{e}_{1'}\} = 1$. We now have two vectors $\mathbf{e}_1$ and $\mathbf{e}_{1'}$ satisfying the conditions

$$\{\mathbf{e}_1\mathbf{e}_1\} = 0, \qquad \{\mathbf{e}_{1'}\mathbf{e}_{1'}\} = 0, \qquad \{\mathbf{e}_1\mathbf{e}_{1'}\} = 1. \tag{10–66}$$

The vectors $\mathbf{e}_1$, $\mathbf{e}_{1'}$ are linearly independent: If $\lambda\mathbf{e}_1 + \mu\mathbf{e}_{1'} = 0$, we find, by taking the skew product with $\mathbf{e}_1$ and $\mathbf{e}_{1'}$ in turn, that $\lambda = \mu = 0$.

The vectors **z** of R_n which satisfy the two independent linear equations

$$\{\mathbf{e}_1\mathbf{z}\} = 0, \qquad \{\mathbf{e}_{1'}\mathbf{z}\} = 0, \tag{10–67}$$

form an $(n - 2)$-dimensional linear subspace of R_n. Every vector **x** of R_n can be written as

$$\mathbf{x} = x_1\mathbf{e}_1 + x_{1'}\mathbf{e}_{1'} + \mathbf{z}, \tag{10–68}$$

where **z** satisfies (10–67). In fact, using (10–66) and (10–67), we have

$$x_1 = \{\mathbf{x}\mathbf{e}_{1'}\}, \qquad x_{1'} = -\{\mathbf{x}\mathbf{e}_1\}.$$

We repeat the argument for the $(n - 2)$-dimensional subspace. By induction, we conclude that we can select a *symplectic coordinate basis* of vectors $\mathbf{e}_1, \ldots, \mathbf{e}_\nu, \mathbf{e}_{\nu'}$ $(n = 2\nu)$, so that

$$\{\mathbf{e}_\alpha \mathbf{e}_\beta\} = \{\mathbf{e}_{\alpha'} \mathbf{e}_{\beta'}\} = 0, \qquad \{\mathbf{e}_\alpha \mathbf{e}_{\beta'}\} = -\{\mathbf{e}_{\alpha'} \mathbf{e}_\beta\} = \delta_{\alpha\beta}. \tag{10–69}$$

The components of a vector $\mathbf{x}$ in this basis are x_α, $x_{\alpha'}$ $(\alpha = 1, \ldots, \nu)$. Evaluating the skew product (10–64) in the basis (10–69), we find

$$\{\mathbf{xy}\} = (x_1 y_{1'} - y_1 x_{1'}) + (x_2 y_{2'} - y_2 x_{2'}) + \cdots + (x_\nu y_{\nu'} - y_\nu x_{\nu'}), \tag{10–70}$$

or

$$\{\mathbf{xy}\} = \epsilon_{ij} x_i y_j; \qquad \epsilon_{ij} = \begin{cases} 1 & \text{for } i = \alpha, j = \alpha', \\ -1 & \text{for } i = \alpha', j = \alpha, \\ 0 & \text{otherwise.} \end{cases} \tag{10–71}$$

The matrix $\mathbf{J} = (\epsilon_{ij})$ of the canonical form (10–71) satisfies the equation

$$\mathbf{J}^2 = -1, \qquad \epsilon_{ij} \epsilon_{jk} = -\delta_{ik}. \tag{10–72}$$

The form of the matrix (10–71) will depend on the order in which we write the indices α, α':

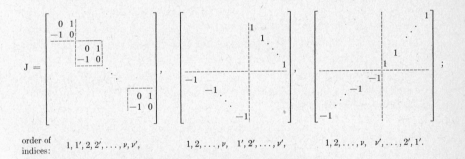

order of indices: $1, 1', 2, 2', \ldots, \nu, \nu',$ $1, 2, \ldots, \nu, \ 1', 2', \ldots, \nu',$ $1, 2, \ldots, \nu, \ \nu', \ldots, 2', 1'.$

The determinant formed from the components of n vectors $\mathbf{x}^{(1)}, \ldots, \mathbf{x}^{(n)}$ can be expressed in terms of skew products. To show this we consider the quantity

$$\frac{1}{2^\nu \cdot \nu!} \sum_p \delta_p \{\mathbf{x}^{(1)} \mathbf{x}^{(2)}\} \{\mathbf{x}^{(3)} \mathbf{x}^{(4)}\} \cdots \{\mathbf{x}^{(n-1)} \mathbf{x}^{(n)}\}, \tag{10–74}$$

where the summation extends over all permutations p of the superscripts (i.e., over all permutations of the vectors $\mathbf{x}^{(1)}, \ldots, \mathbf{x}^{(n)}$), and δ_p is the sign of the permutation p. Writing out the skew products in terms of components by using (10–71), we find that the expression (10–74) becomes

$$\frac{1}{2^{\nu} \cdot \nu!} \sum_{p} \delta_p \sum_{i_1 \ldots i_n} \epsilon_{i_1 i_2} \cdots \epsilon_{i_{n-1} i_n} x_{i_1}^{(1)} x_{i_2}^{(2)} \cdots x_{i_{n-1}}^{(n-1)} x_{i_n}^{(n)}$$

$$= \frac{1}{2^{\nu} \cdot \nu!} \sum_{i_1 \ldots i_n} \epsilon_{i_1 i_2} \cdots \epsilon_{i_{n-1} i_n} \sum_{p} \delta_p x_{i_1}^{(1)} \cdots x_{i_n}^{(n)}. \quad (10\text{–}75)$$

The sum

$$\sum_{p} \delta_p x_{i_1}^{(1)} \cdots x_{i_n}^{(n)}$$

is zero unless the indices $i_1, \ldots, i_n$ are all different. For any set of different $i_1, \ldots, i_n$, the sum is equal to

$$\delta_{\pi} \cdot [\mathbf{x}^{(1)} \ldots \mathbf{x}^{(n)}],$$

where δ_{π} is the sign of the permutation π which brings $1, 2, \ldots, n$ to the order $i_1, \ldots, i_n$, and $[\mathbf{x}^{(1)} \ldots \mathbf{x}^{(n)}]$ is the determinant formed from the components of the vectors $\mathbf{x}^{(1)}, \ldots, \mathbf{x}^{(n)}$. The expression (10–75) is therefore equal to

$$\frac{[\mathbf{x}^{(1)} \ldots \mathbf{x}^{(n)}]}{2^{\nu} \cdot \nu!} \sum_{i_1 \neq i_2 \neq \ldots \neq i_n} \delta_{\pi} \epsilon_{i_1 i_2} \cdots \epsilon_{i_{n-1} i_n}. \quad (10\text{–}76)$$

The product $\epsilon_{i_1 i_2} \cdots \epsilon_{i_{n-1} i_n}$ will be zero unless each pair of indices assumes the values α, α' or α', α. The term $\epsilon_{11'} \epsilon_{22'} \cdots \epsilon_{\nu\nu'}$ is equal to $+1$. Interchanging pairs of indices still gives $+1$, and the corresponding permutation π is even, so that $\delta_{\pi} = 1$. Replacing $\epsilon_{\alpha\alpha'}$ by $\epsilon_{\alpha'\alpha}$ changes the product to -1, but this replacement corresponds to an odd permutation π, with $\delta_{\pi} = -1$. Thus each nonzero term in the sum contributes unity. The number of such terms is $2^{\nu} \cdot \nu!$, so that (10–76) is equal to $[\mathbf{x}^{(1)} \ldots \mathbf{x}^{(n)}]$, and we have the result

$$[\mathbf{x}^{(1)} \ldots \mathbf{x}^{(n)}] = \frac{1}{2^{\nu} \cdot \nu!} \sum_{p} \delta_p \{\mathbf{x}^{(1)} \mathbf{x}^{(2)}\} \cdots \{\mathbf{x}^{(n-1)} \mathbf{x}^{(n)}\}. \quad (10\text{–}77)$$

Under a linear transformation $\mathbf{a}$ in R_n, the determinant $[\mathbf{x}^{(1)} \ldots \mathbf{x}^{(n)}]$ is multiplied by det $\mathbf{a}$. But, if $\mathbf{a}$ is symplectic, it leaves the skew products on the right of (10–77) unchanged, and it must therefore leave $[\mathbf{x}^{(1)} \ldots \mathbf{x}^{(n)}]$ unchanged. Consequently, det $\mathbf{a} = 1$ for symplectic $\mathbf{a}$. The *symplectic transformations are unimodular*. There is no need to distinguish between proper and improper transformations, as we did for $O(n)$.

The procedure for obtaining irreducible representations of the symplectic group $Sp(n)$ is very similar to the method used for the orthogonal group $O(n)$.

When we go from the linear group $GL(n)$ to its symplectic subgroup $Sp(n)$ in an even-dimensional space $(n = 2\nu)$, the representations of $GL(n)$ in terms of tensors of a given symmetry will become reducible. In addition to the operations of permuting the tensor indices, we also have to perform an operation of *contraction* which commutes with the symplectic transformations.

Consider the space of tensors of rank r, with components $F_{i_1 \ldots i_r}$. For symplectic transformations **a**, using (10–65) in the canonical basis, we have

$$\epsilon_{kl} a_{ki} a_{lj} = \epsilon_{ij}. \tag{10-78}$$

From the tensor $F_{i_1 i_2 \ldots i_r}$, we can construct the *(12)-trace (contraction)* by multiplying by $\epsilon_{i_1 i_2}$ and summing over i_1 and i_2:

$$F^{(12)}_{i_3 \ldots i_r} = \epsilon_{i_1 i_2} F_{i_1 i_2 i_3 \ldots i_r}. \tag{10-79}$$

The trace operation gives a tensor of rank $r - 2$ and commutes with the symplectic transformations. If **a** is symplectic, then

$$F'_{i_1 i_2 \ldots i_r} = a_{i_1 j_1} a_{i_2 j_2} \cdots a_{i_r j_r} F_{j_1 j_2 \ldots j_r}, \tag{10-6}$$

$$\begin{aligned}
F'^{(12)}_{i_3 \ldots i_r} = \epsilon_{i_1 i_2} F'_{i_1 i_2 i_3 \ldots i_r} &= \epsilon_{i_1 i_2} a_{i_1 j_1} a_{i_2 j_2} a_{i_3 j_3} \cdots a_{i_r j_r} F_{j_1 j_2 j_3 \ldots j_r} \\
&= \epsilon_{j_1 j_2} a_{i_3 j_3} \cdots a_{i_r j_r} F_{j_1 j_2 j_3 \ldots j_r} \\
&= a_{i_3 j_3} \cdots a_{i_r j_r} F^{(12)}_{j_3 \ldots j_r} \\
&= F^{(12)'}_{i_3 \ldots i_r}. \tag{10-80}
\end{aligned}$$

The contraction process (10–79) can be applied to any pair of indices, so that there are $r(r - 1)/2$ traces $F^{(\alpha \beta)}_{\ldots}$ of the rth-rank tensor.

We select those rth-rank tensors for which all pair traces are zero. Equation (10–80) shows that this subspace of traceless tensors is invariant under transformations induced by $Sp(n)$. Every tensor $F_{i_1 \ldots i_r}$ can be decomposed uniquely into a traceless tensor F^0 plus a tensor of the form

$$\begin{aligned}
\Phi_{i_1 \ldots i_r} = \epsilon_{i_1 i_2} G^{(12)}_{i_3 \ldots i_r} + \cdots \\
+ \epsilon_{i_\alpha i_\beta} G^{(\alpha \beta)}_{i_1 \ldots i_{\alpha-1} i_{\alpha+1} \ldots i_{\beta-1} i_{\beta+1} \ldots i_r} + \cdots \quad \left(\frac{r(r - 1)}{2} \text{ terms} \right).
\end{aligned} \tag{10-81}$$

The proof is identical with the one given in Section 10–5. We get a decomposition of the tensor space into a direct sum of invariant subspaces:

$$F_{i_1 \ldots i_r} = \Phi_{i_1 \ldots i_r} + F^0_{i_1 \ldots i_r}. \tag{10-82}$$

The invariance of the subspace Σ of tensors Φ having the form (10–81) is proved analogously to (10–37). Each term in (10–81) is transformed into a similar term under symplectic transformations:

$$a_{i_1 j_1} a_{i_2 j_2} \cdots a_{i_r j_r} (\epsilon_{j_1 j_2} G_{j_3 \ldots j_r}^{(12)}) = \epsilon_{i_1 i_2} a_{i_3 j_3} \cdots a_{i_r j_r} G_{j_3 \ldots j_r}^{(12)}$$

$$= \epsilon_{i_1 i_2} G_{i_3 \ldots i_r}^{(12)'}. \tag{10–83}$$

The process for finding the traceless part of a tensor is similar to that used for $O(n)$. For example, for $r = 3$, we write $F_{i_1 i_2 i_3}$ as

$$F_{i_1 i_2 i_3} = F_{i_1 i_2 i_3}^0 + H_{i_3} \epsilon_{i_1 i_2} + K_{i_2} \epsilon_{i_3 i_1} + L_{i_1} \epsilon_{i_2 i_3}. \tag{10–84}$$

The requirement that F^0 be traceless gives the equations

$$\text{(12)-trace:} \quad F_{i_3}^{(12)} = n H_{i_3} - K_{i_3} - L_{i_3},$$

$$\text{(31)-trace:} \quad F_{i_2}^{(31)} = -H_{i_2} + n K_{i_2} - L_{i_2},$$

$$\text{(23)-trace:} \quad F_{i_1}^{(23)} = -H_{i_1} - K_{i_1} + n L_{i_1}.$$

Solving, we find

$$H_i = \frac{1}{n^2 - n - 2} [(n-1) F_{..i}^{(12)} + F_{.i.}^{(31)} + F_{i..}^{(23)}],$$

$$K_i = \frac{1}{n^2 - n - 2} [F_{..i}^{(12)} + (n-1) F_{.i.}^{(31)} + F_{i..}^{(23)}],$$

$$L_i = \frac{1}{n^2 - n - 2} [F_{..i}^{(12)} + F_{.i.}^{(31)} + (n-1) F_{i..}^{(23)}]. \tag{10–84a}$$

10–9 The irreducible representations of $Sp(n)$. Decomposition of irreducible representations of $U(n)$ with respect to its symplectic subgroup. Just as for the orthogonal group, permutation of indices takes a traceless tensor into another traceless tensor. If we start from the invariant subspace of traceless tensors F^0 of rank r, we can apply the Young symmetrizers to decompose the space into subspaces of traceless tensors having a definite symmetry type. In this way we arrive at the irreducible representations of $Sp(n)$.

Each irreducible representation of $Sp(n)$ is associated with a Young diagram $[\lambda_1 \ldots \lambda_n]$, with $\lambda_1 + \lambda_2 + \cdots + \lambda_n = r$. However, not all diagrams are admissible. We shall prove that the traceless tensors corresponding to Young diagrams in which there are more than $\nu = n/2$ rows are identically equal to zero. To prove the theorem we show that the tensors corresponding to diagrams with more than $n/2$ rows must have the form (10–81). We need consider only the indices in the first column of the diagram, so we assume that a tensor $F_{i_1 \ldots i_m}$ is antisymmetric in the indices $i_1 \ldots i_m$, where $m > \nu = n/2$. As we showed in Section 10–2, such a tensor is a superposition of tensors of the form (10–14), $[\mathbf{x}^{(1)} \ldots \mathbf{x}^{(m)}]$. We add to this set the vector variables $\mathbf{y}^{(m+1)}, \ldots, \mathbf{y}^{(n)}$, and construct the determinant

$$[\mathbf{x}^{(1)} \ldots \mathbf{x}^{(m)} \mathbf{y}^{(m+1)} \ldots \mathbf{y}^{(n)}]. \tag{10–85}$$

The components of the tensor $[\mathbf{x}^{(1)} \ldots \mathbf{x}^{(m)}]$ will be the coefficients of the monomials $y_{i_{m+1}}^{(m+1)} \cdots y_{i_n}^{(n)}$. From Eq. (10–77), the determinant (10–85) can be expressed as a sum of terms in which the vectors $\mathbf{x}^{(1)}, \ldots, \mathbf{x}^{(m)}$, $\mathbf{y}^{(m+1)}, \ldots, \mathbf{y}^{(n)}$ are paired and contracted. Since $m > \nu$, at least one skew product in each term on the right of (10–77) must contain $\mathbf{x}$'s, giving a factor $\epsilon_{i_\alpha i_\beta} x_{i_\alpha}^{(\alpha)} x_{i_\beta}^{(\beta)}$, so that the term has the form (10–81). Consequently the tensor has zero trace if $m > \nu$. [Actually, we have proved even more: the number of factors on the right of (10–77), in which $\mathbf{x}$'s are paired, must be at least equal to $(m - \nu)$, so that each term on the right of (10–77) has at least $(m - \nu)$ factors ϵ_{ij}.] As a result of this theorem, we can restrict ourselves to patterns with at most ν rows.

The symmetry type for the irreducible representations of $Sp(n)$ will be denoted by $(\sigma_1 \ldots \sigma_\nu)$, where $\sigma_1 \geq \sigma_2 \geq \cdots \geq \sigma_\nu \geq 0$. The basis vectors for this irreducible representation are the traceless tensors of symmetry $[\sigma_1 \ldots \sigma_\nu]$. The dimensionality of the representation $(\sigma_1 \ldots \sigma_\nu)$ of $Sp(n)$ is given by the formula

$$N(\sigma) = \prod_{i=1}^{\nu} \frac{\sigma_i + \nu - i + 1}{\nu - i + 1}$$

$$\times \prod_{k>i}^{\nu} \frac{(\sigma_i - \sigma_k + k - i)(\sigma_i + \sigma_k + 2\nu + 2 - i - k)}{(k - i)(2\nu + 2 - i - k)}. \tag{10–86}$$

We tabulate some of the simpler irreducible representations of $Sp(n)$ for $n = 4$, 6, and 8 in Table 10–5.

The subspace Σ of tensors Φ [Eq. (10–85)] is invariant under symplectic transformations, but it is *not* irreducible. Just as in Section 10–7, we can take repeated traces and decompose the tensor into a sum of terms of the form

$$\epsilon_{i_{\alpha_1} i_{\alpha_1}'} \cdots \epsilon_{i_{\alpha_s} i_{\alpha_s}'} \Phi_{i_{\beta_1} \ldots i_{\beta_\nu}} \qquad (2s + v = r), \tag{10–87}$$

where the tensor $\Phi_{i_1, \ldots, i_v}$ is traceless.

If we start from rth-rank tensors having a definite symmetry, the contraction process will decompose the space into traceless tensors of rank r, $r - 2$, $r - 4$, etc., which have definite symmetry. The decomposition of

TABLE 10–5

IRREDUCIBLE REPRESENTATIONS OF $Sp(n)$

r	$n = 4$ $(\sigma_1\sigma_2)$	$N(\sigma)$	r	$n = 6$ $(\sigma_1\sigma_2\sigma_3)$	$N(\sigma)$	r	$n = 8$ $(\sigma_1\sigma_2\sigma_3\sigma_4)$	$N(\sigma)$
0	(00)	1	0	(000)	1	0	(0000)	1
1	(10)	4	1	(100)	6	1	(1000)	8
2	(20)	10	2	(200)	21	2	(2000)	36
	(11)	5		(110)	14		(1100)	27
3	(21)	16	3	(210)	64	3	(2100)	160
4	(22)	14		(111)	14		(1110)	48
			4	(220)	90	4	(2200)	308
				(211)	70		(2110)	315
			5	(221)	126		(1111)	42
			6	(222)	84	5	(2210)	792
							(2111)	288
						6	(2220)	825
							(2211)	792
						7	(2221)	1056
						8	(2222)	594

a representation $[\lambda_1 \ldots \lambda_n]$ of $U(n)$ into irreducible representations $(\sigma_1 \ldots \sigma_\nu)$ of $Sp(n)$ means that the tensor F with symmetry $[\lambda_1 \ldots \lambda_n]$ is written as a sum of terms (10–87), where the tensor Φ has symmetry $[\sigma_1 \ldots \sigma_\nu]$. Since the factors $\epsilon_{i_{\alpha_\nu} i_{\alpha_\nu}'}$ in (10–87) each have the symmetry [11], we see that the tensor F with symmetry $[\lambda_1 \ldots \lambda_n]$ is obtained from the tensor Φ with symmetry $[\sigma_1 \ldots \sigma_\nu]$ by taking the outer product with s factors, each having the symmetry [11]:

$$[\lambda_1 \ldots \lambda_n] \quad \text{is contained in} \quad [\sigma_1 \ldots \sigma_\nu] \otimes \underbrace{[11] \otimes \cdots \otimes [11]}_{s \text{ factors}}.$$

$$(10\text{–}88)$$

For the special case of $s = 1$ [Eq. (10–85)], we get those patterns $[\sigma_1 \ldots \sigma_\nu]$ for which

$$[\sigma_1 \ldots \sigma_\nu] \otimes [11] \qquad (10\text{–}89)$$

contains $[\lambda_1 \ldots \lambda_n]$. Again we can use the results of Section 7–12 concerning outer products. For (10–89), the pattern $(\sigma_1 \ldots \sigma_\nu)$ is obtained

TABLE 10–6

REDUCTION OF REPRESENTATIONS OF $U(n)$ TO REPRESENTATIONS OF $Sp(n)$

r	$[\lambda]$	(σ)-structure	r	$[\lambda]$	(σ)-structure
	$n = 4$			$n = 6$	
0	[0]	(00)	0	[0]	(000)
1	[1]	(10)	1	[1]	(100)
2	[2]	(20)	2	[2]	(200)
	[11]	(00) (11)		[11]	(000) (110)
3	[21]	(10) (21)	3	[21]	(100) (210)
				[111]	(100) (111)
4	[22]	(00) (11) (22)	4	[22]	(000) (110) (220)
	[211]	(20) (11)		[211]	(200) (110) (211)
			5	[221]	(100) (210) (111) (221)
				[2111]	(100) (111) (210)
			6	[222]	(200) (211) (222)
				[2211]	(000) $(110)^2$ (220) (211)
				[21111]	(200) (110)

r	$[\lambda]$	(σ)-structure
	$n = 8$	
0	[0]	(0000)
1	[1]	(1000)
2	[2]	(2000)
	[11]	(0000) (1100)
3	[21]	(1000) (2100)
	[111]	(1000) (1110)
4	[22]	(0000) (1100) (2200)
	[211]	(2000) (1100) (2110)
	[1111]	(0000) (1100) (1111)
5	[221]	(1000) (2100) (1110) (2210)
	[2111]	(1000) (2100) (1110) (2111)
6	[222]	(2000) (2110) (2220)
	[2211]	(0000) $(1100)^2$ (2200) (2110) (1111) (2211)
	[21111]	(2000) (1100) (2110) (1111)
7	[2221]	(1000) (2100) (1110) (2210) (2111) (2221)
	[22111]	(1000) (2100) $(1110)^2$ (2210) (2111)
	[211111]	(1000) (2100) (1110)
8	[2222]	(0000) (1100) (2200) (1111) (2211) (2222)
	[22211]	(2000) (1100) $(2110)^2$ (1111) (2220) (2211)
	[221111]	(0000) $(1100)^2$ (2200) (2110) (1111)
	[2111111]	(2000) (1100)

from $[\lambda_1 \ldots \lambda_n]$ by regular removal of an antisymmetric pair [11]. This step is followed by a second regular removal of a pair [11], etc.

The pattern [11] is decomposed into (11) + (00). For the pattern [22] we get (22) + (11) + (00). The process becomes complicated for partitions $[\lambda_1 \ldots \lambda_n]$ with more than $n/2$ rows. We shall not discuss the general procedure, but give the reduction tables for the simplest cases (Table 10–6).

CHAPTER 11

APPLICATIONS TO ATOMIC AND NUCLEAR PROBLEMS

11-1 The classification of states of systems of identical particles according to $SU(n)$. One of the main problems of atomic and nuclear physics is the determination of the energy levels of a system of identical (equivalent) particles. Since we cannot solve the problem for a system of interacting particles, we use the methods of perturbation theory. Each particle of the system is assumed to move in some averaged potential field. We determine the eigenstates for this average field and take, as basis functions for the full problem, products of the single-particle eigenfunctions. The perturbation then consists of any part of the single-particle field plus the interactions among the particles. If the particles are identical, the interaction operator will be symmetric in all the particles. Consequently its matrix elements between basis functions will depend sensitively on the symmetry of these functions under interchange of particles.

To keep the argument general, we start from the assumption that we have solved a single-particle problem whose Hamiltonian is invariant under a transformation group G. The functions corresponding to a given eigenvalue $\epsilon^{(j)}$ of the single-particle problem will form the basis for a representation $D^{(j)}$ of the group G. If the dimension of the representation $D^{(j)}$ is n, the n basis functions $\psi_1, \ldots, \psi_n$ correspond to the same energy $\epsilon^{(j)}$. Suppose that we have a system of r identical particles, and that, in zeroth approximation, each of the particles has energy $\epsilon^{(j)}$. Any product wave function $\psi_{i_1}(1)\psi_{i_2}(2) \ldots \psi_{i_r}(r)$ $(i_\nu = 1, \ldots, n)$ will correspond to the same energy $r \cdot \epsilon^{(j)}$ in zeroth approximation. The perturbation will split this degeneracy. Since the total Hamiltonian is symmetric under interchange of identical particles, the proper zeroth-order functions are those linear combinations of product wave functions which have a definite symmetry type. The problem of constructing such functions was solved in Chapter 7.

An alternative point of view is the following: We regard the single-particle wave function ψ as a vector in the n-dimensional space spanned by the basis vectors $\psi_1, \ldots, \psi_n$:

$$\psi = \sum_1^n a_i\psi_i, \qquad \sum_i |a_i|^2 = 1. \tag{11-1}$$

If the basis vectors ψ_i are subjected to a unitary transformation so that

$$\psi_i' = u_{ij}\psi_j, \tag{11-2}$$

we obtain another basis for the same vector space. Moreover, the unitary transformation can be made unimodular by taking out a phase factor $e^{i\delta}$. This phase factor is common to all the ψ_i', so that it will not change any matrix elements. We may therefore regard the space spanned by $\psi_1, \ldots, \psi_n$ as providing us with a basis for the (faithful) representation of the unitary unimodular group $SU(n)$. If we have a system of r identical particles, the products $\psi_{i_1}(1) \cdots \psi_{i_r}(r)$ form the basis for a representation of $SU(n)$ in terms of rth-rank tensors. Our problem is then to construct the irreducible representations of $SU(n)$ in terms of rth-rank tensors of definite symmetry. We have solved this problem in Chapter 10.

From the preceding argument, we see that the space of rth-rank tensors plays two roles:

(1) It is the space of a representation of $SU(n)$.

(2) It is the space of a representation of the group G.

When we decompose the space into subspaces of rth-rank tensors of definite symmetry, we find the irreducible representations of $SU(n)$. At the same time, each of these subspaces provides a representation of the group G. One of our main problems is to decompose the irreducible representations of $SU(n)$ into irreducible representations of G.

11–2 Angular momentum analysis. Decomposition of representations of $SU(n)$ into representations of $O^+(3)$. The simplest example of the preceding discussion is the case where the single-particle Hamiltonian is invariant under the group $O^+(3)$, the rotation group in three dimensions. Then the single-particle energies $\epsilon^{(j)}$ corresponding to different irreducible representations $D^{(j)}$ of $O^+(3)$ are well separated. The wave function for the single particle is a vector in the $(2j + 1)$-dimensional space of the irreducible representation $D^{(j)}$ spanned by the basis vectors

$$\psi_i \ (i = -j, -j + 1, \ldots, j - 1, j).$$

The operator $J_x^2 + J_y^2 + J_z^2$ has the fixed value $j(j + 1)$ for all vectors in this space, and we can choose the ψ_i so that, say, $J_z \psi_i = i\psi_i$. From products of the ψ's, we now wish to construct irreducible tensors [with respect to $SU(2j + 1)$], and determine which representations $D^{(j)}$ of $O^+(3)$ are contained in them.

The simplest procedure is the following: The tensor of rank one corresponds to the Young diagram [1]. It provides an irreducible representation [1] of $SU(2j + 1)$ and an irreducible representation $D^{(j)}$ of $O^+(3)$. For $r = 2$, we get the irreducible representations [2] and [11] of $SU(2j + 1)$. The analysis of these representations in terms of $O^+(3)$ is easily done by means of the results in Section 9–8 on Clebsch-Gordan coefficients. If we set $j_1 = j_2 = j$ in Eq. (9–116), we have

$$A_J = (u_1 v_2 - u_2 v_1)^{2j-J}(u_1 x_1 + u_2 x_2)^J(v_1 x_1 + v_2 x_2)^J. \quad (11-3)$$

If we interchange $u_1 \leftrightarrow v_1$, $u_2 \leftrightarrow v_2$, then A_J is multiplied by $(-1)^{2j-J}$. Thus, if $2j - J$ is even, the second-rank tensor formed from $\psi_{i_1}\psi_{i_2}$ is symmetric, while for $2j - J$ odd, the tensor is antisymmetric. Consequently, for integral j, the representation $\square\square$ contains

$$J = 2j, 2j - 2, \ldots, 0, \tag{11-4}$$

while $\begin{array}{c}\square\\\square\end{array}$ contains

$$J = 2j - 1, 2j - 3, \ldots, 1. \tag{11-4a}$$

For half-integral j, $\square\square$ contains

$$J = 2j, 2j - 2, \ldots, 1, \tag{11-5}$$

while $\begin{array}{c}\square\\\square\end{array}$ contains

$$J = 2j - 1, 2j - 3, \ldots, 0. \tag{11-5a}$$

In particular, we note that the function having $J = 0$ is a symmetric tensor for integral j and an antisymmetric tensor for odd j. The expansion of this function in terms of the ψ_i is obtained by setting $J = 0$ in (11-3):

$$\Psi_{J=0} = (u_1 v_2 - u_2 v_1)^{2j}$$

$$= \sum_{m=-j}^{j} \frac{(2j)!}{(j+m)!(j-m)!} (-1)^{j-m} (u_1 v_2)^{j+m} (u_2 v_1)^{j-m}$$

$$= (2j)! \sum_{m=-j}^{j} (-1)^{j-m} \psi_m^j \phi_{-m}^j,$$

whence the normalized function for $J = 0$ is

$$\Psi_{J=0} = \frac{1}{\sqrt{2j+1}} \sum_{m=-j}^{j} (-1)^{j-m} \psi_m^j \phi_{-m}^j. \tag{11-6}$$

For tensors of higher rank, we illustrate the procedure first for $j = \frac{1}{2}$. Since $n = 2j + 1 = 2$, the Young diagrams can contain at most two rows. Furthermore, from Eq. (10-29), $[\lambda_1 \lambda_2] \equiv [\lambda_1 - \lambda_2, 0]$. Combining this with the results of Eq. (11-5), we see that [2] has $J = 1$, while $[11] \equiv [0]$ has $J = 0$. The pattern [21] is equivalent to [1], so that $J = \frac{1}{2}$. Similarly, any pattern $[a + 1, a] \equiv [1]$ has $J = \frac{1}{2}$. If we take the outer product

$$\square\square \otimes \square = \square\square\square + \begin{array}{c}\square\square\\\square\end{array},$$

we have, on the left, products of functions with $j = 1$ and $j = \frac{1}{2}$. These yield resultants J of $\frac{1}{2}$ and $\frac{3}{2}$. The second pattern on the right has $J = \frac{1}{2}$, so [3] has $J = \frac{3}{2}$. Repeating the argument,

$$[3] \otimes [1] = [4] + [31]$$
$$j = \tfrac{3}{2} \quad \tfrac{1}{2} \quad 2 \quad 1 \; ,$$

we find that [4] has $J = 2$, and in general, $[\lambda_1 \lambda_2]$ has $J = \frac{1}{2}(\lambda_1 - \lambda_2)$. For $j = 1$, $(n = 2j + 1 = 3)$, we already know that

$\square$, $\quad J = 1, \quad$ has dimension 3;

$\square\square$, $\quad J = 2, 0, \quad$ has dimension 6;

$\dfrac{\square}{\square}$, $\quad J = 1, \quad$ has dimension 3;

$\dfrac{\square}{\dfrac{\square}{\square}}$, $\quad J = 0, \quad$ has dimension 1 since $n = 3$.

Next we construct the direct product

$$\dfrac{\square}{\square} \otimes \square = \dfrac{\square}{\dfrac{\square}{\square}} + \dfrac{\square\square}{\square} \; .$$

On the left we combine $j = 1$ and $j' = 1$, and obtain $J = 2, 1, 0$. On the right the representation [111] has $J = 0$, so [21] has $J = 2, 1$. Again, analyzing the outer product

$$\square\square \otimes \square = \square\square\square + \dfrac{\square\square}{\square}$$
$$j: \quad 2, 0 \quad\quad 1 \quad\quad\quad\quad J = 2, 1$$

yields $J = 1, 1, 2, 3$, so $\square\square\square = [3]$ contains $J = 3, 1$. Taking

$$\dfrac{\square}{\square} \otimes \square\square = \dfrac{\square\square\square}{\square} + \dfrac{\square\square}{\dfrac{\square}{\square}}$$

$$\equiv \dfrac{\square\square\square}{\square} + \square \; ,$$

we have, on the left $j = 1, j' = 2, 0$, giving $1, 1, 2, 3$, so that $\dfrac{\square\square\square}{\square} = [31]$ contains $J = 3, 2, 1$. From

$$\square\square\square \otimes \square = \square\square\square\square + \dfrac{\square\square\square}{\square} \; ,$$

we find that $\square\square\square\square = [4]$ contains $J = 4, 2, 0$.

TABLE 11–1

$j = 1$

Outer product	Angular momenta S P D F G $J = 0$ 1 2 3 4					Resolution of outer product
$[1] \otimes [11]$	1	1	1	·	·	$[21]$, $[0]$
$[1] \otimes [2]$	·	2	1	1	·	$[3]$, $[21]$
$[1] \otimes [21]$	1	2	2	1	·	$[31]$, $[2]$, $[1]$
$[1] \otimes [3]$	1	1	2	1	1	$[4]$, $[31]$
$[1] \otimes [31]$	1	2	3	2	1	$[41]$, $[31]$, $[2]$
$[1] \otimes [32](\equiv [31])$	1	2	3	2	1	$[42]$, $[3]$, $[21]$

This recursion method, using Eqs. (11–4), (11–5) and the formulas for resolving outer products, enables us to analyze any representation of $SU(n = 2j + 1)$ into its component angular momenta J. We illustrate the procedure for $j = 1$. We tabulate a sequence of outer products at the left, their component angular momenta in the middle of the table, and the resolution of the outer product at the right (Table 11–1). We have made use of the equivalences

$$[\lambda_1 \lambda_2 \ldots \lambda_n] \equiv [\lambda_1 - \lambda_{n_1} \lambda_2 - \lambda_n, \ldots, \lambda_{n-1} - \lambda_n] \qquad (10\text{–}29)$$

and

$$[\lambda_1 \lambda_2 \ldots \lambda_n] \equiv [\lambda_1 - \lambda_n, \lambda_1 - \lambda_{n-1}, \ldots, \lambda_1 - \lambda_2], \qquad (10\text{–}30)$$

with $n = 2j + 1 = 3$. By combining the results in the table, we find the angular momentum structure of $[\lambda]$. We tabulate the results for $j = 1, \frac{3}{2}, 2, \frac{5}{2}, 3, \frac{7}{2}$ (Tables 11–2 through 11–7, see pp. 418 through 421). We restrict the tables to partitions with $\lambda_1 \leq 4$ for integral j and $\lambda_1 \leq 2$ for half-integral j.

In constructing the tables for higher values of j, we must use the direct method for analyzing partitions $[1^r]$, which is described in detail in Section 11–5.

Problem. Construct the table for $j = \frac{3}{2}$.

11–3 The Pauli principle. Atomic spectra in Russell-Saunders coupling.

Before applying the results of the preceding section to physical problems, we must impose one further condition. The *Pauli principle* states that the *total* wave function of a system of electrons must be *antisymmetric* under any interchange of the particles. The same statement holds for any system of identical particles having half-integral spin [i.e., particles whose internal

TABLE 11–2

ANGULAR-MOMENTUM ANALYSIS OF THE $(j)^r$-CONFIGURATION

$$j = 1, \ SU(3)$$

r	$[\lambda]$	J	$N([\lambda])$
0	[0]	$0 = S$	1
1	[1]	$1 = P$	3
2	[2]	$0, 2 = S \ D$	6
	$[11] \equiv [1]$	$1 = P$	3
3	[3]	$1, 3 = P \ F$	10
	[21]	$1, 2 = P \ D$	8
	$[111] \equiv [0]$	$0 = S$	1
4	[4]	$0, 2, 4 = S \ D \ G$	15
	[31]	$1, 2, 3 = P \ D \ F$	15
	$[22] \equiv [2]$	$0, 2 = S \ D$	6
	$[211] \equiv [1]$	$1 = P$	3
5	[41]	$1, 2, 3, 4 = P \ D \ F \ G$	24
	$[32] \equiv [31]$	$1, 2, 3 = P \ D \ F$	15
	$[311] \equiv [2]$	$0, 2 = S \ D$	6
	$[221] \equiv [1]$	$1 = P$	3
6	[42]	$0, (2)^2, 3, 4 = S \ D^2 \ F \ G$	27
	$[411] \equiv [3]$	$1, 3 = P \ F$	10
	$[33] \equiv [3]$	$1, 3 = P \ F$	10
	$[321] \equiv [21]$	$1, 2 = P \ D$	8
	$[222] \equiv [0]$	$0 = S$	1

TABLE 11–3

ANGULAR-MOMENTUM ANALYSIS OF THE $(j)^r$-CONFIGURATION

$$j = \tfrac{3}{2}, \ SU(4)$$

r	$[\lambda]$	J	$N([\lambda])$
0	[0]	0	1
1	[1]	$\frac{3}{2}$	4
2	[2]	$1, 3$	10
	[11]	$0, 2$	6
3	[21]	$\frac{1}{2}, \frac{3}{2}, \frac{5}{2}, \frac{7}{2}$	20
4	[22]	$0, (2)^2, 4$	20
	[211]	$1, 2, 3$	15

TABLE 11–4

ANGULAR-MOMENTUM ANALYSIS OF THE $(j)^r$-CONFIGURATION
$$j = 2,\ SU(5)$$

r	$[\lambda]$	S $J=0$	P 1	D 2	F 3	G 4	H 5	I 6	K 7	L 8	M 9	N 10	O 11	Q 12	$N([\lambda])$
0	[0]	1													1
1	[1]	·	·	1											5
2	[2]	1	·	1	·	1									15
	[11]	·	1	·	1										10
3	[3]	1	·	1	1	1	·	1							35
	[21]		1	2	1	1	1								40
4	[4]	1	·	2	·	2	1	1	·	1					70
	[31]	·	2	2	3	2	2	1	1						105
	[22]	2	·	2	1	2	·	1							50
	[211]	·	2	1	2	1	1								45
5	[41]	1	2	3	4	4	3	3	2	1	1				224
	[32]	1	2	4	3	4	3	2	1	1					175
	[311]	·	3	2	4	2	3	1	1						126
	[221]	1	1	3	2	2	1	1							75
	[2111]	·	1	1	1	1									24
6	[42]	3	2	7	5	8	5	6	3	3	1	1			420
	[411]	·	4	3	6	4	5	3	3	1	1				280
	[33]	·	3	1	5	2	3	2	2	·	1				175
	[321]	1	4	6	6	6	5	3	2	1					280
	[3111]	1	1	2	2	2	1	1							70
7	[43]	1	4	7	7	8	8	6	5	4	2	1	1		560
	[421]	3	6	10	11	12	10	9	6	4	2	1			700
	[331]	·	5	4	7	5	6	3	3	1	1				315
	[4111]	·	2	3	3	3	3	2	1	1					160
	[322]	2	2	5	4	5	3	3	1	1					210
	[3211]	1	3	4	5	4	3	2	1						175
8	[44]	4	1	6	4	8	4	7	3	4	2	2	·	1	490
	[431]	2	9	12	16	15	15	12	10	6	4	2	1		1050
	[422]	4	3	10	7	11	7	8	4	4	1	1			560
	[4211]	1	6	6	9	8	8	5	4	2	1				450
9	[441]	4	5	11	11	14	11	12	8	7	4	3	1	1	980
	[432]	3	9	14	16	17	16	13	10	7	4	2	1		1120
	[4311]	2	7	10	12	12	11	9	6	4	2	1			720
	[4221]	2	5	8	9	9	8	6	4	2	1				480
10	[442]	6	5	15	12	18	13	15	9	9	4	4	1	1	1176
	[4411]	·	7	7	11	9	11	7	7	4	3	1	1		700
	[4321]	4	10	14	18	18	15	13	9	5	3	1			1024
	[4222]	2	1	5	3	5	3	3	1	1					200

TABLE 11–5

ANGULAR-MOMENTUM ANALYSIS OF THE $(j)^r$-CONFIGURATION

$j = \frac{5}{2}$, $SU(6)$

r	$[\lambda]$	J	$N([\lambda])$
0	[0]	0	1
1	[1]	$\frac{5}{2}$	6
2	[2]	1, 3, 5	21
	[11]	0, 2, 4	15
3	[21]	$\frac{1}{2}$, $\frac{3}{2}$, $(\frac{5}{2})^2$, $(\frac{7}{2})^2$, $\frac{9}{2}$, $\frac{11}{2}$, $\frac{13}{2}$	70
	[111]	$\frac{3}{2}$, $\frac{5}{2}$, $\frac{9}{2}$	20
4	[22]	$(0)^2$, $(2)^3$, 3, $(4)^3$, 5, $(6)^2$, 8	105
	[211]	$(1)^2$, $(2)^2$, $(3)^3$, $(4)^2$, $(5)^2$, 6, 7	105
5	[221]	$(\frac{1}{2})^2$, $(\frac{3}{2})^3$, $(\frac{5}{2})^4$, $(\frac{7}{2})^4$, $(\frac{9}{2})^4$, $(\frac{11}{2})^3$, $(\frac{13}{2})^2$, $\frac{15}{2}$, $\frac{17}{2}$	210
	[2111]	$\frac{1}{2}$, $(\frac{3}{2})^2$, $(\frac{5}{2})^2$, $(\frac{7}{2})^2$, $(\frac{9}{2})^2$, $\frac{11}{2}$, $\frac{13}{2}$	84
6	[222]	$(1)^3$, 2, $(3)^5$, $(4)^2$, $(5)^3$, $(6)^2$, $(7)^2$, 9	175
	[2211]	$(0)^2$, 1, $(2)^5$, $(3)^3$, $(4)^5$, $(5)^2$, $(6)^3$, 7, 8	189
	[21111]	1, 2, 3, 4, 5	35

TABLE 11–6

ANGULAR-MOMENTUM ANALYSIS OF THE $(j)^r$-CONFIGURATION

$j = 3$, $SU(7)$, for $r \leq 4$

r	$[\lambda]$	$J=0$	1	2	3	4	5	6	7	8	9	10	11	12	$N([\lambda])$
0	[0]	1													1
1	[1]	·	·	·	1										7
2	[2]	1	·	1	·	1	·	1							28
	[11]	·	1	·	1	·	1								21
3	[3]	·	1	·	2	1	1	1	1	·	1				84
	[21]	·	1	2	2	2	2	1	1	1					112
	[111]	1	·	1	1	1	·	1							35
4	[4]	2	·	2	1	3	1	3	1	2	1	1	·	1	210
	[31]	·	3	3	5	4	5	4	4	2	2	1	1		378
	[22]	2	·	4	1	4	2	3	1	2	·	1			196
	[211]	·	3	2	4	3	4	2	2	1	1				210

TABLE 11–7

ANGULAR-MOMENTUM ANALYSIS OF THE $(j)^r$-CONFIGURATION

$$j = \tfrac{7}{2},\ SU(8),\ \text{for}\ r \leq 4$$

r	$[\lambda]$	J	$N([\lambda])$
0	[0]	0	1
1	[1]	$\frac{7}{2}$	8
2	[2]	1, 3, 5, 7	36
	[11]	0, 2, 4, 6	28
3	[21]	$\frac{1}{2}, \frac{3}{2}, (\frac{5}{2})^2, (\frac{7}{2})^3, (\frac{9}{2})^2, (\frac{11}{2})^2, (\frac{13}{2})^2, \frac{15}{2}, \frac{17}{2}, \frac{19}{2}$	168
	[111]	$\frac{3}{2}, \frac{5}{2}, \frac{7}{2}, \frac{9}{2}, \frac{11}{2}, \frac{15}{2}$	56
4	[22]	$0^3, 2^4, 3^2, 4^5, 5^2, 6^5, 7^2, 8^3, 9, 10^2, 12$	336
	[211]	$1^3, 2^3, 3^5, 4^4, 5^5, 6^4, 7^4, 8^2, 9^2, 10, 11$	378
	[1111]	$0, 2^2, 4^2, 5, 6, 8$	70

wave functions form the basis for an irreducible representation $D^{(j)}$ of $O^+(3)$, with j half-integral]. For systems of identical particles having *integral* spin, the total wave function must be *symmetric* under any interchange.

For many-electron atoms, we start from the single-particle orbits in some averaged central field. The state of a single electron is characterized by quantum numbers n, l, m_l, m_s. The first quantum number gives the energy, l and m_l label the basis functions of the representation $D^{(l)}$ of the rotation group which is provided by the orbital (external) motion, and m_s labels the basis functions of the representation $D^{(1/2)}$ of the rotation group which is provided by the internal motion (spin). The perturbation consists of the Coulomb interaction between the electrons and of terms involving the electron spins.

For light atoms, the terms containing the electron spins make a contribution to the energy which is small compared to that from the Coulomb repulsion. The approximation procedure which starts from this assumption is called Russell-Saunders coupling (L-S coupling). In L-S coupling, we treat the orbital wave functions and the spin functions of the electrons separately. If there are r electrons in single-particle orbits with angular momentum l, the product of the orbital wave functions will be an rth-rank tensor with respect to $SU(2l + 1)$. Since the Coulomb repulsion is symmetric in the coordinates of all the electrons, the Coulomb energy will depend strongly on the symmetry of the coordinate wave function. Thus the appropriate linear combinations are the irreducible rth-rank tensors.

We see that, in general, states of higher orbital symmetry will have higher energy: if the wave function is symmetric in a pair of particles, the Coulomb repulsion between them will be effective; if the wave function is antisymmetric in a pair of particles, their Coulomb repulsion will be reduced.

At the same time, we must combine the internal (spin) wave functions of the r electrons. For a single electron, the wave functions form the basis for a representation of $SU(2)$ since $j = \frac{1}{2}$, $2j + 1 = 2$. For r electrons, we can construct rth-rank tensors of definite symmetry.

Finally, we must multiply the space and spin wave functions to obtain the total wave function. But, according to the Pauli principle, the total wave function must be completely antisymmetric. From the results of Chapter 7, we know how to construct such products: the space and spin functions must have conjugate diagrams [cf. Eq. (7–211)].

For the spin wave functions we use the results of Section 11–2 for $j = \frac{1}{2}$. The Young diagram contains at most two rows, $\lambda_1 + \lambda_2 = r$, and the resultant angular momentum is

$$S = \tfrac{1}{2}(\lambda_1 - \lambda_2). \tag{11–7}$$

In other words, the Young diagram for r electrons with total spin S is

$$[\lambda_1 \lambda_2] = \left[\frac{r + 2S}{2}, \frac{r - 2S}{2} \right]. \tag{11–8}$$

The dimensionality (multiplicity) of the representation is $2S + 1$.

Since the space wave function must correspond to the conjugate Young diagram (with rows and columns interchanged), the patterns for the orbital functions can have at most two boxes in any row. For the orbital wave functions in atomic problems we can restrict ourselves to partitions with $\lambda_1 \leq 2$.

We start with the configuration $(p)^r$. Here $l = 1$. The total number of different orbital states available for a single particle is $2l + 1 = 3$. The internal wave function (for $s = \frac{1}{2}$) has two basis states. So six different total states are possible, and the p-shell will be filled for $r = 6$. All the results are contained in the tables of the preceding section. We use Eq. (11–7) for the total spin functions, and Table 11–2 for the total orbital wave functions. The results are given in Table 11–8.

We need not go beyond $r = 3$ (halfway through the shell), because the rest of the table can be obtained by using the equivalences (10–29) and (10–30). For example, for $r = 4$, [22] ≡ [2] with $L = 2, 0$; [22] has $S = 0$, so we obtain the same results as for $r = 2$: $[2] \cdot [1]^2$.

For the configuration $(d)^r$ with $l = 2$, we use Table 11–4 for $j = 2$. The shell contains 10 states. The resultant terms are given in Table 11–9.

TABLE 11–8

TERMS FOR ELECTRON CONFIGURATION $(p)^r$

	Orbital	Spin	Multiplet
$r = 1$	$[1]\ L = 1$	$[1]\ S = \frac{1}{2}$	^2P
$r = 2$	$[2]\ L = 2, 0$	$[1^2]\ S = 0$	$^1S,\ ^1D$
	$[1^2]\ L = 1$	$[2]\ S = 1$	^3P
$r = 3$	$[21]\ L = 2, 1$	$[21]\ S = \frac{1}{2}$	$^2P,\ ^2D$
	$[1^3]\ L = 0$	$[3]\ S = \frac{3}{2}$	^4S

TABLE 11–9

TERMS FOR ELECTRON CONFIGURATION $(d)^r$

	Orbital	Spin	Multiplet
$r = 1$	$[1]\ L = 2$	$[1]\ S = \frac{1}{2}$	^2D
$r = 2$	$[2]\ L = 4, 2, 0$	$[1^2]\ S = 0$	$^1S,\ ^1D,\ ^1G$
	$[1^2]\ L = 3, 1$	$[2]\ S = 1$	$^3P,\ ^3F$
$r = 3$	$[21]\ L = 5, 4, 3, (2)^2, 1$	$[21]\ S = \frac{1}{2}$	$^2P,\ (^2D)^2,\ ^2F,\ ^2G,\ ^2H$
	$[1^3]\ L = 3, 1$	$[3]\ S = \frac{3}{2}$	$^4P,\ ^4F$
$r = 4$	$[1^4]\ L = 2$	$[4]\ S = 2$	^5D
	$[21^2]\ L = 5, 4, (3)^2, 2, (1)^2$	$[31]\ S = 1$	
	$[2^2]\ L = 6, (4)^2, 3, (2)^2, (0)^2$	$[2^2]\ S = 0$	
$r = 5$	$[1^5]\ L = 0$	$[5]\ S = \frac{5}{2}$	^6S
	$[21^3]\ L = 4, 3, 2, 1$	$[41]\ S = \frac{3}{2}$	
	$[2^21]\ L = 6, 5, (4)^2, (3)^2, (2)^3, 1, 0$	$[32]\ S = \frac{1}{2}$	

Problems. (1) Use Table 11–6 to construct the table of terms arising from the configuration $(f)^r$ $(l = 3)$.

(2) The arguments used here can also be applied to the classification of rotational states of homonuclear diatomic molecules. Apply them to the classification of the states of ortho- and para-hydrogen and ortho- and para-deuterium. Write the partition function for these molecules in their ground-electronic and vibrational state.

The next step in the perturbation procedure would be to include the spin-dependent terms in the Hamiltonian. The product of the representations D^J and D^S then splits into representations D^J. The spin-dependent terms cause a splitting of the multiplets.

11–4 Seniority in atomic spectra. For the configuration $(p)^r$, the symmetry pattern $[\lambda]$ and the angular momentum L completely characterized the state: a multiplet occurred only once for a given symmetry.

This was not the case for the configuration $(d)^r$. For three particles, there are two ^2D-states with the same symmetry [21]. Many of the multiplets occur several times for the same [λ] when $r = 4, 5$. It would be advantageous to have some additional quantum number for classification of the states. We would like to find a group G which is contained in $SU(2l + 1)$, and which contains the rotation group $O^+(3)$ as a subgroup. Then a state would be characterized by its symmetry pattern [λ] for $SU(2l + 1)$, the irreducible representation of G to which it belongs, and its angular momentum L. Such groups G can be found, but the resulting classification will be useful only if the perturbation Hamiltonian commutes (or commutes approximately) with the transformations of G. If this is the case, the label of the irreducible representation of G provides us with an additional quantum number having physical significance. For the spectra of many-electron atoms, the group $O^+(2l + 1)$ provides such a subgroup G.

Suppose that we have a system of two electrons in the configuration $(l)^2$. The orbital wave functions $\psi_i^{(1)}$, $\psi_j^{(2)}$ of the two electrons are vectors in the $(2l + 1)$-dimensional space of the representation $D^{(l)}$ of the rotation group. The product functions $\psi_i^{(1)}\psi_j^{(2)}$ form the basis for a representation of the unitary group $SU(2l + 1)$ and, at the same time, form the basis for a representation of the rotation group $O^+(3)$. For $j = l$ (an $integer$), Eq. (11–6) shows that the $symmetric$ bilinear form (scalar product)

$$\Psi_{L=0} = \frac{1}{\sqrt{2l + 1}} \sum_{m=-l}^{l} \psi_m^{(1)}\psi_{-m}^{(2)} \equiv \boldsymbol{\psi}^{(1)} \cdot \boldsymbol{\psi}^{(2)} \tag{11–9}$$

couples the angular momenta of the two particles to give a resultant $L = 0$. The three-dimensional rotations will induce linear transformations in the tensor space, but the function $\Psi_{L=0}$ of (11–9) will be left unchanged. But there is a much larger group of transformations which leaves the scalar product (11–9) invariant. The scalar product (11–9) is a symmetric bilinear form in the vectors of a $(2l + 1)$-dimensional space. It is therefore invariant under the orthogonal transformations $O^+(2l + 1)$, which we studied in Chapter 10. The process (11–9) of coupling the orbital functions of two electrons to give a resultant having $L = 0$ reduces the rank of the tensor $\psi_i^{(1)}\psi_j^{(2)}$ by 2, and is invariant under the transformations of the group $O^+(2l + 1)$. The operation (11–9) for taking the scalar product is just the contraction which we introduced in Section 10–5.

We now see that the product functions $\psi_i^{(1)}\psi_j^{(2)}$ serve as the representation space for three groups: $SU(2l + 1) \supset O^+(2l + 1) \supset O^+(3)$. The products $\psi_i^{(1)}\psi_j^{(2)}$ can first be reduced to give the irreducible representations [2] and [11] of $SU(2l + 1)$. For the functions of the representation [2], we can use the trace operation provided by (11–9) to decompose the space into two parts, a tensor of rank zero, $(\boldsymbol{\psi}^{(1)} \cdot \boldsymbol{\psi}^{(2)})$, and a traceless tensor of rank two. These parts were called Φ and F^0 in Section 10–5. Thus the representation [2] of $SU(2l + 1)$ decomposes into representations (00) and (20) of $O^+(2l + 1)$.

The scalar product (11–9) is symmetric and has $L = 0$. The corresponding spin function for the two electrons must be antisymmetric and have the symmetry [11], with $S = 0$. The contraction process (11–9) therefore consists in coupling a pair of electrons to give a ^1S-state.

The next step in the classification is to decompose the representations of $O^+(2l + 1)$ into representations of the rotation group $O^+(3)$. The angular-momentum analysis $SU(2l + 1) \to O^+(2l + 1) \to O^+(3)$ is easily done: [2] contains $L = 2l, 2l - 2, \ldots, 2, 0$. The subspace (00) contains only $L = 2$, so the subspace (20) contains $L = 2l, 2l - 2, \ldots, 2$. The functions belonging to [11] give zero when contracted, and hence they belong to the representation (11) of $O^+(2l + 1)$. The possible values of L are $L = 2l - 1, 2l - 3, \ldots, 3, 1$.

For two particles, we have the classification:

$SU(2l + 1)$	$O^+(2l + 1)$	$O^+(3)$	
[2]	(20)	$L = 2l, \quad 2l - 2, \ldots, 2;$	
[2]	(00)	$L = 0;$	
[11]	(11)	$L = 2l - 1, \quad 2l - 3, \ldots, 3, 1.$	(11–10)

Next we consider a system of four electrons in the configuration $(l)^4$. The operation (11–9) applied to the functions $\psi_{i_1}^{(1)}\psi_{i_2}^{(2)}\psi_{i_3}^{(3)}\psi_{i_4}^{(4)}$ gives terms of the form

$$(\psi^{(1)} \cdot \psi^{(2)})\psi_{i_3}^{(3)}\psi_{i_4}^{(4)}. \tag{11–11}$$

In such terms, a ^1S-state of one pair of electrons is combined with the second-rank tensor formed from the other pair. If we apply the trace operation again, we obtain terms of the form

$$(\psi^{(1)} \cdot \psi^{(2)})(\psi^{(3)} \cdot \psi^{(4)}), \tag{11–12}$$

a combination of ^1S-states which is a tensor of zero rank and has $L = 0$.

If we start from the irreducible representation [22] of $SU(2l + 1)$, the trace operation will decompose the space into the representations (00), (20), and (22) of $O^+(2l + 1)$. The functions for (00) have the form (11–12). Those for (20) have the form (11–11), and are the product of a traceless tensor in two of the particles with a ^1S-state of the other pair. The angular-momentum analysis follows from these statements and the results for two particles. The functions (00) must have $L = 0$. The functions (20) have $L = 2l, 2l - 2, \ldots, 2$. To find the angular momenta for (22), we use the fact that the subspaces (00), (20), and (22) add to give the space of the representation [22] of $SU(2l + 1)$. Since the angular momenta contained in [22] can be found by the methods of Section 11–2, the L-values for (22) can be found by subtraction.

In this example, we note that the space of fourth-rank tensors with symmetry [22] was decomposed into a part (00) consisting of tensors of rank $v = 4 - 4 = 0$, a part (20) containing tensors of rank $v = 4 - 2 = 2$, and a part (22) consisting of traceless tensors of rank $v = 4$. The tensors (00) already occurred for *zero* particles (identity representation). The tensors (20) already occurred above for a configuration of *two* particles. The *smallest number* of particles for which the traceless tensor can occur is called its *seniority*. The tensors (00) have seniority $v = 0$, those for (20) have seniority $v = 2$, those for (22) have seniority $v = 4$.

In general, the tensors of rank r corresponding to the partition

$$[\lambda_1 \ldots \lambda_{2l+1}]$$

can be decomposed into traceless tensors of type $(\mu_1, \ldots, \mu_l)$ having rank $v \leq r$. For even r, the lowest possible seniority is $v = 0$, corresponding to the irreducible representation $(00 \cdots)$ of $O^+(2l + 1)$; for odd r, the lowest possible seniority is $v = 1$, corresponding to the irreducible representation $(10 \cdots)$ of $O^+(2l + 1)$. The decomposition $SU(2l + 1) \to O^+(2l + 1)$ can be done for each l, by means of Section 10–7 and Tables 10–3 and 10–4. For the decomposition $O^+(2l + 1) \to O^+(3)$ we use the tables in Section 11–2.

For the configuration $(p)^r$, $l = 1$, and $O^+(2l + 1)$ is just the rotation group in three dimensions. The analysis in Table 11–8 is complete.

For the configuration $(d)^r$, we combine the results in Tables 10–3 and 11–9.

For $r = 1$:

[1], $L = 2 \to (10)$, so (10) contains $L = 2$.

For $r = 2$:

[2], $L = 4, 2, 0 \to (20) + (00)$, so (20) contains $L = 4, 2$.

[11], $L = 3, 1 \to (11)$, so (11) contains $L = 3, 1$.

For $r = 3$:

[21], $L = 5, 4, 3, 2^2, 1 \to (21) + (10)$. Since (10) contains $L = 2$, (21) contains $L = 5, 4, 3, 2, 1$.

$[1^3] \equiv [1^2] \to (11)$.

By this procedure we obtain the decomposition $O^+(5) \to O^+(3)$. The results for the $(d)^r$-configuration are given in Table 11–10, which shows that the terms for the $(d)^r$-configuration are uniquely characterized by $[\lambda]$, v, and L.

The same method yields Table 11–11 for the $(f)^r$-configuration.

For the $(f)^r$-configuration, the addition of the seniority quantum number is not sufficient to characterize the terms uniquely. It is possible to find a subgroup of $O^+(7)$, which contains the rotation group, and thus obtain a further refinement of the classification, but we shall not attempt this.

TABLE 11–10

r	Orbital $[\lambda]$ $(\mu_1\mu_2)$	Seniority v	Angular momentum L	Spin $[\tilde{\lambda}]$ S	Multiplet
$r = 0$	[0] (00)	0	0	[0] 0	^1S
$r = 1$	[1] (10)	1	2	[1] $\frac{1}{2}$	^2D
$r = 2$	[2] (00)	0	0	[11] 0	^1S
	(20)	2	4, 2	[11] 0	$^1D, {}^1G$
	[11] (11)	2	3, 1	[2] 1	$^3P, {}^3F$
$r = 3$	[21] (10)	1	2	[21] $\frac{1}{2}$	^2D
	(21)	3	5, 4, 3, 2, 1	[21] $\frac{1}{2}$	$^2P, {}^2D, {}^2F, {}^2G, {}^2H$
	[111] (11)	2	3, 1	[3] $\frac{3}{2}$	$^4P, {}^4F$
$r = 4$	[1111] (10)	1	2	[4] 2	^5D
	[211] (11)	2	3, 1	[31] 1	$^3P, {}^3F$
	(21)	3	5, 4, 3, 2, 1	[31] 1	$^3P, {}^3D, {}^3F, {}^3G, {}^3H$
	[22] (00)	0	0	[22] 0	^1S
	(20)	2	4, 2	[22] 0	$^1D, {}^1G$
	(22)	4	6, 4, 3, 2, 0	[22] 0	$^1S, {}^1D, {}^1F, {}^1G, {}^1I$
$r = 5$	[11111] (00)	0	0	[5] $\frac{5}{2}$	^6S
	[2111] (11)	2	3, 1	[41] $\frac{3}{2}$	$^4P, {}^4F$
	(20)	2	4, 2	[41] $\frac{3}{2}$	$^4D, {}^4G$
	[221] (10)	1	2	[32] $\frac{1}{2}$	^2D
	(21)	3	5, 4, 3, 2, 1	[32] $\frac{1}{2}$	$^2P, {}^2D, {}^2F, {}^2G, {}^2H$
	(22)	4	6, 4, 3, 2, 0	[32] $\frac{1}{2}$	$^2S, {}^2D, {}^2F, {}^2G, {}^2I$

TABLE 11–11

r	Orbital $[\lambda]$ $(\mu_1\mu_2\mu_3)$	Seniority v	Angular momentum L	Spin $[\tilde{\lambda}]$ S
$r=0$	$[0]$ (000)	0	0	$[0]$ 0
$r=1$	$[1]$ (100)	1	3	$[1]$ $\tfrac{1}{2}$
$r=2$	$[2]$ (000)	0	0	$[11]$ 0
	$[2]$ (200)	2	$6, 4, 2$	$[11]$ 0
	$[11]$ (110)	2	$5, 3, 1$	$[2]$ 1
$r=3$	$[21]$ (100)	1	3	$[21]$ $\tfrac{1}{2}$
	$[21]$ (210)	3	$8, 7, 6, 5^2, 4^2, 3, 2^2, 1$	$[21]$ $\tfrac{1}{2}$
	$[111]$ (111)	3	$6, 4, 3, 2, 0$	$[3]$ $\tfrac{3}{2}$
$r=4$	$[22]$ (000)	0	0	$[22]$ 0
	$[22]$ (200)	2	$6, 4, 2$	$[22]$ 0
	$[22]$ (220)	4	$10, 8^2, 7, 6^2, 5^2, 4^3, 3, 2^3, 0$	$[22]$ 0
	$[211]$ (110)	2	$5, 3, 1$	$[31]$ 1
	$[211]$ (211)	4	$9, 8, 7^2, 6^2, 5^3, 4^3, 3^3, 2^2, 1^2$	$[31]$ 1
	$[1111]\equiv[111]$ (111)	3	$6, 4, 3, 2, 0$	$[4]$ 2
$r=5$	$[221]$ (100)	1	3	$[32]$ $\tfrac{1}{2}$
	$[221]$ (210)	3	$8, 7, 6, 5^2, 4^2, 3, 2^2, 1$	$[32]$ $\tfrac{1}{2}$
	$[221]$ (221)	5	$11, 10, 9^2, 8^2, 7^4, 6^4, 5^5, 4^4, 3^5, 2^3, 1^3$	$[32]$ $\tfrac{1}{2}$
	$[2111]$ (111)	3	$6, 4, 3, 2, 0$	$[41]$ $\tfrac{3}{2}$
	$[2111]$ (211)	4	$9, 8, 7^2, 6^2, 5^3, 4^3, 3^3, 2^2, 1^2$	$[41]$ $\tfrac{3}{2}$
	$[11111]\equiv[11]$ (110)	2	$5, 3, 1$	$[5]$ $\tfrac{5}{2}$

$r = 6$	[222]	(000)	0	0	[33]	0
		(200)	2	6, 4, 2	[33]	0
		(220)	4	10, 8², 7, 6², 5², 4³, 3, 2³, 0	[33]	0
		(222)	6	12, 10, 9², 8², 7², 6⁴, 5², 4⁴, 3³, 2², 1, 0²	[33]	0
	[2211]	(110)	2	5, 3, 1	[42]	1
		(211)	4	9, 8, 7², 6², 5³, 4³, 3, 2², 1²	[42]	1
		(221)	5	11, 10, 9², 8², 7⁴, 6⁴, 5⁴, 4⁵, 3⁵, 2³, 1³	[42]	1
	[21111]	(111)	3	6, 4, 3, 2, 0	[51]	2
		(210)	3	8, 7, 6, 5², 4², 3, 2², 1	[51]	2
	[1⁶] ≡ [1]	(100)	1	3	[6]	3
$r = 7$	[2221]	(100)	1	3	[43]	$\frac{1}{2}$
		(210)	3	8, 7, 6, 5², 4², 3, 2², 1	[43]	$\frac{1}{2}$
		(221)	5	11, 10, 9², 8², 7⁴, 6⁴, 5⁵, 4⁵, 3⁵, 2³, 1³	[43]	$\frac{1}{2}$
		(222)	6	12, 10, 9², 8², 7², 6⁴, 5², 4⁴, 3³, 2², 1, 0²	[43]	$\frac{1}{2}$
	[22111]	(111)	3	6, 4, 3, 2, 0	[52]	$\frac{3}{2}$
		(211)	4	9, 8, 7², 6², 5³, 4³, 3³, 2², 1²	[52]	$\frac{3}{2}$
		(220)	4	10, 8², 7, 6², 5², 4³, 3, 2³, 0	[52]	$\frac{3}{2}$
	[211111]	(110)	2	5, 3, 1	[61]	$\frac{5}{2}$
		(200)	2	6, 4, 2	[61]	$\frac{5}{2}$
	[1⁷] ≡ [0]	(000)	0	0	[7]	$\frac{7}{2}$

Problems. (1) Construct Table 11–11.

(2) For electrons in an $(l)^r$-configuration, the symbol $(\mu_1, \ldots, \mu_l)$ can contain only the values $\mu_i = 0, 1, 2$. Let α be the number of 2's in the symbol and let β be the number of 1's. Prove that

$$\alpha = \frac{v}{2} - S,$$

$$\beta = \min \begin{cases} 2S, \\ 2l + 1 - v, \end{cases} \tag{11-13}$$

where v is the seniority and S the spin.

11–5 Atomic spectra in jj-coupling. In heavy atoms, when the contribution of spin-dependent interactions to the energy becomes important, we may obtain a better description by using jj-coupling instead of Russell-Saunders coupling.

In the jj-approximation, we first combine the one-electron orbital wave function with the internal (spin) wave function of the electron. These functions form the bases for representations $D^{(l)}$ and $D^{(1/2)}$ of the rotation group. We resolve the product $D^{(l)} \times D^{(1/2)}$ into irreducible representations $D^{(j)}$ of the rotation group $(j = l \pm \frac{1}{2})$. Because of the strong spin-orbit interaction, the single-particle functions with different angular momenta j have well-separated energies. The single-particle wave functions corresponding to a given energy form the basis $\psi_j, \psi_{j-1}, \ldots, \psi_{-j+1}, \psi_{-j}$ for the representation $D^{(j)}$ of the rotation group.

If there are r equivalent electrons in a configuration $(j)^r$, we must construct the total wave function from products $\psi_{i_1}^{(1)} \cdots \psi_{i_r}^{(r)}$ $(i_v = -j, \ldots, j)$. But, since we are now considering total wave functions for the identical particles, the Pauli principle requires that we take only the completely antisymmetric tensor $[1^r]$.

The angular-momentum analysis of the antisymmetric tensor $[1^r]$ is straightforward, but tedious. Since the tensor is antisymmetric, the only nonzero components are those for which $i_1, \ldots, i_r$ are all different. Since the index i on the function ψ_i labels the row of $D^{(j)}$ to which ψ_i belongs (i.e., the value of $J_z^{(1)}$ in the state $\psi_i^{(1)}$), the sum of the (different) indices, $i_1 + \cdots + i_r$, gives the row to which $\psi_{i_1}^{(1)} \cdots \psi_{i_r}^{(r)}$ belongs (i.e., the value of $J_z^{(1)} + \cdots + J_z^{(r)}$ in this state). To find the values of J for the configuration $(j)^r$, we simply write all possible sets of indices $i_1, \ldots, i_r$, where $i_1 > i_2 > \cdots > i_r$, and take the sum $\sum_1^r i_v$. If the largest sum is J_1, the resolution of $[D^{(j)}]^r$ must contain the representation D^{J_1} of the rotation group, with its $2J_1 + 1$ basis functions having $\sum i_v = J_1, J_1 - 1, \ldots, -J_1$. We strike these entries from the table of sums, and proceed to the highest remaining value of $\sum i_v$. Continuing this process, we find the

TABLE 11–12

i_1	i_2	$i_1 + i_2$
$\frac{3}{2}$	$\frac{1}{2}$	2
$\frac{3}{2}$	$-\frac{1}{2}$	1
$\frac{3}{2}$	$-\frac{3}{2}$	0
$\frac{1}{2}$	$-\frac{1}{2}$	0
$\frac{1}{2}$	$-\frac{3}{2}$	-1
$-\frac{1}{2}$	$-\frac{3}{2}$	-2

angular-momentum resolution. We illustrate the procedure with a few examples.

For $j = \frac{3}{2}$, there are four states (four values of the indices) available. We need only go up to $r = 2$. For $r = 1$, $J = \frac{3}{2}$. For $r = 2$, we tabulate the possible values of i_1, i_2, and $i_1 + i_2$ (Table 11–12). The highest value of $\sum i_\nu$ is 2. We strike out the entries 2, 1, 0, -1, -2. This leaves the single entry 0. Thus the configuration $(\frac{3}{2})^2$ contains $J = 2, 0$.

For the configuration $(\frac{5}{2})^r$, we have:

$r = 1$, $J = \frac{5}{2}$;

$r = 2$, $J = 4, 2, 0$;

$r = 3$: We tabulate all possible sets of indices i_1, i_2, i_3 and their sums $i_1 + i_2 + i_3$ (Table 11–13). The highest entry is $\sum i_\nu = \frac{9}{2}$, so the representation contains $J = \frac{9}{2}$. We strike out the entries $\frac{9}{2}, \frac{7}{2}, \frac{5}{2}, \ldots, -\frac{7}{2}, -\frac{9}{2}$. The highest remaining entry is $\frac{5}{2}$, so the representation contains $J = \frac{5}{2}$. We strike out the entries $\frac{5}{2}, \frac{3}{2}, \ldots, -\frac{5}{2}$. The highest remaining entry is

TABLE 11–13

i_1	i_2	i_3	Σi_ν	i_1	i_2	i_3	Σi_ν
$\frac{5}{2}$	$\frac{3}{2}$	$\frac{1}{2}$	$\frac{9}{2}$	$\frac{3}{2}$	$\frac{1}{2}$	$-\frac{1}{2}$	$\frac{3}{2}$
		$-\frac{1}{2}$	$\frac{7}{2}$			$-\frac{3}{2}$	$\frac{1}{2}$
		$-\frac{3}{2}$	$\frac{5}{2}$			$-\frac{5}{2}$	$-\frac{1}{2}$
		$-\frac{5}{2}$	$\frac{3}{2}$	$\frac{3}{2}$	$-\frac{1}{2}$	$-\frac{3}{2}$	$-\frac{1}{2}$
$\frac{5}{2}$	$\frac{1}{2}$	$-\frac{1}{2}$	$\frac{5}{2}$			$-\frac{5}{2}$	$-\frac{3}{2}$
		$-\frac{3}{2}$	$\frac{3}{2}$	$\frac{3}{2}$	$-\frac{3}{2}$	$-\frac{5}{2}$	$-\frac{5}{2}$
		$-\frac{5}{2}$	$\frac{1}{2}$	$\frac{1}{2}$	$-\frac{1}{2}$	$-\frac{3}{2}$	$-\frac{3}{2}$
$\frac{5}{2}$	$-\frac{1}{2}$	$-\frac{3}{2}$	$\frac{1}{2}$			$-\frac{5}{2}$	$-\frac{5}{2}$
		$-\frac{5}{2}$	$-\frac{1}{2}$	$\frac{1}{2}$	$-\frac{3}{2}$	$-\frac{5}{2}$	$-\frac{7}{2}$
$\frac{5}{2}$	$-\frac{3}{2}$	$-\frac{5}{2}$	$-\frac{3}{2}$	$-\frac{1}{2}$	$-\frac{3}{2}$	$-\frac{5}{2}$	$-\frac{9}{2}$

TABLE 11-14

TOTAL ANGULAR MOMENTA IN THE CONFIGURATION $(j)^r$

		J
$j = \frac{3}{2}$	$r = 1$	$\frac{3}{2}$
	2	2, 0
$j = \frac{5}{2}$	$r = 1$	$\frac{5}{2}$
	2	4, 2, 0
	3	$\frac{9}{2}, \frac{5}{2}, \frac{3}{2}$
$j = \frac{7}{2}$	$r = 1$	$\frac{7}{2}$
	2	6, 4, 2, 0
	3	$\frac{15}{2}, \frac{11}{2}, \frac{9}{2}, \frac{7}{2}, \frac{5}{2}, \frac{3}{2}$
	4	$8, 6, 5, (4)^2, (2)^2, 0$
$j = \frac{9}{2}$	$r = 1$	$\frac{9}{2}$
	2	8, 6, 4, 2, 0
	3	$\frac{21}{2}, \frac{17}{2}, \frac{15}{2}, \frac{13}{2}, \frac{11}{2}, (\frac{9}{2})^2, \frac{7}{2}, \frac{5}{2}, \frac{3}{2}$
	4	$12, 10, 9, (8)^2, 7, (6)^3, 5, (4)^3, 3, (2)^2, (0)^2$
	5	$\frac{25}{2}, \frac{21}{2}, \frac{19}{2}, (\frac{17}{2})^2, (\frac{15}{2})^2, (\frac{13}{2})^2, (\frac{11}{2})^2, (\frac{9}{2})^3,$ $(\frac{7}{2})^2, (\frac{5}{2})^2, \frac{3}{2}, \frac{1}{2}$

$\frac{3}{2}$, and hence $J = \frac{3}{2}$ is contained in the representation. Striking out $\frac{3}{2}, \frac{1}{2},$ $-\frac{1}{2}, -\frac{3}{2}$, we exhaust the table, so the configuration $(\frac{5}{2})^3$ contains $J = \frac{9}{2},$ $\frac{5}{2}, \frac{3}{2}.$

There is a simple check on the angular-momentum analysis. The tensor of symmetry $[1^r]$ with respect to $SU(2j + 1)$ has $\binom{2j+1}{r}$ independent components. (This is the number of ways of selecting the r indices $i_1 > i_2 > \cdots > i_r$ from the $2j + 1$ values $j, j - 1, \ldots, -j$.) Since a representation $D^{(J)}$ of the rotation group has dimension $2J + 1$, we must have

$$\sum_{J \text{ in } [1^r]} (2J + 1) = \binom{2j + 1}{r}. \tag{11-14}$$

In the last example,

$$\binom{2j + 1}{r} = \binom{6}{3} = 20 = \left(2 \cdot \frac{9}{2} + 1\right) + \left(2 \cdot \frac{5}{2} + 1\right) + \left(2 \cdot \frac{3}{2} + 1\right).$$

We tabulate the results through $j = \frac{9}{2}$ (Table 11-14).

Problems. (1) Prove that the largest angular momentum in the configuration $(j)^r$ is $J_{max} = r/2[2j - r + 1]$.

(2) Prove that, in the configuration $(j)^r$, the angular momentum $J = J_{max} - 1$ cannot occur, and that the maximum multiplicity for the angular momentum

$$J = J_{max} - 2 \quad \text{is 1,}$$
$$= J_{max} - 3 \quad \text{is 1,}$$
$$= J_{max} - 4 \quad \text{is 2,}$$
$$= J_{max} - 5 \quad \text{is 2,}$$
$$= J_{max} - 6 \quad \text{is 4.}$$

11–6 Nuclear structure. Isotopic spin. Perturbation procedures similar to those for the many-electron problem can be applied to nuclei. The nuclear problem is complicated by the fact that the system is built up from two kinds of particles, neutrons and protons. (In addition, we have no definite knowledge of the nuclear interaction. The comparison of calculated and observed nuclear structures provides us with information concerning the nuclear Hamiltonian.) The neutron and proton have (approximately) the same mass and spin ($S = \frac{1}{2}$), and transform into each other in beta-decay. The neutron is neutral, while the proton has charge $+e$, so only the protons will be subjected to Coulomb forces. However, the Coulomb forces are small compared to the specifically nuclear forces. In addition, the available experimental evidence shows that the specifically nuclear forces between two particles in the nucleus do not depend on whether the particles are neutrons or protons—the nuclear forces are *charge-independent*. It is therefore useful to regard neutron and proton as *states* of a single fundamental entity which we call a *nucleon*.

The nucleon can be in either of two *charge states*, which we label $\psi_{1/2}$ and $\psi_{-1/2}$, corresponding to the neutron state and proton state, respectively. The operator t_ζ has these states as eigenstates:

$$t_\zeta \psi_{1/2} = \tfrac{1}{2}\psi_{1/2} \qquad \text{(neutron),}$$
$$t_\zeta \psi_{-1/2} = -\tfrac{1}{2}\psi_{-1/2} \qquad \text{(proton).} \tag{11–15}$$

By applying t_ζ to the charge function, we determine whether it is a neutron or a proton state. The operator t_ζ can be expressed as a 2×2 matrix:

$$t_\zeta = \begin{pmatrix} \tfrac{1}{2} & 0 \\ 0 & -\tfrac{1}{2} \end{pmatrix}. \tag{11–15a}$$

To account for the beta transformations between the neutron and proton states, we must have operators of the form

$$\begin{pmatrix} 0 & 1 \\ 0 & 0 \end{pmatrix}, \quad \begin{pmatrix} 0 & 0 \\ 1 & 0 \end{pmatrix}. \tag{11–15b}$$

The operators (11–15a and b) give an algebra which is formally identical with that of the angular-momentum operators J_x, J_y, J_z. Since we are dealing here with a two-dimensional space, the states $\psi_{\pm 1/2}$ are the basis of a representation analogous to the representation $D^{(1/2)}$ of the rotation group. The space of the representation spanned by $\psi_{\pm 1/2}$ is called the *isotopic spin space*. We have operators t_ξ, t_η, and t_ζ whose properties are analogous to those of the angular-momentum operators J_x, J_y, J_z. If we have a system of several nucleons, the operator

$$T_\zeta = \sum_i t_\zeta^{(i)} \tag{11–16}$$

is the infinitesimal operator for the simultaneous "rotation about the ζ-axis in isotopic spin space" for all particles. We can similarly define operators T_ξ and T_η.

For a system of two nucleons, the function $\psi_{1/2}(1)\psi_{1/2}(2)$ has both nucleons in the neutron state, and $T_\zeta(\psi_{1/2}(1)\psi_{1/2}(2)) = \psi_{1/2}(1)\psi_{1/2}(2)$. Thus $T_\zeta = 1$ for this state. The function $\psi_{-1/2}(1)\psi_{-1/2}(2)$ describes a system of two protons, and $T_\zeta = -1$. The functions $\psi_{1/2}(1)\psi_{-1/2}(2)$ and $\psi_{-1/2}(1)\psi_{1/2}(2)$ describe states containing one neutron and one proton, with $T_\zeta = 0$. We see that

$$T_\zeta = \tfrac{1}{2}(N - Z), \tag{11–17}$$

where N and Z are the numbers of neutrons and protons in the system. The total number of nucleons in the system is $N + Z$. So far as the nuclear forces are concerned, the value of T_ζ is irrelevant.

For a system of identical nucleons, we must construct states having a definite symmetry with respect to interchange of identical particles. For two particles, the first two functions listed above are symmetric under interchange of nucleons 1 and 2. The other two functions can be combined to give the symmetric and antisymmetric combinations $\psi_{1/2}(1)\psi_{-1/2}(2) \pm \psi_{-1/2}(1)\psi_{1/2}(2)$. Only the symmetry of the wave functions can affect the energy of the system, since we are neglecting the Coulomb forces. For two nucleons we see that the number of symmetric states is three, while the number of antisymmetric states is one. The first symmetry type corresponds to the partition [2] with dimension 3, the second to the partition [11] with dimension 1.

In analogy to the total spin S, we introduce the *isotopic spin* (isobaric spin, I-spin) T, where $(2T + 1)$ gives the multiplicity of the states of a given charge symmetry. For two nucleons we have:

$$[2]: \quad T = 1, \quad T_{\zeta} = 1, 0, -1;$$
$$[11]: \quad T = 0, \quad T_{\zeta} = 0. \tag{11-18}$$

The symmetric state, with $T = 1$, is an *isobaric triplet*. The energy will be the same for the three states with $T_{\zeta} = 1, 0, -1$ ($N - Z = 2, 0, -2$). Similarly, for three nucleons, we have:

$$[3]: \quad T = \tfrac{3}{2}, \quad 2T + 1 = 4, \quad T_{\zeta} = \tfrac{3}{2}, \tfrac{1}{2}, -\tfrac{1}{2}, -\tfrac{3}{2},$$
$$N - Z = 3, 1, -1, -3;$$
$$[21]: \quad T = \tfrac{1}{2}, \quad 2T + 1 = 2, \quad T_{\zeta} = \tfrac{1}{2}, -\tfrac{1}{2},$$
$$N - Z = 1, -1. \tag{11-19}$$

The first charge multiplet has four isobars, and the second has two.

The results for any number of nucleons are identical with the corresponding results for spin $j = \tfrac{1}{2}$. For a state with charge-symmetry pattern $[\lambda_1 \lambda_2]$, $T = \tfrac{1}{2}(\lambda_1 - \lambda_2)$.

11-7 Nuclear spectra in *L-S* coupling. Supermultiplets. If the nuclear forces do not depend strongly on the spins, we can, as in the atomic problem, write the wave function as the product of an orbital function and a function of the spin and charge variables. The interaction Hamiltonian is symmetric in the space coordinates of the nucleons, so the orbital wave functions should be combined to give a total orbital function of definite symmetry. The energy of the state will depend critically on this symmetry. Since the nuclear forces are primarily attractive, the energy will be lowered if the symmetry of the orbital wave function is increased. Thus we may expect that the state whose orbital function has the highest symmetry will have the lowest energy. Since the total wave function of the system of identical nucleons is required by the Pauli principle to be completely antisymmetric, we must construct charge-spin functions of definite symmetry and obtain the total wave function by taking the product of the orbital function with a charge-spin function having the conjugate symmetry. Since the energy of the state is determined only by the orbital function, while the multiplicity depends on the charge-spin function, each energy level will be a *supermultiplet*.

For a single nucleon, four charge-spin states are possible. If we label the states by the values of t_{ζ} and s_z, we have four basis functions:

$$\psi_{1/2, 1/2}, \quad \psi_{1/2, -1/2}, \quad \psi_{-1/2, 1/2}, \quad \psi_{-1/2, -1/2}. \tag{11-20}$$

The symmetry character for a single nucleon is [1], and has $T = \tfrac{1}{2}, S = \tfrac{1}{2}$, with multiplicities $2T + 1 = 2, 2S + 1 = 2$.

For two nucleons, there are sixteen charge-spin states. If the charge-spin function is symmetric, [2], then either both the spin function and the charge function are symmetric so that $2T + 1 = 3, 2S + 1 = 3$, or they are both antisymmetric with $2T + 1 = 1, 2S + 1 = 1$. Thus the Young pattern [2] contains the charge-spin multiplicities (11), singlet charge, singlet spin; and (33), triplet charge, triplet spin. Similarly, the antisymmetric charge-spin function [11] must be a product of charge and spin functions of opposite symmetry, so it contains (13), singlet charge, triplet spin; and (31), triplet charge, singlet spin.

The Young pattern $[1^4]$ represents four nucleons in a completely antisymmetric charge-spin state. Since all four sets of indices must be different, each of the functions (11–20) appears once, so that $T = S = 0$. Thus $[1^4] \equiv [0]$ has the multiplicity (11), singlet charge and singlet spin. Similarly, the pattern $[1^3]$ is equivalent to [1] and has multiplicity (22).

Now we can find the charge-spin functions for three nucleons by using the outer product and analyzing for S and T:

$$\square \otimes \square\square = \square\square\square + \begin{array}{c}\square\square \\ \square\end{array}. \tag{11–21}$$

The partition [1] has multiplicity (22), while [2] contains the multiplicities (11) and (33). The left side of (11–21) is $[1](22) \otimes [2]((11) + (33))$. We analyze this product and find

$$
\left.
\begin{array}{l}
[1](22) \otimes [2](11) \rightarrow (22) \\
[1](22) \otimes [2](33) \rightarrow (22) + (24) + (42) + (44)
\end{array}
\right\} [3] + [21]. \tag{11–21a}
$$

Similarly, we have

$$\square \otimes \begin{array}{c}\square \\ \square\end{array} = \begin{array}{c}\square\square \\ \square\end{array} + \begin{array}{c}\square \\ \square \\ \square\end{array} \equiv \begin{array}{c}\square\square \\ \square\end{array} + \square, \tag{11–22}$$

$$
\left.
\begin{array}{l}
[1](22) \otimes [11](13) \rightarrow (22) + (24) \\
[1](22) \otimes [11](31) \rightarrow (22) + (42)
\end{array}
\right\} [21] + [1]. \tag{11–22a}
$$

Subtracting the known structure [1](22) from Eq. (11–22a), we find that [21] contains the multiplicities (22), (24), and (42). Subtracting this last result from Eq. (11–21a), we find the structure of [3]. Finally we have the multiplet structures for three particles (Table 11–15).

The same ladder process takes us from $r = 3$ to $r = 4$. For $r = 4$, the possible Young patterns are [4], [31], [22], and [211] (omitting $[1^4] \equiv [0]$). To carry out the ladder process, we take the outer product of [1](22) with each structure of Table 11–15, and obtain Table 11–16. From the last entry, since $[1^4] \equiv [0]$ has multiplicity (11), we find the structure of [211].

TABLE 11–15

[λ]	(22)	(24)	(42)	(44)	Dimensionality
		(2T + 1, 2S + 1)			
[3]	1	0	0	1	20
[21]	1	1	1	0	20

TABLE 11–16

	(11)	(13) (31)	(15) (51)	(33)	(35) (53)	(55)	
		(2T + 1, 2S + 1)					
[1] (22) ⊗ [3] (22)	1	1	0	1	0	0	[4] + [31]
[1] (22) ⊗ [3] (44)	0	0	0	1	1	1	
[1] (22) ⊗ [21] (22)	1	1	0	1	0	0	[31] + [22] + [211]
[1] (22) ⊗ [21] $\binom{24}{42}$	0	1	1	2	1	0	
[1] (22) ⊗ [1³] (22)	1	1	0	1	0	0	[211] + [1⁴]

TABLE 11–17

[λ]	(11)	$\binom{13}{31}$	$\binom{15}{51}$	(33)	$\binom{35}{53}$	(55)	Dimension N
		(2T + 1, 2S + 1)					
[4]	1	0	0	1	0	1	35
[31]	0	1	0	1	1	0	45
[22]	1	0	1	1	0	0	20
[211]	0	1	0	1	0	0	15

TABLE 11–18

STRUCTURE OF ALL CHARGE-SPIN SUPERMULTIPLETS FOR $r \leq 10$

$[\lambda]$	(P, P', P'')	$(2T + 1, 2S + 1)$-structure	Dimension N
[0]	(000)	(11)	1
[1]	$(\tfrac{1}{2}\tfrac{1}{2}\tfrac{1}{2})$	(22)	4
[2]	(111)	$(11)(33)$	10
[11]	(100)	$\binom{13}{31}$	6
[3]	$(\tfrac{3}{2}\tfrac{3}{2}\tfrac{3}{2})$	$(22)(44)$	20
[21]	$(\tfrac{3}{2}\tfrac{1}{2}\tfrac{1}{2})$	$(22)\binom{24}{42}$	20
[4]	(222)	$(11)(33)(55)$	35
[31]	(211)	$\binom{13}{31}(33)\binom{35}{53}$	45
[22]	(200)	$(11)\binom{15}{51}(33)$	20
[211]	(110)	$\binom{13}{31}(33)$	15
[5]	$(\tfrac{5}{2}\tfrac{5}{2}\tfrac{5}{2})$	$(22)(44)(66)$	56
[41]	$(\tfrac{5}{2}\tfrac{3}{2}\tfrac{3}{2})$	$(22)\binom{24}{42}(44)\binom{46}{64}$	84
[32]	$(\tfrac{5}{2}\tfrac{1}{2}\tfrac{1}{2})$	$(22)\binom{24}{42}\binom{26}{62}(44)$	60
[311]	$(\tfrac{3}{2}\tfrac{3}{2}\tfrac{1}{2})$	$(22)\binom{24}{42}(44)$	36

[6]	(333)	$(11)(33)(55)(77)$	84
[51]	(322)	$\binom{13}{31}\binom{33}{53}\binom{35}{55}\binom{57}{75}$	140
[42]	(311)	$(11)\binom{15}{51}\binom{33}{51}^2\binom{35}{53}\binom{37}{73}(55)$	126
[411]	(221)	$\binom{13}{31}(33)\binom{35}{53}(55)$	70
[33]	(300)	$\binom{13}{31}\binom{17}{71}\binom{35}{53}$	50
[321]	(210)	$\binom{13}{31}\binom{15}{51}\binom{33}{51}^2\binom{35}{53}$	64
[7]	$(\tfrac{7}{2}\tfrac{7}{2}\tfrac{7}{2})$	$(22)(44)(66)(88)$	120
[61]	$(\tfrac{7}{2}\tfrac{5}{2}\tfrac{5}{2})$	$(22)\binom{24}{42}(44)\binom{46}{64}(66)\binom{68}{86}$	216
[52]	$(\tfrac{7}{2}\tfrac{3}{2}\tfrac{3}{2})$	$(22)\binom{24}{42}\binom{26}{62}(44)^2\binom{46}{64}\binom{48}{84}(66)$	224
[511]	$(\tfrac{5}{2}\tfrac{5}{2}\tfrac{3}{2})$	$(22)\binom{24}{42}(44)\binom{46}{64}(66)$	120
[43]	$(\tfrac{7}{2}\tfrac{1}{2}\tfrac{1}{2})$	$(22)\binom{24}{42}\binom{26}{62}\binom{28}{82}(44)\binom{46}{64}$	140
[421]	$(\tfrac{5}{2}\tfrac{3}{2}\tfrac{1}{2})$	$(22)\binom{24}{42}^2\binom{26}{62}(44)^2\binom{46}{64}$	140
[8]	(444)	$(11)\binom{33}{53}\binom{55}{75}^2(77)(99)$	165
[71]	(433)	$\binom{13}{31}\binom{33}{53}\binom{35}{53}\binom{55}{75}\binom{57}{75}\binom{77}{97}\binom{79}{97}$	315
[62]	(422)	$(11)\binom{33}{53}^2\binom{55}{51}^2(77)\binom{15}{51}\binom{35}{53}\binom{37}{73}\binom{57}{75}\binom{59}{95}$	360

(continued)

TABLE 11–18 (*continued*)

[λ]	(P, P', P'')	(2T + 1, 2S + 1)-structure	Dimension N
[611]	(332)	$(33)(55)(77)\binom{13}{31}\binom{35}{53}\binom{57}{75}$	189
[53]	(411)	$(33)(55)\binom{13}{31}\binom{17}{71}\binom{35}{53}^2\binom{37}{73}\binom{39}{93}\binom{57}{75}$	280
[521]	(321)	$(33)^2(55)^2\binom{13}{31}\binom{15}{51}\binom{35}{53}^2\binom{37}{73}\binom{57}{75}$	256
[44]	(400)	$(11)(33)(55)\binom{15}{51}\binom{19}{91}\binom{37}{73}$	105
[431]	(310)	$\binom{13}{31}\binom{15}{51}\binom{17}{71}\binom{33}{53}^2\binom{35}{53}\binom{37}{73}(55)$	175
[422]	(220)	$(11)\binom{15}{51}(33)^2\binom{35}{53}(55)$	84
[9]	$\left(\frac{999}{222}\right)$	$(22)(44)(66)(88)(10,10)$	220
[81]	$\left(\frac{977}{222}\right)$	$(22)(44)(66)(88)\binom{24}{42}\binom{46}{64}\binom{68}{86}\binom{8,10}{10,8}$	440
[72]	$\left(\frac{955}{222}\right)$	$(22)(44)^2(66)^2(88)\binom{24}{42}\binom{26}{62}\binom{46}{64}\binom{48}{84}\binom{68}{86}\binom{6,10}{10,6}$	540
[711]	$\left(\frac{775}{222}\right)$	$(22)(44)(66)(88)\binom{24}{42}\binom{46}{64}\binom{68}{86}$	280
[63]	$\left(\frac{933}{222}\right)$	$(22)(44)^2(66)\binom{24}{42}\binom{26}{62}\binom{28}{82}\binom{46}{64}^2\binom{48}{84}\binom{4,10}{10,4}\binom{68}{86}$	480
[621]	$\left(\frac{753}{222}\right)$	$(22)(44)^2(66)^2\binom{24}{42}^2\binom{26}{62}\binom{46}{64}^2\binom{48}{84}\binom{68}{86}$	420
[54]	$\left(\frac{911}{222}\right)$	$(22)(44)(66)\binom{24}{42}\binom{26}{62}\binom{28}{82}\binom{2,10}{10,}\binom{46}{}(48)$	280

[531]	$\left(\tfrac{731}{222}\right)$	$(22)(44)^3(66)\binom{24}{42}^2\binom{26}{62}^2\binom{28}{82}\binom{46}{64}^2\binom{48}{84}$	360
[522]	$\left(\tfrac{551}{222}\right)$	$(22)(44)^2(66)\binom{24}{42}\binom{26}{62}\binom{46}{64}$	160
[10]	(555)	$(11)(33)(55)(77)(99)(11,11)$	286
[91]	(544)	$(33)(55)(77)(99)\binom{13}{31}\binom{35}{53}\binom{57}{75}\binom{79}{97}\binom{9,11}{11,9}$	594
[82]	(533)	$(11)(33)^2(55)^2(77)^2(99)\binom{15}{51}\binom{35}{53}\binom{37}{73}\binom{57}{75}\binom{59}{95}\binom{79}{97}\binom{7,11}{11,7}$	770
[811]	(443)	$(33)(55)(77)(99)\binom{13}{31}\binom{35}{53}\binom{57}{75}\binom{79}{97}$	396
[73]	(522)	$(33)(55)^2(77)\binom{13}{31}\binom{17}{71}\binom{35}{53}^2\binom{37}{73}\binom{39}{93}\binom{57}{75}^2\binom{59}{95}\binom{5,11}{11,5}\binom{79}{97}$	750
[721]	(432)	$(33)^2(55)^2(77)^2\binom{13}{31}\binom{15}{51}\binom{35}{53}^2\binom{37}{73}\binom{57}{75}^2\binom{59}{95}\binom{79}{97}$	640
[64]	(511)	$(11)(33)^2(55)^2(77)\binom{15}{51}\binom{19}{91}\binom{35}{53}\binom{37}{73}^2\binom{39}{93}\binom{3,11}{11,3}\binom{57}{75}^2\binom{59}{95}$	540
[631]	(421)	$(33)^2(55)^3(77)\binom{13}{31}\binom{15}{51}\binom{17}{71}\binom{35}{53}^3\binom{37}{73}^2\binom{39}{93}\binom{57}{75}^2\binom{59}{95}$	630
[622]	(331)	$(11)(33)^2(55)^2(77)\binom{15}{51}\binom{35}{53}\binom{37}{73}\binom{57}{75}$	270
[55]	(600)	$\binom{13}{31}\binom{17}{71}\binom{1,11}{11,1}\binom{35}{53}\binom{39}{93}\binom{57}{75}$	196
[541]	(410)	$(33)^2(55)^2\binom{13}{31}\binom{15}{51}\binom{17}{71}\binom{19}{91}\binom{35}{53}^2\binom{37}{73}\binom{39}{93}\binom{57}{75}$	384
[532]	(320)	$(33)^2(55)^2\binom{13}{31}\binom{15}{51}\binom{17}{71}\binom{35}{53}^3\binom{37}{73}\binom{57}{75}$	300

The analysis of the completely symmetric structure [4] can be found from the following theorem:

THEOREM.   To obtain a completely symmetric charge-spin function, we must take direct products of charge and spin functions both of which have the same symmetry. Thus the charge-spin function for the partition $[n]$ is a sum of products $[n] \otimes [n]$, $[n-1, 1] \otimes [n-1, 1]$, ..., etc., where the sequence ends with $[\nu + 1, \nu] \otimes [\nu + 1, \nu]$ if $n = 2\nu + 1$ is odd; and with $[\nu, \nu] \otimes [\nu, \nu]$ if $n = 2\nu$ is even.

The values of S (or T) for these simple charge and spin structures are given by Eqs. (11–4) and (11–5), so we find that

for n even, $n = 2\nu$, the charge-spin function $[n]$ contains the multiplicities $(2\nu + 1, 2\nu + 1)$, $(2\nu - 1, 2\nu - 1)$, ..., (55), (33), (11);

for n odd, $n = 2\nu + 1$, the charge-spin function $[n]$ contains the multiplicities $(2\nu + 2, 2\nu + 2)$, $(2\nu, 2\nu)$, ... (44), (22).

$$(11\text{–}23)$$

Thus the structure of $[4]$ is $(55) + (33) + (11)$. Combining this with the first entry in Table 11–16, we find that $[31]$ contains $\binom{13}{31} + \binom{35}{53} + (33)$. Using these results, we can then determine the structure of $[22]$ (Table 11–17).

Continuing this procedure, we can construct the charge-spin functions for $r = 5$ by adding a particle to the structures for $r = 4$, and so on.

In general, the symmetry pattern of the charge-spin function will have four rows $[\lambda_1 \lambda_2 \lambda_3 \lambda_4]$. Using Eqs. (10–29) and (10–30), we find that this pattern is equivalent to $[\lambda_1' \lambda_2' \lambda_3' 0]$, where

$$\lambda_1' = \lambda_1 - \lambda_4, \qquad \lambda_2' = \lambda_2 - \lambda_4, \qquad \lambda_3' = \lambda_3 - \lambda_4 \quad (11\text{–}24)$$

are the *reduced partition numbers*. The λ''s completely characterize the symmetry of the charge-spin function. In place of these one can also use the numbers P, P', P'' defined by

$$(P, P', P'') = \left(\frac{\lambda_1' + \lambda_2' - \lambda_3'}{2}, \frac{\lambda_1' - \lambda_2' + \lambda_3'}{2}, \frac{\lambda_1' - \lambda_2' - \lambda_3'}{2} \right).$$

$$(11\text{–}25)$$

Since $\lambda_1' \geq \lambda_2' \geq \lambda_3'$, $P \geq P' \geq P''$, and P and P' are necessarily positive. The significance of the quantum numbers P, P', P'' is the following: P is the largest value of T_ζ contained in the supermultiplet. P' is the largest value of S_z for a state having $T_\zeta = P$. (At the same time, P is the largest value of S_z contained in the supermultiplet, and P' is the largest value of

T_ζ for a state with $S_z = P$.) Finally, P'' is the largest value of $\sum_{i=1}^{r} s_z^{(i)} t_\zeta^{(i)}$ for a state with $T_\zeta = P$ and $S_z = P'$ (or $S_z = P$ and $T_\zeta = P'$).

Problems. (1) Prove the last statements.

(2) Prove that $P + P' + P'' + \frac{1}{2}r$ is always an even positive integer.

Finally, we present Table 11–18 (p. 438) listing the charge-spin structures for $r \leq 10$.

11–8 The L-S coupling shell model. Seniority. The charge-spin functions found in the last section must be combined with orbital functions to give the total wave function of the system. In order to satisfy the Pauli principle, the two factors must have conjugate symmetries. Since the Young pattern for the charge-spin function has at most four rows, the pattern for the orbital function can have at most four columns.

The L-S shell model of the nucleus is very similar to the corresponding atomic model which we discussed in Section 11–3. The single-particle wave functions are calculated in some averaged central potential and are labeled by quantum numbers n and l. The wave function for a single nucleon is a vector in a $(2l + 1)$-dimensional space. If there are r nucleons in the (n, l)-orbit, the orbital wave function will be an rth-rank tensor. To obtain functions of definite symmetry we must reduce the space of rth-rank tensors into its component irreducible representations of $SU(2l + 1)$. The angular-momentum analysis will be similar to that of Sections 11–2 and 11–3.

We illustrate the procedure for $l = 1$. For the p-shell nuclei, configuration $(p)^r$, the maximum number of particles in the shell is $4(2l + 1) = 12$. All the results are available in Tables 11–2 and 11–18. For $r = 0$ or 12, we obtain a ^{11}S-term $(L = 0, S = T = 0)$. For $r = 1$, the orbital and charge-spin patterns are both [1], and we have ^{22}P.

TABLE 11–19

Orbital		Charge-spin			
[λ]	L	[λ̃]	(P, P', P'')	$(2T + 1, 2S + 1)$	
[2]	S, D	[11]	(100)	(13)	(31)
[11] ≡ [1]	P	[2]	(111)	(11)	(33)

TABLE 11-20

STRUCTURE OF THE NUCLEAR $(p)^r$-CONFIGURATION

r	[λ]	Orbit		Charge-spin	
		L	[λ̃]	(P, P', P'')	(2T + 1, 2S + 1)
0	[0]	S	[0]	(000)	(11)
1	[1]	P	[1]	$(\frac{1}{2}\frac{1}{2}\frac{1}{2})$	(22)
2	[2]	SD	[11]	(100)	(13)(31)
	[11] ≡ [1]	P	[2]	(111)	(11)(33)
3	[3]	PF	[111] ≡ [1]	$(\frac{1}{2}\frac{1}{2}-\frac{1}{2})$	(22)
	[21]	PD	[21]	$(\frac{3}{2}\frac{1}{2}\frac{1}{2})$	(22)(24)(42)
	[111] ≡ [0]	S	[3]	$(\frac{3}{2}\frac{3}{2}\frac{3}{2})$	(22)(44)
4	[4]	SDG	[1111] ≡ [0]	(000)	(11)
	[31]	PDF	[211]	(110)	(13)(31)(33)
	[22] ≡ [2]	SD	[22]	(200)	(11)(15)(51)(33)
	[211] ≡ [1]	P	[31]	(211)	(13)(31)(33)(35)(53)
5	[41]	PDFG	[2111] ≡ [1]	$(\frac{1}{2}\frac{1}{2}\frac{1}{2})$	(22)
	[32] ≡ [31]	PDF	[221] ≡ [21]	$(\frac{3}{2}\frac{1}{2}-\frac{1}{2})$	(22)(24)(42)
	[311] ≡ [2]	SD	[311]	$(\frac{3}{2}\frac{3}{2}\frac{1}{2})$	(22)(24)(42)(44)
	[221] ≡ [1]	P	[32]	$(\frac{5}{2}\frac{1}{2}\frac{1}{2})$	(22)(24)(42)(26)(62)(44)
6	[42]	SD²FG	[2211] ≡ [11]	(100)	(13)(31)
	[411] ≡ [3]	PF	[3111] ≡ [2]	(111)	(11)(33)
	[33] ≡ [3]	PF	[222] ≡ [2]	(11 − 1)	(11)(33)
	[321] ≡ [21]	PD	[321]	(210)	(13)(31)(33)²(15)(51)(35)(53)
	[222] ≡ [0]	S	[33]	(300)	(13)(31)(35)(53)(17)(71)

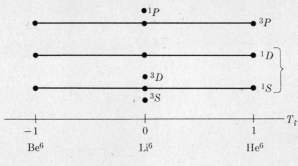

FIGURE 11-1

For $r = 2$, the results are given in Table 11-19. Because the nuclear forces are attractive, the states with higher orbital symmetry will be lower in energy. In addition, one finds that the energy increases with increasing L. A diagram of levels for p-shell nuclei with $r = 2$ is given in Fig. 11-1. The spin multiplicities are written explicitly, while the charge multiplicity is shown by dots above the corresponding values of T_ζ. Thus the ^{13}D-term in Table 11-19 is shown on the diagram as the ^3D in Li^6; the ^{33}P-term is shown as the ^3P-level in Be^6, Li^6, and He^6. When we include the perturbations due to spin-orbit interactions, we obtain a further splitting of the levels. The effect of the Coulomb interaction will be to make the horizontal lines in the diagram slope up toward the left.

The structure for $r = 0, 1, \ldots, 6$ is given in Table 11-20.

Problem. Draw the level diagrams for the $(p)^3$- and $(p)^4$-configurations.

The same analysis can be applied to configurations of nucleons in the $d, f, \ldots$ shells. As in atomic spectra, one can improve the classification by introducing the group $O^+(2l + 1)$. The arguments are almost identical with those of Section 11-3. The only difference is that the contraction process which couples a pair of l-nucleons to $L = 0$ gives their orbital function the symmetry [2]. The corresponding charge-spin function has the symmetry [11] and contains the multiplets (13) and (31). The results for $r \leq 4$ in the d- and f-shells are given in Tables 11-21 and 11-22. The tables are constructed by combining the results of Tables 11-10 and 11-11 with those of Table 11-18.

TABLE 11-21

STATES ARISING FROM FILLING OF THE NUCLEAR d-SHELL ($r \leq 4$)

r	Orbit			Charge-spin		
	$[\lambda]$	$(\mu_1\mu_2)$	$[L]$	$[\tilde{\lambda}]$	(P, P', P'')	$(2T+1, 2S+1)$
0	$[0]$	(00)	S	$[0]$	(000)	(11)
1	$[1]$	(10)	D	$[1]$	$(\tfrac{1}{2}\tfrac{1}{2}\tfrac{1}{2})$	(22)
2	$[2]$	(00) (20)	S GD	$[11]$	(100)	$\binom{13}{31}$
	$[11]$	(11)	FP	$[2]$	(111)	$(11)(33)$
3	$[3]$	(10) (30)	D $IGFS$	$[111] \equiv [1]$	$(\tfrac{1}{2}\tfrac{1}{2} - \tfrac{1}{2})$	(22)
	$[21]$	(10) (21)	D $HGFDP$	$[21]$	$(\tfrac{3}{2}\tfrac{1}{2}\tfrac{1}{2})$	$(22) \binom{24}{42}$
	$[111] \equiv [11]$	(11)	FP	$[3]$	$(\tfrac{3}{2}\tfrac{3}{2}\tfrac{3}{2})$	$(22)(44)$

4						
[4]	(00) (20) (40)	S GD $LIHGD$	[1111] ≡ [0]	(000)	(11)
[31]	(11) (20) (31)	FP GD KIH^2GF^2DP	[211]	(110)	$\binom{13}{31}$ (33)
[22]	(00) (20) (22)	S GD $IGFDS$	[22]	(200)	(11) (33) $\binom{15}{51}$
[211]	(11) (21)	FP $HGFDP$	[31]	(211)	$\binom{13}{31}$ (33) $\binom{35}{53}$
[1111] ≡ [1]	(10)	D	[4]	(222)	(11) (33) (55)

TABLE 11–22

STATES ARISING FROM FILLING OF THE NUCLEAR f-SHELL ($r \leq 4$)

r	Orbit		Charge spin		
	$[\lambda]$	$(\mu_1\mu_2\mu_3)$	$[\tilde\lambda]$	(P, P', P'')	$(2T + 1, 2S + 1)$
0	[0]	(000)	[0]	(000)	(11)
1	[1]	(100)	[1]	$(\frac{1}{2}\frac{1}{2}\frac{1}{2})$	(22)
2	[2]	(000)(200)	[11]	(100)	$\binom{13}{31}$
	[11]	(110)	[2]	(111)	(11)(33)
3	[3]	(100)(300)	[111] ≡ [1]	$(\frac{1}{2}\frac{1}{2} - \frac{1}{2})$	(22)
	[21]	(100)(210)	[21]	$(\frac{3}{2}\frac{1}{2}\frac{1}{2})$	(22) $\binom{24}{42}$
	[111]	(111)	[3]	$(\frac{3}{2}\frac{3}{2}\frac{3}{2})$	(22)(44)
4	[4]	(000)(200)(400)	[1111] ≡ [0]	(000)	(11)
	[31]	(110)(200)(310)	[211]	(110)	$\binom{13}{31}$ (33)
	[22]	(000)(200)(220)	[22]	(200)	(11) $\binom{15}{51}$ (33)
	[211]	(110)(211)	[31]	(211)	$\binom{13}{31}$ (33) $\binom{35}{53}$
	[1111] ≡ [111]	(111)	[4]	(222)	(11)(33)(55)

11–9 The jj-coupling shell model.   Seniority in jj-coupling.   The jj-coupling scheme in nuclei is analogous to the approximation scheme used in Section 11–5 for atoms. The individual nucleons move in some averaged potential. The spin-orbit interaction is assumed to be large, so that the energy levels of a single nucleon can be specified by quantum numbers n, l, j, m_j, where $j = l + \frac{1}{2}$ or $l - \frac{1}{2}$. If there are r nucleons in the system, the charge state of the nucleus will be described by the charge functions which we constructed and labeled with the value of the isotopic spin T in Section 11–6. If the r nucleons are equivalent (i.e., if they are in the same nlj-shell), we construct the spin-orbit function for the system by taking products of r single-particle functions. The single-particle functions form the basis for a representation of the group $SU(2j + 1)$. The spin-orbit wave functions of a definite symmetry type will form the basis for an irreducible representation of $SU(2j + 1)$. To obtain the total wave function we must multiply the spin-orbit function by a charge function of conjugate symmetry. Thus the spin-orbit functions will be labeled by the value of T. Since the Young pattern for the charge function has at most two rows, the pattern for the spin-orbit function contains no partitions with $\lambda_i > 2$. In addition, the number of rows in the pattern for the spin-orbit function does not exceed $2j + 1$. For example, for $r = 5$, we may have a spin-orbit function with $[\lambda] = [221]$. Then the charge function has $[\tilde\lambda] = [32]$, so that $T = \frac{1}{2}$.

The angular-momentum analysis of the $(j)^r$-configuration was given in Section 11–2, Tables 11–3, 11–5, and 11–7.

The concept of seniority can be introduced in the jj-coupling scheme by a procedure similar to that used in Section 11–4. We would like to find some additional quantum numbers to characterize the terms arising from a given nuclear configuration. To do this, we look for a group G which is a subgroup of $SU(2j + 1)$ and which contains the rotation group $O^+(3)$ as a subgroup. The terms will then be characterized by the partition $[\lambda]$, the angular momentum J, and the irreducible representation of G to which the term belongs.

Suppose that we have two nucleons in the configuration $(j)^2$. The spin-orbit wave functions $\psi_{i_1}^{(1)}, \psi_{i_2}^{(2)}$ $(i_1, i_2 = -j, -j + 1, \ldots, j - 1, j)$ of the two nucleons are vectors in the even-dimensional space (dimension $2j + 1$) of the representation $D^{(j)}$ of the rotation group. The product functions $\psi_{i_1}^{(1)}\psi_{i_2}^{(2)}$ form the basis for a representation of the unitary group $SU(2j + 1)$ and, at the same time, form the basis for a representation of the rotation group $O^+(3)$. Equation (11–6), for half-integral j, shows that the *antisymmetric* bilinear form (skew product)

$$\Psi_{J=0} = \frac{1}{\sqrt{2j + 1}} \sum_{m=-j}^{j} (-1)^{j-m} \psi_m^{(1)}\psi_{-m}^{(2)} \equiv \{\psi^{(1)}\psi^{(2)}\} \quad (11\text{–}26)$$

couples the angular momenta of the two particles to give a resultant $J = 0$. The three-dimensional rotations induce linear transformations in the tensor space, but they leave the function $\Psi_{J=0}$ of (11–26) unchanged. But there is a larger group G of transformations in the $(2j + 1)$-dimensional space which leave (11–26) invariant. The skew product (11–26) is an antisymmetric bilinear form in the vectors of an even-dimensional space. It is therefore invariant under the symplectic transformations $Sp(2j + 1)$, which we studied in Chapter 10. The operation (11–26) for taking the skew product is precisely the contraction which we introduced in Section 10–8.

The skew product (11–26) is antisymmetric and has $J = 0$. The corresponding charge function for the two nucleons must be symmetric, with partition [2], so $T = 1$.

The decomposition $SU(2j + 1) \to Sp(2j + 1) \to O^+(3)$ is carried out by methods similar to those of Section 11–4, using the results of Chapter 10 on the symplectic group.

For $j = \frac{3}{2}$, the angular-momentum analysis $Sp(4) \to O^+(3)$ is obtained from Tables 10–5 and 11–3. For $r = 1$, $[\lambda] = [1]$ contains $J = \frac{3}{2}$ (Table 11–3), and [1] contains $(\sigma_1\sigma_2) = (10)$ (Table 10–5), so the representation (10) of $Sp(4)$ contains $J = \frac{3}{2}$. For $r = 2$, the partition [11] contains $J = 0, 2$ (Table 11–3) and $(\sigma_1\sigma_2) = (00), (11)$ (Table 10–5). Since the representation (00) of $Sp(4)$ has $J = 0$, (11) has $J = 2$. Continuing in this fashion, we obtain the resolution $Sp(4) \to O^+(3)$ (see Table 11–23).

TABLE 11-23

ANALYSIS OF REPRESENTATIONS OF $Sp(2j+1)$ INTO
REPRESENTATIONS OF $O^+(3)$

	r	(σ)	J-structure
(a) $j = \frac{3}{2}$	0	(00)	0
	1	(10)	$\frac{3}{2}$
	2	(20)	$1\ 3$
		(11)	2
	3	(21)	$\frac{1}{2}\ \frac{5}{2}\ \frac{7}{2}$
	4	(22)	$2\ 4$
(b) $j = \frac{5}{2}$	0	(000)	0
	1	(100)	$\frac{5}{2}$
	2	(200)	$1\ 3\ 5$
		(110)	$2\ 4$
	3	(210)	$\frac{1}{2}\ \frac{3}{2}\ \frac{5}{2}\ (\frac{7}{2})^2\ \frac{9}{2}\ \frac{11}{2}\ \frac{13}{2}$
		(111)	$\frac{3}{2}\ \frac{9}{2}$
	4	(220)	$0\ 2^2\ 3\ 4^2\ 5\ 6^2\ 8$
		(211)	$1\ 2\ 3^2\ 4\ 5\ 6\ 7$
	5	(221)	$\frac{1}{2}\ \frac{3}{2}\ (\frac{5}{2})^2\ (\frac{7}{2})^2\ (\frac{9}{2})^2\ (\frac{11}{2})^2\ \frac{13}{2}\ \frac{15}{2}\ \frac{17}{2}$
	6	(222)	$1\ 3^2\ 4\ 5\ 6\ 7\ 9$

(c) $j = \frac{7}{2}$

n		J
0	(0000)	0
1	(1000)	$\frac{7}{2}$
2	(2000)	$1\ 3\ 5\ 7$
	(1100)	$2\ 4\ 6$
3	(2100)	$\frac{1}{2}\ \frac{3}{2}\ (\frac{5}{2})^2\ (\frac{7}{2})^2\ (\frac{9}{2})^2\ (\frac{11}{2})^2\ (\frac{13}{2})^2\ \frac{15}{2}\ \frac{17}{2}\ \frac{19}{2}$
	(1110)	$\frac{3}{2}\ \frac{5}{2}\ \frac{9}{2}\ \frac{11}{2}\ \frac{15}{2}$
4	(2200)	$0^2\ 2^3\ 3^2\ 4^4\ 5^2\ 6^4\ 7^2\ 8^3\ 9\ 10^2\ 12$
	(2110)	$1^2\ 2^2\ 3^4\ 4^3\ 5^4\ 6^3\ 7^3\ 8^2\ 9^2\ 10\ 11$
	(1111)	$2\ 4\ 5\ 8$
5	(2210)	$(\frac{1}{2})^2\ (\frac{3}{2})^4\ (\frac{5}{2})^5\ (\frac{7}{2})^7\ (\frac{9}{2})^9\ (\frac{11}{2})^7\ (\frac{13}{2})^7\ (\frac{15}{2})^6\ (\frac{17}{2})^5\ (\frac{19}{2})^4\ (\frac{21}{2})^3\ (\frac{23}{2})^2\ \frac{25}{2}\ \frac{27}{2}$
	(2111)	$\frac{1}{2}\ (\frac{3}{2})^2\ (\frac{5}{2})^2\ (\frac{7}{2})^3\ (\frac{9}{2})^3\ (\frac{11}{2})^3\ (\frac{13}{2})^3\ (\frac{15}{2})^2\ (\frac{17}{2})^2\ (\frac{19}{2})^2\ \frac{21}{2}\ \frac{23}{2}$
6	(2220)	$1^4\ 2^2\ 3^7\ 4^5\ 5^7\ 6^6\ 7^7\ 8^4\ 9^6\ 10^3\ 11^3\ 12^2\ 13^2\ 15$
	(2211)	$0^2\ 1^2\ 2^5\ 3^5\ 4^7\ 5^6\ 6^7\ 7^6\ 8^6\ 9^4\ 10^4\ 11^2\ 12^2\ 13\ 14$
7	(2221)	$(\frac{1}{2})^3\ (\frac{3}{2})^3\ (\frac{5}{2})^6\ (\frac{7}{2})^7\ (\frac{9}{2})^7\ (\frac{11}{2})^8\ (\frac{13}{2})^8\ (\frac{15}{2})^7\ (\frac{17}{2})^6\ (\frac{19}{2})^6\ (\frac{21}{2})^4\ (\frac{23}{2})^3\ (\frac{25}{2})^3\ \frac{27}{2}\ \frac{29}{2}\ \frac{31}{2}$
8	(2222)	$0\ 1\ 2^3\ 3^2\ 4^5\ 5^6\ 6^7\ 7^8\ 8^4\ 9^3\ 10^3\ 11^2\ 12^2\ 13\ 14\ 16$

TABLE 11-24

CLASSIFICATION OF THE STATES OF THE NUCLEAR CONFIGURATION $(j)^r$

(a) $j = \frac{3}{2}$

r	$[\lambda]$	T	(σ)	(s, t)	J-structure
0	[0]	0	(00)	(0, 0)	0
1	[1]	$\frac{1}{2}$	(10)	$(1, \frac{1}{2})$	$\frac{3}{2}$
2	[2]	0	(20)	(2, 0)	1 3
	[1]	1	(00)	(0, 0)	0
			(11)	(2, 1)	2
3	[21]	$\frac{1}{2}$	(10)	$(1, \frac{1}{2})$	$\frac{3}{2}$
	[111] $\equiv$ [1]	$\frac{3}{2}$	(21)	$(3, \frac{3}{2})$	$\frac{1}{2}\ \frac{5}{2}\ \frac{7}{2}$
			(10)	$(1, \frac{1}{2})$	$\frac{3}{2}$
4	[22]	0	(00)	(0, 0)	0
			(11)	(2, 1)	2
			(22)	(4, 0)	2 4
	[211]	1	(20)	(2, 0)	1 3
			(11)	(2, 1)	2
	[1111] $\equiv$ [0]	2	(00)	(0, 0)	0

(b) $j = \frac{5}{2}$

r	$[\lambda]$	T	(σ)	(s, t)	J-structure
0	[0]	0	(000)	(0, 0)	0
1	[1]	$\frac{1}{2}$	(100)	$(1, \frac{1}{2})$	$\frac{5}{2}$
2	[2]	0	(200)	(2, 0)	1 3 5
	[11]	1	(000)	(0, 0)	0
			(110)	(2, 1)	2 4
3	[21]	$\frac{1}{2}$	(100)	$(1, \frac{1}{2})$	$\frac{5}{2}$
		$\frac{3}{2}$	(210)	$(3, \frac{3}{2})$	$\frac{1}{2}\ \frac{3}{2}\ \frac{5}{2}\ (\frac{7}{2})^2\ \frac{9}{2}\ \frac{11}{2}\ \frac{13}{2}$

n	$[f]$		$(\lambda\mu\nu)$	(x,y)	J
4	[111]	$\tfrac{3}{2}$	(100)	$(1,\tfrac{1}{2})$	$\tfrac{5}{2}\ \tfrac{3}{2}$
			(111)	$(3,\tfrac{3}{2})$	$\tfrac{9}{2}\ \tfrac{3}{2}$
	[22]	0	(000)	(0,0)	0
	[211]	1	(110)	(2,1)	2 4
			(220)	(4,0)	0 2² 3 4² 5 6² 8
			(200)	(2,0)	1 3 5
	[1111]$\equiv$[11]	2	(110)	(2,1)	2 4
			(211)	(4,1)	1 2 3² 4 5 6 7
5	[221]	$\tfrac{1}{2}$	(000)	(0,0)	0
			(110)	(2,1)	2 4
	[2111]	$\tfrac{3}{2}$	(100)	$(1,\tfrac{1}{2})$	$\tfrac{3}{2}\ \tfrac{9}{2}$
			(210)	$(3,\tfrac{3}{2})$	$\tfrac{1}{2}\ \tfrac{3}{2}\ \tfrac{5}{2}\ (\tfrac{7}{2})^2\ \tfrac{9}{2}\ \tfrac{11}{2}\ \tfrac{13}{2}$
			(111)	$(3,\tfrac{3}{2})$	$\tfrac{3}{2}\ \tfrac{9}{2}$
			(221)	$(5,\tfrac{1}{2})$	$\tfrac{1}{2}\ \tfrac{3}{2}\ (\tfrac{5}{2})^2\ (\tfrac{7}{2})^2\ (\tfrac{9}{2})^2\ (\tfrac{11}{2})^2\ \tfrac{13}{2}\ \tfrac{15}{2}\ \tfrac{17}{2}$
	[11111]$\equiv$[1]	$\tfrac{5}{2}$	(100)	$(1,\tfrac{1}{2})$	$\tfrac{1}{2}\ \tfrac{3}{2}\ \tfrac{5}{2}\ \tfrac{7}{2}\ \tfrac{9}{2}\ \tfrac{11}{2}\ \tfrac{13}{2}$
			(210)	$(3,\tfrac{3}{2})$	$\tfrac{3}{2}\ \tfrac{5}{2}\ (\tfrac{7}{2})^2\ \tfrac{9}{2}\ \tfrac{11}{2}\ \tfrac{13}{2}$
			(111)	$(3,\tfrac{3}{2})$	$\tfrac{3}{2}\ \tfrac{9}{2}$
			(100)	$(1,\tfrac{1}{2})$	$\tfrac{5}{2}\ \tfrac{3}{2}\ \tfrac{5}{2}$
6	[222]	0	(200)	(2,0)	1 3 5
			(211)	(4,1)	1 2³ 4 5 6 7
			(222)	(6,0)	1 3² 4 5 6 7 9
	[2211]	1	(000)	(0,0)	0
			(110)²	(2,1)²	(2 4)²
	[21111]	2	(220)	(4,0)	0 2² 3 4² 5 6² 8
			(211)	(4,1)	1 2³ 4 5 6 7
			(200)	(2,0)	1 3 5
	[111111]$\equiv$[0]	3	(110)	(2,1)	2 4
			(000)	(0,0)	0

(continued)

TABLE 11-24 (continued)

r	$[\lambda]$	T	(σ)	(s,t)	J-structure
(c) $j=\frac{7}{2}$					
0	[0]	0	(0000)	(0,0)	0
1	[1]	$\frac{1}{2}$	(1000)	$(1,\frac{1}{2})$	$\frac{7}{2}$
2	[2]	0	(2000)	(2,0)	1 3 5 7
	[11]	1	(0000)	(0,0)	0
			(1100)	(2,1)	2 4 6
3	[21]	$\frac{1}{2}$	(1000)	$(1,\frac{1}{2})$	$\frac{7}{2}$
			(2100)	$(3,\frac{1}{2})$	$\frac{1}{2}\ \frac{3}{2}\ (\frac{5}{2})^2\ (\frac{7}{2})^2\ (\frac{9}{2})^2\ (\frac{11}{2})^2\ (\frac{13}{2})^2\ \frac{15}{2}\ \frac{17}{2}\ \frac{19}{2}$
	[111]	$\frac{3}{2}$	(1000)	$(1,\frac{1}{2})$	$\frac{7}{2}$
			(1110)	$(3,\frac{3}{2})$	$\frac{3}{2}\ \frac{5}{2}\ \frac{9}{2}\ \frac{11}{2}\ \frac{15}{2}$
4	[22]	0	(0000)	(0,0)	0
			(1100)	(2,1)	2 4 6
			(2200)	(4,0)	$0^2\ 2^3\ 3^2\ 4^4\ 5^2\ 6^4\ 7^2\ 8\ 3\ 9\ 10^2\ 12$
	[211]	1	(2000)	(2,0)	1 3 5 7
			(1100)	(2,1)	2 4 6
			(2110)	(4,1)	$1^2\ 2^2\ 3^4\ 4^3\ 5^4\ 6^3\ 7^3\ 8^2\ 9^2\ 10\ 11$
	[1111]	2	(0000)	(0,0)	0
			(1100)	(2,1)	2 4 6
			(1111)	(4,2)	2 4 5 8
5	[221]	$\frac{1}{2}$	(1000)	$(1,\frac{1}{2})$	$\frac{7}{2}$
			(2100)	$(3,\frac{1}{2})$	$\frac{1}{2}\ \frac{3}{2}\ (\frac{5}{2})^2\ (\frac{7}{2})^2\ (\frac{9}{2})^2\ (\frac{11}{2})^2\ (\frac{13}{2})^2\ \frac{15}{2}\ \frac{17}{2}\ \frac{19}{2}$
			(1110)	$(3,\frac{3}{2})$	$\frac{3}{2}\ \frac{5}{2}\ \frac{9}{2}\ \frac{11}{2}\ \frac{15}{2}$
			(2210)	$(5,\frac{1}{2})$	$(\frac{1}{2})^2\ (\frac{3}{2})^4\ (\frac{5}{2})^5\ (\frac{7}{2})^7\ (\frac{9}{2})^7\ (\frac{11}{2})^7\ (\frac{13}{2})^7\ (\frac{15}{2})^6\ (\frac{17}{2})^5\ (\frac{19}{2})^4$ $(\frac{21}{2})^3\ (\frac{23}{2})^2\ \frac{25}{2}\ \frac{27}{2}$
	[2111]	$\frac{3}{2}$	(1000)	$(1,\frac{1}{2})$	$\frac{7}{2}$
			(2100)	$(3,\frac{1}{2})$	$\frac{1}{2}\ \frac{3}{2}\ (\frac{5}{2})^2\ (\frac{7}{2})^2\ (\frac{9}{2})^2\ (\frac{11}{2})^2\ (\frac{13}{2})^2\ \frac{15}{2}\ \frac{17}{2}\ \frac{19}{2}$
			(1110)	$(3,\frac{3}{2})$	$\frac{3}{2}\ \frac{5}{2}\ \frac{9}{2}\ \frac{11}{2}\ \frac{15}{2}$

(continued)

[f]	n		(abcd)	(v,t)	J
[11111] ≡ [111]	6		(1000)	(1, ½)	$\frac{7}{2}$
			(1110)	(3, 3/2)	$\frac{3}{2}\ \frac{5}{2}\ \frac{9}{2}\ \frac{11}{2}\ \frac{15}{2}$
[222]		0	(2000)	(2, 0)	1 3 5 7
			(2110)	(4, 1)	$1^2\ 2^2\ 3^4\ 4^3\ 5^4\ 6^3\ 7^3\ 8^2\ 9^2\ 10\ 11$
			(2220)	(6, 0)	$1^4\ 2^2\ 3^7\ 4^5\ 5^7\ 6^6\ 7^7\ 8^4\ 9^6\ 10^3\ 11^3\ 12^2\ 13^2\ 15$
[2211]		1	(0000)	(0, 0)	0
			(1100)²	(2, 1)²	$(2\ 4\ 6)^2$
			(2200)	(4, 0)	$0^2\ 2^3\ 3^2\ 4^4\ 5^2\ 6^4\ 7^2\ 8^3\ 9\ 10^2\ 12$
			(2110)	(4, 1)	$1^2\ 2^2\ 3^4\ 4^3\ 5^4\ 6^3\ 7^3\ 8^2\ 9^2\ 10\ 11$
			(1111)	(4, 2)	2 4 5 8
			(2211)	(6, 1)	$0^2\ 1^2\ 2^5\ 3^5\ 4^7\ 5^6\ 6^7\ 7^6\ 8^6\ 9^4\ 10^4\ 11^2\ 12^2\ 13\ 14$
[21111]		2	(2000)	(2, 0)	1 3 5 7
			(1100)	(2, 1)	2 4 6
			(2110)	(4, 1)	$1^2\ 2^2\ 3^4\ 4^3\ 5^4\ 6^3\ 7^3\ 8^2\ 9^2\ 10\ 11$
			(1111)	(4, 2)	2 4 5 8
[111111] ≡ [11]		3	(0000)	(0, 0)	0
			(1100)	(2, 1)	2 4 6
[2221]	7	½	(1000)	(1, ½)	$\frac{7}{2}$
			(2100)	(3, ½)	$(\tfrac{1}{2})^2\ (\tfrac{3}{2})2\ (\tfrac{5}{2})2\ (\tfrac{7}{2})3\ (\tfrac{9}{2})3\ (\tfrac{11}{2})2\ (\tfrac{13}{2})2\ \frac{15}{2}\ \frac{17}{2}$
			(1110)	(3, 3/2)	$(\tfrac{3}{2})^2\ (\tfrac{5}{2})\ (\tfrac{7}{2})2\ (\tfrac{9}{2})2\ (\tfrac{11}{2})\ (\tfrac{13}{2})\ \frac{15}{2}$
			(2210)	(5, ½)	$(\tfrac{1}{2})^2\ (\tfrac{3}{2})4\ (\tfrac{5}{2})5\ (\tfrac{7}{2})7\ (\tfrac{9}{2})7\ (\tfrac{11}{2})7\ (\tfrac{13}{2})7\ (\tfrac{15}{2})6\ (\tfrac{17}{2})5\ (\tfrac{19}{2})4\ (\tfrac{21}{2})^2\ (\tfrac{23}{2})^2\ \frac{25}{2}\ \frac{27}{2}$
			(2111)	(5, 3/2)	$1\ (\tfrac{3}{2})^2\ (\tfrac{5}{2})^2\ (\tfrac{7}{2})3\ (\tfrac{9}{2})3\ (\tfrac{11}{2})3\ (\tfrac{13}{2})3\ (\tfrac{15}{2})^2\ (\tfrac{17}{2})^2\ (\tfrac{19}{2})\ (\tfrac{21}{2})$
			(2221)	(7, ½)	$(\tfrac{1}{2})3\ (\tfrac{3}{2})4\ (\tfrac{5}{2})5\ (\tfrac{7}{2})7\ (\tfrac{9}{2})7\ (\tfrac{11}{2})8\ (\tfrac{13}{2})8\ (\tfrac{15}{2})7\ (\tfrac{17}{2})6\ (\tfrac{19}{2})6\ (\tfrac{21}{2})4\ (\tfrac{23}{2})3\ (\tfrac{25}{2})3\ \frac{27}{2}\ \frac{29}{2}\ \frac{31}{2}$
[22111]		3/2	(1000)	(1, ½)	$\frac{7}{2}$
			(2100)	(3, ½)	$\tfrac{1}{2}\ 3\ (\tfrac{5}{2})^2\ (\tfrac{7}{2})^2\ (\tfrac{9}{2})^2\ (\tfrac{11}{2})^2\ (\tfrac{13}{2})^2\ \frac{15}{2}\ \frac{17}{2}\ \frac{19}{2}$
			(1110)²	(3, 3/2)²	$(\tfrac{3}{2}\ \tfrac{5}{2}\ \tfrac{9}{2}\ \tfrac{11}{2}\ \tfrac{15}{2})^2$

TABLE 11-24 (continued)

r	$[\lambda]$	T	(σ)	(s,t)	J-structure
8	[211111]	$\frac{5}{2}$	(2210)	$(5,\frac{1}{2})$	$(\frac{1}{2})^2 (\frac{3}{2})^4 (\frac{5}{2})^5 (\frac{7}{2})^7 (\frac{9}{2})^7 (\frac{11}{2})^7 (\frac{13}{2})^7 (\frac{15}{2})^6 (\frac{17}{2})^5 (\frac{19}{2})^4$ $(\frac{21}{2})^3 (\frac{23}{2})^2 (\frac{25}{2}) (\frac{27}{2})$
	[1111111] ≡ [1]	$\frac{7}{2}$	(2111)	$(5,\frac{3}{2})$	$\frac{1}{2} (\frac{3}{2})^2 (\frac{5}{2})^2 (\frac{7}{2})^3 (\frac{9}{2})^3 (\frac{11}{2})^3 (\frac{13}{2})^3 (\frac{15}{2})^2 (\frac{17}{2})^2 \frac{15}{2} \frac{17}{2} \frac{19}{2}$
			(1000)	$(1,\frac{1}{2})$	$\frac{7}{2}$
			(2100)	$(3,\frac{1}{2})$	$\frac{1}{2} \frac{3}{2} (\frac{5}{2})^2 (\frac{7}{2})^2 (\frac{9}{2})^2 (\frac{11}{2})^2 (\frac{13}{2})^2 \frac{15}{2} \frac{17}{2} \frac{19}{2}$
			(1110)	$(3,\frac{3}{2})$	$\frac{3}{2} \frac{5}{2} \frac{9}{2} \frac{11}{2} \frac{15}{2}$
			(1000)	$(1,\frac{1}{2})$	$\frac{7}{2}$
	[2222]	0	(0000)	(0,0)	0
			(1100)	(2,1)	2 4 6
			(2200)	(4,0)	$0^2\ 2^3\ 3^2\ 4^4\ 5^2\ 6^4\ 7^2\ 8^3\ 9\ 10^2\ 12$
			(1111)	(4,2)	2 4 5 8
			(2211)	(6,1)	$0^2\ 1^2\ 2^5\ 3^5\ 4^7\ 5^6\ 6^7\ 7^6\ 8^6\ 9^4\ 10^4\ 11^2\ 12^2\ 13\ 14$
			(2222)	(8,0)	$0\ 1\ 2^3\ 3^2\ 4^5\ 5^3\ 6^5\ 7^3\ 8^4\ 9^3\ 10^3\ 11^2\ 12^2\ 13\ 14\ 16$
	[22221]	1	(2000)	(2,0)	1 3 5 7
			(1100)	(2,1)	2 4 6
			(2110)²	$(4,1)^2$	$(1^2\ 2^2\ 3^4\ 4^3\ 5^4\ 6^3\ 7^3\ 8^2\ 9^2\ 10\ 11)^2$
			(1111)	(4,2)	2 4 5 8
			(2220)	(6,0)	$1^4\ 2^2\ 3^7\ 4^5\ 5^7\ 6^6\ 7^7\ 8^4\ 9^6\ 10^3\ 11^3\ 12^2\ 13^2\ 15$
			(2211)	(6,1)	$0^2\ 1^2\ 2^5\ 3^5\ 4^7\ 5^6\ 6^7\ 7^6\ 8^6\ 9^4\ 10^4\ 11^2\ 12^2\ 13\ 14$
	[221111]	2	(0000)	(0,0)	0
			(1100)²	$(2,1)^2$	$(2\ 4\ 6)^2$
			(2200)	(4,0)	$0^2\ 2^3\ 3^2\ 4^4\ 5^2\ 6^4\ 7^2\ 8^3\ 9\ 10^2\ 12$
			(2110)	(4,1)	$1^2\ 2^2\ 3^4\ 4^3\ 5^4\ 6^3\ 7^3\ 8^2\ 9^2\ 10\ 11$
			(1111)	(4,2)	2 4 5 8
	[2111111]	3	(2000)	(2,0)	1 3 5 7
			(1100)	(2,1)	2 4 6
	[1111111] ≡ [0]	4	(0000)	(0,0)	0

Problem. Construct the tables for $j = \frac{3}{2}, \frac{5}{2}$ in Table 11–23.

By combining the results in Table 11–23 with the results of Table 10–5 we obtain the classification of the spin-orbit functions. The corresponding charge function is completely determined by giving the value of T. The representations of $Sp(2j + 1)$ are characterized by symbols $(\sigma_1 \sigma_2 \ldots \sigma_j)$, where $\sigma_i \leq 2$. Since the Young pattern for this symbol has two columns, it can be described by giving the sum and difference of the lengths of the two columns. We express the lengths of the two columns as $\frac{1}{2}s \pm t$. The *seniority number* s is the smallest number of particles for which the representation of $Sp(2j + 1)$ can occur; the quantum number t is called the *reduced isotopic spin.* The results of the analysis are tabulated in Table 11–24.

Problem. Construct Table 11–24 for $j = \frac{3}{2}, j = \frac{5}{2}$.

CHAPTER 12

RAY REPRESENTATIONS. LITTLE GROUPS

Throughout the previous chapters we have considered the representation of groups by linear operators acting on vectors in a Hilbert space. To each element g of the group G, we associated a linear operator $D(g)$, and required that

$$D(g_1)D(g_2) = D(g_3) \quad \text{if} \quad g_1 g_2 = g_3. \tag{12-1}$$

The operators $D(g)$ of the representation act on the vectors ψ in the Hilbert space. However, a pure state of a physical system is described in quantum mechanics not by a normalized vector ψ, but rather by a *ray* $\epsilon\psi$, where ϵ is an arbitrary phase factor ($|\epsilon| = 1$). To each element g of the symmetry group G we should therefore associate an operator $D(g)$ which acts on the *rays* in Hilbert space and maps *rays* into *rays*. In this way we associate with the group element g a process of "translation" which takes each physical state of the system into another possible physical state of the system. We should therefore, *a priori*, not restrict ourselves to requiring that the operators of a representation satisfy (12-1). Instead we should require only that

$$D(g_1)D(g_2) = \epsilon D(g_3) \quad \text{if} \quad g_1 g_2 = g_3, \tag{12-2}$$

where ϵ is a phase factor which depends on g_1 and g_2. Such representations are called *ray representations* (or *projective representations*), while the representations defined by (12-1) are called *vector representations*. In this chapter we shall consider the problem of finding the ray representations of a group. At the same time, we shall examine the conditions under which the ray representation (12-2) can be replaced by a vector representation (12-1).

12-1 Projective representations of finite groups. It is remarkable that the problem of finding the ray representations of finite groups was stated and completely solved long before the advent of quantum mechanics. In a series of papers, Schur gave the general method for finding the irreducible representations of a finite group in terms of *fractional linear transformations* (*projective transformations, collineations*).

By a *projective representation* we mean the following:

We associate with each element $A, B, \ldots$ of a finite group a *fractional linear transformation*

$$A \to \{A\}: \quad x'_i = \frac{a_{i1}x_1 + a_{i2}x_2 + \cdots + a_{i,n-1}x_{n-1} + a_{in}}{a_{n1}x_1 + a_{n2}x_2 + \cdots + a_{n,n-1}x_{n-1} + a_{nn}}$$

$$(i = 1, \ldots, n - 1), \quad (12\text{–}3)$$

where the matrix $D(A)$ of the coefficients a_{ij} is nonsingular. The fractional linear transformations give a representation of the group H if

$$\{A\}\{B\} = \{AB\}. \tag{12–4}$$

The degree of the representation is the degree n of the matrices $D(A)$.

The transformations (12–3) can be regarded as linear transformations

$$y'_i = \sum_{j=1}^{n} a_{ij}y_j \quad (i = 1, \ldots, n), \tag{12–5}$$

where

$$x_i = \frac{y_i}{y_n} \quad (i = 1, \ldots, n - 1) \tag{12–5a}$$

are homogeneous coordinates.

From (12–3) or (12–5a) we see that the multiplication of the elements a_{ij} by a common factor does not change the transformation $\{A\}$. If we write out the matrices corresponding to (12–4), we find

$$D(A)D(B) = \omega_{A,B}D(AB), \tag{12–6}$$

where the $\omega_{A,B}$ are a set of constants depending on the choice of $D(A)$ and $D(B)$. Conversely, if we are given a set of nonsingular matrices satisfying (12–6), we can find a corresponding representation (12–4) in terms of fractional linear transformations. Thus the problem of finding representations of a group by fractional linear transformations is equivalent to finding representations by linear transformations $D(A)$ which satisfy (12–6).

We shall call the representation (12–6) a *projective representation* belonging to the *factor system* $\omega_{A,B}$. If, in particular, all factors $\omega_{A,B}$ are equal to unity, Eq. (12–6) defines the usual representations, which we shall call *vector representations*.

If the matrices of two representations differ only by a factor,

$$D'(A) = aD(A), \quad D'(B) = bD(B), \ldots, \tag{12–7}$$

they will correspond to the *same* set of fractional linear transformations $\{A\}, \{B\}, \ldots$ and are said to be *associated*.

The concepts of equivalence and reducibility are the same as before. The projective representation $D'(A)$ is *equivalent* to the representation $D(A)$ of (12–6) if there exists a nonsingular matrix S such that $D'(A) = SD(A)S^{-1}$ for all A in H. Then

$$D'(A)D'(B) = \omega_{A,B}D'(AB), \tag{12-8}$$

so that equivalent representations belong to the same factor system. If we can find a matrix S which gives all the representatives $D'(A)$ the reduced form

$$D'(A) = \begin{bmatrix} D'_1(A) & 0 \\ 0 & D'_2(A) \end{bmatrix}, \tag{12-9}$$

the representation $D(A)$ is *reducible* to the direct sum of the projective representations $D'_1 + D'_2$. From (12–8) and (12–9) we see that the factor system $\omega_{A,B}$ is the same for $D(A)$, $D'(A)$, $D'_1(A)$, and $D'_2(A)$.

The projective representation (12–6) is *irreducible* if there is no equivalent representation having the form (12–9).

The elements $\omega_{A,B}$ of the factor system in (12–6) cannot be chosen arbitrarily. Suppose that the finite group H consists of h elements $H_0 = E$, $H_1, \ldots, H_{h-1}$. If the matrices $D(H_i)$ give a projective representation of H belonging to the factor system $\omega_{A,B}$, then for any three elements P, Q, R of the group H,

$$D(P)D(Q) = \omega_{P,Q}D(PQ),$$
$$D(P)D(Q)D(R) = \omega_{P,Q}D(PQ)D(R) = \omega_{P,Q}\omega_{PQ,R}D(PQR); \tag{12-10}$$

$$D(Q)D(R) = \omega_{Q,R}D(QR),$$
$$D(P)D(Q)D(R) = \omega_{Q,R}D(P)D(QR) = \omega_{Q,R}\omega_{P,QR}D(PQR); \tag{12-10a}$$

$$\omega_{P,Q}\omega_{PQ,R} = \omega_{P,QR}\omega_{Q,R} \qquad (P, Q, R = H_0, \ldots, H_{h-1}). \tag{12-11}$$

Thus the associative law for group multiplication forces the h^2 constants $\omega_{A,B}$ to satisfy the h^3 equations (12–11).

Conversely we now show that for every system of h^2 nonvanishing constants $\omega_{A,B}$ which satisfy (12–11), there exists a projective representation of H belonging to the factor system $\omega_{A,B}$. To prove this we introduce h independent variables $x_{H_0}, x_{H_1}, \ldots, x_{H_{h-1}}$, and construct the h-by-h matrix X with matrix elements

$$X_{P,Q} = \omega_{PQ^{-1},Q}x_{PQ^{-1}} \qquad (P, Q = H_0, \ldots, H_{h-1}). \tag{12-12}$$

The rows and columns of X are labeled by the group elements H_i. The matrix X clearly can be written as

$$X = \sum_{R \subset H} D(R)x_R, \tag{12–13}$$

where the sum extends over all elements R of H, and the matrix $D(R)$ can be found from (12–12) by setting $x_R = 1$ and all the other variables equal to zero. We shall show that the matrices $D(R)$ satisfy (12–6).

Let $y_{H_0}, y_{H_1}, \ldots, y_{H_{h-1}}$ be a second set of independent variables. Define the h variables $z_{H_0}, z_{H_1}, \ldots, z_{H_{h-1}}$ by the equations

$$z_P = \sum_{RS=P} \omega_{R,S} \, x_R y_S, \tag{12–14}$$

where the summation extends over all elements R and S for which $RS = P$. We denote the matrix (12–12) by Y or Z if the variables x_R are replaced by y_R or z_R. The P, Q-matrix element of the product XY is

$$(XY)_{P,Q} = \sum_R \omega_{PR^{-1},R} \omega_{RQ^{-1},Q} \, x_{PR^{-1}} y_{RQ^{-1}}. \tag{12–15}$$

But, for the elements (PR^{-1}), (RQ^{-1}), Q, we have, from Eq. (12–11),

$$\omega_{PR^{-1},R} \omega_{RQ^{-1},Q} = \omega_{PR^{-1},RQ^{-1}} \omega_{PQ^{-1},Q},$$

so that (12–15) becomes

$$(XY)_{P,Q} = \omega_{PQ^{-1},Q} \sum_R \omega_{PR^{-1},RQ^{-1}} \, x_{PR^{-1}} y_{RQ^{-1}}. \tag{12–16}$$

Comparing the sum in (12–16) with (12–14), we see that it is equal to $z_{PQ^{-1}}$, so that

$$(XY)_{P,Q} = \omega_{PQ^{-1},Q} \, z_{PQ^{-1}} = Z_{P,Q}, \tag{12–17}$$

or

$$XY = Z. \tag{12–17a}$$

Substituting from (12–13) and (12–14), we find

$$\sum_{R,S} D(R)D(S)x_R y_S = \sum_T D(T)z_T = \sum_{R,S} \omega_{R,S} D(RS)x_R y_S, \quad (12–18)$$

and, when the coefficients of the independent variables $x_R y_S$ are equated, the h-by-h matrices $D(R)$, $D(S)$ satisfy

$$D(R)D(S) = \omega_{R,S} D(RS). \tag{12–19}$$

From (12–12) and (12–13), the matrix $D(R)$ contains only one nonzero element in each row and each column and is therefore nonsingular. Thus the matrices $D(R)$ constructed in Eqs. (12–12) and (12–13) form a representation belonging to the factor system $\omega_{A,B}$. We have thus shown that the necessary and sufficient condition for the existence of representations belonging to a factor system $\omega_{A,B}$ is that the h^2 numbers $\omega_{A,B}$ satisfy the h^3 equations (12–11).

The equations (12–11) have an infinite number of solutions. If $\omega_{A,B}$ is a solution and $c_{H_0}, c_{H_1}, \ldots, c_{H_{h-1}}$ are any constants not equal to zero, then

$$\omega'_{A,B} = \frac{c_A c_B}{c_{AB}} \omega_{A,B} \tag{12–20}$$

is also a solution. Solutions $\omega_{A,B}$ and $\omega'_{A,B}$ for which one can find constants c so that (12–20) is satisfied are called *associated* or *equivalent* factor systems. If we find a representation $D(A)$ belonging to the factor system $\omega_{A,B}$, the matrices

$$D'(A) = c_A D(A) \tag{12–21}$$

satisfy

$$D'(A)D'(B) = c_A c_B \omega_{A,B} D(AB) = \frac{c_A c_B}{c_{AB}} \omega_{A,B} D'(AB) = \omega'_{A,B} D'(AB). \tag{12–22}$$

Thus, if two factor systems are equivalent, the representations belonging to one are immediately determined from the representations belonging to the other by means of (12–20). Representations belonging to nonequivalent factor systems are said to be of *different type*.

We now distribute all solutions of (12–11) into *classes* of equivalent factor systems. The number of classes is *finite:* Taking the determinant of (12–19) and letting $\det D(R) = d_R$, we have

$$(\omega_{R,S})^h = \frac{d_R d_S}{d_{RS}}. \tag{12–23}$$

If we choose the constants c_A in (12–20) to be $c_A = d_A^{-1/h}$ (so that c_A is one of the h solutions of the equation $c_A^{-h} = d_A$), we can obtain from the factor system $\omega_{R,S}$ an equivalent system

$$\omega'_{R,S} = \frac{c_R c_S}{c_{RS}} \omega_{R,S}$$

for which, according to (12–23),

$$(\omega'_{R,S})^h = 1. \tag{12–24}$$

Thus a class of equivalent factor systems necessarily contains a factor system in which the h^2 quantities $\omega_{A,B}$ are hth roots of unity. Therefore the number of classes cannot exceed h^{h^2}, and hence the number of classes is finite.

Let us denote the m classes by $K_0, K_1, \ldots, K_{m-1}$, where K_0 is the class containing the factor system $\omega_{A,B} \equiv 1$. If the factor systems $\omega_{A,B}^{(\lambda)}$ and $\omega_{A,B}^{(\mu)}$ are solutions of (12–11) and belong to the classes K_λ and K_μ, their products $\omega_{A,B}^{(\lambda)} \cdot \omega_{A,B}^{(\mu)}$ also satisfy (12–11). The class K_ν, to which this new solution belongs, does not depend on the particular choice of the solutions $\omega_{A,B}^{(\lambda)}$ and $\omega_{A,B}^{(\mu)}$ within the classes K_λ and K_μ, but is determined by the classes themselves. We write $K_\lambda K_\mu = K_\nu$, and note that $K_\mu K_\lambda = K_\lambda K_\mu = K_\nu$. We thus have defined a multiplication of classes which is commutative. Furthermore, we easily see that if $K_\alpha K_\beta = K_\alpha K_\gamma$, then $K_\beta = K_\gamma$. The class K_0 serves as the unit element. Finally it is clear that this class multiplication is associative, and the classes $K_0, K_1, \ldots, K_{m-1}$ therefore form an abelian group of order m, which is called the *multiplicator* of the group H.

Schur developed general methods for finding the multiplicator and for the construction of the irreducible projective representations of large classes of finite groups. The argument is too long to be given here. Instead we illustrate the procedure for some simple cases.

12–2 Examples of projective representations of finite groups. The cyclic group generated from an element A of order n, $A^0 = E, A, A^2, \ldots,$ A^{n-1}, can be defined by the algebraic condition

$$A^n = E. \tag{12–25}$$

For a projective representation, the matrix $D(A)$ must satisfy

$$[D(A)]^n = \omega 1. \tag{12–26}$$

If we replace the representative $D(A)$ by $D'(A) = \omega^{-1/n} D(A)$, we obtain a vector representation, $D'(A^m) = [D'(A)]^m$. Thus for cyclic groups, the multiplicator consists of the unit element K_0, and all projective representations are equivalent to the usual vector representations. The irreducible representations are all one-dimensional.

Let the abelian group H be the direct product of two cyclic groups generated by elements A with order n and B with order n', respectively. The algebraic equations characterizing the group are

$$A^n = E, \quad B^{n'} = E, \quad AB = BA. \tag{12–27}$$

The representatives will satisfy the equations

$$[D(A)]^n = a1, \quad [D(B)]^{n'} = b1, \quad D(A)D(B) = cD(B)D(A), \tag{12–28}$$

with constants a, b, c. The constants a and b can be incorporated into $D(A)$ and $D(B)$, so we are left with only the last constant c in (12–28).

Since $D(A)$ and $D(B)$ appear on both sides of the last equation in (12–28), the constant c cannot be changed by altering the matrices. However, we must determine the restrictions imposed on c by (12–28). Multiplying the last equation on the right by $D(B)$, we have

$$D(A)[D(B)]^2 = cD(B)[D(A)D(B)] = c^2[D(B)]^2D(A).$$

Repeating this process, we obtain

$$D(A)[D(B)]^3 = c^3[D(B)]^3D(A)$$

and, finally,

$$D(A)[D(B)]^{n'} = c^{n'}[D(B)]^{n'}D(A),$$

or

$$c^{n'} = 1. \tag{12–29}$$

Similarly, by repeated left multiplication with $D(A)$ we find

$$c^n = 1. \tag{12–29a}$$

Thus the constant c must be both an nth and an n'th root of unity. If d is the greatest common divisor of n and n', c must satisfy

$$c^d = 1. \tag{12–29b}$$

If n and n' are relatively prime, $d = 1$, and therefore $c = 1$. In this case, all representations are equivalent to vector representations, and all irreducible representations are one-dimensional. This is the case, for example, for $n = 3$, $n' = 2$.

If $n = n' = 2$, we get the four-group

$$A^2 = E, \qquad B^2 = E, \qquad C = AB = BA. \tag{12–30}$$

The argument given above shows that $c^2 = 1$, $c = \pm 1$. For $c = +1$, the operators satisfy the equations

$$[D(A)]^2 = [D(B)]^2 = 1, \qquad D(A)D(B) = D(B)D(A), \tag{12–31}$$

and we obtain the usual one-dimensional irreducible representations. If $c = -1$, we have

$$[D(A)]^2 = [D(B)]^2 = 1, \qquad D(A)D(B) = -D(B)D(A). \tag{12–32}$$

If we choose $D(C) = iD(A)D(B)$, then

$$[D(C)]^2 = -[D(A)D(B)][D(A)D(B)] = [D(A)D(B)][D(B)D(A)] = 1.$$

The matrices $D(A)$, $D(B)$, $D(C)$ have their squares equal to one, and anti-commute. The irreducible representation is given by the Pauli matrices.

Problem. Prove that the irreducible representations of (12–32) have dimension two.

The dihedral groups D_n are generated from two elements A and S satisfying the equations

$$A^n = E, \qquad S^2 = E, \qquad SAS = A^{-1}. \tag{12–33}$$

We denote the matrix representatives of A and S by B and T. They satisfy the equations

$$B^n = b1, \qquad T^2 = t1, \qquad TBT = cB^{-1}. \tag{12–34}$$

The constants b and t can be incorporated into the matrices B and T ($B' = b^{-1/n}B$, $T' = t^{-1/2}1$). After this is done, our equations are

$$B^n = 1, \qquad T^2 = 1, \qquad TBT = cB^{-1}. \tag{12–34a}$$

Taking the nth power of the last equation in (12–34a) and using the first two equations, we find $c^n = 1$, so the constant c must be an nth root of unity. Let $c = \epsilon^m$, $\epsilon = \exp(2\pi i/n)$. If we multiply B by $\epsilon^{m'}$, the first two equations in (12–34a) are still valid, while the last is replaced by $TBT = \epsilon^{m-2m'}B^{-1}$. If n is odd, we can always choose m' so that $m - 2m' = 0$, or $m - 2m' = n$. Then $TBT^{-1} = B$, and we obtain only the usual vector representations. If n is even, we can eliminate c if m is even, and we can replace it by ϵ if m is odd. Thus, for even n, there are two classes K_0 and K_1. For the class K_0, $c = 1$, the operator equations are

$$B^n = 1, \qquad T^2 = 1, \qquad TBT = B^{-1}, \tag{12–35}$$

and we obtain the usual irreducible vector representations. For the class K_1, $c = \epsilon$, and the operator equations are

$$B^n = 1, \qquad T^2 = 1, \qquad TBT = \epsilon B^{-1} \qquad (n \text{ even}). \tag{12–36}$$

For $n = 2$, we get the four-group which was treated above. It is easy to show that the irreducible representations of (12–36) are two-dimensional.

Problem. Show that the inequivalent irreducible representations of (12–36) have the form

$$B = \begin{bmatrix} \epsilon^r & 0 \\ 0 & \epsilon^{1-r} \end{bmatrix}, \qquad T = \begin{bmatrix} 0 & 1 \\ 1 & 0 \end{bmatrix}, \tag{12–37}$$

where $r = 1, 2, \ldots, n/2$.

The symmetric group S_n is generated by the $n - 1$ transpositions

$$T_1 = (12), \quad T_2 = (23), \quad \ldots, \quad T_{n-1} = (n - 1, n)$$

which satisfy the equations

$$T_i^2 = E, \quad (T_j T_{j+1})^3 = E, \quad T_r T_s = T_s T_r \quad \begin{bmatrix} i = 1, \ldots, n - 1 \\ j = 1, \ldots, n - 2 \\ r = 1, 2, \ldots, n - 3 \\ s = r + 2, \ldots, n - 1 \end{bmatrix}.$$

$$(12\text{–}38)$$

Conversely, one can show that the abstract group defined by the equations (12–38) is isomorphic to S_n.

Let us assume that we have a projective representation of S_n in which the elements T_i are represented by matrices A_i. The operator equations corresponding to (12–38) are

$$A_i^2 = a_i 1, \tag{12–39}$$

$$(A_j A_{j+1})^3 = b_j 1, \tag{12–39a}$$

$$A_r A_s = c_{rs} A_s A_r, \tag{12–39b}$$

where the a_i, b_j, c_{rs} are constants different from zero. The constants c_{rs} occur only for $n \geq 4$ and, as we see from (12–39b), cannot be altered by multiplying the matrices A_r by factors. Thus the constants c_{rs} are determined by the fractional linear transformations.

From (12–39b), we have $A_r A_s A_r^{-1} = c_{rs} A_s$. Squaring and using (12–39), we find

$$c_{rs}^2 = 1. \tag{12–40}$$

The indices r, $r + 1$, s, $s + 1$, which appear in the permutations T_r and T_s of (12–38), are all different. If we take another pair of permutations $T_{r'}$, $T_{s'}$ for which r', $r' + 1$, s', $s' + 1$ are all different, there exists a permutation T which takes the indices r, $r + 1$, s, $s + 1$ into r', $r' + 1$, s', $s' + 1$, so that

$$T T_r T^{-1} = T_{r'}, \quad T T_s T^{-1} = T_{s'}. \tag{12–41}$$

If the permutation T is represented by the operator A, the corresponding equations for the operators of the representation are

$$A A_r A^{-1} = c A_{r'}, \quad A A_s A^{-1} = d A_{s'}, \tag{12–41a}$$

where c and d are constants different from zero. From (12–39b),

$$A A_r A^{-1} A A_s A^{-1} = c_{rs} A A_s A^{-1} A A_r A^{-1},$$

and, using (12–41a), we have

$$cd\, A_{r'}A_{s'} = cd\, c_{rs}A_{s'}A_{r'} \quad \text{or} \quad A_{r'}A_{s'} = c_{rs}A_{s'}A_{r'}.$$

Comparing with (12–39b), that is,

$$A_{r'}A_{s'} = c_{r's'}A_{s'}A_{r'},$$

we find that $c_{rs} = c_{r's'}$. Thus according to (12–40) all constants c_{rs} are the same and equal to ± 1. Letting $j = \pm 1$, we have $c_{rs} = j$ for all r, s.

From (12–39a) we have

$$A_j A_{j+1} A_j = b_j A_{j+1}^{-1} A_j^{-1} A_{j+1}^{-1}.$$

Squaring, we get

$$A_j A_{j+1} A_j^2 A_{j+1} A_j = b_j^2 A_{j+1}^{-1} A_j^{-1} A_{j+1}^{-2} A_j^{-1} A_{j+1}^{-1},$$

$$a_j^2 a_{j+1} = \frac{b_j^2}{a_j a_{j+1}^2},$$

$$b_j^2 = a_j^3 a_{j+1}^3. \tag{12–42}$$

Since the matrices A_i can be multiplied by arbitrary constants, we can choose the constants a_i in (12–39) arbitrarily. We do this in two alternative, but equivalent, ways.

First we set

$$a_1 = a_2 = \cdots = a_{n-1} = j. \tag{12–43}$$

From (12–42), $b_j = \pm 1$. We define matrices $B_1, \ldots, B_{n-1}$ by the equations

$$B_1 = A_1, \quad B_2 = j b_1 A_2, \quad B_3 = b_1 b_2 A_3, \quad B_4 = j b_1 b_2 b_3 A_4, \ldots \tag{12–44}$$

and find that they satisfy the equations

$$B_j^2 = j1, \quad (B_j B_{j+1})^3 = j1, \quad B_r B_s = j B_s B_r. \tag{12–45}$$

Alternatively, we set

$$a_1 = a_2 = \cdots = a_{n-1} = 1, \tag{12–43a}$$

and introduce matrices $C_1, \ldots, C_{n-1}$:

$$C_1 = A_1, \quad C_2 = b_1 A_2, \quad C_3 = b_1 b_2 A_3, \quad C_4 = b_1 b_2 b_3 A_4, \ldots \tag{12–44a}$$

which satisfy

$$C_i^2 = 1, \quad (C_j C_{j+1})^3 = 1, \quad C_r C_s = j C_s C_r. \tag{12–45a}$$

For $j = +1$, both systems reduce to (12–39), and we obtain the class K_0 of vector representations of the permutation group. For $j = -1$, we obtain the class K_1 of projective representations, with operator equations

$$B_i^2 = -1, \qquad (B_j B_{j+1})^3 = -1, \qquad B_r B_s = -B_s B_r, \qquad (12\text{–}46)$$

or

$$C_i^2 = 1, \qquad (C_j C_{j+1})^3 = 1, \qquad C_r C_s = -C_s C_r. \qquad (12\text{–}46a)$$

The last two sets are interchangeable. If we multiply all matrices B_i by $\sqrt{-1}$, the resulting set satisfies (12–46a).

For $n < 4$, the constants c_{rs} do not occur in (12–38). Thus for $n < 4$, the symmetric group has only the usual irreducible vector representations which we found in Chapter 7. For $n \geq 4$, there are additional irreducible representations for which the operators satisfy (12–46). Schur showed that for $n \geq 4$ the number of inequivalent irreducible representations of this second type is equal to the number of partitions

$$n = \nu_1 + \nu_2 + \cdots + \nu_m \qquad (\nu_1 > \nu_2 > \cdots > \nu_m > 0) \qquad (12\text{–}47)$$

of n into unequal integers. To the partition (12–47) there corresponds an irreducible representation of degree

$$f_{\nu_1, \ldots, \nu_m} = 2^{[(n-m)/2]} \frac{n!}{\nu_1! \nu_2! \cdots \nu_m!} \prod_{\alpha < \beta} \frac{\nu_\alpha - \nu_\beta}{\nu_\alpha + \nu_\beta}, \qquad (12\text{–}48)$$

where $[(n - m)/2]$ denotes the largest integer which is $\leq (n - m)/2$. For $n = 4$, we have two new irreducible representations of degree $f_4 = 2$ and $f_{3,1} = 4$. For $n = 4$, the equations (12–46a) are:

$$C_1^2 = C_2^2 = C_3^2 = 1, \qquad (C_1 C_2)^3 = (C_2 C_3)^3 = 1, \qquad C_1 C_3 = -C_3 C_1. \qquad (12\text{–}49)$$

The representation of degree two is obtained by setting

$$C_1 = \sigma_x, \qquad C_2 = \frac{\sqrt{3}}{2}\sigma_y - \frac{1}{2}\sigma_x, \qquad C_3 = \sqrt{\frac{2}{3}}\sigma_z - \sqrt{\frac{1}{3}}\sigma_y, \qquad (12\text{–}50)$$

where $\sigma_x = \left(\begin{smallmatrix} 0 & 1 \\ 1 & 0 \end{smallmatrix}\right)$, $\sigma_y = \left(\begin{smallmatrix} 0 & -i \\ i & 0 \end{smallmatrix}\right)$, $\sigma_z = \left(\begin{smallmatrix} 1 & 0 \\ 0 & -1 \end{smallmatrix}\right)$ are the Pauli matrices.

For the complete theory, we refer the reader to the third paper of Schur which contains many remarkable results which were later rediscovered independently.

Problem. For $n = 4$, assume the existence of an irreducible two-dimensional representation of (12–46a) and derive the result (12–50).

12–3 Ray representations of Lie groups. The states of a quantum-mechanical system are described by unit vectors ψ in a Hilbert space. [An inner product (ϕ, ψ) is defined for the Hilbert space, and this product is positive definite, so that $\|\phi\| = (\phi, \phi) > 0$ for any vector in the space. In particular, for unit vectors ψ, $\|\psi\| = 1$.] To every unit vector ψ there corresponds a unique physical state of the system. The converse is not true. A given physical state is describable by any of the set Ψ of unit vectors $\epsilon\psi$, where ϵ is complex and $|\epsilon| = 1$. There is a one-to-one correspondence between physical states and *rays* Ψ. Any vector in the ray Ψ can serve as a representative of the ray.

The transition probability between physical states corresponding to rays Φ and Ψ is given by $|(\phi, \psi)|^2$, where ϕ is any vector in the ray Φ and ψ is any vector in the ray Ψ. We may therefore define an inner product of rays,

$$(\Phi, \Psi) = |(\phi, \psi)|, \qquad \phi \text{ in } \Phi, \quad \psi \text{ in } \Psi, \tag{12–51}$$

which is independent of the particular choice of ϕ and ψ.

To each transformation of the symmetry group of the physical system there corresponds a one-to-one mapping of the physical states. Each ray Ψ is mapped into a ray Ψ':

$$\Psi \to \Psi'. \tag{12–52}$$

Since the new description must give the same observable results as the original one, the mapping must preserve transition probabilities:

$$|(\Phi', \Psi')|^2 = |(\Phi, \Psi)|^2. \tag{12–53}$$

The remarkable result, first proved by Wigner, is that the ray correspondence (12–52) can always be replaced by a vector correspondence: Representatives ψ, ψ' can be selected from the rays Ψ, Ψ' so that the vector mapping

$$\psi \to \psi' = U\psi \tag{12–54}$$

satisfies

$$U(\phi + \psi) = U\phi + U\psi, \tag{12–55}$$

and either

$$\text{(a):} \qquad (U\phi, U\psi) = (\phi, \psi), \tag{12–56a}$$

or

$$\text{(b):} \qquad (U\phi, U\psi) = (\psi, \phi). \tag{12–56b}$$

In other words, we can replace the ray mapping by an additive operation on the vectors of the Hilbert space. Two alternatives may occur. The operator U is either (a) linear and unitary, or (b) antiunitary. There is no *a priori* reason for selecting one of the alternatives (a) or (b).

The operator U is still not defined uniquely, since multiplication by any phase factor ϵ will not alter the preceding results. Thus we may regard the operator U as a *representative* of the *operator ray* ϵU. We can select the representative U in any convenient way.

If r and s are two elements of the symmetry group G of the quantum-mechanical system, we associate with each of them an operator representative $U(r)$, $U(s)$. To the product rs, which is also in G, there will correspond an operator $U(rs)$. But the requirements imposed above lead only to the relation

$$U(r)U(s) = \omega(r, s)U(rs), \qquad (12\text{-}57)$$

where $|\omega| = 1$, and the phase factor ω depends on the elements r and s of G. Equation (12-57) is identical with (12-6), which defined the ray representations of finite groups.

We now wish to consider continuous groups and, in particular, Lie groups. In our discussion in Chapter 8, we showed that the parameter space of a Lie group will in general consist of several disconnected pieces, one of which contains the identity element e of the group G. For those elements r of G which are contained in the sheet of the identity element e (so that they can be reached by a continuous path from the identity), we can prove that only the alternative (a) of Eq. (12-56) occurs, so that $U(r)$ is a unitary operator. In any neighborhood of the identity e, an element r can be written as the square of some element s, so $r = s^2$. From (12-57)

$$U(r) = \frac{1}{\omega(s, s)} U(s)U(s). \qquad (12\text{-}58)$$

Since the square of a unitary or antiunitary operator is unitary, we conclude that $U(r)$ is unitary for any r in the neighborhood of the identity. Finally, every element r in the sheet of the identity can be expressed as a product $r_1 r_2 \cdots r_n$ of elements in a given neighborhood of the identity. Since each of the operators $U(r_i)$ is unitary, and a product of a finite number of unitary operators is unitary, we conclude that $U(r)$ is unitary.

The transformations in the sheet of the identity form a subgroup, and we shall restrict ourselves to this group G'. For any elements r, s of G', the operators $U(r)$, $U(s)$ are unitary and satisfy Eqs. (12-55) and (12-56a). So far, we have imposed no conditions of continuity on the operator $U(r)$ as r varies over G. It can be shown that, in the neighborhood of the identity e, the operator representatives $U(r)$ can be selected so that they are continuous; i.e., for any $\epsilon > 0$ and any vector ψ, there exists a neighborhood $\mathfrak{N}$ of r such that

$$\|U(r)\psi - U(s)\psi\| < \epsilon \qquad (12\text{-}59)$$

for s in $\mathfrak{N}$. From this result it follows that the factor $\omega(r, s)$ in (12-57) is a continuous function of r and s. [Use the identity

$$\omega(r, s)[U(rs) - U(r's')]\psi + U(r')[U(s') - U(s)]\psi$$
$$+ [U(r') - U(r)]U(s)\psi = [\omega(r', s') - \omega(r, s)]U(r's')\psi \quad (12\text{--}60)$$

to prove the continuity of $\omega(r, s)$.]

We conclude that, in the neighborhood of the identity, the elements r of the group G' can be represented by continuous unitary operators $U(r)$, and that $\omega(r, s)$ in (12–57) is a continuous function. The function $\omega(r, s)$ is called a *local factor*. It can be written as

$$\omega(r, s) = \exp[i\xi(r, s)], \quad (12\text{--}61)$$

where ξ is a real, continuous function of r and s, which we call a *local exponent*.

The operators $U(r)$ can be multiplied by any function $\phi(r)$ which is continuous and has absolute value unity. The operators

$$U'(r) = \phi(r)U(r) = \exp[i\zeta(r)] \cdot U(r), \quad (12\text{--}62)$$

where $\zeta(r)$ is a real, continuous function, will also be continuous. If we substitute (12–62) in (12–57), we find

$$U'(r)U'(s) = \phi(r)\phi(s)U(r)U(s)$$
$$= \phi(r)\phi(s)\omega(r, s)U(rs)$$
$$= \frac{\phi(r)\phi(s)}{\phi(rs)}\omega(r, s)U'(rs),$$

so that

$$U'(r)U'(s) = \omega'(r, s)U'(rs), \quad (12\text{--}63)$$

$$\omega'(r, s) = \frac{\phi(r)\phi(s)}{\phi(rs)}\omega(r, s). \quad (12\text{--}64)$$

This procedure is completely analogous to that for finite groups. [Compare Eqs. (12–21) and (12–22).] The factors $\omega(r, s)$ and $\omega'(r, s)$ are *equivalent local factors*. The local exponents $\xi(r, s)$ and $\xi'(r, s)$ are related by

$$\xi'(r, s) = \xi(r, s) + \zeta(r) + \zeta(s) - \zeta(rs). \quad (12\text{--}65)$$

We shall choose $U(e)$ to be the identity operator 1. If we set $r = s = e$ in (12–57), we find

$$U(e)U(e) = \omega(e, e)U(e),$$

so that $\omega(e, e) = 1$ and $\xi(e, e) = 0$. If we set $s = e$ in (12–57), we find

$$U(r)U(e) = \omega(r, e)U(r),$$

so that

$$\omega(r, e) = 1, \qquad \xi(r, e) = 0. \tag{12–66}$$

Similarly,

$$\omega(e, r) = 1, \qquad \xi(e, r) = 0. \tag{12–66a}$$

Using (12–57) and the associative law $(rs)t = r(st)$, we find

$$\omega(r, s)\omega(rs, t) = \omega(r, st)\omega(s, t) \tag{12–67}$$

or

$$\xi(r, s) + \xi(rs, t) = \xi(r, st) + \xi(s, t). \tag{12–67a}$$

If the Hilbert space of the representation (12–57) is finite-dimensional, the operators $U(r)$ are n-by-n unitary matrices. If we let det $U(r) = D(r)$, then $D(r)$ is continuous and $|D(r)| = 1$. Taking the determinant of (12–57), we find

$$D(r)D(s) = e^{in\xi(r,s)}D(rs), \tag{12–68}$$

where n is the dimension of the representation. If we let

$$U'(r) = \frac{U(r)}{[D(r)]^{1/n}}, \tag{12–69}$$

and divide (12–57) by (12–68), we find

$$U'(r)U'(s) = U'(rs). \tag{12–70}$$

Thus finite-dimensional ray representations are *always* equivalent to vector representations. In deriving this result, we assume that r and s are near the identity e. In (12–69), we must select a particular nth root of $D(r)$. So long as we remain in the neighborhood of the identity, the result (12–70) is valid. If the parameter space of the group is simply connected, any closed curve in the space can be contracted to a point. Then the root $[D(r)]^{1/n}$ can be defined uniquely along any path and will return to its initial value when r describes a closed path. The argument fails for the three-dimensional rotation group. The parameter space is doubly connected, and for even n we obtain instead of (12–70) the result

$$U'(r)U'(s) = \pm U'(rs). \tag{12–70a}$$

Two local exponents ξ and ξ', which satisfy (12–65), are said to be *equivalent*,

$$\xi \equiv \xi'. \tag{12–71}$$

If $\xi \equiv \xi'$, then $\xi' \equiv \xi$. If $\xi \equiv \xi'$ and $\xi' \equiv \xi''$, so that

$$\xi'(r, s) = \xi(r, s) + \zeta_1(r) + \zeta_1(s) - \zeta_1(rs),$$

$$\xi''(r, s) = \xi'(r, s) + \zeta_2(r) + \zeta_2(s) - \zeta_2(rs),$$

then

$$\xi''(r, s) = \xi(r, s) + \zeta_1(r) + \zeta_2(r) + \zeta_1(s) + \zeta_2(s) - \zeta_1(rs) - \zeta_2(rs)$$
$$= \xi(r, s) + \zeta(r) + \zeta(s) - \zeta(rs),$$

where

$$\zeta(r) = \zeta_1(r) + \zeta_2(r),$$

so that $\xi \equiv \xi''$. We therefore get a separation of the local exponents into equivalence classes.

If ξ_1 and ξ_2 are local exponents, then $\xi = \alpha\xi_1 + \beta\xi_2$, with real α, β, is also a local exponent. Moreover, it is clear that the class to which ξ belongs is independent of the choice of ξ_1 and ξ_2 within their classes. Thus the nonequivalent classes of local exponents form a linear vector space with real coefficients. The dimensionality of this vector space determines the number of essentially different types of ray representations. If the dimensionality is zero, all local exponents are equivalent to $\xi = 0$, so that the ray representations all reduce to vector representations.

Next we consider a one-parameter subgroup of the Lie group G. From Section 8–11 we know that we can introduce a canonical parameter θ so that

$$r(0) = e, \qquad r(\theta_1)r(\theta_2) = r(\theta_1 + \theta_2). \qquad (12\text{–}72)$$

We denote the unitary operator $U(r)$ for $r = r(\theta)$ by $U(\theta)$. The infinitesimal Hermitian operator u which generates $U(\theta)$ is

$$iu = \frac{dU(\theta)}{d\theta}\bigg|_{\theta=0} \qquad (12\text{–}73)$$

If r and s in Eq. (12–57) correspond to parameter values θ_1 and θ_2, we can write Eq. (12–57) as

$$U(\theta_1)U(\theta_2) = \omega(\theta_1, \theta_2)U(\theta_1 + \theta_2). \qquad (12\text{–}74)$$

Differentiating with respect to θ_1 and setting $\theta_1 = 0$, we have

$$iuU(\theta_2) = \frac{d\omega(\theta_1, \theta_2)}{d\theta_1}\bigg|_{\theta_1=0} \cdot U(\theta_2) + \omega(0, \theta_2)\frac{dU(\theta_2)}{d\theta_2}$$

$$= f(\theta_2)U(\theta_2) + \frac{dU(\theta_2)}{d\theta_2}.$$

From (12–66a), $\omega(0, \theta_2) = 1; \omega(\theta_1, \theta_2) = \exp[i\xi(\theta_1, \theta_2)]$, so

$$\frac{d\omega}{d\theta_1} = i \frac{d\xi}{d\theta_1} \omega(\theta_1, \theta_2).$$

Setting $\theta_1 = 0$, we find

$$f(\theta_2) = i \frac{d\xi}{d\theta_1}\bigg]_{\theta_1=0}.$$

Thus f is a pure imaginary, continuous function of θ, and $f(0) = 0$. Dropping the subscript on θ_2 and setting $f = ig$, where g is real, we find that our differential equation for $U(\theta)$ is

$$\frac{dU}{d\theta} + igU = iuU. \tag{12–75}$$

Introducing the unitary operator

$$U'(\theta) = \exp\left[-i\int_0^\theta g(\theta')\,d\theta'\right] \cdot U(\theta), \tag{12–76}$$

we obtain

$$\frac{dU'}{d\theta} = iuU'. \tag{12–77}$$

Integration of this equation gives

$$U'(\theta) = e^{iu\theta} \tag{12–78}$$

which satisfies

$$U'(\theta_1)U'(\theta_2) = U'(\theta_1 + \theta_2). \tag{12–79}$$

Thus, for a one-parameter subgroup, the ray representation can always be replaced by an equivalent vector representation. This result is the analog of the result we found in Section 12–1 for cyclic groups.

It can be shown that it is possible to construct *canonical exponents*; i.e., for any $\xi'(r, s)$ we can find an equivalent exponent $\xi(r, s)$ for which $\xi(r, s) = 0$ if r and s are in the same one-parameter subgroup.

An n-parameter Lie group is generated by the infinitesimal elements of n one-parameter subgroups. In Chapter 8, we showed that the n infinitesimal elements X_i ($i = 1, \ldots, n$) form the basis for the Lie algebra of the group. The multiplication law for the algebra is

$$[X_i, X_j] = c_{ij}^k X_k, \tag{12–80}$$

where the real numbers c_{ij}^k are the structure constants of the group. The finite transformations of the one-parameter subgroup generated by X_i

are exp (θX_i). In Chapter 8, we showed that the representations of the group G can be determined, by integration, from the representations of its Lie algebra. If the Hermitian operators D_i represent the basis elements X_i in a vector representation, the operators satisfy the commutation equations

$$i[D_i, D_j] = c_{ij}^k D_k. \tag{12-81}$$

The operators for the finite transformations will then be unitary operators exp $(i\theta D_i)$.

The phase factors $\omega(r, s)$ which occur in the ray representations of the group will manifest themselves by adding constants to the commutator equations (12–81). The Hermitian operators D_i for a unitary ray representation will satisfy the equations

$$i[D_i, D_j] = c_{ij}^k D_k + \beta_{ij} \cdot 1, \tag{12-82}$$

where the β_{ij} are $n(n - 1)/2$ real constants. From (12–82) we see that $\beta_{ij} = -\beta_{ji}$. If we replace the operators D_i by

$$D_i' = D_i + \alpha_i \cdot 1 \qquad (\alpha_i \text{ real}), \tag{12-83}$$

then the operators for finite transformations, exp $(i\theta D_i)$, are replaced by

$$\exp\{i\theta(D_i + \alpha_i)\} = \exp(i\alpha_i\theta) \cdot \exp(i\theta D_i),$$

so that they are multiplied by phase factors. The effect on (12–82) of the shift (12–83) to an equivalent set of phase factors for the finite transformations is to replace it by

$$i[D_i', D_j'] = c_{ij}^k D_k' + \beta_{ij}' \cdot 1, \tag{12-82a}$$

where

$$\beta_{ij}' = \beta_{ij} - \alpha_k c_{ij}^k. \tag{12-84}$$

Two sets of constants β_{ij}, β_{ij}' which are related by (12–84) are equivalent. Their representations are transformed into each other by (12–83).

The constants β_{ij} in (12–82) are in general not independent. If we write the Jacobi identity (8–57) for the operators D_i, D_j, D_k,

$$[D_i, [D_j, D_k]] + [D_j, [D_k, D_i]] + [D_k, [D_i, D_j]] = 0, \tag{12-85}$$

and substitute from (12–82), we find the conditions

$$c_{ij}^k \beta_{lk} + c_{jl}^k \beta_{ik} + c_{li}^k \beta_{jk} = 0. \tag{12-86}$$

If the sets of constants $\beta_{ij}^{(1)}$ and $\beta_{ij}^{(2)}$ satisfy (12–86), then

$$\beta_{ij} = \alpha_1 \beta_{ij}^{(1)} + \alpha_2 \beta_{ij}^{(2)} \tag{12-87}$$

also satisfies (12–86). Thus the admissible β_{ij}'s are vectors in a real vector space of dimension $\leq n(n-1)/2$. The ray representations (12–82) will all be equivalent to vector representations if the admissible β_{ij}'s are equivalent to the null vector, i.e., if the dimensionality of the vector space of the β_{ij}'s is zero.

First we consider some simple examples. For an n-parameter abelian group, the structure constants c_{ij}^k are all zero. The infinitesimal elements satisfy the equations

$$[X_i, X_j] = 0 \qquad (i, j = 1, \ldots, n). \tag{12–88}$$

The corresponding equations (12–82) for the operators of the representation are

$$i[D_i, D_j] = \beta_{ij} \cdot 1. \tag{12–89}$$

Since the structure constants vanish, Eq. (12–86) imposes no restriction on the β_{ij}'s. The dimensionality of the vector space of the β_{ij}'s is $n(n-1)/2$. Any vector β_{ij} can appear in (12–89). The positive multiples of a given β_{ij} correspond to equivalent representations, since

$$i[mD_i, mD_j] = m^2 \beta_{ij} \cdot 1, \tag{12–90}$$

and the Hermitian operator mD_i will give finite transformations

$$\exp(im\theta D_i),$$

so that the representations differ only in a scale change in the variable θ. Only the null vector $\beta_{ij} = 0$ corresponds to the vector representations.

For the rotation group in three dimensions, the Hermitian angular-momentum operators satisfy

$$i[J_x, J_y] = J_z + \beta_z \cdot 1, \qquad i[J_y, J_z] = J_x + \beta_x \cdot 1,$$
$$i[J_z, J_x] = J_y + \beta_y \cdot 1. \tag{12–91}$$

The β's are equivalent to zero since the transformation (12–83),

$$J_s' = J_s + \beta_s \cdot 1,$$

reduces (12–91) to

$$i[J_x', J_y'] = J_z', \qquad i[J_y', J_z'] = J_x', \qquad i[J_z', J_x'] = J_y'. \tag{12–92}$$

So we have another proof that the ray representations of the three-dimensional rotation group are all equivalent to vector representations.

Instead of considering the quantities β_{ij}, we can introduce a function which is defined for all elements of the Lie algebra. The elements of the

Lie algebra are linear combinations of the basis elements X_i with real coefficients,

$$a = a_m X_m, \qquad b = b_m X_m, \tag{12–93}$$

and the operators representing these elements are

$$A = a_m D_m, \qquad B = b_m D_m. \tag{12–94}$$

For the elements a and b, Eq. (12–80) becomes

$$[a, b] = c_{lm}^k a_l b_m X_k, \tag{12–95}$$

and Eq. (12–82) yields

$$
\begin{aligned}
i[A, B] &= c_{lm}^k a_l b_m D_k + \beta_{lm} a_l b_m \cdot 1 \\
&= c_{lm}^k a_l b_m D_k + F(a, b) \cdot 1,
\end{aligned}
\tag{12–96}
$$

where

$$F(a, b) = \beta_{lm} a_l b_m = -F(b, a). \tag{12–97}$$

In particular, if $a = X_i$ and $b = X_j$,

$$F(X_i, X_j) = \beta_{ij}. \tag{12–98}$$

If the operators D_m are replaced by the equivalent operators $D'_m = D_m + \alpha_m \cdot 1$, so that A and B in (12–94) are correspondingly changed, Eq. (12–96) becomes

$$
\begin{aligned}
i[A', B'] &= c_{lm}^k a_l b_m D'_k + [F(a, b) - \alpha_k c_{lm}^k a_l b_m] \cdot 1 \\
&= c_{lm}^k a_l b_m D'_k + F'(a, b) \cdot 1,
\end{aligned}
\tag{12–99}
$$

where

$$F'(a, b) = F(a, b) - \alpha_k c_{lm}^k a_l b_m. \tag{12–100}$$

The shift from one representation to another equivalent representation is described by the change from F to F', where α_k are arbitrary real constants. If the α_k can be chosen so that $F'(a, b) = 0$ for all a and b, the representation is equivalent to a vector representation.

Equation (12–100) can be simplified as follows: Consider a linear function Λ defined on the Lie algebra so that

$$\Lambda(X_i) = \alpha_i. \tag{12–101}$$

Then

$$\Lambda(a) = \alpha_i a_i, \tag{12–102}$$

and using (12–95), we obtain

$$\Lambda([a, b]) = \alpha_k c_{lm}^k a_l b_m. \tag{12–103}$$

Comparing with (12–100), we have

$$F'(a, b) = F(a, b) - \Lambda([a, b]). \qquad (12\text{–}104)$$

Thus, two systems of exponents will be equivalent if the difference between the corresponding functions $F'(a, b)$ and $F(a, b)$ is a linear function of the commutator $[a, b]$.

We can find conditions for the function F corresponding to the conditions (12–86) on the constants β_{ij} by writing the Jacobi identity

$$[A, [B, C]] + [B, [C, A]] + [C, [A, B]] = 0$$

and substituting from (12–96). The result is

$$F(a, [b, c]) + F(b, [c, a]) + F(c, [a, b]) = 0. \qquad (12\text{–}105)$$

12–4 Ray representations of the pseudo-orthogonal groups. The real homogeneous linear transformations which leave the real, nonsingular quadratic form

$$F(\mathbf{x}) = g_{ik}x_i x_k \qquad (g_{ik} = g_{ki}) \qquad (12\text{–}106)$$

in the n real variables x_i $(i = 1, \ldots, n)$ invariant are called *pseudo-orthogonal transformations*. By a suitable change of coordinate basis, the form F can be brought to principal axes, so that

$$F_p(\mathbf{x}) = \sum_{i=1}^{p} x_i^2 - \sum_{i=p+1}^{n} x_i^2 = \epsilon_i\, \delta_{ik} x_i x_k$$

$$\left[\begin{array}{ll} \epsilon_i = +1 & \text{for } i = 1, \ldots, p \\ \epsilon_i = -1 & \text{for } i = p+1, \ldots, n \end{array} \right].$$

$$(12\text{–}107)$$

The group of pseudo-orthogonal transformations $\mathbf{W}$,

$$\mathbf{x}' = \mathbf{W}\mathbf{x}, \qquad x_i' = w_{ik}x_k \qquad (i = 1, \ldots, n), \qquad (12\text{–}108)$$

which leave the form F_p invariant is denoted by G_n^p. From (12–107), we see that G_n^p and G_n^{n-p} are the same group.

If $p = n$, the form is positive definite, and G_n^n is the orthogonal group in n dimensions, O_n. For $p \neq n$, we get groups analogous to the Lorentz group (for which $n = 4$, $p = 3$).

The transformations of G_n^p which can be reached continuously from the identity form a subgroup $G_n'^p$ of *proper* pseudo-orthogonal transformations. A transformation $\mathbf{W}$ belongs to $G_n'^p$ if its determinant is equal to 1, and if the determinant of its first p rows and columns is positive.

Problem. Prove the preceding statement.

We can also discuss the groups I_n^p of *inhomogeneous pseudo-orthogonal transformations,*

$$\mathbf{x}' = \mathbf{W}\mathbf{x} + \mathbf{u}, \qquad x_i' = w_{ij}x_j + u_i \qquad (i = 1, \ldots, n), \quad (12\text{–}109)$$

where $\mathbf{u}$ is an arbitrary real vector (translation). If $\mathbf{W}$ leaves the form $F_p(\mathbf{x})$ of Eq. (12–107) invariant, the transformation (12–109) leaves the form $F_p(\mathbf{x} - \mathbf{y})$ invariant, where $\mathbf{x}$ and $\mathbf{y}$ are arbitrary vectors. If we restrict $\mathbf{W}$ to $G_n''^p$, I_n^p is restricted to its proper subgroup $I_n'^p$.

If the transformation (12–109) is followed by a second transformation

$$\mathbf{x}'' = \mathbf{W}'\mathbf{x}' + \mathbf{u}',$$

the resultant transformation is

$$\mathbf{x}'' = \mathbf{W}'(\mathbf{W}\mathbf{x} + \mathbf{u}) + \mathbf{u}'$$
$$= \mathbf{W}'\mathbf{W}\mathbf{x} + (\mathbf{u}' + \mathbf{W}'\mathbf{u}).$$

Thus, if we denote the transformation (12–109) by $(\mathbf{W}, \mathbf{u})$, the group multiplication in I_n^p is defined by

$$(\mathbf{W}', \mathbf{u}')(\mathbf{W}, \mathbf{u}) = (\mathbf{W}'\mathbf{W}, \mathbf{W}'\mathbf{u} + \mathbf{u}'). \qquad (12\text{–}110)$$

It is convenient to interpret this law of combination as the matrix multiplication of matrices of order $n + 1$: We associate with the transformation $(\mathbf{W}, \mathbf{u})$ the matrix

$$\begin{bmatrix} W & u \\ 0 & 1 \end{bmatrix}, \qquad (12\text{–}111)$$

where the last column consists of the components of $\mathbf{u}$ followed by the constant 1. The element $(1, 0)$ is the identity element of I_n^p.

Problems. (1) Verify the group-multiplication law (12–110) for the matrices (12–111).

(2) Show that the inverse of $(W, \mathbf{u})$ is $(W^{-1}, -W^{-1}\mathbf{u})$.

(3) Show that the translations $(1, u)$ form an invariant subgroup T_n in I_n^p, and that

$$\frac{I_n^p}{T_n} \simeq G_n^p.$$

We wish to construct the Lie algebras of the groups G_n^p and I_n^p. This is done most simply in terms of the matrices (12–111). The "rotations" in the ij-plane form a complete set of one-parameter subgroups of G_n^p. The "rotation" $W^{(ij)}$ in the ij-plane is a transformation acting only on the variables x_i, x_j:

$$W^{(ij)}: \qquad \begin{aligned} x_i' &= w_{ii}x_i + w_{ij}x_j \\ x_j' &= w_{ji}x_i + w_{jj}x_j \\ x_r' &= x_r \quad \text{for } r \neq i, j; \end{aligned} \qquad \text{(no summation!),}$$

$$\tag{12-112}$$

$$\epsilon_i x_i'^2 + \epsilon_j x_j'^2 = \epsilon_i x_i^2 + \epsilon_j x_j^2. \tag{12-113}$$

If $\epsilon_i = \epsilon_j = \pm 1$, then $W^{(ij)}$ is a rotation:

$$\begin{aligned} x_i' &= x_i \cos\theta + x_j \sin\theta, \\ x_j' &= -x_i \sin\theta + x_j \cos\theta, \\ x_r' &= x_r \quad \text{for } r \neq i, j. \end{aligned} \tag{12-114}$$

The corresponding matrix (12–111) is

$$\begin{bmatrix} W^{(ij)} & 0 \\ 0 & 1 \end{bmatrix}, \tag{12-115}$$

where the matrix $W^{(ij)}$ is

$$
\begin{array}{cc}
 & \quad i \qquad\qquad j \\
\begin{array}{c} \\ \\ \\ i \\ \\ \\ j \\ \\ \\ \\ \end{array}
&
\left[
\begin{array}{ccccccc}
1 & & & \vdots & & \vdots & \\
 & 1 & & \vdots & & \vdots & \\
 & & \ddots & & & & \\
\cdots & \cos\theta & \cdots & & \sin\theta & & \\
 & & \ddots & \ddots & \vdots & & \\
 & & & & \vdots & & \\
\cdots & -\sin\theta & \cdots & & \cos\theta & & \\
 & & & & & \ddots & \\
 & & & & & & 1 \\
 & & & & & & \quad 1
\end{array}
\right]
\end{array}
\tag{12-116}
$$

To find the corresponding infinitesimal matrix we evaluate $dW^{(ij)}/d\theta$ at $\theta = 0$:

$$w^{(ij)} = \frac{dW^{(ij)}}{d\theta}\bigg|_{\theta=0} \quad \text{has} \quad \begin{array}{l} +1 \text{ at the } (ij)\text{-position,} \\ -1 \text{ at the } (ji)\text{-position,} \\ 0 \text{ elsewhere.} \end{array} \tag{12-117}$$

If $\epsilon_i = -\epsilon_j$, the "rotation" $W^{(ij)}$ is a Lorentz transformation:

$$x'_i = x_i \cosh \theta - x_j \sinh \theta,$$
$$x'_j = -x_i \sinh \theta + x_j \cosh \theta,$$
$$x'_r = x_r \quad \text{for } r \neq i, j, \tag{12-118}$$

and the corresponding infinitesimal matrix has -1 at the (ij)- and (ji)-positions, and zero elsewhere.

Both cases can be taken care of by choosing the infinitesimal matrices in the form

$$
\begin{aligned}
(w^{(ij)})_{kl} &= -\epsilon_i \, \delta_{il} \, \delta_{jk} + \epsilon_j \, \delta_{jl} \, \delta_{ik} \\
&= -g_{il} \, \delta_{jk} + g_{jl} \, \delta_{ik}, \tag{12-119}
\end{aligned}
$$

where

$$g_{il} = \epsilon_i \, \delta_{il}, \qquad g_{il}g_{im} = \delta_{lm}, \qquad g_{il}g_{il} = n. \tag{12-120}$$

We denote by a_{ij} the element of the Lie algebra which corresponds to the matrix $w^{(ij)}$. We can evaluate the commutator $[a_{ij}, a_{kl}]$ by using the corresponding matrices (12-119). We find

$$[a_{ij}, a_{kl}] = g_{jk}a_{il} - g_{ik}a_{jl} + g_{il}a_{jk} - g_{jl}a_{ik},$$
$$a_{ji} = -a_{ij} \qquad (i, j, k, l = 1, \ldots, n). \tag{12-121}$$

The infinitesimal matrix for a translation along the i-axis will have a one in the $(i, n + 1)$-position and zero elsewhere:

$$(t^{(i)})_{kl} = \delta_{ik} \, \delta_{l,n+1}. \tag{12-122}$$

We denote the corresponding element of the Lie algebra by b_i. The translations commute with one another, so that

$$[b_i, b_k] = 0 \qquad \text{for all } i, k, \tag{12-123}$$

and, using (12-119) and (12-122), we find

$$[a_{ij}, b_k] = g_{jk}b_i - g_{ik}b_j. \tag{12-124}$$

Equation (12-121) describes completely the Lie algebra of G_n^p. Adding Eqs. (12-123) and (12-124), we obtain the Lie algebra of I_n^p.

Multiplying (12-121) by g_{jk} and using (12-120), we find

$$g_{jk}[a_{ij}, a_{kl}] = (n - 2)a_{il} \tag{12-125}$$

($g_{jk}a_{jk} = 0$ because g_{jk} is symmetric and a_{jk} is antisymmetric). Similarly, from (12-124),

$$g_{jk}[a_{ij}, b_k] = (n - 1)b_i. \tag{12-126}$$

First we consider the homogeneous groups G_n^p. For $n = 2$, the Lie algebra has a single element a_{12}. The group is cyclic, and all representations are vector representations. For $n > 2$, we use (12–125) to transform the function F defined in Eq. (12–97):

$$\begin{aligned}
(n - 2)F(a_{ij}, a_{kl}) &= F([g_{rh}a_{ir}, a_{hj}], a_{kl}) \\
&= F([g_{rh}a_{ir}, a_{kl}], a_{hj}) - F([a_{hj}, a_{kl}], g_{rh}a_{ir}) \\
&= F(g_{rh}a_{jh}, [a_{ir}, a_{kl}]) + F(g_{rh}a_{ir}, [a_{hj}, a_{kl}]).
\end{aligned}$$

Now substitute for the commutators from (12–121):

$$(n - 2)F(a_{ij}, a_{kl}) = g_{rh}\left\{\begin{array}{l}
g_{rk}F(a_{jh}, a_{il}) - g_{ik}F(a_{jh}, a_{rl}) \\
+ g_{il}F(a_{jh}, a_{rk}) - g_{rl}F(a_{jh}, a_{ik}) \\
+ g_{jk}F(a_{ir}, a_{hl}) - g_{hk}F(a_{ir}, a_{jl}) \\
+ g_{hl}F(a_{ir}, a_{jk}) - g_{jl}F(a_{ir}, a_{hk})
\end{array}\right\}.$$

The first and seventh terms, and the fourth and sixth terms cancel, leaving

$$\begin{aligned}
(n - 2)F(a_{ij}, a_{kl}) = {} & g_{jk}F(g_{rh}a_{ir}, a_{hl}) - g_{ik}F(g_{rh}a_{jh}, a_{rl}) \\
& + g_{il}F(g_{rh}a_{jh}, a_{rk}) - g_{jl}F(g_{rh}a_{ir}, a_{hk}). \quad (12\text{–}127)
\end{aligned}$$

We define a linear function on the Lie algebra by setting

$$\Lambda(a_{st}) = g_{rh}F(a_{sr}, a_{ht}). \quad (12\text{–}128)$$

Then, according to (12–121),

$$\begin{aligned}
(n - 2)F(a_{ij}, a_{kl}) &= g_{jk}\Lambda(a_{il}) - g_{ik}\Lambda(a_{jl}) \\
& \quad + g_{il}\Lambda(a_{jk}) - g_{jl}\Lambda(a_{ik}) \\
&= \Lambda([a_{ij}, a_{kl}]). \quad (12\text{–}129)
\end{aligned}$$

Thus $F(a_{ij}, a_{kl})$ is a linear function of the commutator $[a_{ij}, a_{kl}]$, and is therefore equivalent to zero.

Next we consider the inhomogeneous groups I_n^p for $n > 2$. We first use Eq. (12–126) to transform the expression for $F(a_{ij}, b_k)$:

$$\begin{aligned}
(n - 1)F(a_{ij}, b_k) &= g_{hl}F(a_{ij}, [a_{kl}, b_h]) \\
&= g_{hl}\{F([a_{ij}, a_{kl}], b_h) + F(a_{kl}, [a_{ij}, b_h])\};
\end{aligned}$$

then we use Eqs. (12–121) and (12–124) to expand the commutators which appear as arguments of F:

$$(n - 1)F(a_{ij}, b_k) = g_{hl}\left\{\begin{array}{l}
g_{jk}F(a_{il}, b_h) - g_{ik}F(a_{jl}, b_h) \\
+ g_{il}F(a_{jk}, b_h) - g_{jl}F(a_{ik}, b_h) \\
+ g_{jh}F(a_{kl}, b_i) - g_{ih}F(a_{kl}, b_j)
\end{array}\right\}.$$

The third and fifth terms, and the fourth and sixth terms cancel, leaving

$$(n - 1)F(a_{ij}, b_k) = g_{jk}F(g_{hl}a_{il}, b_h) - g_{ik}F(g_{hl}a_{jl}, b_h). \quad (12\text{-}130)$$

If we define the linear function Λ for the basis elements b_i so that

$$\Lambda(b_i) = g_{hl}F(a_{il}, b_h), \quad (12\text{-}131)$$

we have, from Eq. (12-124),

$$(n - 1)F(a_{ij}, b_k) = g_{jk}\Lambda(b_i) - g_{ik}\Lambda(b_j)$$
$$= \Lambda([a_{ij}, b_k]). \quad (12\text{-}132)$$

Since $F(a_{ij}, b_k)$ is a linear function of the commutator $[a_{ij}, b_k]$,

$$F(a_{ij}, b_k) \equiv 0.$$

Finally we must show that $F(b_i, b_j) \equiv 0$. The linear function Λ is already prescribed by (12-128) and (12-131), so we can only check to see whether $F(b_i, b_j) \equiv 0$. We transform $F(b_i, b_j)$ by using (12-124):

$$(n - 1)F(b_i, b_j) = g_{hk}F([a_{ih}, b_k], b_j)$$
$$= g_{hk}\{F([a_{ih}, b_j], b_k) + F([b_j, b_k], a_{ih})\}.$$

The second term is zero since $[b_j, b_k] = 0$. In the first term we use (12-124):

$$(n - 1)F(b_i, b_j) = g_{hk}\{g_{hj}F(b_i, b_k) - g_{ij}F(b_h, b_k)\}$$
$$= F(b_i, b_j) - g_{ij}g_{hk}F(b_h, b_k) = F(b_i, b_j)$$

$[g_{hk}F(b_h, b_k) = 0$ since it is the product of a symmetric factor g_{hk} and an antisymmetric factor $F(b_h, b_k)]$. Thus,

$$(n - 2)F(b_i, b_j) = 0$$

and, for $n > 2$,

$$F(b_i, b_j) = 0. \quad (12\text{-}133)$$

We have shown that, for $n > 2$, all ray representations of I_n^p are equivalent to vector representations. Just as for the rotation group, there may still be multivalued representations.

For $n = 2$, the Lie algebra of I_2^p is defined by

$$[a_{12}, b_1] = -\epsilon_1 b_2,$$
$$[a_{12}, b_2] = \epsilon_2 b_1, \quad (12\text{-}134)$$
$$[b_1, b_2] = 0.$$

The Jacobi condition yields nothing in this special case. If we denote the operators of a representation by A_{12}, B_1, B_2, their commutators are

$$i[A_{12}, B_1] = -\epsilon_1 B_2 + \beta' \cdot 1, \qquad i[A_{12}, B_2] = \epsilon_2 B_1 + \beta'' \cdot 1,$$
$$i[B_1, B_2] = \beta \cdot 1, \tag{12-135}$$

where β, β', β'' are real constants. We use the change of phase (12–83) and introduce operators

$$B_1' = B_1 + \frac{\beta''}{\epsilon_2} \cdot 1, \qquad B_2' = B_2 - \frac{\beta'}{\epsilon_1} \cdot 1. \tag{12-136}$$

Then

$$i[A_{12}, B_1'] = -\epsilon_1 B_2', \qquad i[A_{12}, B_2'] = \epsilon_2 B_1', \tag{12-137}$$

but we are left with the constant β in the equation

$$i[B_1', B_2'] = \beta \cdot 1. \tag{12-138}$$

Thus, for $n = 2$, there is a one-dimensional manifold of inequivalent ray representations. For $\beta = 0$, we get the vector representations of I_2^p.

12–5 Ray representations of the Galilean group. The Galilean group G is the group of transformations which connect inertial frames of reference. Such frames of reference can be displaced from one another, shifted in time, rotated in space, and finally, they can move with constant relative velocity. Thus Galilean transformations change the coordinate vector $\mathbf{x}$ and the time t to $\mathbf{x}'$ and t', where

$$\mathbf{x}' = W\mathbf{x} + \mathbf{v}t + \mathbf{u},$$
$$t' = t + \tau. \tag{12-139}$$

In (12–139) W is a proper orthogonal transformation in three-dimensional space, $\mathbf{v}$ is the constant velocity vector, $\mathbf{u}$ is the vector describing the translation of the system, and τ is the time displacement. *Pure* Galilean transformations are the transformations

$$\mathbf{x}' = \mathbf{x} + \mathbf{v}t, \qquad t' = t. \tag{12-140}$$

Such a transformation is sometimes referred to as an "acceleration."

Since the rotations W are described by three parameters, $\mathbf{v}$ by three, $\mathbf{u}$ by three, and τ by one, the Galilean group G is a 10-parameter Lie group. If we denote the element (12–139) of G by $(W, \tau, \mathbf{v}, \mathbf{u})$, the multiplication law is

$$(W', \tau', \mathbf{v}', \mathbf{u}') \cdot (W, \tau, \mathbf{v}, \mathbf{u})$$
$$= (W'W, \tau + \tau', W'\mathbf{v} + \mathbf{v}', W'\mathbf{u} + \mathbf{u}' + \mathbf{v}'\tau), \tag{12-141}$$

and can be represented by the multiplication of the 5-by-5 matrices

$$\begin{bmatrix} W & \mathbf{v} & \mathbf{u} \\ 0 & 1 & \tau \\ 0 & 0 & 1 \end{bmatrix}. \tag{12-142}$$

The unit element of G is $(1, 0, 0, 0)$, and the inverse of $(W, \tau, \mathbf{v}, \mathbf{u})$ is $(W^{-1}, -\tau, -W^{-1}\mathbf{v}, -W^{-1}\{\mathbf{u} - \mathbf{v}\tau\})$. The infinitesimal matrices have the form

$$\begin{bmatrix} S & \mathbf{k}' & \mathbf{k} \\ 0 & 0 & \phi \\ 0 & 0 & 0 \end{bmatrix}, \tag{12-143}$$

where S is a 3-by-3 skew-symmetric matrix, $\mathbf{k}'$ and $\mathbf{k}$ are arbitrary column vectors, and ϕ is a real number. We denote the basis elements of the Lie algebra by a_{ij}, b_i (as in the preceding section), the pure Galilean transformations by d_i, and the time displacement by f. The commutator equations are:

$$[a_{ij}, a_{kl}] = \delta_{jk}a_{il} - \delta_{ik}a_{jl} + \delta_{il}a_{jk} - \delta_{jl}a_{ik}, \tag{12-144}$$

$$[a_{ij}, b_k] = \delta_{jk}b_i - \delta_{ik}b_j, \qquad [b_i, b_j] = 0; \tag{12-145}$$

$$[a_{ij}, d_k] = \delta_{jk}\, d_i - \delta_{ik}\, d_j, \qquad [d_i, d_j] = 0; \qquad [d_i, b_j] = 0; \tag{12-146}$$

$$[a_{ij}, f] = 0, \qquad [b_i, f] = 0, \qquad [d_i, f] = b_i. \tag{12-147}$$

Problem. Verify Eqs. (12–144) through (12–147).

In investigating the ray representations of the Galilean group, we can make use of the results of the preceding section for $n = 3$. The rotations plus translations, and the rotations plus pure Galilean transformations form subgroups of G which are isomorphic to G_3^3. From the results of the last section we can always adjust the function F of Eq. (12–97) so that

$$F(a_{ij}, a_{kl}) = F(a_{ij}, b_k) = F(a_{ij}, d_k) = F(b_i, b_j)$$
$$= F(d_i, d_j) = 0. \tag{12-148}$$

From Eq. (12–125), for $n = 3$, and $g_{jk} = \delta_{jk}$,

$$a_{il} = [a_{ik}, a_{kl}], \tag{12-149}$$

so that

$$F(a_{il}, f) = F([a_{ik}, a_{kl}], f)$$
$$= F(a_{ik}, [a_{kl}, f]) + F([a_{ik}, f], a_{kl}) = 0, \tag{12-150}$$

since the arguments are zero according to Eq. (12–147). Similarly, from Eq. (12–126),

$$2b_i = [a_{ik}, b_k], \qquad (12\text{–}151)$$

so that

$$F(b_i, f) = \tfrac{1}{2}F([a_{ik}, b_k], f).$$

Again, using the Jacobi relation (12–105) for F, and Eqs. (12–147), we find $F(b_i, f) = 0$. Similarly, $F(d_i, f) = 0$.

The only remaining quantity to be examined is $F(b_i, d_k)$. From (12–126),

$$2F(b_i, d_k) = F([a_{ij}, b_j], d_k)$$
$$= F([a_{ij}, d_k], b_j) + F([d_k, b_j], a_{ij}).$$

Using Eqs. (12–146), we have

$$2F(b_i, d_k) = \delta_{jk}F(d_i, b_j) - \delta_{ik}F(d_j, b_j)$$
$$= F(d_i, b_k) - \delta_{ik}F(d_j, b_j)$$
$$= -F(b_k, d_i) - \delta_{ik}F(d_j, b_j). \qquad (12\text{–}152)$$

Interchanging i and k, we obtain

$$2F(b_k, d_i) = -F(b_i, d_k) - \delta_{ik}F(d_j, b_j),$$

and substituting on the right of (12–152), we find

$$F(b_i, d_k) = -\tfrac{1}{3}\, \delta_{ik}F(d_j, b_j) = \gamma\, \delta_{ik}, \qquad (12\text{–}153)$$

where

$$\gamma = -\tfrac{1}{3}F(d_j, b_j) \qquad (12\text{–}154)$$

is a constant. For $\gamma \neq 0$, we cannot eliminate the constant from the corresponding operator equations

$$i[B_i, D_k] = \gamma\, \delta_{ik} \cdot 1. \qquad (12\text{–}155)$$

Thus the set of equivalence classes for the Galilean group is one-dimensional.

12–6 Irreducible representations of translation groups.

The groups which we discussed in the preceding sections included the Galilean group, the group I_3^3 of euclidean motions (rotations and translations) in three dimensions, and the inhomogeneous Lorentz group, I_4^3. In all these groups, the translations form an abelian invariant subgroup T_n. The simplest procedure for finding the representations of the full group is to consider first the representations of the translation group T_n.

T_n is the group of transformations

$$\mathbf{x}' = \mathbf{x} + \mathbf{u}, \qquad x_\nu' = x_\nu + u_\nu \qquad (\nu = 1, \ldots, n). \qquad (12\text{–}156)$$

We shall restrict ourselves to finding the irreducible unitary vector representations of T_n. If the operator representative of the translation $\mathbf{u}$ is $D(\mathbf{u})$, then

$$D(\mathbf{u})D(\mathbf{u}') = D(\mathbf{u} + \mathbf{u}'). \qquad (12\text{–}157)$$

Since the translation group T_n is abelian, all its irreducible vector representatives are one-dimensional. If ψ is the basis function of an irreducible unitary representation, then

$$D(\mathbf{u})\psi = e^{i\delta(\mathbf{u})}\psi, \qquad (12\text{–}158)$$

where the real number δ depends on the translation vector $\mathbf{u}$. If $\mathbf{n}_1$ is a unit translation along the x_1-axis,

$$D(\mathbf{n}_1)\psi = e^{ik_1}\psi, \qquad k_1 = \delta(\mathbf{n}_1),$$

and if $u_1\mathbf{n}_1$ is a translation of amount u_1 along the x_1-direction, then

$$D(u_1\mathbf{n}_1)\psi = e^{ik_1 u_1}\psi.$$

Proceeding similarly for $x_2, x_3, \ldots, x_n$, we find

$$D(\mathbf{u})\psi = e^{i\mathbf{k}\cdot\mathbf{u}}\psi \qquad \text{for } \mathbf{u} = \sum_1^n u_\nu\mathbf{n}_\nu, \qquad (12\text{–}159)$$

where $\mathbf{k}$ is a real vector.

Thus an irreducible unitary representation of T_n is characterized by a real vector $\mathbf{k}$ in the n-dimensional space. We shall indicate the representation by writing the basis function as $\psi(\mathbf{k})$ or $\psi_\mathbf{k}$. The general representation of T_n will be a direct sum of representations (12–159) for different $\mathbf{k}$'s.

In particular, we can realize the representations in terms of space functions $\psi(\mathbf{r})$ by defining the operators $D(\mathbf{u})$ to be

$$D(\mathbf{u})\psi(\mathbf{r}) = \psi(\mathbf{r} + \mathbf{u}), \qquad (12\text{–}160)$$

so that, with $\psi_\mathbf{k}(\mathbf{r}) = e^{i\mathbf{k}\cdot\mathbf{r}}$,

$$D(\mathbf{u})\psi_\mathbf{k}(\mathbf{r}) = \psi_\mathbf{k}(\mathbf{r} + \mathbf{u}) = e^{i\mathbf{k}\cdot\mathbf{u}}\psi_\mathbf{k}(\mathbf{r}). \qquad (12\text{–}161)$$

The resolution of $\psi(\mathbf{r})$ into basis functions $\psi_\mathbf{k}(\mathbf{r})$ is just the Fourier analysis

$$\psi(\mathbf{r}) = \int d\mathbf{k} a_\mathbf{k} e^{i\mathbf{k}\cdot\mathbf{r}} = \int d\mathbf{k} a_\mathbf{k}\psi_\mathbf{k}(\mathbf{r}), \qquad (12\text{–}162)$$

and

$$D(\mathbf{u})\psi(\mathbf{r}) = \int d\mathbf{k} a_\mathbf{k} e^{i\mathbf{k}\cdot\mathbf{u}}\psi_\mathbf{k}(\mathbf{r}). \qquad (12\text{–}162a)$$

Alternatively, we can consider the Hermitian infinitesimal operators p_ν of the representation which satisfy $[p_\nu, p_{\nu'}] = 0$. Since the operators p_ν all commute, we can determine simultaneous eigenfunctions $\psi_{\mathbf{k}}$ such that $\mathbf{p}\psi_{\mathbf{k}} = \mathbf{k}\psi_{\mathbf{k}}$, where $\mathbf{k}$ is any real vector.

In considering space groups in crystals, we find that the continuous translation group T_3 is replaced by a discrete group of translations along three noncoplanar directions. The translations which describe the lattice are given by the vectors

$$\mathbf{a} = n_1\mathbf{a}_1 + n_2\mathbf{a}_2 + n_3\mathbf{a}_3 \qquad (n_1, n_2, n_3 = 0, \pm1, \ldots), \qquad (12\text{--}163)$$

where $\mathbf{a}_1$, $\mathbf{a}_2$, $\mathbf{a}_3$ are not coplanar. The group is abelian and therefore has only one-dimensional irreducible vector representations. The argument given above shows that for each real vector $\mathbf{k}$, we obtain a one-dimensional representation,

$$D(\mathbf{a})\psi_{\mathbf{k}} = e^{i\mathbf{k}\cdot\mathbf{a}}\psi_{\mathbf{k}}.$$

But here, because of the discreteness of the group, the representations corresponding to different $\mathbf{k}$'s may be equivalent. Two vectors $\mathbf{k}$ and $\mathbf{k}'$ which differ by an amount

$$\mathbf{b} = m_1\mathbf{b}_1 + m_2\mathbf{b}_2 + m_3\mathbf{b}_3 \qquad (m_1, m_2, m_3 = 0, \pm1, \ldots), \qquad (12\text{--}164)$$

where

$$\mathbf{b}_1 = 2\pi \frac{\mathbf{a}_2 \times \mathbf{a}_3}{\mathbf{a}_1 \cdot \mathbf{a}_2 \times \mathbf{a}_3}, \qquad \mathbf{b}_2 = 2\pi \frac{\mathbf{a}_3 \times \mathbf{a}_1}{\mathbf{a}_1 \cdot \mathbf{a}_2 \times \mathbf{a}_3},$$

$$\mathbf{b}_3 = 2\pi \frac{\mathbf{a}_1 \times \mathbf{a}_2}{\mathbf{a}_1 \cdot \mathbf{a}_2 \times \mathbf{a}_3}, \qquad (12\text{--}165)$$

give equivalent representations, since $e^{i\mathbf{b}\cdot\mathbf{a}} = 1$. The vectors $\mathbf{b}_1, \mathbf{b}_2, \mathbf{b}_3$ are the basis vectors of the *reciprocal lattice*, and $\mathbf{b}$ is called a *reciprocal lattice vector*. To find a complete set of nonequivalent representations, we find the set of vectors $\mathbf{k}$ such that the length of the vector $\mathbf{k}$ is less than or equal to the length of any vector $\mathbf{k} + \mathbf{b}$. The vectors $\mathbf{k}$ found in this way fill the *first Brillouin zone*.

The basis functions $\psi_{\mathbf{k}}$ can be realized as coordinate functions,

$$\psi_{\mathbf{k}}(\mathbf{r}) = e^{i\mathbf{k}\cdot\mathbf{r}}\phi_{\mathbf{k}}(\mathbf{r}), \qquad (12\text{--}166)$$

where $\phi_{\mathbf{k}}(\mathbf{r})$ has the periodicity of the lattice,

$$\phi_{\mathbf{k}}(\mathbf{r} + \mathbf{a}) = \phi_{\mathbf{k}}(\mathbf{r}). \qquad (12\text{--}167)$$

We obtain a complete set by letting $\mathbf{k}$ run through the first Brillouin zone.

Problem. Draw the first Brillouin zone for a plane lattice in which the angle θ between $\mathbf{a}_1$ and $\mathbf{a}_2$ is 90°. Do the same for $|\mathbf{a}_1| = |\mathbf{a}_2|$ and $\theta = 120°$.

12–7 Little groups. We consider the transformation group I_n^p,

$$\mathbf{x}' = W\mathbf{x} + \mathbf{u}, \tag{12–109}$$

$$(W', \mathbf{u}')(W, \mathbf{u}) = (W'W, W'\mathbf{u} + \mathbf{u}'), \tag{12–110}$$

which leaves the form $F_p(\mathbf{x} - \mathbf{y})$ invariant, where

$$F_p(\mathbf{x}) = \sum_{i=1}^{n} \epsilon_i x_i^2. \tag{12–107}$$

Then the bilinear form ("scalar product")

$$\{\mathbf{x}, \mathbf{y}\} = \sum_{i=1}^{n} \epsilon_i x_i y_i$$

is left invariant by the homogeneous transformations W:

$$\{W\mathbf{x}, W\mathbf{y}\} = \{\mathbf{x}, \mathbf{y}\}. \tag{12–168}$$

If we have found a representation of I_n^p and then restrict ourselves to the subgroup T_n, the representation will decompose into a direct sum of one-dimensional representations of T_n. Let us consider a particular irreducible representation of T_n which is contained in the representation of I_n^p:

$$D(\mathbf{u})\psi(\mathbf{k}_0, \zeta) = e^{i\{\mathbf{k}_0, \mathbf{u}\}}\psi(\mathbf{k}_0, \zeta), \tag{12–169}$$

where ζ is an additional set of indices which may be needed to label the basis functions of the representation of I_n^p. The basis function $\psi(\mathbf{k}_0, \zeta)$ gives a one-dimensional subspace of the representation of I_n^p. We apply to (12–169) the operator $D(W)$ which represents the "rotation" $(W, 0)$. Since $(W, 0)(1, \mathbf{u}) = (1, W\mathbf{u})(W, 0)$,

$$D(W)D(\mathbf{u})\psi(\mathbf{k}_0, \zeta) = e^{i\{\mathbf{k}_0, \mathbf{u}\}}[D(W)\psi(\mathbf{k}_0, \zeta)]$$
$$= D(W\mathbf{u})[D(W)\psi(\mathbf{k}_0, \zeta)].$$

Letting $W\mathbf{u} = \mathbf{u}'$, we have

$$D(\mathbf{u}')[D(W)\psi(\mathbf{k}_0, \zeta)] = e^{i\{\mathbf{k}_0, W^{-1}\mathbf{u}'\}}[D(W)\psi(\mathbf{k}_0, \zeta)]$$
$$= e^{i\{W\mathbf{k}_0, \mathbf{u}'\}}[D(W)\psi(\mathbf{k}_0, \zeta)], \tag{12–170}$$

because of (12–168). Thus $D(W)\psi(\mathbf{k}_0, \zeta)$ is in the subspace of the representation of T_n which corresponds to the vector $W\mathbf{k}_0$. The vector $W\mathbf{k}_0$ is obtained by applying the transformation W to the components of the vector $\mathbf{k}_0$. According to (12–168), $\{W\mathbf{k}_0, W\mathbf{k}_0\} = \{\mathbf{k}_0, \mathbf{k}_0\}$. Consequently, if the space of an irreducible representation of I_n^p contains the representation of T_n corresponding to $\mathbf{k}_0$, it must also contain the representations of T_n corresponding to all vectors $\mathbf{k}$ having the same "length" as $\mathbf{k}_0$, that is, all vectors $\mathbf{k}$ with $\{\mathbf{k}, \mathbf{k}\} = \{\mathbf{k}_0, \mathbf{k}_0\}$ which can be reached by transformations W.

We now consider the subgroup $G_{\mathbf{k}_0}$ which consists of all "rotations" $W_{\mathbf{k}_0}$ that leave the vector $\mathbf{k}_0$ unchanged:

$$W_{\mathbf{k}_0}\mathbf{k}_0 = \mathbf{k}_0. \tag{12–171}$$

The group $G_{\mathbf{k}_0}$ is called the *group of the wave vector* $\mathbf{k}_0$, or the *little group*.

From Eq. (12–170), we see that the functions $D(W_{\mathbf{k}_0})\psi(\mathbf{k}_0, \zeta)$ also belong to the vector $\mathbf{k}_0$. We consider this subspace and find an irreducible representation of the little group $G_{\mathbf{k}_0}$:

$$D(W_{\mathbf{k}_0})\psi(\mathbf{k}_0, \zeta) = \sum_\eta \psi(\mathbf{k}_0, \eta)[D(W_{\mathbf{k}_0})]_{\eta\zeta}, \tag{12–172}$$

where η labels the basis functions of the irreducible representation of $G_{\mathbf{k}_0}$.

We now show that an irreducible representation of the entire group I_n^p is automatically determined by the selection of the irreducible representation (12–172) of the little group.

Let $W_{\mathbf{k}}$ be a "rotation" which takes $\mathbf{k}_0$ into $\mathbf{k}$:

$$W_{\mathbf{k}}\mathbf{k}_0 = \mathbf{k}. \tag{12–173}$$

For each $\mathbf{k}$ which satisfies $\{\mathbf{k}, \mathbf{k}\} = \{\mathbf{k}_0, \mathbf{k}_0\}$, we select one $W_{\mathbf{k}}$. Any other rotation which takes $\mathbf{k}_0$ into $\mathbf{k}$ can be expressed in the form $W_{\mathbf{k}}W_{\mathbf{k}_0}$. Corresponding to the basis functions $\psi(\mathbf{k}_0, \zeta)$ we define sets of basis functions for each $\mathbf{k}$:

$$\psi(\mathbf{k}, \zeta) = D(W_{\mathbf{k}})\psi(\mathbf{k}_0, \zeta). \tag{12–174}$$

If a rotation W takes $\mathbf{k}$ into $\mathbf{k}'$, then

$$W\mathbf{k} = \mathbf{k}',$$

$$WW_{\mathbf{k}}\mathbf{k}_0 = W_{\mathbf{k}'}\mathbf{k}_0,$$

$$W_{\mathbf{k}'}^{-1}WW_{\mathbf{k}}\mathbf{k}_0 = \mathbf{k}_0,$$

so that $W_{\mathbf{k}'}^{-1}WW_{\mathbf{k}}$ is in $G_{\mathbf{k}_0}$, and

$$W = W_{\mathbf{k}'}W_{\mathbf{k}_0}W_{\mathbf{k}}^{-1}, \tag{12–175}$$

where $W_{\mathbf{k}_0}$ is some element of $G_{\mathbf{k}_0}$.

The representation of W is now completely determined:

$$D(W)\psi(\mathbf{k},\, \zeta) = D(W_{\mathbf{k}'})D(W_{\mathbf{k}_0})D(W_{\mathbf{k}}^{-1})\psi(\mathbf{k},\, \zeta)$$

$$= D(W_{\mathbf{k}'})D(W_{\mathbf{k}_0})\psi(\mathbf{k}_0,\, \zeta)$$

$$= D(W_{\mathbf{k}'}) \sum_{\eta} \psi(\mathbf{k}_0,\, \eta)[D(W_{\mathbf{k}_0})]_{\eta\zeta}$$

$$= \sum_{\eta} \psi(\mathbf{k}',\, \eta)[D(W_{\mathbf{k}_0})]_{\eta\zeta};$$

$$D(W)\psi(\mathbf{k},\, \zeta) = \sum_{\eta} \psi(W\mathbf{k},\, \eta)[D(W_{\mathbf{k}_0})]_{\eta\zeta}. \qquad (12\text{–}176)$$

Equation (12–176) and Eq. (12–170),

$$D(\mathbf{u})\psi(\mathbf{k},\, \zeta) = e^{i\{\mathbf{k},\mathbf{u}\}}\psi(\mathbf{k},\, \zeta), \qquad (12\text{–}170)$$

completely determine the irreducible representation of I_n^p.

Problem. Prove that the choice of the vector $\mathbf{k}_0$, for which the little group was selected, is irrelevant. Show that the representations obtained by starting from the little group $G_{\mathbf{k}}$, where $\mathbf{k} = W_{\mathbf{k}}\mathbf{k}_0$, are equivalent to the representations obtained by starting from $G_{\mathbf{k}_0}$.

For the euclidean group in three dimensions, the "scalar product" $\{\mathbf{x}, \mathbf{y}\}$ is the usual scalar product $\mathbf{x} \cdot \mathbf{y}$. The vector $\mathbf{k}_0$ can be transformed by a rotation W into all vectors $\mathbf{k}$ having the same length $\mathbf{k}_0 \cdot \mathbf{k}_0 = \mathbf{k}_0^2$. The little group $G_{\mathbf{k}_0}$ is the group of rotations around the direction of $\mathbf{k}_0$. Since this is an abelian group, it has only one-dimensional representations. The index ζ takes on a single value j_0 which is either integral or half-integral.

For the Lorentz group, we have

$$\{x, x\} = x_1^2 + x_2^2 + x_3^2 - x_4^2 = g_{ij}x_i x_j,$$

$$g_{ij} = \epsilon_i\, \delta_{ij} \quad \left[\epsilon_i = \begin{matrix} +1, & i = 1, 2, 3 \\ -1, & i = 4 \end{matrix}\right], \qquad (12\text{–}177)$$

where $x_4 = ct$. The transformations $\mathbf{x}' = W\mathbf{x}$ leave $\{\mathbf{x}, \mathbf{x}\}$ unchanged:

$$\{W\mathbf{x}, W\mathbf{x}\} = \{\mathbf{x}, \mathbf{x}\},$$

$$\widetilde{Wx}\, gWx = \tilde{x}gx,$$

$$\widetilde{W}gW = g. \qquad (12\text{–}178)$$

Taking the determinant of (12–178), we find that det $W = \pm 1$. The proper Lorentz transformations are those for which det $W = +1$.

Writing out the (4, 4)-component of (12–178), we find

$$w_{4i}g_{ij}w_{j4} = g_{44},$$

or

$$w_{44}^2 - w_{14}^2 - w_{24}^2 - w_{34}^2 = 1. \tag{12–179}$$

If we take the inverse of (12–178) and replace W^{-1} by W, we find

$$Wg\widetilde{W} = g. \tag{12–178a}$$

Taking the (4, 4)-component yields

$$w_{44}^2 - w_{41}^2 - w_{42}^2 - w_{43}^2 = 1. \tag{12–179a}$$

From the last equation we see that

$$w_{44}^2 \geq 1 : w_{44} \geq +1 \qquad \text{or} \qquad w_{44} \leq -1. \tag{12–180}$$

The Lorentz transformations for which $w_{44} \geq +1$ are said to be *orthochronous*. The *proper orthochronous Lorentz transformations* satisfy the conditions

$$\det W = +1, \qquad w_{44} \geq +1, \tag{12–181}$$

and can be reached continuously from the identity. They form the group which we designated by $G_4'^3$.

A vector **x** is said to be *timelike* if

$$\{\mathbf{x}, \mathbf{x}\} < 0; \tag{12–182}$$

spacelike if

$$\{\mathbf{x}, \mathbf{x}\} > 0; \tag{12–182a}$$

and a *null vector* if

$$\{\mathbf{x}, \mathbf{x}\} = 0. \tag{12–182b}$$

The speed of a material particle must be less than the velocity of light, c, so its space and time displacements Δx_i, Δt must satisfy the condition

$$(\Delta x_1)^2 + (\Delta x_2)^2 + (\Delta x_3)^2 < c^2 (\Delta t)^2. \tag{12–183}$$

Thus the space-time displacement of a material particle is a timelike vector.

A timelike vector is said to be *positive* (*future, in the positive light cone*) if $x_4 > 0$, and *negative* (*past, in the negative light cone*) if $x_4 < 0$.

A null vector is said to be *on the positive light cone* if $x_4 > 0$, and *on the negative light cone* if $x_4 < 0$.

We now show that the proper orthochronous Lorentz transformations take positive timelike (or null) vectors into positive timelike (or null) vectors; i.e., if $x_1^2 + x_2^2 + x_3^2 \le x_4^2$ and $x_4 > 0$, then the transformed vector $\mathbf{x}'$ satisfies the same conditions.

The first part follows from the condition $\{\mathbf{x}', \mathbf{x}'\} = \{\mathbf{x}, \mathbf{x}\}$. The fourth component of $\mathbf{x}'$ is

$$x_4 = w_{41}x_1 + w_{42}x_2 + w_{43}x_3 + w_{44}x_4. \qquad (12\text{–}184)$$

The Schwartz inequality applied to the first three terms on the right gives

$$|w_{41}x_1 + w_{42}x_2 + w_{43}x_3|^2 \le (w_{41}^2 + w_{42}^2 + w_{43}^2)(x_1^2 + x_2^2 + x_3^2)$$
$$\le (w_{44}^2 - 1)x_4^2 < w_{44}^2 x_4^2. \qquad (12\text{–}185)$$

This equation shows that the last term in (12–184) dominates the first three. Thus x_4' has the same sign as $w_{44}x_4$. If $w_{44} \ge 1$, then x_4' has the same sign as x_4.

In finding the irreducible representations of the proper orthochronous Lorentz group, we choose a vector $\mathbf{k}_0$ and find the irreducible representations of the little group $G_{\mathbf{k}_0}$. There will be four principal types of representations:

    (1) $\mathbf{k}_0$ is timelike, $\{\mathbf{k}_0, \mathbf{k}_0\} < 0$;
    (2) $\mathbf{k}_0$ is a null vector, $\{\mathbf{k}_0, \mathbf{k}_0\} = 0$, but $\mathbf{k}_0 \ne 0$;
    (3) $\mathbf{k}_0 = 0$;
    (4) $\mathbf{k}_0$ is spacelike, $\{\mathbf{k}_0, \mathbf{k}_0\} > 0$.

Type 1. Starting from $\mathbf{k}_0$, we obtain functions corresponding to all vectors $\mathbf{k} = W\mathbf{k}_0$. As shown above, if k_{04} is positive (negative), then k_4 is positive (negative). We denote these two subtypes as 1_+ and 1_-. If $\{\mathbf{k}_0, \mathbf{k}_0\} = -m^2$, we can find a transformation W such that the vector $\mathbf{k} = W\mathbf{k}_0$ has $k_1 = k_2 = k_3 = 0$, and $k_4^2 = m^2$. If $\mathbf{k}_0$ is a positive time-like vector, $k_4 = +\sqrt{m^2}$; if $\mathbf{k}_0$ is a negative timelike vector, $k_4 = -\sqrt{m^2}$. We choose to construct the little group for the vector of the form

$$(0, 0, 0, k_4).$$

The little group is then the group of those Lorentz transformations W which leave the fourth component of a vector unchanged; i.e., the group O_3^+ of rotations in three-dimensional space. We know the irreducible representations of this little group: they are the representations $D^{(j)}$, where j is integral or half-integral. For each j we obtain an irreducible representation of the little group and a corresponding irreducible representation of the Lorentz group. The constants $k_4 = \pm\sqrt{m^2}$ and j determine the mass and spin of the system.

The irreducible representations have the *same* form for *all* values of $\{k_0, k_0\}$. This can be shown by using the isomorphism (automorphism)

$$W \to W, \quad \mathbf{u} \to \alpha\mathbf{u} \quad (\alpha \text{ real}),$$

$$(W, \mathbf{u}) \to (W, \alpha\mathbf{u}). \tag{12–186}$$

This is an isomorphism, since

$$(W', \mathbf{u}')(W, \mathbf{u}) = (W'W, W'\mathbf{u} + \mathbf{u}'),$$

$$(W', \alpha\mathbf{u}')(W, \alpha\mathbf{u}) = (W'W, \alpha W'\mathbf{u} + \alpha\mathbf{u}')$$

$$= (W'W, \alpha[W'\mathbf{u} + \mathbf{u}']).$$

Thus, if we are given any representation $D(W)$, $D(\mathbf{u})$ of the group, the operators $D(W)$, $D'(\mathbf{u}) = D(\alpha\mathbf{u})$ also give a representation. But

$$D'(\mathbf{u})\psi(\mathbf{k}_0, \zeta) = D(\alpha\mathbf{u})\psi(\mathbf{k}_0, \zeta) = e^{i\{\mathbf{k}_0, \alpha\mathbf{u}\}}\psi(\mathbf{k}_0, \zeta)$$

$$= e^{i\{\alpha\mathbf{k}_0, \mathbf{u}\}}\psi(\mathbf{k}_0, \zeta).$$

So the new representation corresponds to the vector $\alpha\mathbf{k}_0$, with $\{\alpha\mathbf{k}_0, \alpha\mathbf{k}_0\} = \alpha^2\{\mathbf{k}_0, \mathbf{k}_0\}$ and $(\alpha\mathbf{k}_0)_4 = \alpha\mathbf{k}_{04}$.

Type 2. Again we have two subtypes: 0_+ when k_{04} is positive and 0_- when k_{04} is negative. The rest mass m of the particles is zero. The argument given under Type 1 shows that the representations are similar for all values of k_{04}. We can always find a W to make our typical $\mathbf{k}$ have the form $(0, 0, 1, 1)$. The little group $G_\mathbf{k}$ is the set of transformations W which leave the vector $\mathbf{k}:(0, 0, 1, 1)$ unchanged. These transformations are of three types: The first is the group of rotations in the (1-2)-plane:

$$R_{12}(\theta) = \begin{bmatrix} \cos\theta & -\sin\theta & 0 & 0 \\ \sin\theta & \cos\theta & 0 & 0 \\ 0 & 0 & 1 & 0 \\ 0 & 0 & 0 & 1 \end{bmatrix}; \tag{12–187}$$

the other two are:

$$T_1(a) = \begin{bmatrix} 1 & 0 & -a & a \\ 0 & 1 & 0 & 0 \\ a & 0 & 1 - \dfrac{a^2}{2} & \dfrac{a^2}{2} \\ a & 0 & -\dfrac{a^2}{2} & 1 + \dfrac{a^2}{2} \end{bmatrix} \tag{12–187a}$$

and

$$
T_2(b) = \begin{bmatrix} 1 & 0 & 0 & 0 \\ 0 & 1 & -b & b \\ 0 & b & 1 - \dfrac{b^2}{2} & \dfrac{b^2}{2} \\ 0 & b & -\dfrac{b^2}{2} & 1 + \dfrac{b^2}{2} \end{bmatrix}, \qquad (12\text{–}187b)
$$

where a and b are arbitrary real numbers. The transformations T_1 and T_2 form one-parameter abelian groups. $T_1(a)$ and $T_2(b)$ commute, and we find that the little group is isomorphic to the group of euclidean motions in two dimensions.

The result is easily derived by noting that the group is a three-parameter subgroup of the homogeneous Lorentz group. We then try to find those infinitesimal matrices which leave $\mathbf{k}$ invariant, namely the matrices

$$
a_{12}, \qquad a_{14} + a_{34}, \qquad a_{24} + a_{34} \qquad (12\text{–}188)
$$

which generate the three subgroups listed above.

Type 3. If $\mathbf{k}_0 = 0$, the basis functions are invariant under all translations. The little group is the group of proper orthochronous homogeneous Lorentz transformations G'^3_4.

Type 4. We choose as the typical vector $\mathbf{k}_0 = (1, 0, 0, 0)$. The little group is the group which leaves the first coordinate unchanged. It is therefore the group G^2_3 of Lorentz transformations with two space coordinates and one time coordinate.

Only the representations of Types 1 and 2 appear to have physical significance.

Problem. Show that the infinitesimal matrices (12–188) generate the finite transformations (12–187).

BIBLIOGRAPHY AND NOTES

Bibliography and Notes

BOERNER, H., *Group Representations*. Springer, Berlin, 1955.

LANDAU, L., and E. LIFSHITZ, *Quantum Mechanics*. Addison-Wesley, Reading, Mass., 1958.

LITTLEWOOD, D. E., *The Theory of Group Characters*. Oxford University Press, Oxford, 1950.

LOMONT, J. S., *Applications of Finite Groups*. Academic Press, New York, 1959. (Contains a very complete bibliography.)

LYUBARSKII, G. YA., *The Application of Group Theory to Physics*, Pergamon Press, Oxford, 1960.

PONTRIAGIN, L. S., *Topological Groups*. Teubner, Leipzig, 1957–58.

VAN DER WAERDEN, B. L., *The Group-Theoretic Method in Quantum Mechanics*. Springer, Berlin, 1932.

WEYL, H., *Theory of Groups and Quantum Mechanics*. Princeton University Press, Princeton, 1931.

WEYL, H., *The Classical Groups*. Princeton University Press, Princeton, 1946.

WIGNER, E. P., *Group Theory and its Application to the Quantum Mechanics of Atomic Spectra*. Academic Press, New York, 1959.

CHAPTER 2

Section 2–10: The derivation of the magnetic symmetry groups follows the paper of B. Tavger and V. Zaitsev, *Soviet Phys.-JETP*, **3**, 430 (1956). Cf. also Landau and Lifshitz, *Statistical Physics*, Addison-Wesley, Reading, Mass., 1958, p. 426.

CHAPTER 5

Section 5–5: The discussion of real representations is based on the classic paper of Schur and Frobenius, *Berliner Berichte*, 1906, p. 186.

Section 5–8: Cf. E. P. Wigner, *Am. J. Math.*, **63**, 57 (1941).

Section 5–9: Cf. E. P. Wigner, *On the Matrices which Reduce the Kronecker Products of Representations of S. R. Groups*, Princeton, 1951.

CHAPTER 7

Sections 7–1 through 7–5: Cf. Littlewood, *op. cit.*

Sections 7–6 and 7–7: T. Yamanouchi, *Proc. Phys.-Math. Soc., Japan* **19**, 436 (1937)

The result in Eq. (7–114) has not been noted previously, though it follows immediately from the paper of Schur and Frobenius cited above.

Section 7–8: F. Hund, *Z. Physik*, **43**, 788 (1927).

Sections 7–9 and 7–10: Boerner, *op. cit.* and Weyl's *Classical Groups*.

Section 7–11: V. Fock, *JETP*, **10**, 961 (1940); Yu. N. Demkov, *Soviet Phys.-JETP*, **34**, 491 (1958).

Section 7–12: Cf. Littlewood, *op. cit.*

Section 7–13: The graphical procedure for evaluating inner products is an extension and simplification of a paper by Gamba and Radicati, *Rend. Acad. Lincei*, VIII, **14**, 632 (1953).

Section 7–14: This section is the content of unpublished work of E. A. Crosbie and M. Hamermesh, *Bull. Am. Phys. Soc.*, **1**, 209 (1956).

CHAPTER 8

For the entire chapter, cf. Pontriagin, *op. cit.*, and G. Racah, "Group Theory and Spectroscopy," *Notes from Princeton*, 1951.

Section 8–13: H. B. G. Casimir, *Proc. Roy. Ac. Amsterdam*, **34**, 844 (1931); G. Racah, *Rend. Acad. Lincei*, VIII, **8**, 108 (1950).

CHAPTER 9

Section 9–7: H. A. Bethe, *Ann. Physik.*, **3**, 133 (1929).

Section 9–8: This method for calculating the CG-coefficients is due to Van der Waerden, *op. cit.*, p. 68. The normalization procedure was given by G. Ludwig, *Grundlagen der Quantenmechanik*, Springer, Berlin, 1954.

CHAPTER 10

Sections 10–1 through 10–4: Boerner, *op. cit.*, and Weyl's *Classical Groups*.

Sections 10–5 through 10–9: Weyl's *Classical Groups*.

Section 10–7: The tables are from H. A. Jahn, *Proc. Roy. Soc.* (London), A **201**, 516 (1950), and B. H. Flowers, *Proc. Roy. Soc.* (London), A **210**, 497, (1952).

Section 10–9: The tables are from B. H. Flowers, *Proc. Roy. Soc.* (London), A **212**, 248 (1952).

CHAPTER 11

Section 11–2: Cf. the material cited for Chapter 10.

Section 11–4: G. Racah, *Phys. Rev.*, **76**, 1352 (1949).

Section 11–7: E. P. Wigner, *Phys. Rev.*, **51**, 106 (1937); F. Hund, *Z. Physik*, **105**, 202 (1937).

Sections 11–7 through 11–9: The tables are from the papers of Jahn and Flowers cited above.

CHAPTER 12

Sections 12–1 and 12–2: I. Schur, *J. Reine u. Ang. Math.*, **127**, 20 (1904), **132**, 85 (1907), and **139**, 155 (1911). Cf. also Weyl's *Theory of Groups*.

Section 12–3: The theorem of Eqs. (12–52) through (12–56) was first stated by Wigner, but the proof given in his book is incomplete. Cf. R. Hagedorn, *Nuovo cimento*, Supplemento 1, 1959, p. 73.

Sections 12–3 through 12–5: V. Bargmann, *Ann. Math.*, **59**, 1, (1954).

Section 12–5: The vector representations of the Galilean group were studied by E. Inonu and E. P. Wigner, *Nuovo cimento*, **9**, 705 (1952). Cf. also M. Hamermesh, *Ann. Phys.*, **9**, 518 (1960).

Sections 12–6 and 12–7: Bouckaert, Smoluchowski, and Wigner, *Phys. Rev.*, **50**, 58 (1936); F. Seitz, *Ann. Math.*, **37**, 17 (1936). For a derivation of the space groups, cf. F. Seitz, *Z. Krist*, **90**, 289 (1935), **91** 336 (1935), and **94**, 100 (1936).

Section 12–7: A summary of the work by Gelfand and Naimark on the representations of the homogeneous Lorentz group is given in M. A. Naimark, *Uspekhi Mat. Nauk*, **9**, No. 4, 19 (1954). The classification of the representations of the inhomogeneous Lorentz group was given by E. P. Wigner, *Ann. Math.*, **40**, 149 (1939). A very readable discussion, which includes the treatment of the full Lorentz group, is by Yu. M. Shirokov, *Soviet Phys.-JETP*, **6**, 664, 919, 929 (1958) **7**, 493 (1958), and **8**, 703 (1959); also *Nuclear Phys.*, **15**, 1, 13 (1960).

INDEX

INDEX

Abelian group, 7
Abstract group, 7
Addition, of angular momenta, 367
  of representations, 80
Adjoint, 90
Adjoint operator, 91
Adjoint representation, 135
  characters of, 135
Adjunction to a group, 52
A-form, 234
Alternating group, 14, 182
Ambivalent classes, 146, 152
Angular-momentum analysis, 414
  of $(l)^r$-configuration, 418
  of $(j)^r$-configuration, 432
Antisymmetrized product, 132
Antisymmetrizer, 231
Antisymmetry type, 234
Antiunitary operator, 469
Associate diagrams, 396
Associated partitions, 200
Associative algebra, 241
Associative law, for mappings, 2
  for an abstract group, 7
Axial distance, 220
Axial vector, 169
Axioms for a group, 7

Basis, 71
Basis vector, 71
Bethe, H. A., 122
Bisymmetric transformation, 379
Branching laws, 208
Brillouin zones, 488

Canonical exponents, 474
Canonical parameter, 295
Cartan, E., 311
Casimir operator, 317
Cayley's theorem, 15
Characters, 79
  compound, 104, 184
  primitive, 104

simple, 104
  of the symmetric group, 189
Character table, 111
  for crystal point groups, 125
  for S_n $(n = 3, \ldots, 7)$, 276
Charge states, 433
Classes, 23
Classification, of states, 413
  of atomic $(d)^r$-configuration, 427
  of atomic $(f)^r$-configuration, 428
  of nuclear d-shell, 446
  of nuclear f-shell, 448
  of nuclear $(j)^r$-configuration, 452
  of spectral terms, 161
Clebsch-Gordan coefficients, 148
  for crystal groups, 180
  for the rotation group, 367
  for S_n, 260
Clebsch-Gordan series, 147
  for S_n, 254, 256
Closed group manifold, 282
Collineation, 458
Color groups, 63
Commutator, 297
Compact groups, 309
Complete space, 94
Complex, 28
Complex conjugate representation, 135
  characters of, 135
Complex extension, 304
Composition of mappings, 2
Compound characters, 104, 184
Conformal group, 290
Conjugate classes, 23
Conjugate elements, 23
Conjugate partitions, 200
Conjugate permutations, 24
Conjugate representations, 206
Conjugate subgroup, 28
Conjugate symmetry type, 238
Conjugate transformations, 24
Conjugate transposed, 90

Construction of wave functions of a given symmetry, 246
Continuous groups, 283
Continuous operator, 94
Contraction of a tensor, 391
Coordinates, 72
Correspondence, 1
Coset, 20
  left (right), 21
Coupled systems, 178, 367
Cover group, 321
Cross ratio, 4
Crystal eigenfunctions, 342
Crystal groups, 161
Crystalline fields, 179, 337
Crystal systems, 60
Cycle of a permutation, 13
Cycle structure, 25
Cyclic group, 20

Decomposable representation, 97
Decomposition of $U(n)$, with respect to $O^+(n)$, 399
  with respect to $Sp(n)$, 408
  reduction tables, 411
Decrement of a permutation, 14
Degree, of a permutation, 4
  of a representation, 78
Density function, on a group, 314
  for the rotation group, 328
Dihedral groups, 41, 465
Direct factor, 30
Direct product, 30
Direct sum of representations, 97
Distance, 93
Double group, 358
Double-valued representations of crystal groups, 357
  character tables, 362

Electric dipole transitions, 170
Electric quadrupole transitions, 175
Enantiomorphism, 35
Equivalence relation, 23
Equivalent axes and planes, 38
Equivalent matrices, 77
Equivalent representations, 79

Essential parameters, 284
Euler angles, 331
Euler's theorem, 49
Even application, 202
Even representation, 155
Expansion of eigenfunctions, 111

Factor group, 29
Factor systems, 459
  associated, 461
  classes of, 461
  equivalent, 461
Faithful representation, 78
Finite continuous group, 282
Fock's cyclic symmetry conditions, 246
Four-group, 12, 464
Fractional linear transformation, 458
Frobenius' formula, 189
Fully reducible representation, 97
Fundamental sequence, 94

Galilean group, 484
Gamba, 255
  and Radicati, 257
Generator, 242
$GL(n)$, 288, 377
  irreducible representations of, 384
Graphical methods, for determining characters, 201
  for S_n, 198
Graphs, 198
Group algebra, 106, 239
Group axioms, 7
Group manifold, 352
Group of the wave vector, 489
Group table, 10

Half-integer representation, 143
Hemihedry, 61
Hermitian, 90
Hermitian conjugate, 90
Hermitian form, 90
Hermitian invariant, 136
Hilbert space, 93
Holohedry, 61
Homomorphism, 30

Homotopy, 330
Horizontal permutation of a Young tableau, 243
Hund, F., 224

Icosahedral group, 51
Idempotent, 242
Identical particles, 413
Improper subgroup, 15
Index of a subgroup, 21
Infinite continuous group, 287
Infinite discrete group, 281
Infinitesimal transformations, 293
Inner product of representations, 249, 254
Integer representations, 143
International notation, 63
Intersection of sets, 16
Invariance, 86
Invariant integration, 313
Invariants, 136
Invariant subgroup, 28
Inverse of a composite, 3
Inverse mapping, 2
Inversion, 36
Irreducible representations, 94
Irreducible tensors, 377
Isobaric spin, 433
Isomorphism, 9
Isotopic spin, 433
Isotopic spin space, 434
I-spin, 433

Jacobian, 75
jj-coupling, in atomic spectra, 430
in nuclear spectra, 448

Kronecker product, 128

Lagrange's theorem, 20
Lattice permutation, 198
Left ideal, 241
Left-invariant measure, 314
Lie algebra, 301
Lie group, 283
Line of nodes, 331
Linear dependence, 70

Linear group in n dimensions, 288, 377
Linear molecules, states of, 323
Linear operator, 75
Linear representation, 78
Linear transformation in n-dimensional space, 4
Linear vector space, 68
Little groups, 489
Local exponents, 471
equivalent, 472
Local factors, 471
equivalent, 471
Lorentz group, 307, 491
types of representations, 493
L-S coupling, 435

Magnetic classes, 66
Magnetic dipole transitions, 170
Magnetic symmetry groups, 63
Mappings, 1
Matrix, 5
Matrix elements in crystalline fields, 166
Matrix product, 6
Matrix representative, 76
for S_n, 214
Measure on a group, 314
Metric, 88
Metric matrix, 89
Minimal ideal, 242
Mixed continuous group, 291
Multiple-valued representation, 319
Multiplicator, 463
Murnaghan, F., 255

n-fold rotation axis, 33
Negative representations, 337
Noncommutative operations, 2
Normal antisymmetric form, 234
Normal symmetric form, 233
Nonequivalent representations, 103
Nonsingular matrix, 5
Normal divisor, 28
Normalized vector, 89
Notation for point groups, 62
Nuclear structure, 433
Null vector, 492

Octahedral group, 51
Odd application, 202
Odd representation, 155
$O(n)$, 289, 391
    irreducible representations, 394, 398
One-j symbol, 160
One-parameter groups, 293
Operator, 75
Operator ray, 470
Order, of a group, 7
    of a group element, 8
Orthochronous Lorentz transformation, 492
Orthogonal group, 289, 391
Orthogonal matrices for S_n, 224
Orthogonal vectors, 90
Orthogonality relations, 101
Orthonormal basis, 90
Outer products for S_n, 249

Partition of n, 26
Pauli principle, 417
Peirce resolution, 242
Period of a group element, 21
Permutation, 3
Perturbation theory, 162
Point groups, 33
Polar vector, 168
Pole figures, 34
Positive definite form, 90
Positive representations, 337
Primitive characters, 104
Primitive idempotent, 243
Product of mappings, 2
Product representation, 128
Projection operator, 113
Projective representation, 458
Projective transformation, 4, 458
Proper orthogonal transformation, 325
Proper subgroup, 15
Pseudo-orthogonal groups, 478

Quaternion group, 28, 145, 152, 156
Quotient group, 29

Rank of a tensor, 378
Rational indices, law of, 45

Rays, 469
Ray representations, 458
    of Lie groups, 469
Real representations, 138
Reciprocal lattice, 488
    vector, 488
Reciprocal symmetry type, 238
Recursion formulas, for CG-coefficients of S_n, 260
    for characters, 208
Reduced isotopic spin, 457
Reduced partition numbers, 442
Reduction of Kronecker products, 147
Regular application of nodes, 198, 203
Regular graphs, 198
Regular permutations, 19
Regular polyhedra, 48
Representation, 78
    of a direct product, 114
    matrix, 76
    reducible, 94
    regular, 107
Rest mass, 494
Right-invariant measure, 316
Rigid motions, 290
Rotation group, in two dimensions, 322
    in three dimensions, 325
    in n dimensions, 394
Rotation-reflection symmetry, 36
Russell-Saunders coupling, 417
r-parameter continuous group, 283

Scalar product, 88
Schoenflies notation, 61
Schur, I., 458, 463, 468
Schur's lemmas, 98
Selection rules, 166
    for diagonal matrix elements, 173
Self-adjoint, 90
Self-conjugate partition, 200
Self-conjugate subgroup, 28
Semisimple groups, 28, 309
Seniority, 426
    in atomic spectra, 423
    in nuclear spectra, 443
    number s, 457

S-form, 233
Shubnikov notation, 61
Similitude group, 290
Simple characters, 104
Simple group, 28
Simply reducible groups, 151
Skew product, 404
S_n, 182
$SL(n)$, 288
  irreducible representations of, 388
Spacelike vector, 492
Space groups, 488
$Sp(n)$, 293, 403
Spherical harmonics, table of, 334
Splitting of atomic levels in crystals,
  337
Square roots of group elements, 143
Standard tableau, 200
Stereographic projection, 34
Structure constants, 299
Subalgebra, 241
Subgroup, 15
Sum of representations, 84
Supermultiplets, 435
  structure tables of, 438
Symmetric group, 13, 182, 466
Symmetric tensor, 175
Symmetrized product, 132
Symmetrizer, 231
Symmetry group, 32
Symmetry properties of CG-coeffi-
  cients for S_n, 260
Symmetry types, 233
Symplectic coordinate basis, 405
Symplectic group, 293, 403
  irreducible representations of, 409

Tensors, 175, 377
Terms for electron configurations, 423
Tetartohedry, 61
Tetrahedral group, 50
Three-j symbols, 156
Timelike vector, 492
Time reversal, 118
  operator, 64

Traceless tensor, 391
Transform, 23
  of a matrix, 76
Transitive relation, 15
Translation, 4
Translation groups, 486
Transpose, 73
Transposition, 13
Two-sided axis, 38
Two-valued representations of the
  rotation group, 348
Types of representations, 139

$U(n)$, 290
  irreducible representations of, 388
Uniaxial groups, 41
Unimodular group, 14, 288, 388
Unit antisymmetric tensor, 384
Unitary group, 290, 388
Unitary matrix, 91
Unitary representations, 92
Unitary space, 89
Unitary unimodular group in two
  dimensions, 348
Universal covering group, 321

Vector representations, 458
Vector space, 68
Vector addition coefficients, 148
Vertical permutation of a Young
  tableau, 243
Volume on a group, 314

Wigner coefficients, 148

Yamanouchi, 214
  symbols, 221
Young antisymmetrizer, 244
Young operators, 243
Young patterns, 198
Young symmetrizer, 244
Young tableau, 198
Y-symbols, 221

Zeroth-order functions, 413